Machining Fundamentals

From Basic to Advanced Techniques

by

John R. Walker

Publisher

The Goodheart-Willcox Company, Inc.

Tinley Park, Illinois

Copyright 2000
by
THE GOODHEART-WILLCOX COMPANY, INC.
Previous Editions Copyright 1998, 1993, 1989, 1981, 1977, 1973

Library of Congress Card Catalog Number 99-17776
International Standard Book Number 1-56637-662-9

1 2 3 4 5 6 7 8 9 10   00   03 02 01 00 99

Cover photo ©Westlight (K. Tiedge)

Library of Congress Cataloging in Publication Data

Walker, John R.
    Machining Fundamentals: from basic to advanced techniques /
by John R. Walker

    p. cm.
Includes index.
ISBN 1-56637-662-9
1. Machine-shop practice. 2. Machining.
I. Title.
TJ1160.W25 2000
671.3'5--dc21

                                                99-17776
                                                CIP

Introduction

Machinists are highly skilled men and women. They use drawings, hand tools, precision measuring tools, drilling machines, grinders, lathes, milling machines, and other specialized machine tools to shape and finish metal and nonmetal parts. Machinists must have a sound understanding of basic and advanced machining technology, which includes:

- Proficiency in safely operating machine tools of various types (manual, automatic, and computer controlled).
- Knowledge of the working properties of metals and nonmetals.
- The academic skills (math, science, English, print reading, metallurgy, etc.) needed to make precision layouts and machine set-ups.

Machining Fundamentals provides an introduction to this important area of manufacturing technology. The text explains the "how, why, and when" of numerous machining operations, set-ups, and procedures. Through it, you will learn how machine tools operate and when to use one particular machine instead of another. The advantages and disadvantages of various machining techniques are discussed, along with their suitability for particular applications.

Machining Fundamentals details the many common methods of machining and shaping parts to meet given specifications. It also covers newer processes such as laser machining and welding, water-jet cutting, high-energy-rate forming (HERF), cryogenics, chipless machining, electrical discharge machining (EDM), electrochemical machining (ECM), robotics, and rapid prototyping. The importance of computer numerical control (CNC) in the operation of most machine tools, and its role in automated manufacturing is explored thoroughly.

This new edition of *Machining Fundamentals* has many features that make it easy to read and understand. A numbering system for headings has been adopted to make it easier to locate information in a chapter. Learning objectives are presented at the beginning of each chapter, along with a list of selected technical terms important to understanding the material in that chapter. Throughout the book, technical terms are highlighted in bold italic type as they are introduced and defined. Several hundred of these terms are also listed and defined in a *Glossary of Technical Terms* at the end of this text. Review questions covering the content taught are presented at the end of each chapter.

Color is employed extensively in this new edition to enhance understanding and to emphasize safety precautions. A consistent color coding has been employed in the hundreds of line illustrations (most made especially for the text) to help you visualize more clearly the machining operations and procedures. Many of the black and white photographs in the text have been replaced with new, full-color photos showing the most current types of equipment and processes.

Machining Fundamentals is a valuable guide to anyone interested in machining, since the procedures and techniques presented have been drawn from all areas of machining technology.

John R. Walker

Machining Fundamentals Color Key

Colors are used throughout *Machining Fundamentals* to indicate various materials or equipment features. The following key shows what each color represents.

Metals (surfaces)

Metals (in section)

Machines/machine parts

Tools

Cutting edges

Work-holding and tool-holding devices

Rulers and measuring devices

Direction or force arrows, dimensional information

Fasteners

Abrasives

Fluids

Miscellaneous

IMPORTANT SAFETY NOTICE

Work procedures and shop practices described in this book are effective, but general, methods of performing given operations. Always use special tools and equipment as recommended. Carefully follow all safety warnings and cautions (they are printed in red type for greater legibility). Note that these warnings are *not* exhaustive. Proceed with care and under proper supervision to minimize the risk of personal injury or injury to others. Also follow specific equipment operating instructions.

This book contains the most complete and accurate information that could be obtained from various authoritative sources at the time of publication. Goodheart-Willcox Publisher cannot assume responsibility for any changes, errors, or omissions.

Contents

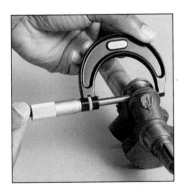

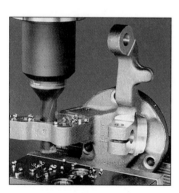

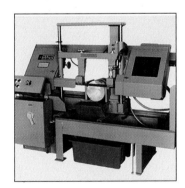

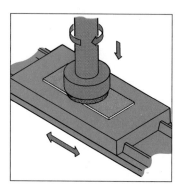

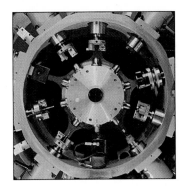

Chapter 1

An Introduction to Machining Technology

LEARNING OBJECTIVES

After studying this chapter, you will be able to:
- ○ Discuss how modern machine technology affects the workforce.
- ○ Give a brief explanation of the evolution of machine tools.
- ○ Provide an overview of machining processes.
- ○ Explain how CNC machining equipment operates.
- ○ Describe the role of the machinist.

IMPORTANT TERMS

band machining
computer numerical
   control (CNC)
drill press
lathe
machine tools

machinist
milling machine
numerical control (NC)
precision grinding
skill standards

Figure 1-1. *Machine tools have made it possible to manufacture parts with the precision and speed necessary for low-cost mass production. Without machine tools, most products on the market today would not be available or affordable. (Courtesy of SURFCAM by Surfware)*

A study of technology will show that industry has progressed from the time when everything was made by hand to the present fully automated manufacturing of products. Machine tools have played an essential role in all technological advances.

Without machine tools, **Figure 1-1**, there would be no airplanes, automobiles, television sets, or computers. Many of the other industrial, medical, recreational, and domestic products we take for granted would not have been developed. For example, if machine tools were not available to manufacture tractors and farming implements, farmers might still be plowing with oxen and hand-forged plowshares.

It is difficult to name a product that does not require, either directly or indirectly, the use of a machine tool somewhere in its manufacture. Today,

no country can hope to compete successfully in a global economy without making use of the most advanced machine tools.

There is one very important point that must be emphasized concerning modern manufacturing technology. The high-paying skilled jobs in manufacturing, such as tool-and-die making and precision machining, *require aptitudes comparable to those of college graduates*. Jobs that require few or no skills have almost disappeared.

1.1 THE EVOLUTION OF MACHINE TOOLS

Machine tools are the class of machines which, taken as a group, can reproduce themselves (manufacture other machine tools). There are many variations of each type of machine tool, and they are available in many sizes. Tools range from those small enough to fit on a bench top to machines weighing several hundred tons.

The evolution of machine tools is somewhat akin to the old question, "Which came first, the chicken or the egg?" You could also ask, "How could there be machine tools when there were no machine tools to make them?"

1.1.1 Early Machine Tools

The first machine tools, the bow lathe and bow drill, were hand-made. They have been dated back to about 1200 BC. Until the end of the 17th Century, the lathe could only be used to turn softer materials, such as wood, ivory, or at most, soft metals like lead or copper. All of them were human-powered. Eventually, the bow lathe with its *reciprocating* (back-and-forth) *motion* gave way to treadle power, which made possible work rotation that was continuous in one direction. Later, machines were powered by a "great wheel" turned by flowing water or by a person or animal walking on a treadmill. Power was transmitted from the wheel to one or more machines by a belt and pulley system.

When inventor James Watt first experimented with his steam engine, the need for perfectly bored cylinders soon became apparent. This brought about the development of the first *true* machine tool. It was a form of the lathe and was called a "boring mill," **Figure 1-2**. The water-powered tool was developed in 1774 by Englishman John Wilkinson.

This machine was capable of turning a cylinder 36" in diameter to an accuracy of a "thin-worn shilling"(an English coin). However, operation of the boring mill, like all metal cutting lathes at the time, was hampered by the lack of tool control. The "mechanic" (the first *machinist*) had to unbolt and reposition the cutting tool after each cut.

About 1800, the first lathe capable of cutting accurate screw threads was designed and constructed by Henry Maudslay, an English master mechanic and machine toolmaker. As shown in **Figure 1-3,** a hand-made screw thread was geared to the spindle and moved a cutting tool along the work. Maudslay also devised a slide rest and fitted it to his lathe. It allowed the cutting tool to be accurately repositioned after each cut. Maudslay's lathe is considered the "granddaddy" of all modern chip-making machine tools.

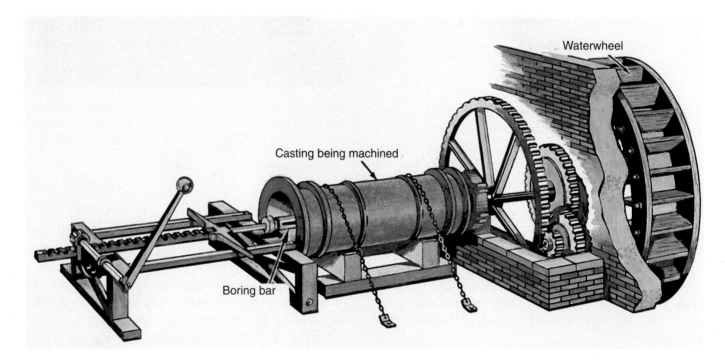

Figure 1-2. The first true machine tool is thought to be the boring mill invented by John Wilkinson in 1774. It enabled James Watt to complete the first successful steam engine. The boring bar was rigidly supported at both ends, and was rotated by waterpower. It could bore a 36" diameter cylinder to an accuracy of less than 1/16". (DoALL Co.)

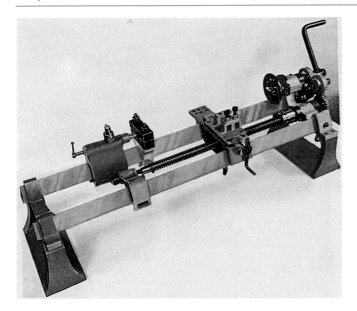

Figure 1-3. Henry Maudslay's screw-cutting lathe. This machine tool, constructed on a heavy frame, combined a master lead screw and a movable slide rest. The lead screw had to be changed when a different thread pitch was required. (DoALL Co.)

Figure 1-4. One of the first practical milling machines manufactured in America. Eli Whitney used it and similar machines to mass-produce musket parts that were interchangeable. (DoALL Co.)

In retrospect, the Industrial Revolution could not have taken place if there had not been a cheap, convenient source of power: the *steam engine*. Until the advent of the steam engine, industry had to locate near sources of water power. This was often some distance from raw materials and workers. With cheap power, industry could locate where workers were plentiful and where the products they produced were needed. The steam engine, in turn, would not have been possible without machine tools. Until the boring mill and lathe were developed to the point where metal could be machined with some degree of accuracy, there could be no steam engine.

The milling machine was the next important development in machine tools. It also evolved from the lathe. In 1820, Eli Whitney, an American inventor and manufacturer, devised a system to mass produce muskets (guns). Whitney began using a milling machine, **Figure 1-4**, to make interchangeable musket parts. Until then, muskets were made individually by hand, so parts from one musket would not fit in another. Whitney's milling machine even had power feed, but it had one defect. There was no provision to raise the worktable. The part had to be raised by shimming after each cut. Since each machine was used to produce the same part again and again, this shortcoming was not a great problem. It wasn't too much later that this problem was corrected.

Whitney had another problem, however. His ideas were used in several armories producing gun parts. There was no standard of measurement at that time, so parts made in one armory were not interchangeable with parts in another armory. It was not until the mid-1860s that the United States adopted a standard measuring system.

By 1875, basic machine tools such as the lathe, the milling machine, and the drill press, **Figure 1-5**, were capable of attaining accuracies of one one-thousandth of an inch. America was well on its way to becoming the greatest industrial nation in the world.

1.1.2 Power Sources

As machine tools were improved, so was the way they were powered. At first, the changes were very slow, taking hundreds of years. The great changes have come only in the last 150 years or so.

- *Hand power.* The bow lathe and bow drill are examples. Direction of rotation changed at each stroke of the bow.
- *Foot power.* A treadle or a treadmill made possible continuous rotation of the work in one direction.
- *Animal power.* Treadmills were used to power early devices for boring cannon barrels. Human foot power was not sufficiently strong for this work.

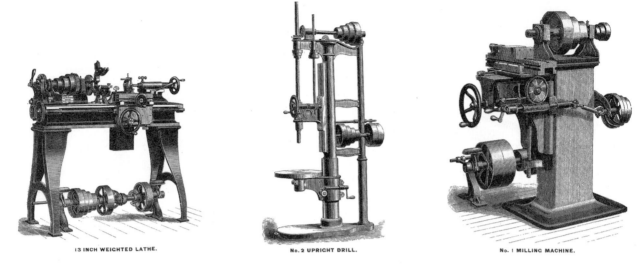

13 INCH WEIGHTED LATHE.

No. 2 UPRIGHT DRILL.

No. 1 MILLING MACHINE.

Figure 1-5. *Illustrations of Pratt & Whitney machine tools from an 1876 advertisement. Built from heavy iron castings, the machines were driven by overhead pulleys and belting. A central steam engine or large electric motor powered the overhead pulleys in factories until the 1920s.*

- *Water power.* Not always dependable as a power source, because of lack of water during dry seasons.
- *Steam power.* The first real source of dependable power. A centrally located steam engine turned shafts and overhead pulleys that were belted to the individual machines.
- *Central electrical power.* Large electric motors simply replaced the steam engines. Power transmission to the machines did not change.
- *Individual electrical power.* Motors were built into the individual machine tools. Overhead belting was eliminated.

1.2 BASIC MACHINE TOOL OPERATION

Almost all machine tools have evolved from the *lathe*, **Figure 1-6**. This machine tool performs one of the most important machining operations. It operates on the principle of work being rotated against the edge of a cutting tool, **Figure 1-7**. Many other operations—drilling, boring, threadcutting, milling, and grinding—can also be performed on a lathe. The most advanced version of the lathe is the *CNC turning center*, **Figure 1-8**. See Chapters 13-15 for basic lathe operations, and Chapters 21 and 22 for automated machining.

Figure 1-6. *A modern lathe using digital technology to perform operations such as maintaining a constant surface speed, automatic threading cycle, automatic radius cutting, and taper turning. Note the safety shield that moves with the carriage. Except those tools that perform nontraditional machining operations, all machine tools have evolved from the lathe. (Harrison/Rem Sales, Inc.)*

1.2.1 Drill Press

A *drill press*, **Figure 1-9**, rotates a cutting tool (*drill*) against the material with sufficient pressure

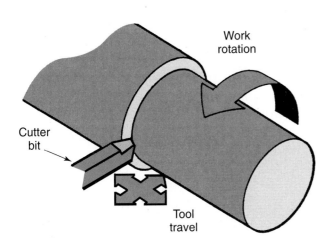

Figure 1-7. *The lathe operates on the principle of the work being rotated against the edge of a cutting tool.*

Figure 1-8. *Slant bed CNC lathe with hydraulic chucking and an electronically indexed 12 station turret. (Clausing Industrial, Inc.)*

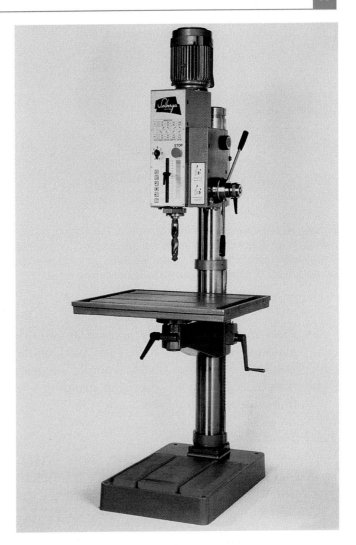

Figure 1-9. *A typical 20″ variable-speed gear head drill press with power feed. It can drill holes up to 1 1/2″ in diameter in cast iron. (Willis Machinery and Tools Corp.)*

to cause the tool to penetrate the material. It is primarily used for cutting round holes. See **Figure 1-10**. Drill presses are available in many versions. Some are designed to machine holes as small as 0.0016″ (0.04 mm) in diameter. See Chapter 10.

1.2.2 Grinding Machines

Grinding, **Figure 1-11**, is an operation that removes metal by rotating a grinding wheel or abrasive belt against the work. The process falls into two basic categories:

- *Offhand grinding.* Work that does not require great accuracy is hand-held and manipulated until ground to the desired shape. See Chapter 11.

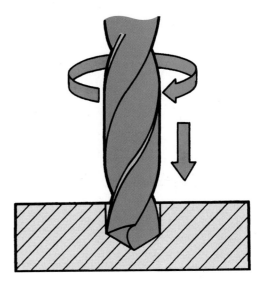

Figure 1-10. *A drill press operates by rotating a cutting tool (drill) against the material with sufficient pressure to cause the tool to penetrate the material.*

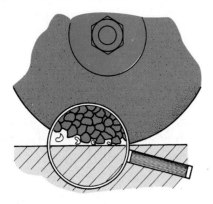

Figure 1-11. Grinding is a cutting operation, like turning, drilling, milling, or sawing. However, instead of the one, two, or multiple-edge cutting tools used in other applications, grinding employs an abrasive tool composed of thousands of cutting edges.

- *Precision grinding.* Only a small amount of material is removed with each pass of the grinding wheel, so that a smooth, accurate surface is generated. Precision grinding is a finishing operation. See Chapter 19.

1.2.3 Band Machines

Band machining, **Figure 1-12**, is a widely employed technique that makes use of a continuous saw blade. Chip removal is rapid and accuracy can

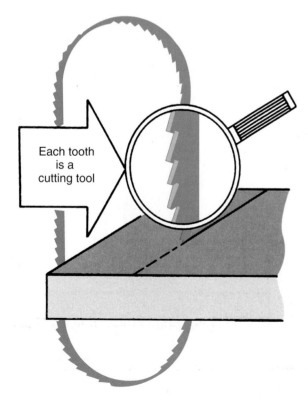

Figure 1-12. Band machining makes use of a continuous saw blade, with each tooth functioning as a precision cutting tool.

be held to close tolerances, eliminating or minimizing many secondary machining operations. See Chapter 20.

1.2.4 Milling Machine

A *milling machine* rotates a multitoothed cutter into the work, **Figure 1-13**. A wide variety of cutting operations can be performed on milling machines. See Chapters 17 and 18.

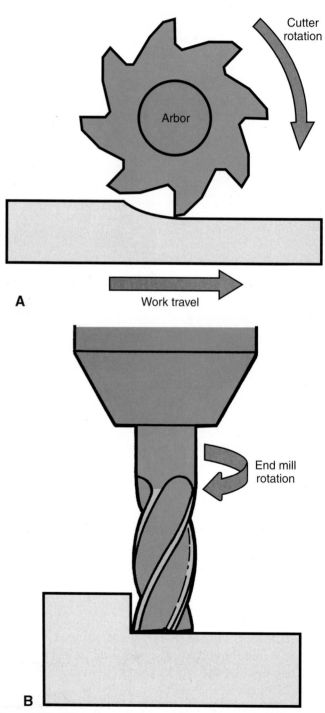

Figure 1-13. Milling removes material by rotating a multitoothed cutter into the work. A—With peripheral milling, the surface being machined is parallel to periphery of the cutter. B—End mills have cutting edges on the circumference and the end.

1.2.5 Broaching Machines

Broaching machines are designed to push or pull a multitoothed cutter across the work, **Figure 1-14**. Each tooth of the *broach* (cutting tool) removes only a small amount of the material being machined.

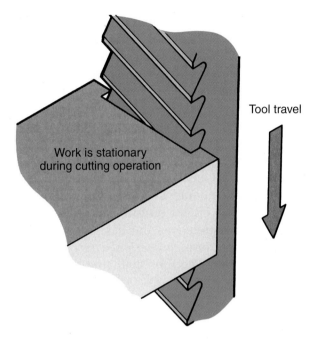

Figure 1-14. A broach is a multitoothed cutting tool that moves against the work. Each tooth removes only a small portion of the material being machined. The cutting operation may be on a vertical or horizontal plane.

1.3 NONTRADITIONAL MACHINING PROCESSES

There are a number of machining operations that have not evolved from the lathe. They are classified as *nontraditional machining processes*. These processes include:

- *Electrical discharge machining (EDM).* An advanced machining process that uses a fine, accurately controlled electrical spark to erode metal.
- *Electrochemical machining (ECM).* A method of material removal that shapes a workpiece by removing electrons from its surface atoms. In effect, ECM is exactly the opposite of electroplating.
- *Chemical milling.* A process in which chemicals are employed to etch away selected portions of metal.

- *Chemical blanking.* A material removal method in which chemicals are employed to produce small, intricate, ultrathin parts by etching away unwanted material.
- *Hydrodynamic machining (HDM).* A computer-controlled technique that uses a 55,000 psi water jet to cut complex shapes with minimum waste. The work can be accomplished with or without abrasives added to the jet.
- *Ultrasonic machining.* A method that uses ultrasonic sound waves and an abrasive slurry to remove metal.
- *Electron beam machining (EBM).* A thermoelectric process that focuses a high-speed beam of electrons on the workpiece. The heat that is generated vaporizes the metal.
- *Laser machining.* The laser produces an intense beam of light that can be focused onto an area only a few microns in diameter. It is useful for cutting and drilling.
- *Hexapods.* CNC has made possible unconventional machine tools that use new work-positioning and tool-positioning concepts. See **Figure 1-15**. These tools, already available to industry, utilize the same movement principles developed for the flight simulators that train aircraft pilots. They offer basic advantages in stiffness, accuracy, speed, dexterity, and scaling (making larger or smaller versions of the same part).

1.4 AUTOMATING THE MACHINING PROCESS

In the late 1940s, the United States Air Force was searching for ways to increase production on complex parts for the new jet aircraft and missiles then going into production.

The Parsons Corporation, a manufacturer of aircraft parts, had developed a two-axis technique for generating data to check helicopter blade airfoil patterns. This system used punched-card tabulating equipment. To determine the accuracy of the data, a pattern was mounted on a Bridgeport milling machine. With a dial indicator in place, the X and Y points were called out to a machinist operating the machine's X-axis handwheel and another machinist who controlled the Y-axis handwheel. With enough reference points established, the generated data proved accurate to ±0.0015″ (0.038 mm).

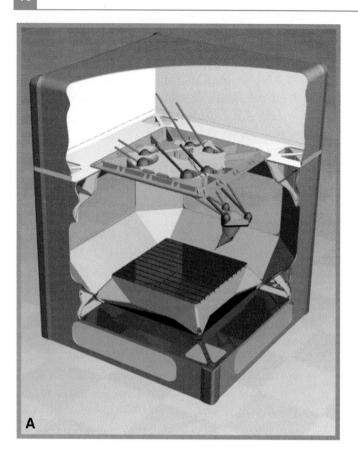

A

1.4.1 The Development of Numerical Control

Parsons realized that the technique might also be developed into a two-axis, or even three-axis, machining system. With an Air Force contract to manufacture a contoured integrally stiffened aircraft wing section, the Parsons Corporation subcontracted with the Servomechanism Laboratory at the Massachusetts Institute of Technology to design a three-axis machining system. MIT eventually took over the entire NC development project.

By 1952, MIT had designed a control system and mounted it on a vertical spindle machine tool. The system operated on instructions coded in the binary number system on punched (perforated) tape. Programming required the use of an early computer upon which MIT was also experimenting.

Later in that year, MIT demonstrated the first machine tool capable of executing simultaneous cutting tool movement on three axes. Since mathematical information was the basis of the concept, MIT coined the term *numerical control (NC)*. The first NC machines became available to industry by late 1955.

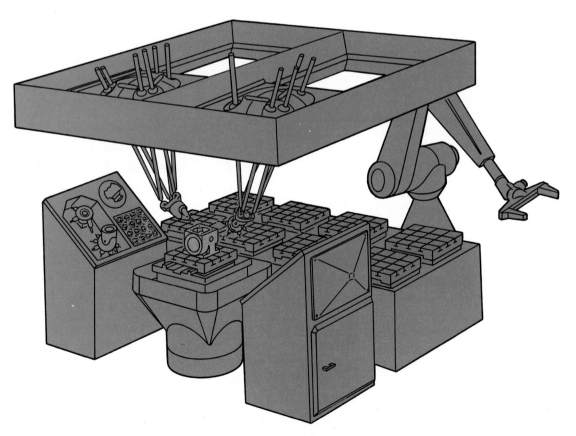

B

Figure 1-15. New machine tool concept, the hexapod, seems to defy almost every preconception about what a machine tool should be. A—The hexapod uses an entirely new concept for cutting tool movement and work positioning, with six degrees of freedom provided by a framework of variable length struts. (Renaissance Design, Inc.) B—The hexapod can be configured to perform multiple functions such as milling, drilling, tapping, polishing, grinding, welding, and even mechanical assembly.

1.4.2 Computer Numerical Control

In the mid-1970s, with the introduction of the microchip, the use of onboard computers on individual machine tools became possible. This led to the introduction of *computer numerical control (CNC)*, **Figure 1-16**.

CNC machine tools are much easier to use than manually controlled machines. They have menu-selectable displays, advanced graphics (the multi-function screen displays the full operational data as a part is being machined), and a word address format for programming. The program is made up of sentence-like commands. Programs can be entered at the machine, or may be downloaded by direct line from an external computer. Programs on punched tapes are rarely used. A modern CNC *horizontal machining center (HMC)* is shown in **Figure 1-17**.

A CNC machine tool offers:

- *Accuracy.* It is capable of producing consistent and accurate workpieces.
- *Repeatability.* It is able to produce any number of identical workpieces once a program is verified.

Figure 1-16. *CNC machine tools are equipped with on-board computers that permit computer-aided or manual programming. All controls needed for complete machine operation are in one location. A CRT screen displays all important machining information, such as the tool path. (Giddings & Lewis, Inc.)*

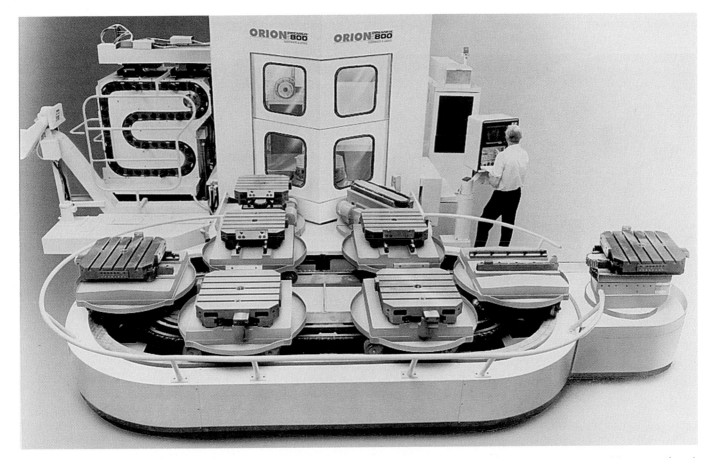

Figure 1-17. *A state-of-the-art CNC horizontal machining center (HMC) with multiaxis capabilities. It can handle a wide range of workpiece sizes and materials. The center is fitted with a multiple pallet work storage system (foreground) that automatically transfers workpieces into and out of the machine upon command from the CNC unit. (Giddings & Lewis, Inc.)*

- *Flexibility.* Changeover to running another type of part requires only a short period of nonproductive machine downtime.

The use of **robotic systems** for loading and unloading permits some machine tools to operate unattended during the entire machining cycle. Robots also have many other industrial applications, **Figure 1-18.** They can:

Figure 1-18. *A robot is a programmable, multifunctional manipulator designed to move material, tools, or specialized devices through programmed motions for the performance of a variety of tasks. This robot is deburring a complex part following machining operations. (Fanuc Robotics)*

- Operate in hazardous and harsh environments.
- Perform operations that would be tedious for a human operator.
- Handle heavy materials.
- Position parts with great repetitive precision.

The automotive industry makes extensive use of robots in the manufacture and assembly of motor vehicles, **Figure 1-19.**

1.5 THE EVOLVING ROLE OF THE MACHINIST

In recent years, the number of highly skilled machinists has been in decline. CNC machine tools have compensated for this trend to some degree. Since these machines operate under programmed control, the men and women who use them do not require the same level of skill or training as a skilled machinist.

However, because of these same CNC machine tools, the demand for machinists has not diminished. Machinists understand machining technology and what machine tools are capable of accomplishing. For these reasons, they make the best programmers and setup personnel.

Figure 1-19. *The automotive industry makes extensive use of robots for positioning parts, welding, painting, and performing quality control tasks.*

There is still another reason for the high demand for machinists: although CNC equipment is found in almost all machine shops, surveys consistently show that there is still considerable work being produced on conventional manually operated machine tools.

Whether planning an NC program or preparing to produce work on a conventional machine tool, a machinist must make many decisions and determinations on how to manufacture a part in the most economical way. The machinist must:

- Make a thorough study of the print.
- Determine the machining that must be done.
- Ascertain tolerance requirements.
- Plan the machining sequence.
- Determine how the setup will be made.
- Select the machine tool, cutter(s), and other tools and equipment that will be needed.
- Calculate cutting speeds and feeds.
- Select a proper cutting fluid for the material being machined.

All of this is possible because of the skill, knowledge, and experience of the machinist. Essentially, a machinist is able to *visualize the machining program.* When NC and CNC came along, most machinists quickly adapted to the new technology because they were already experienced in machining technology.

1.5.1 Acquiring Machining Skills and Knowledge

The skills and knowledge needed by the machinist are not acquired in a short time. It normally requires taking part in a multiyear salaried apprentice program. In addition to machine tool training under an experienced machinist, the program also involves related subjects such as English, algebra, geometry, trigonometry, print reading, safety, production techniques, and CNC principles and programming. Refer to Chapter 30 for additional information on machining technology occupations.

The *National Tooling and Machining Association,* with the aid of the metalworking industry, has developed three levels of *skill standards* reflecting industry skill requirements. A major goal of the Metalworking Skills Standards program is performance testing. The standards will provide skilled workers with certification that will afford them industry recognition.

TEST YOUR KNOWLEDGE

Please do not write in this text. Write your answers on a separate sheet of paper.

1. One of the first machine tools, the bow lathe:
   a. Could only turn softer materials.
   b. Has been dated back to about 1200 BC.
   c. Eventually gave way to treadle power.
   d. None of the above.
   e. All of the above.

2. The Industrial Revolution could not have taken place without the cheap, convenient power of the _____ _____.

3. List seven power sources in the order they have evolved over the last 150 years or so.

4. Almost all machine tools have evolved from the _____.

5. Jobs such as tool-and-diemaking and precision machining require aptitudes comparable to those of _____ .
   a. High school graduates.
   b. College graduates.
   c. High school equivalency graduates.
   d. All of the above
   e. None of the above.

6. Eli Whitney's mass-production system for muskets had a major problem because _____.
   a. There were no skilled workers
   b. There was no good source of power.
   c. There was no standard of measurement.
   d. All of the above.
   e. None of the above.

7. What occurred in the mid-1860s that was very important to the development of machining technology in the United States?

8. List four types of nontraditional machining processes and briefly describe their operation.

9. The introduction of the microchip in the mid-1970s led to the introduction of _____ machine tools.

10. List four industrial applications of robots.

The role of the computer in manufacturing has expanded greatly in recent years. In addition to computer numerical control of machine tools, many production operations include computer-controlled robotic assembly lines like this one. (Giddings & Lewis, Inc.)

Chapter 2

Shop Safety

LEARNING OBJECTIVES

After studying this chapter, you will be able to:
○ Give reasons why shop safety is important.
○ Explain why it is important to develop safe work habits.
○ Recognize and correct unsafe work practices.
○ Apply safe work practices when employed in a machine shop.
○ Select the appropriate fire extinguisher for a particular type of fire.

IMPORTANT TERMS

adequate ventilation
approved-type respirator
back injuries
combustible materials
electrocution

OSHA
protective clothing
safety equipment
unsafe practice
warning signs

Shop safety is *not* something to be studied at the start of a training program and then forgotten; most accidents are caused by carelessness or by not observing safety rules. Remember this when your instructor insists on safe work practices. If you are diligent and follow instructions with care, machining operations can be safe and enjoyable. Safe work practices should become a force of habit!

Since it is not possible to include *every* safety precaution, the safety practices in this chapter are general. Safety precautions for specific tools and machines are described in the text where they apply, along with the description and operation of the equipment. Refer to **Figure 2-1**.

Study all safety rules carefully and constantly apply them. When in doubt about any task, get help! *Do not* take chances!

2.1 SAFETY IN THE SHOP

Keep the shop *clean*. Metal scraps should be placed in the scrap bin. Never allow them to remain on the bench or floor.

Exercise extreme care when you are machining unfamiliar materials. For example, magnesium chips burn with great intensity under certain conditions. Applying water to the burning magnesium chips only intensifies the fire. Machining equipment can be damaged beyond repair and very serious burns can result.

Inhaling fumes or dust from some of the newer space-age and exotic materials can cause serious respiratory ailments. *Do not* machine a material until you *know what it is* and how it should be handled.

An approved-type respirator and special **protective clothing** must be worn when machining some materials. Machines must be fitted with effective vacuum systems as needed.

The shop is a place to work, not play. It is *not* a place for horseplay. A "joker" in a machine shop is

Figure 2-1. *None of the pilots in this precision flying team would think of taking off to give a flight demonstration until all the plane's systems were in safe operating condition. In the same way, you should never operate a machine tool until you have determined that it is in safe operating condition.*

a *walking hazard* to everyone. Daydreaming also increases your chances of injury.

If you have been ill and are using *medication*, check with your doctor or school clinic to determine whether it is safe for you to operate machinery. For example, many cold remedies recommend that you do not operate machinery while taking the medication because of possible drowsiness.

Avoid using *compressed air* to remove chips and cutting oil from machines. Flying chips can cause serious eye injuries. Also, oil that has been vaporized by the stream of air can ignite, resulting in painful burns and property damage.

Oily rags must be placed in an *approved safety container* (a metal can with metal lid). See **Figure 2-2**. Rags or waste used to clean machines will also have metal slivers embedded in them, posing an additional hazard. Placing them in a safety container will help make sure they will not be used again. Dispose of the rags daily. This will minimize the possibility of spontaneous combustion (ignition by rapid oxidation or burning of oil without an external source of heat).

Keep *hand tools* in good condition. Store tools in such a way that people cannot be injured while they are removing the tools from the tool panel or storage rack.

Use care when handling long sections of metal stock — accidentally contacting a light fixture with the stock, for example, could cause severe electrical burns or even death by *electrocution* (electric current passing through body tissues). An electric shock has been compared to "being hit by a truck."

When moving heavy machine accessories or large pieces of metal stock, always secure help. The *back injuries* that result from improper lifting are usually long-term injuries!

Dress properly for working around machinery — severe injuries or even death can result if clothing, hair, or jewelry gets caught in moving parts. Avoid wearing loose-fitting sweaters or similar clothing that could catch in machinery. A snug-fitting shop coat or apron can be worn to protect your street clothes, **Figure 2-3**. Keep sleeves rolled up. Rings and other jewelry should be removed before working around machinery. If you have long hair, wear a cap or use other means of containing it.

For jobs where dust and fumes are a hazard, ensure *adequate ventilation*. Return solvents and oils to proper storage after use. Wipe up spilled oil or solvent right away. If the spill area is extensive, use an approved-type oil absorbent. See **Figure 2-4**.

Figure 2-2. Oily rags used for cleaning machines or soaking up spills should be placed in an approved safety container like this one to minimize fire hazards. The container should be emptied daily and contents disposed of properly. (Justrite Manufacturing Company)

Figure 2-3. This trainee is properly dressed for the job she is doing. She is wearing approved eye protection and a snug-fitting apron. The machine was carefully checked before she began to operate it. (Millersville University)

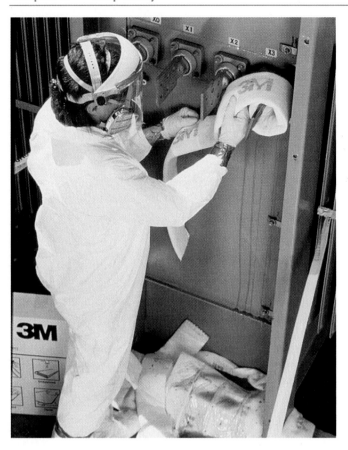

Figure 2-4. *This worker in a manufacturing plant is using a special oil-absorbent material to wrap a leaking machine component. Note that the material, which is packaged in roll form, has also been used to soak up oil on the machine base and floor. (3M Company)*

Wear appropriate *safety equipment*, **Figure 2-5**. In noisy areas, use earplugs or another type of hearing protection. Disposable plastic gloves will protect your hands when handling oils, cutting fluids, or solvents. Wear a dust mask when machining produces airborne particles, such as those from

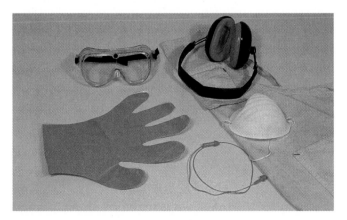

Figure 2-5. *Wear appropriate safety equipment. Shown are approved eye protection, an apron to protect clothing, plastic gloves for handling oils and solvents, a hearing protector, earplugs, and a dust mask.*

sand castings, plastics, and some grinding operations. A disposable dust mask is *not* suitable in areas where machining operations produce a mist of oil or coolants. An *approved-type respirator* must be worn in such situations. See **Figure 2-6**. Suitable personal protective equipment must also be worn when handling sharp, hot, or contaminated materials.

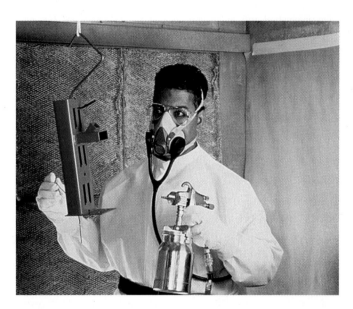

Figure 2-6. *Whenever fine airborne mists of oil, coolant, or other materials are present, an approved respirator is required. This spray painter is wearing a respirator supplied with clean air through a tube from a central source. It is also important to use proper eye protection, such as the safety glasses worn by this worker. (3M Company)*

Take no chances! Always protect your eyes. Eyesight that has been damaged or destroyed cannot be replaced. Wear safety glasses, goggles, or face shields approved by *OSHA* (the United States Occupational Safety and Health Administration).

Wear eye protection whenever you are in the shop. Protective eyewear should be used when conditions call for it. It is good practice to have your own personal safety glasses. The cost is reasonable. Your instructor can help you determine the style best suited for your needs, **Figure 2-7**. If you wear glasses, special safety lenses are available that can be ground to your prescription. Your eye doctor or optician can help get them.

Know your job! It is foolish and disastrous to operate machines without first receiving proper instruction. If you are not sure what must be done, or how a task should be performed, get help.

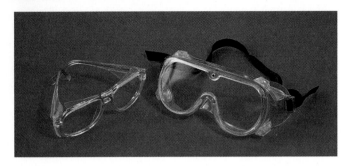

Figure 2-7. Safety glasses are available in a number of styles. The model at left is similar to regular eyeglasses, but has "wings" at each side and on top to guard against flying particles. The goggle-style model at right fits tightly against the face and can be worn over regular eyeglasses.

2.1.1 Safety Aids

Awareness barriers, while they all do not provide physical protection from machine hazards, serve to remind the operator of an area that is dangerous. In simplest form, a barrier may be nothing more than red or yellow lines painted on the floor. More complex barriers stop the machine when a light beam or electronic beam is broken by someone entering the danger area.

Machine shields, **Figure 2-8**, provide protection from flying chips and splashing cutting fluids or coolants. Many CNC machine tools are fitted with large sliding shields that cover the entire machining area, **Figure 2-9**.

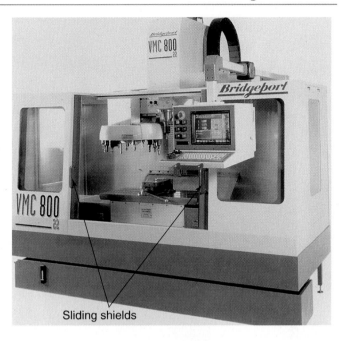

Sliding shields

Figure 2-9. Many CNC machine tools use sliding shields over the machining area to protect the machinist from flying chips and vaporized coolant. (Bridgeport Machines, Inc.)

Many different types of *warning signs*, **Figure 2-10**, are used to notify workers of potential hazards. *No Smoking* signs must be posted in areas where inflammable or combustible materials are used and stored.

2.2 GENERAL MACHINE SAFETY

- Never operate a machine until all guards are in place!
- Always stop your machine to make adjustments or measurements! Resist the urge, while the machine is running, to touch a surface that has been machined. Severe lacerations can result.
- Keep the floor around your machine clear of oil, chips, and metal scrap.
- It is considered an *unsafe practice* to talk to anyone while you are operating a machine. You might become distracted and injure yourself, or someone else.
- *Never* attempt to remove chips or cuttings with your hands or while the machine is operating. Use a brush, **Figure 2-11**. Pliers are one of the safest ways to remove long, stringy chips from a lathe. Better still, learn how to grind the cutting tool to break chips off in shorter pieces. This is explained later in the text.

Figure 2-8. Specially designed safety shields are available for almost all machine tools. They should always be in place before the machine is operated.

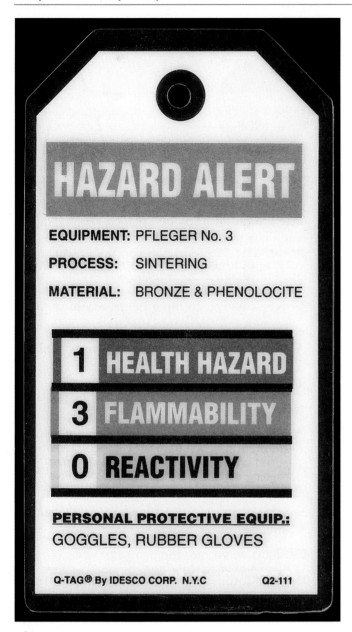

HAZARD ALERT

EQUIPMENT: PFLEGER No. 3

PROCESS: SINTERING

MATERIAL: BRONZE & PHENOLOCITE

1	HEALTH HAZARD
3	FLAMMABILITY
0	REACTIVITY

PERSONAL PROTECTIVE EQUIP.:
GOGGLES, RUBBER GLOVES

Q-TAG® By IDESCO CORP. N.Y.C   Q2-111

Figure 2-10. Signs can remind workers to be alert for potential hazards. (Idesco Corp.)

Figure 2-11. Use a brush, not your hands, to remove accumulated chips!

• Secure prompt medical attention for any cut, bruise, scratch, burn, or other injury. No matter how minor the injury may appear, report it to your instructor!

2.3 GENERAL TOOL SAFETY

• *Never carry sharp-pointed tools in your pockets!* When using sharp tools, lay them on the bench in such a way that you will not injure yourself when you reach for them, **Figure 2-12.**
• Make sure tools are properly sharpened, in good condition, and fitted with suitable handles.

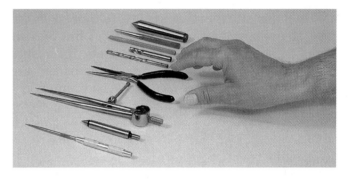

Figure 2-12. Arrange sharp pointed tools on the bench in a way that they will not injure you when you reach to pick them up.

2.4 FIRE SAFETY

Combustible materials are classified into four categories, **Figure 2-13.** Extinguishers should have color-coded symbols to identify their appropriateness for a particular type of fire.

Class A Fires. Those involving ordinary combustible materials—paper, wood, textiles, etc. They require the cooling and quenching effect of water, or solutions containing a large percentage of water. Do not use Class A extinguishers on Class C and D fires.

Class B Fires. Flammable liquid and grease fires require the blanketing or smothering effects of dry chemicals or carbon dioxide.

Class C Fires. Electrical equipment fires require nonconducting extinguishing agents that will smother the flames. Do not use Class A extinguishers on electrical fires.

Class D Fires. Extinguishers containing specially prepared heat-absorbing dry powder are used on flammable metals, such as magnesium and lithium. Do not use Class A extinguishers on flammable metal fires.

Ordinary combustibles	Flammable liquids	Electrical equipment	Combustible metals

Figure 2-13. Symbols coded by color and shape identify the four classifications of fire and the extinguishers that can be used on them.

2.4.1 Dealing with a Fire

Know what to do in case of a fire! Be familiar with the location of your building's fire exits and how they are opened. Be aware of alternate escape routes.

In some situations, students are trained in the use of fire extinguishers by the local fire department. If you are one of those students, make sure that you know where the fire extinguishers are located. If you have *not* received such training, get out of the fire area immediately.

SPECIAL SAFETY NOTE: *Think before acting!* It costs nothing, and you may be saved from painful injury that could result in a permanent disability. If you think it tiring to sit through an hour-long class in school, think what it would be like to spend your entire life in a wheelchair!

TEST YOUR KNOWLEDGE

Please do not write in this text. Write your answers on a separate sheet of paper.

1. Why is shop safety so important?

2. Most shop accidents are caused by _____.

3. Safety glasses should be worn:
   a. most of the time.
   b. only when working on machines.
   c. the entire time you are in the shop.
   d. none of the above.

4. Oily rags should be placed in a safety container to prevent _____.

5. Why should compressed air *not* be used to clean chips from machine tools?

6. Never attempt to operate a machine until _____.

7. Always stop machine tools before making _____ and _____.

8. Use a _____ to remove chips and shavings, *not* your _____.

9. When working in an area contaminated with dust or solvent fumes, be sure there is _____. A(n) _____ should also be worn when working in a dusty area.

10. Secure prompt _____ for any cut, bruise, scratch, or burn.

11. Get help when moving _____.

12. What should you do before operating a machine tool if you are taking medication of any sort?

13. Why is it necessary to take special precautions when handling long sections of metal stock?

Chapter 3

Understanding Drawings

LEARNING OBJECTIVES

After studying this chapter, you will be able to:
- ○ Read drawings that are dimensioned in fractional inches, decimal inches, and in metric units.
- ○ Explain the information found on a typical drawing.
- ○ Describe how detail, subassembly, and assembly drawings differ.
- ○ Point out why drawings are numbered.
- ○ Explain the basics of geometric dimensioning and tolerancing.

IMPORTANT TERMS

actual size
American National
   Standards Institute
bill of materials
dual dimensioning
geometric dimensioning
   and tolerancing

revisions
scale drawings
SI Metric
US Conventional
working drawings

Many products manufactured today are an assembly of parts supplied by a number of different industries. These industries may be in distant geographic locations, **Figure 3-1.**

It would not be possible for industry to manufacture a complex product without using drawings. *Drawings* show the craft worker what to make and identify the standards that must be followed so the various parts will fit together properly. The resulting parts will also be interchangeable with similar components on equipment already in service.

Drawings range from a simple freehand sketch, **Figure 3-2,** to detailed drawings for complex products, **Figure 3-3.**

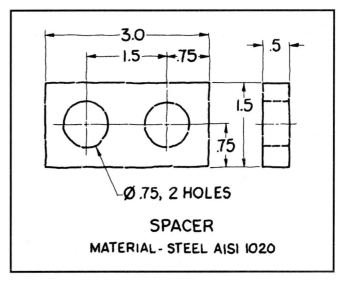

Figure 3-2. *Some drawings are as simple as this freehand sketch.*

Figure 3-1. *Thousands of drawings were required in the design and construction of this vertical takeoff and landing aircraft. Standards and specifications had to be exact because components were manufactured in several geographic locations.*
(Bell Helicopter Textron/Boeing Helicopters)

Symbols, lines, and figures are employed to give drawings meaning, **Figure 3-4.** They have been standardized so they have the same meaning wherever drawings are made and used.

These symbols, lines, and figures have been devised by the *American National Standards Institute,* better known as ANSI. The symbols, lines, and figures on drawings are known as the *"language of industry."*

Periodically, ANSI changes standard drawing symbols. Craft workers must be familiar with past and present practices because only recently made

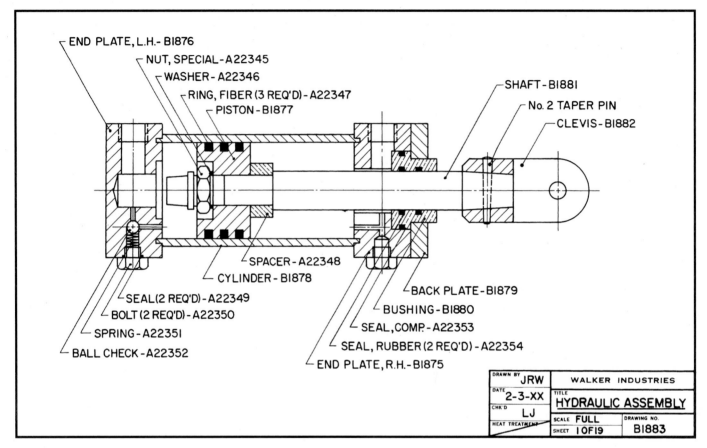

Figure 3-3. *Each manufactured product may require dozens of drawings, one for each part. Even the smallest screw, washer, or pin may require a drawing.*

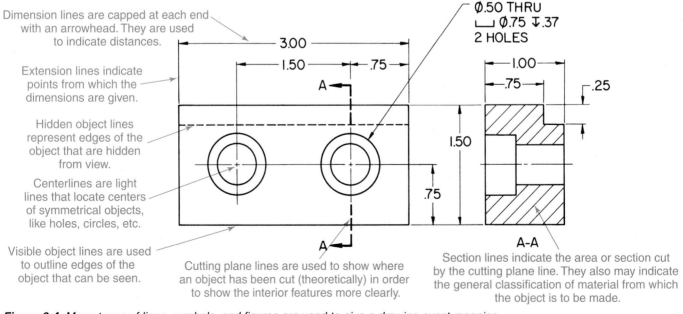

Figure 3-4. *Many types of lines, symbols, and figures are used to give a drawing exact meaning.*

drawings will follow the new standards. It is too expensive to revise the millions of drawings made before the new standards were devised. **Figure 3-5** shows past and present metalworking symbols.

Lines are used to draw views that fully describe the object to be manufactured. In addition, the drawing usually includes other information needed to make the product. Details often show threads, for example. **Figure 3-6** shows several methods of showing threads on a drawing.

3.1 DIMENSIONS

A proper drawing includes all *dimensions* (sizes or measurements) in proper relation to one another. The dimensions are needed to produce the part or object.

Until recently, drawings were only dimensioned in decimal or fractional parts of an inch, **Figure 3-7**. However, some industries in the United States are in the process of converting to the metric system of measurement. During this transition period, craft workers will need to understand drawings dimensioned in more than one system.

3.1.1 Fractional Dimensioning

Drawings using *fractional dimensioning* usually show objects that do not require a high degree of precision in their manufacture. Greater precision is indicated when dimensions are given in decimal parts of an inch.

3.1.2 Dual Dimensioning

Dual dimensioning is a system that employs the **US Conventional** ("English") system of fraction or decimal dimensions and metric dimensions on the same drawing, **Figure 3-8**. If the drawing is intended primarily for use in the United States, the decimal inch will appear above the metric dimension, as in **Figure 3-9A**. The reverse is true if the drawing is to be used in a metric-oriented country, as in **Figure 3-9B**. Some companies place the metric dimension within brackets, as in **Figure 3-9C**.

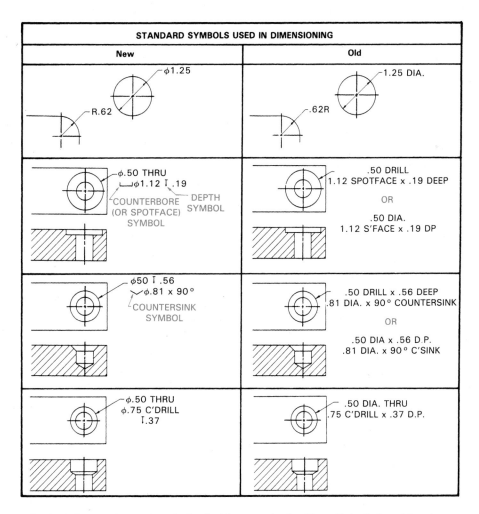

Figure 3-5. *Standard ANSI symbols are changed periodically. You must be familiar with both the old and new symbols because either may be used on the drawings. Compare these examples.*

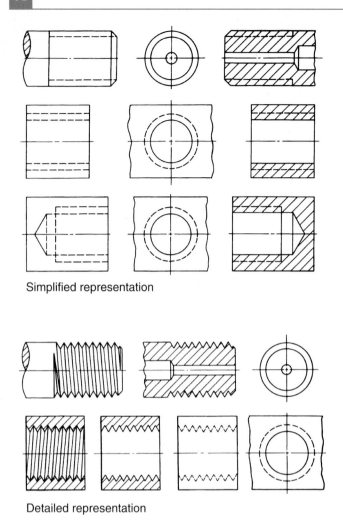

Simplified representation

Detailed representation

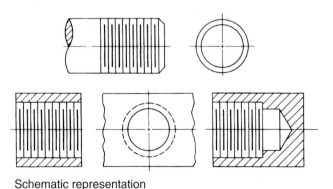

Schematic representation

Figure 3-6. Methods employed to depict threads on drawings. Only one type will be found on a drawing. The simplified version is the most common style.

3.1.3 Metric Dimensioning

With *metric dimensioning,* all of the dimensions on the drawing are in the *SI Metric* system, usually in millimeters. However, until there is full conversion to metric in the United States, a *conversion chart* (equivalents for millimeter and inch dimensions) will appear on the drawing, **Figure 3-10.**

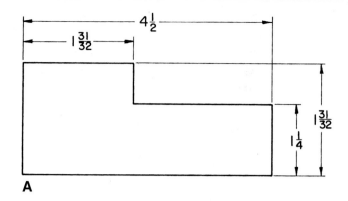

A

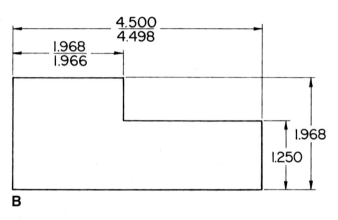

B

Figure 3-7. Dimensional styles. A—Fractional dimensions do not require tolerances closer than ±1/64″. B—Decimal dimensions normally have tolerances of ±0.010″ or 0.015″.

3.2 INFORMATION INCLUDED ON DRAWINGS

Drawings contain additional information to inform the machinist of the material to be used, required surface finish, tolerances, etc. It is important for you to be familiar with this information.

3.2.1 Materials

The general classification of materials to be used in the manufacture of an object may be indicated by the type of section line on the drawing or plan, **Figure 3-11.** Exact material specification is included in a section of the title block, **Figure 3-12(A).** Sometimes, the material specification may be found in the notes shown elsewhere on the drawing.

3.2.2 Surface Finishes

The quality of the *surface finish* (degree of surface smoothness) is important in the manufacture of many products. The smoothness of the bore of an engine cylinder is an example. Usually, the more

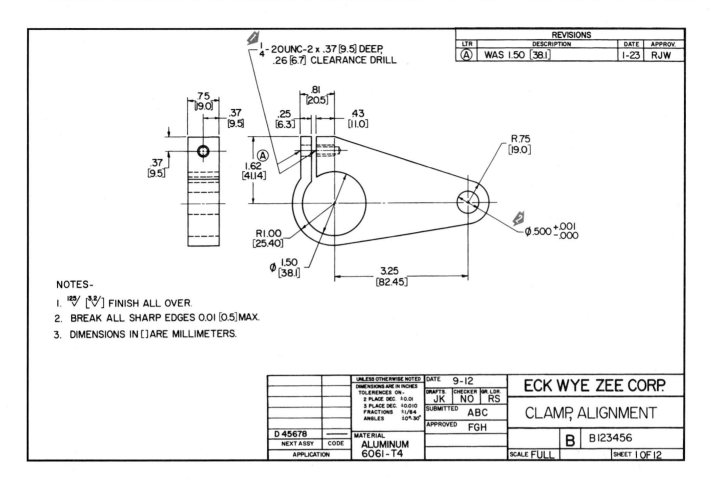

Figure 3-8. *Dual-dimensioned drawing. 1—A metric thread size has not been given. There is no metric thread that is equal to this size fractional thread. 2—There is no metric reamer equal to this size.*

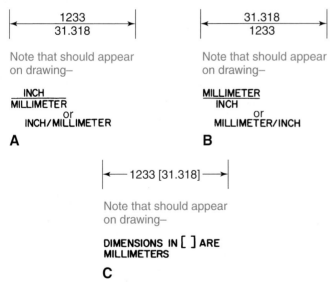

Figure 3-9. *Indicating inch and millimeter dimensions on a dual-dimensioned drawing. A—When the drawing is to be used in the United States, the inch value appears on top. B—When the drawing is to be used primarily in a country that uses the metric system, the millimeter value appears on top. C—Sometimes, brackets are used to indicate the metric equivalent on a drawing to be used in the United States.*

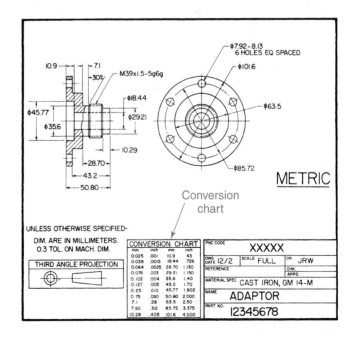

Figure 3-10. *A drawing using metric units of measurement. Until the United States has fully converted to metric measurement, a conversion chart will usually appear on the drawing.*

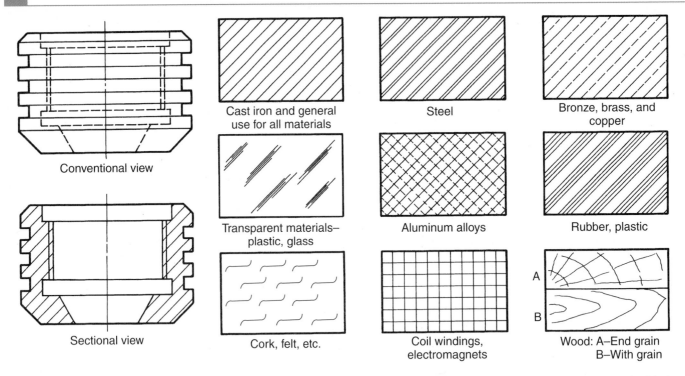

Conventional view

Sectional view

Cast iron and general use for all materials

Transparent materials– plastic, glass

Cork, felt, etc.

Steel

Aluminum alloys

Coil windings, electromagnets

Bronze, brass, and copper

Rubber, plastic

Wood: A–End grain B–With grain

Figure 3-11. *Sectional views make a drawing easier to understand because internal details are shown more clearly. Various materials are identified with unique section lines. However, many section views use general section lining regardless of the material being used.*

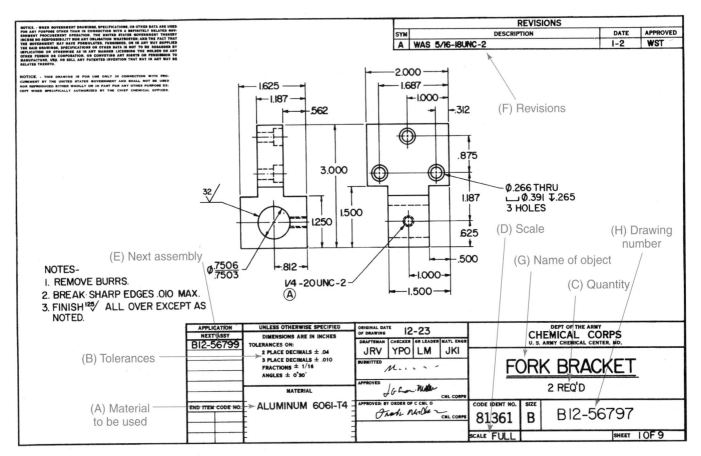

Figure 3-12. *A great deal of information is contained in the drawing's title block. The components highlighted here are standard on most drawings.*

superior (smoother) the finish of a machined surface, the more expensive it is to manufacture.

In the past, symbols were employed to indicate machined surfaces, **Figure 3-13.** These symbols may still be found on some older drawings. With so many machining techniques now in use, symbols such as these do not indicate, in sufficient detail, the quality of the surface finish required on a part.

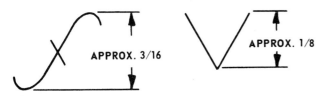

Figure 3-13. *These are older-style finish marks. They do not indicate the degree of smoothness required; they simply specify that the surface is machined. Finish marks of these types are still found on older drawings.*

The method presently used provides more complete surface information. Shown in **Figure 3-14,** a *check mark* and *number* are used to indicate surface roughness in microinches or micrometers. A *microinch* is one-millionth of an inch (0.000001″). A *micrometer* (micron) is one-millionth of a meter (0.000001 m) and is abbreviated μm.

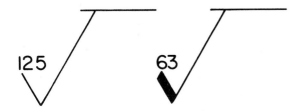

Figure 3-14. *Current surface finish marks. The number indicates the degree of smoothness in microinches—the larger the number, the rougher the finish.*

A machinist compares surface finishes to required specifications by using a *surface roughness comparison standard* as a guide, **Figure 3-15.** If the surface finish is critical, as it is in some jet engine components, the finish is measured electronically with a device called a *profilometer,* **Figure 3-16.**

3.2.3 Tolerances

The control of dimensions to achieve interchangeable manufacturing is known as tolerancing. This controls the size of the features of a part. A standard system of geometric dimensioning and tolerancing has been established.

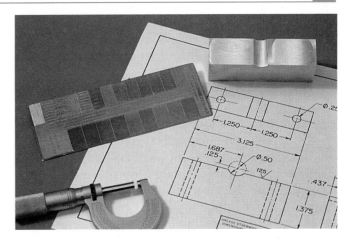

Figure 3-15. *A surface roughness comparison standard is used to check whether a milled surface meets the required specifications.*

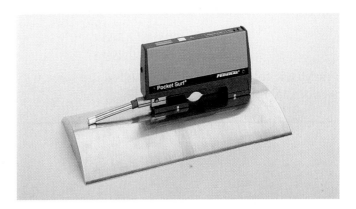

Figure 3-16. *Surface roughness is best determined with a profilometer or electronic surface roughness gage. The probe on the unit is moved across the work and measures surface roughness electronically. A digital display presents the measured roughness value in microinches or micrometers. (Federal Products Co.)*

Tolerances are allowances, either oversize or undersize, that are permitted when machining or making a part. Refer again to **Figure 3-12(B).** Acceptable tolerances are shown on drawings in several different ways.

When the dimension is given in fractional inch units, the permissible tolerances can be assumed to be ±1/64″, unless otherwise indicated.

The symbol "±" means that the machined surface can be *plus* (larger) or *minus* (smaller) by the dimension that follows and still be acceptable. In the example above, the dimension may be up to 1/64″ larger or smaller than the dimension given on the drawing. This "plus *and* minus" tolerance is called a *bilateral tolerance.*

If it is permissible to machine the part larger, but not smaller, the dimension on the drawing might

read
$$2\,1/2\,^{+1/64}$$.

If only a minus tolerance is permitted, the dimension might read
$$2\,1/2\,^{-1/64}$$.

When the tolerance is only plus or only minus (one direction), it is called a **unilateral dimension.**

Drawings dimensioned in decimal inches usually indicate that the work must be machined more precisely than dimensioning in fractional inches. The part can be used as long as the machined dimensions measure within these limits. Unless otherwise indicated, the tolerances can be assumed to be ±0.001".

A *plus* tolerance may be shown as:

$$2.500\,^{+\,.001} \quad\text{or}\quad \frac{2.501}{2.500}$$

A *minus* tolerance may be shown as:

$$2.500\,^{-\,.001} \quad\text{or}\quad \frac{2.500}{2.499}$$

Metric tolerances are presented in the same way as decimal tolerances, **Figure 3-17.**

3.2.4 Quantity of Units

Also shown on the drawing is the *number of parts* (quantity) needed in each assembly. Refer to **Figure 3-12(C).** A work order, included with the job information received by the shop, gives the total number of units to be manufactured. This facilitates ordering the necessary materials, and will help in determining the most economical way to manufacture the pieces.

3.2.5 Drawing Scale

Drawings made other than actual size (1:1) are called *scale drawings.* The scale is usually shown in a section of the title block, **Figure 3-12(D).** A drawing made one-half size would have a scale of 1:2 (one-to-two). A scale of 2:1 (two-to-one) would mean that the drawing is twice the size of the actual part.

3.2.6 Assembly or Subassembly

Assembly or *subassembly* information is necessary to correctly fit the various parts together. The

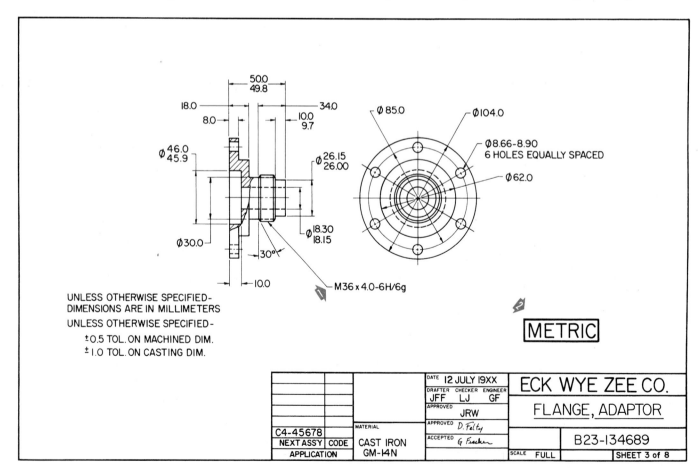

Figure 3-17. *A metric detail drawing. 1—Note that metric thread specifications are different from the more familiar UNC (coarse) and UNF (fine) series threads. The letter "M" denotes standard metric screw threads. The 36 indicates the nominal thread diameter in millimeters. The 4.0 denotes thread pitch in millimeters. The 6H and 6g are tolerance class designations. 2—To avoid possible misunderstanding, metric is shown on the drawing in large letters.*

term *application* is sometimes used in place of the term *next assembly*, **Figure 3-12(E)**.

3.2.7 Revisions

Revisions indicate what changes were made to the original drawing and when they were made. Refer again to **Figure 3-12(F)**.

3.2.8 Name of the Object

A portion of the title block provides this information. It tells the machinist the correct name of the piece, **Figure 3-12(G)**.

3.3 TYPES OF PRINTS

The original drawings are seldom used in the shop because they might be lost, damaged, or destroyed. On many jobs, several sets of plans are required. There are several methods of duplicating original drawings:

- *Blueprints.* The term *blueprint* is often used to refer to all types of prints. An *actual* blueprint has white lines on a blue background. However, the blueprint process is seldom used today because of the time required to make a print.
- *Diazo process.* These are direct positive copies (dark lines on a white background) of the original drawing. They are often referred to as *whiteprints* or *bluelines.*
- *Xerographic (electrostatic) process.* This process makes a copy of the original drawing. The print can be enlarged or reduced in size if necessary. Full color copies can be made on some xerographic machines.

- *Microfilm process.* This is a technique in which the original drawing is reduced by photographic means. Finished negatives can be stored in roll form or on cards, **Figure 3-18**. To produce a working print, the microfilm image is retrieved from files and enlarged onto photographic paper. The print is discarded or destroyed when it is no longer needed. Microfilms can also be viewed on a *reader* (machine for making enlarged projection on display screen). This technique is still widely employed for the storage of older drawings that would be too expensive to convert to computer data.

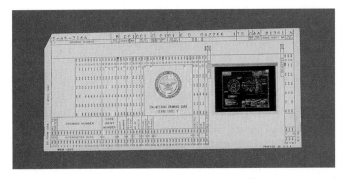

Figure 3-18. *The small negative on the microfilm aperture card is enlarged by a photographic process to the desired print size on a microfilm reader/printer. The enlarged print can be verified or confirmed on the view screen.*

- *Computer-generated prints.* The prints are generated on a plotter (automatic drafting machine) from information stored electronically in computer memory, **Figure 3-19**.

Figure 3-19. *Computer-generated print. Left—Many companies now use computer-aided design and drafting (CAD) techniques to prepare drawings. (Autodesk, Inc.) Right—Prints are generated on a plotter from CAD-developed information. (Hewlett-Packard Marketing Communications)*

This same information can also be used to control machine tools, using *CAM* (computer-aided manufacturing). When these methods are used, the overall manufacturing technique is called *CIM* (computer-integrated manufacturing). More information on computers in manufacturing will be included in later chapters.

3.4 TYPES OF DRAWINGS USED IN THE SHOP

Working drawings, also called *prints*, establish the standards for the product and show the craft worker what to make. There are two major kinds of working drawings:

- *Detail drawings.* These consist of a drawing (usually multiview) of the part with dimensions and other information for making the part, **Figure 3-20**.
- *Assembly drawings.* These drawings show where and how the parts, described on detail drawings, fit into the completed assembly. See **Figure 3-21**.

On large or complex products, *subassembly drawings* are used to show the assembly of a small portion of the completed object, **Figure 3-22**.

Some assembly and subassembly drawings are shown as *exploded pictorial drawings* (a drawing with parts separated, but in proper relationship). One is shown in **Figure 3-23**.

In most instances, a detail drawing provides information on just one item. However, if the mechanism is small in size or if it is composed of only a few parts, the detail and assembly drawings may appear on the same sheet, **Figure 3-24**.

3.5 PARTS LIST

Parts are identified by *circled numbers* and are listed in a note. Some drawings will also include a *parts list* or *bill of materials* listing all of the parts used in the assembly. See **Figure 3-25**.

3.6 DRAWING SIZES

Most firms centralize the preparation and storage of drawings in the engineering department. Generally, engineers and drafters prepare drawings

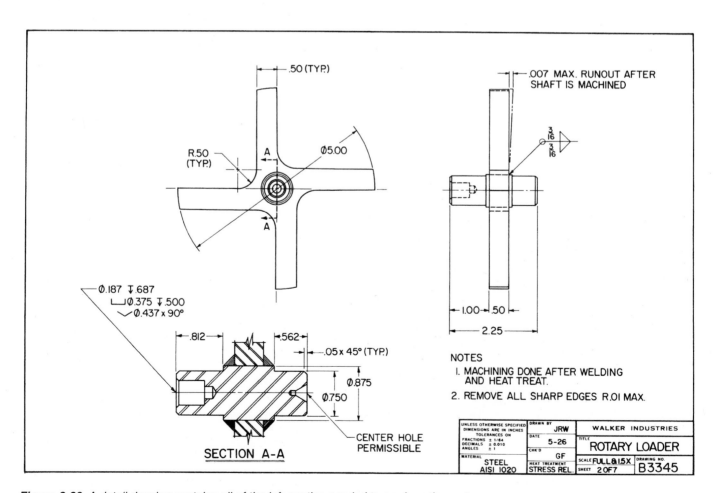

Figure 3-20. A detail drawing contains all of the information needed to produce the part.

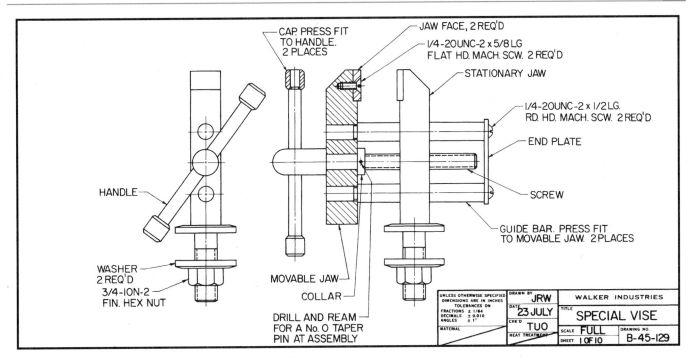

Figure 3-21. *An assembly drawing shows how various parts fit together.*

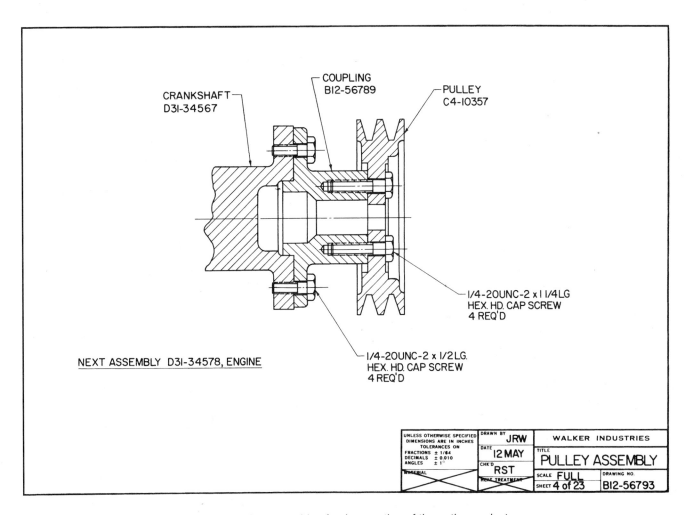

Figure 3-22. *A subassembly drawing contains the assembly of only a portion of the entire product.*

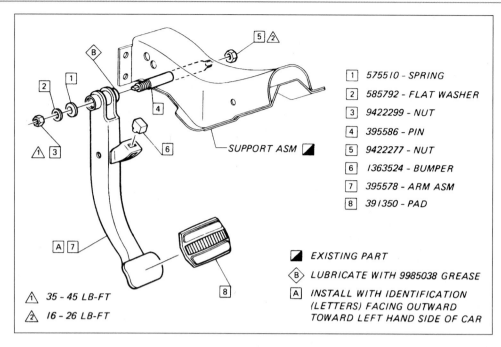

Figure 3-23. *Exploded pictorial drawings are often used with semiskilled workers who have received a minimum of training in print reading. (General Motors Corp.)*

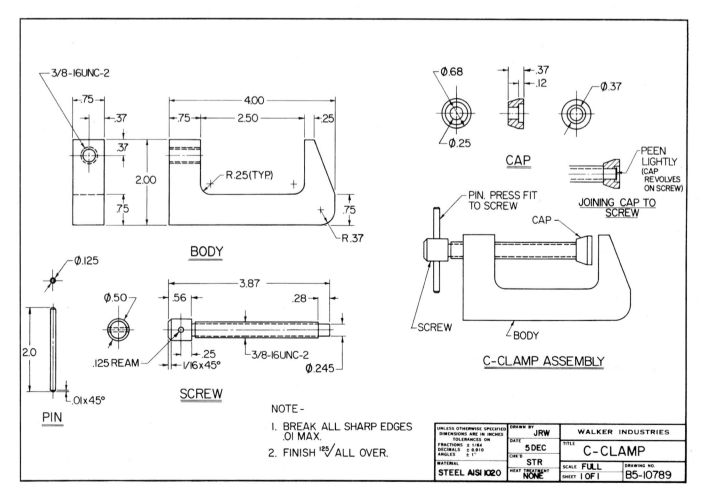

Figure 3-24. *A detail and assembly drawing on the same sheet.*

PARTS LIST

No.	Name	Quan.
1	CRANKCASE	1
2	CRANKSHAFT	1
3	CRANKCASE COVER	1
4	CYLINDER	2
5	PISTON	2

A

A1776	NUT	BRASS	6
A1985	BOLT	BRASS	6
B1765	PLATE	ALUMINUM	2
B1767	CYLINDER	CAST IRON	2
Pt. No.	Name	Material	Quan.
BILL OF MATERIALS			

B

Figure 3-25. A list of parts and materials is normally included with the drawings for a project. A—A typical, but partial, parts list. B—An example of a partial bill of materials.

Figure 3-26. A high degree of precision is needed to produce the parts used in this engine. The tolerances allowed for the shape and location of features on the parts must not be exceeded. (Buick Div. of GMC)

on standard-size sheets. This simplifies the stocking, handling, and storage of the completed drawings.

Standard sizes for drawing sheets include the following:

US CONVENTIONAL SHEET SIZES
A size = 8 1/2″ × 11″
B size = 11″ × 17″
C size = 17″ × 22″
D size = 22″ × 34″
E size = 34″ × 44″

SI METRIC SHEET SIZES
A4 size = 210 × 297 mm
A3 size = 297 × 420 mm
A2 size = 420 × 594 mm
A1 size = 594 × 841 mm
A0 size = 841 × 1189 mm

Also, for convenience in filing and locating drawings in storage, each drawing has an identifying number, **Figure 3-12(H).**

3.7 GEOMETRIC DIMENSIONING AND TOLERANCING

Conventional tolerancing is appropriate for many products. However, for accurately machined parts, the amount of variation (tolerances) in form (shape and size) and position (location) may need to be more strictly defined. This definition provides the precision needed to allow for the most economical manufacture of parts. See **Figure 3-26.**

Geometric dimensioning and tolerancing is a system that provides additional precision compared to conventional dimensioning. It ensures that parts can be easily interchanged.

Only a brief introduction to geometric dimensioning and tolerancing is included in this text. Detailed information can be found in the publication ASME Y14.5M-1994.

3.7.1 Definitions

Geometric characteristic symbols are employed to provide clarity and precision in communicating design specifications. See **Figure 3-27.** These symbols are standardized by the American Society of Mechanical Engineers (ASME). *Geometric tolerance* is a general term that refers to tolerances which control form, profile, orientation, location, and runout.

A *basic dimension* is a numerical value denoting the exact size, profile, orientation, or location of a feature. The *true position* of a feature is its theoretically exact location as established by basic dimensions. A *reference dimension* is a dimension provided for information only. It is not used for production or inspection purposes. See **Figure 3-28.**

Datum is an exact point, axis, or plane. It is the origin from which the location or geometric characteristic of features of a part is established. It is identified by a solid triangle with an identifying letter. See **Figure 3-29.** *Feature* is a general term applied to a physical portion of a part, such as a surface, pin,

Symbol for:	ASME Y14.5
Straightness	—
Flatness	▱
Circularity	○
Cylindricity	⌭
Profile of a line	⌒
Profile of a surface	⌓
All-around profile	⌾
Angularity	∠
Perpendicularity	⊥
Parallelism	//
Position	⊕
Concentricity/coaxiality	◎
Symmetry	⌯
Circular runout	* ↗
Total runout	* ↗↗
At maximum material condition	Ⓜ
At least material condition	Ⓛ
Regardless of feature size	NONE
Projected tolerance zone	Ⓟ
Diameter	⌀
Basic dimension	[30]
Reference dimension	(30)
Datum feature	[A]◄
Datum target	Ⓓ⑥/A1
Target point	✕
Dimension origin	⊕►
Feature control frame	⊕⌀0.5ⓂA B C
Conical taper	▷
Slope	◁
Counterbore/spotface	⌴
Countersink	⌵
Depth/deep	⌇
Square (shape)	□
Dimension not to scale	15
Number of times/places	8X
Arc length	⌒105
Radius	R
Spherical radius	SR
Spherical diameter	S⌀

* May be filled

Figure 3-27. Symbols used to specify positional and form tolerances in geometric dimensioning. (American National Standards Institute)

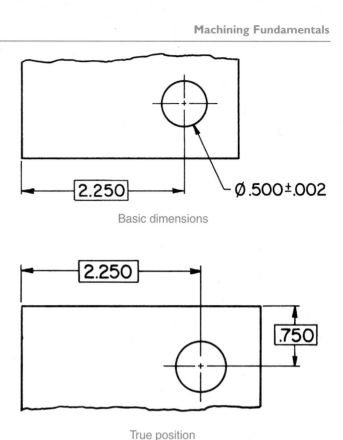

Basic dimensions

True position

Reference dimension

Figure 3-28. Basic dimensions are usually indicated by being enclosed in a rectangular frame. They are not toleranced. True position is the theoretical exact location of feature. It is established by basic dimensions. Reference dimensions are not used for production or inspection purposes. On a drawing, they are shown enclosed in parentheses.

hole, or slot. A *datum feature* is the actual feature of a part used to establish a datum. See **Figure 3-30.**

Maximum material condition (MMC) is the condition in which the size of a feature contains the maximum amount of material within the stated limits of size. Examples include a minimum hole diameter and maximum shaft diameter, both of which result in the greatest possible amount of material

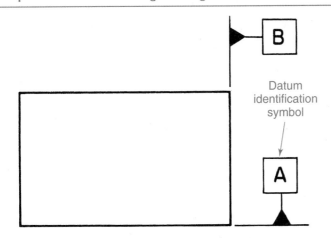

Figure 3-29. *Datums are exact points, axes, or planes from which features of a part are located.*

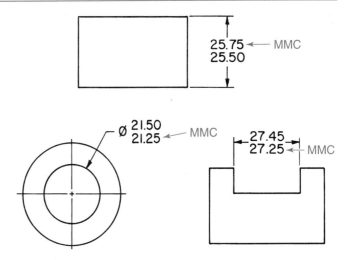

Figure 3-31. *Maximum material condition (MMC) indicates that the size of a feature contains the maximum amount of material within the stated tolerance limits.*

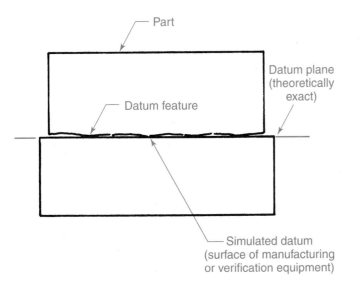

Figure 3-30. *A datum feature is a physical feature on a part used to establish a datum.*

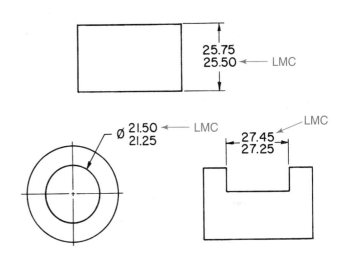

Figure 3-32. *Least material condition (LMC) indicates that the size of a feature contains the least amount of material within the stated limits of size.*

being used. See **Figure 3-31.** MMC is indicated by an *M* within a circle.

Least material condition (LMC) is the condition in which the size of a feature contains the least amount of material within the stated tolerance limits. Examples include a maximum hole diameter and a minimum shaft diameter. See **Figure 3-32.** LMC is indicated by an *L* within a circle.

Regardless of feature size (RFS) specifies that the size of a feature tolerance must not be exceeded. RFS is assumed for all geometric tolerances unless otherwise specified.

The maximum and minimum sizes of a feature are called the *limits of size.* See **Figure 3-33.** The measured size of a part after it is manufactured is the *actual size.*

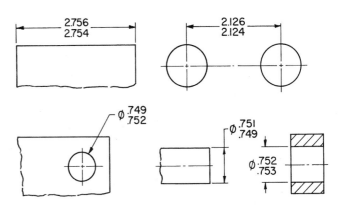

Figure 3-33. *Limits of size are the maximum and minimum sizes of a feature.*

3.7.2 Application of Geometric Dimensioning and Tolerancing

Datum identification symbol. A datum identifying symbol, **Figure 3-34,** consists of a square frame

that contains the *datum reference letter.* All letters but *I, O,* and *Q* may be used. A rectangular frame with the datum reference letter preceded and followed by a dash may be found on older drawings.

Feature control frame. A feature control frame is used when a location or form tolerance is related to a datum. It contains the geometric symbol, allowable tolerance, and the datum reference letter(s). It is connected to an extension line of the feature, a leader running to the feature, or below a leader-directed note of the feature, **Figure 3-35.**

Datum references indicated on the right end of the feature control frame are read from left to right. The letters signify datum preference. They establish three mutually perpendicular planes, **Figure 3-36.**

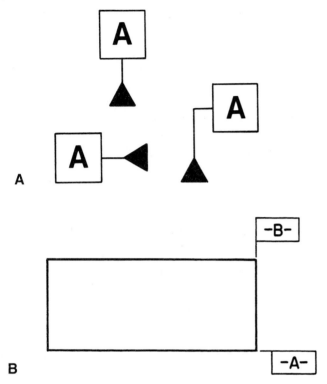

A

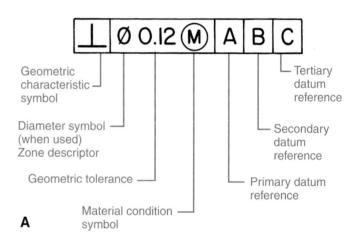

A

Figure 3-34. *Datum points and surfaces are identified by a datum identification symbol. A—Datum identification symbols used on new drawings. B—This type of datum symbol is not used currently, but is still found on old drawings.*

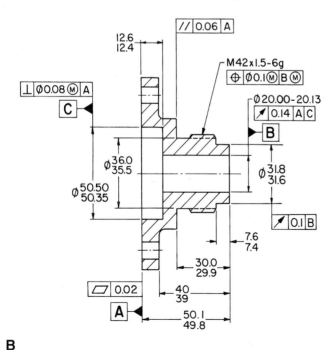

B

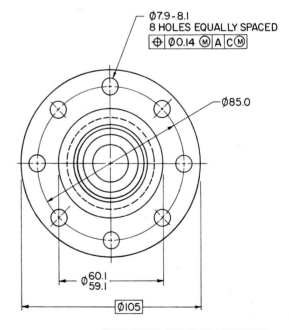

DIMENSIONS ARE IN MILLIMETERS

Figure 3-35. *A feature control frame is employed when a location or form tolerance is related to a datum. A—Components of a feature control frame. B—Feature control frames are used to specify tolerances on this drawing.*

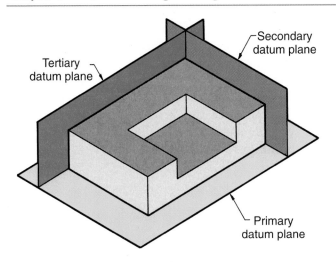

Figure 3-36. *Datum references are perpendicular planes. The first datum referenced is the primary datum, followed by the secondary and tertiary datums.*

3.7.3 Form Geometric Tolerances

Form geometric tolerances control flatness, straightness, circularity (roundness), and cylindricity. They are indicated by the symbols shown in **Figure 3-37.** Form tolerances control only the variation permitted on a single feature and are used when form variation is less than that permitted by size tolerance.

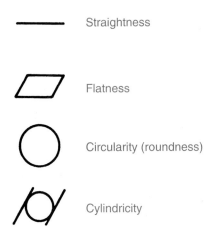

Figure 3-37. *Form geometric symbols.*

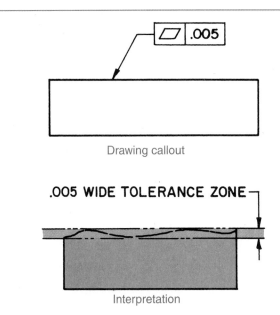

Figure 3-38. *The flatness geometric form tolerance specifies the two parallel planes within which a surface must lie.*

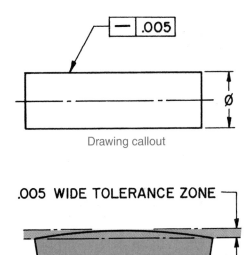

Figure 3-39. *A straightness geometric form tolerance establishes a tolerance zone of uniform width along a straight line. All elements of the surface must lie within this zone.*

Flatness is a measure of the variation of a surface perpendicular to its plane. The flatness geometric tolerance specifies the two parallel planes within which all points of a surface must lie, **Figure 3-38.**

Straightness describes how closely the surface of an object is to a line. A straightness geometric tolerance establishes a tolerance zone of uniform width along a line. All elements of the surface must lie within this zone, **Figure 3-39.**

Circularity is characterized by any given cross section taken perpendicular to the axis of a cylinder or a cone, or through the common center of a sphere. A circularity (roundness) geometric tolerance specifies a tolerance zone bounded by two concentric circles, indicated on a plane perpendicular to the axis of a cylinder or a cone, within which each circular element must lie. It is a single cross-sectional tolerance. See **Figure 3-40.**

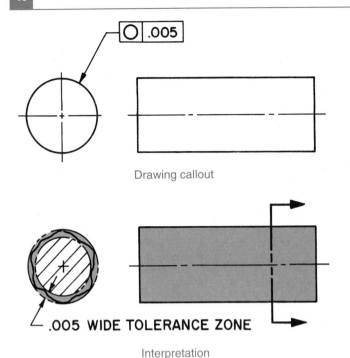

Drawing callout

.005 WIDE TOLERANCE ZONE

Interpretation

Figure 3-40. A circularity geometric tolerance specifies a tolerance zone bounded by two concentric circles on a plane perpendicular to the axis of a cylinder or cone, within which each circular element must lie.

Cylindricity represents a surface in which all points are an equal distance from a common center. The cylindricity geometric tolerance establishes a tolerance zone that controls the diameter of a cylinder throughout its entire length. It consists of two concentric cylinders within which the actual surface must lie. This tolerance covers both the circular and longitudinal elements. See **Figure 3-41.**

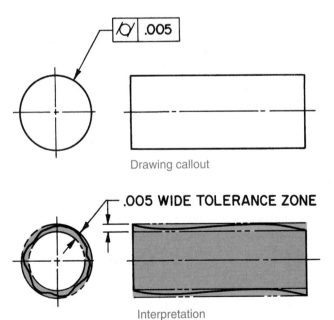

Drawing callout

.005 WIDE TOLERANCE ZONE

Interpretation

Figure 3-41. The cylindricity geometric tolerance establishes a tolerance zone that controls the diameter of a cylinder throughout its entire length.

3.7.4 Profile Geometric Tolerances

A *profile geometric tolerance* controls the outline or contour of an object and can be represented by an external view or by a cross section through the object. It is a boundary along the true profile in which elements of the surface must be contained. The symbols used to indicate profile tolerances are shown in **Figure 3-42.**

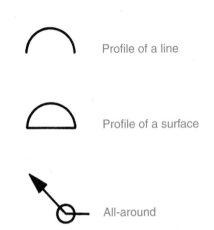

Profile of a line

Profile of a surface

All-around

Figure 3-42. Profile geometric tolerance symbols. When a tolerance is specified for all sides of an object, the "all-around" symbol is used.

A *profile line geometric tolerance* is a two-dimensional (cross-sectional) tolerance zone extending along the length of the element. It is located using basic dimensions, **Figure 3-43.**

The *profile surface geometric tolerance* is three-dimensional and extends along the length and width of the surface. For proper orientation of the profile, a datum reference is usually required, **Figure 3-44.**

3.7.5 Orientation Geometric Tolerances

Orientation geometric tolerances control the degree of parallelism, perpendicularity, or angularity of a feature with respect to one or more datums. There are three orientation tolerances, **Figure 3-45.**

Angularity is concerned with the position of a surface or axis at a specified angle to a datum plane or axis. The specified angle must be other than 90°. An *angularity geometric tolerance* establishes a tolerance zone defined by two parallel lines, planes, or a cylindrical zone at a specified basic angle other than 90°. The line elements, surface, or axis of the considered feature must lie within this zone, **Figure 3-46.**

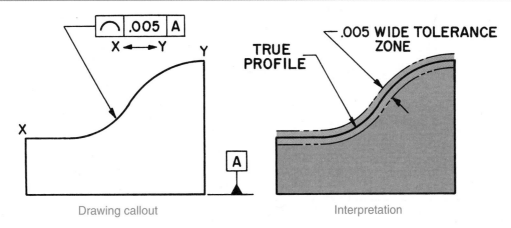

Figure 3-43. *A profile line geometric tolerance is a two-dimensional tolerance zone extending along the length of the considered element.*

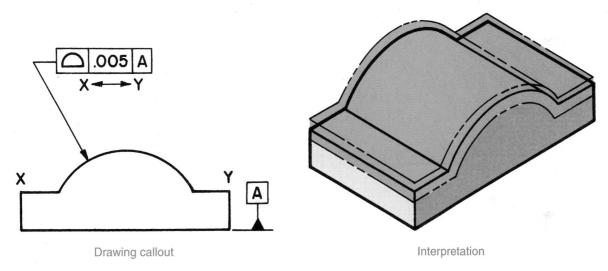

Figure 3-44. *The profile surface geometric tolerance is three-dimensional and extends along the length and width of the surface.*

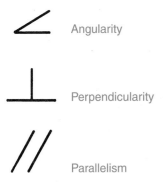

Figure 3-45. *Orientation geometric tolerance symbols.*

A *perpendicularity geometric tolerance* specifies a tolerance zone at right angles to a given datum or axis. It is described by two parallel lines, planes, or a cylindrical tolerance zone. The line, surface, or axis of the considered feature must lie within this zone, **Figure 3-47.**

Parallelism describes how close all elements of a line or surface are to being parallel (equidistant) to a given datum plane or axis. A *parallelism geometric tolerance* is a tolerance zone defined by two lines parallel to a datum within which the elements of a surface or axis must lie, **Figure 3-48.**

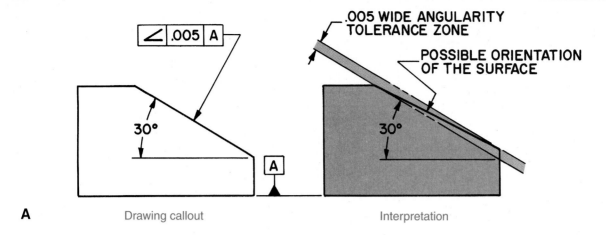

A   Drawing callout   Interpretation

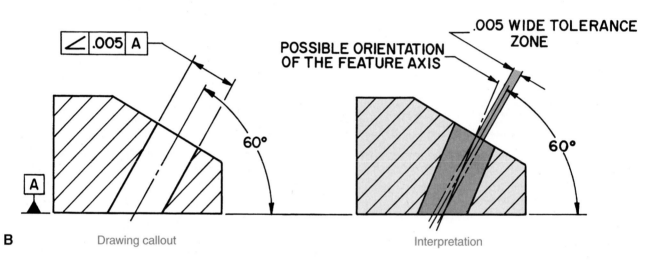

B   Drawing callout   Interpretation

Figure 3-46. An angularity geometric tolerance establishes a tolerance zone defined by two parallel lines, planes, or a cylindrical zone at a specified basic angle other than 90°. A—Angularity of a surface. B—Angularity of an axis.

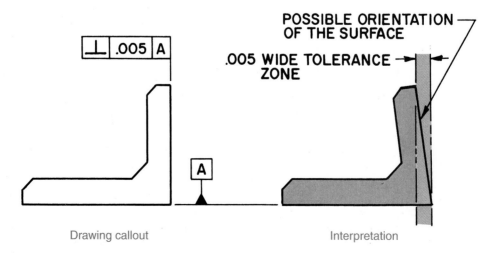

          Drawing callout   Interpretation

Figure 3-47. The line, surface, or axis of a considered feature must lie within the perpendicularity geometric tolerance zone.

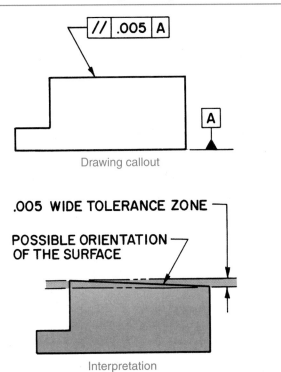

Figure 3-48. *A parallelism geometric tolerance is a tolerance zone defined by two lines parallel to a datum within which the elements of a surface or axis must lie.*

3.7.6 Location Geometric Tolerances

Location geometric tolerances are employed to establish the location of features and datums. They define the zone within which the center, axis, or center plane of a feature may vary from a true (theoretically exact) position. Location tolerances are also known as *positional tolerances* and include position, concentricity, and symmetry. See **Figure 3-49**.

Basic dimensions establish the true position of a feature from specified datums and related features. A *positional geometric tolerance* establishes how far a feature may vary from its true position, **Figure 3-50**.

Concentricity defines the relationship between the axes of two or more of an object's cylindrical features. A *concentricity geometric tolerance* is expressed

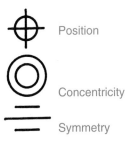

Figure 3-49. *Location or positional tolerance symbols.*

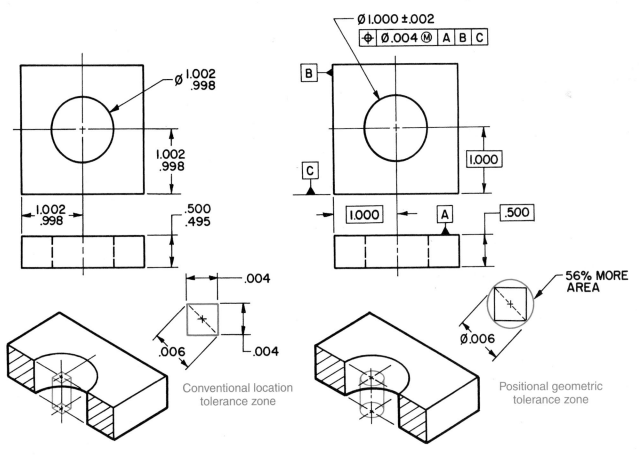

Figure 3-50. *A positional geometric tolerance establishes how far a feature may vary from its true position.*

as a cylindrical tolerance zone. The axis or center point of this zone coincides with a datum axis, **Figure 3-51.**

Since this tolerance is sometimes difficult and time-consuming to verify, runout or positional geometric tolerances are often used instead.

Symmetry indicates equal or balanced proportions on either side of a central plane or datum, **Figure 3-52.** A *symmetry geometric tolerance* is a zone within which the symmetrical surfaces align with the datum of a center plane or axis, **Figure 3-53.**

3.7.7 Runout Geometric Tolerances

There are two types of *runout geometric tolerances*—total runout and circular runout. These tolerances are indicated by the symbols shown in **Figure 3-54.** Runout tolerances are used to control runout of surfaces around or perpendicular to a datum axis.

Total runout controls circularity, straightness, angularity, and cylindricity of a part when applied to surfaces rotated around a datum axis, **Figure 3-55.** The entire surface must lie within the tolerance zone.

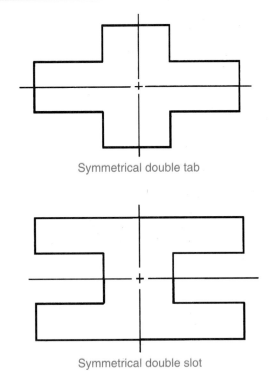

Symmetrical double tab

Symmetrical double slot

Figure 3-52. Symmetry indicates equal or balanced proportions on either side of a central plane.

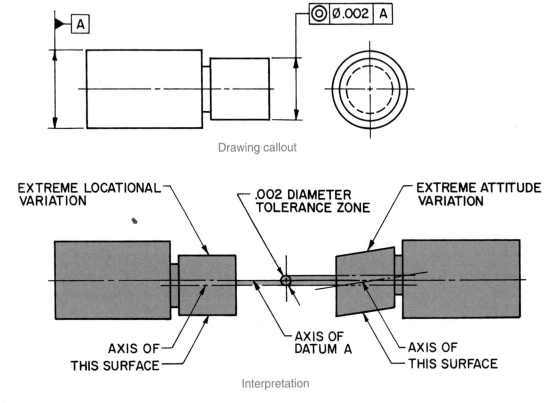

Drawing callout

EXTREME LOCATIONAL VARIATION

.002 DIAMETER TOLERANCE ZONE

EXTREME ATTITUDE VARIATION

AXIS OF THIS SURFACE

AXIS OF DATUM A

AXIS OF THIS SURFACE

Interpretation

Figure 3-51. A concentricity geometric tolerance is expressed as a cylindrical tolerance zone. The axis or center point of this zone coincides with a datum axis.

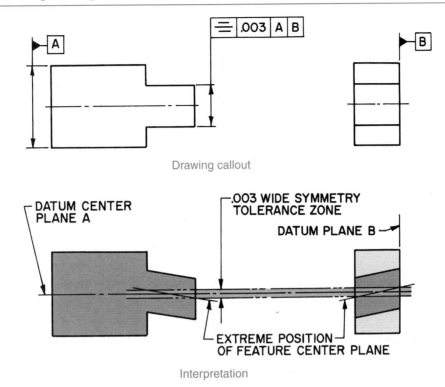

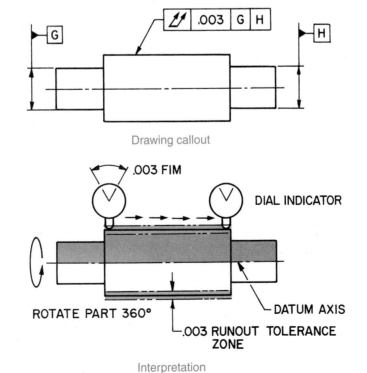

Figure 3-53. *A symmetry geometric tolerance is a zone within which the symmetrical surfaces align with the datum of a center plane or axis.*

Circular runout   Total runout

Figure 3-54. *Runout geometric tolerance symbols. Arrows may be filled or unfilled.*

Figure 3-55. *Total runout controls circularity, straightness, angularity, and cylindricity of a part when applied to surfaces rotated around a datum. The entire surface must lie within the tolerance zone.*

Circular runout is applied to features independently and controls circularity of a single circular cross section, **Figure 3-56.** The tolerance is measured by the *full indicator movement (FIM)* of a dial indicator when it is placed at several positions as the part is rotated.

3.7.8 Summary

Geometric dimensioning and tolerancing is far more involved than described on the preceding pages. As you progress in machining technology, you should consider purchasing a text on the subject, studying a copy of ASME Y14.5M–1994, or enrolling in a class on geometric dimensioning and tolerancing.

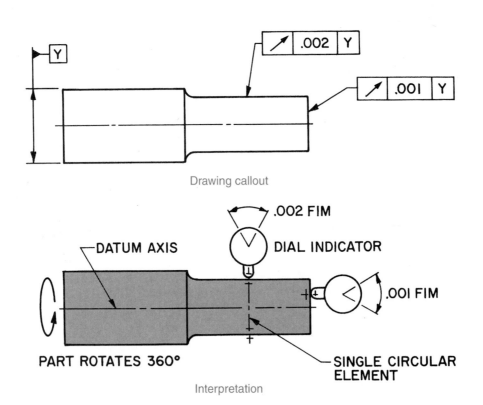

Figure 3-56. *Circular runout controls circularity of a single circular cross section.*

TEST YOUR KNOWLEDGE

Please do not write in the text. Write your answers on a separate sheet of paper.

1. Drawings are used to:
   a. Show, in multiview, what an object looks like before it is made.
   b. Standardize parts.
   c. Show what to make and the sizes to make it.
   d. All of the above.
   e. None of the above.

2. The symbols, lines, and figures that make up a drawing are frequently called the _____.

3. A microinch is _____ of an inch.

4. A micrometer is _____ of a meter.

5. How can surface roughness of a machined part be checked against specifications on the drawing? How can it be measured electronically?

6. When tolerances are plus and minus, it is called a _____ tolerance.

7. When tolerances are only plus or only minus, it is called a _____ tolerance.

8. Tolerances are:
   a. The different materials that can be used.
   b. Allowances in either oversize or undersize that a part can be made and still be acceptable.
   c. Dimensions.
   d. All of the above.
   e. None of the above.

9. Drawings made other than actual size are called _____.

10. A subassembly drawing differs from an assembly drawing by:
    a. Showing only a small portion of the complete object.
    b. Making it possible to use smaller drawings.
    c. Showing the object without all needed dimensions.
    d. All of the above.
    e. None of the above.

11. Why are prints used in place of the original drawings?

12. The craft worker is given all of the information needed to make a part on a _____ drawing.

13. What does an assembly drawing show?

14. Why are standard size drawing sheets used?

15. All dimensions have a tolerance except _____ dimensions.

16. Dimensions placed between parentheses are _____ dimensions.

17. When is a feature control frame employed?

18. Sketch the form geometric tolerance symbols and indicate what they mean.

19. Define the term *maximum material condition* (*MMC*). Use a sketch if necessary.

20. Define the term *least material condition* (*LMC*). Use a sketch if necessary.

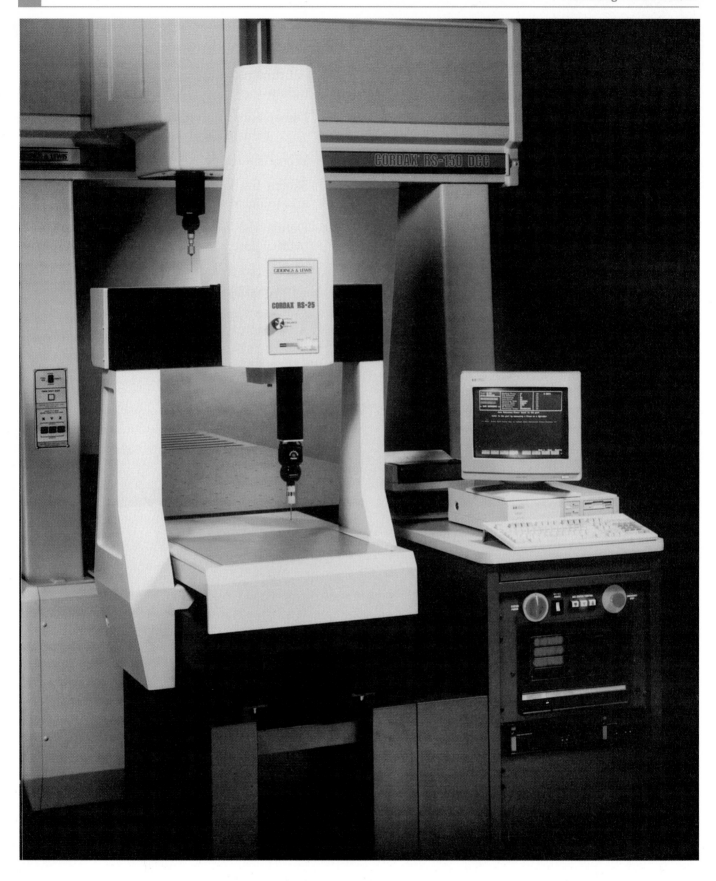

Extremely accurate measurements of small parts (up to 30″ × 30″ × 27″ or 762 mm × 762 mm × 686 mm) can be done on the shop floor by a bridge-type coordinate measuring machine like this one. The CMM is computer-controlled, and can make repeated measurements with an accuracy of 0.00012″ (0.003 mm). (Giddings & Lewis, Inc.)

Chapter 4

Measurement

IMPORTANT TERMS

dial indicators
gage blocks
gaging
graduations
helper measuring tools
International System of Units
metrology
micrometer caliper
steel rule
Vernier caliper

Without some form of accurate measurement, modern industry could not exist. The science that deals with systems of measurement is called *metrology.* Today, industry can make measurements accurate to one microinch (one-millionth of an inch).

If a microinch were as thick as a dime, one inch would be as high as *four* Empire State Buildings (about 5000′ total). An engineer once estimated, with tongue in cheek, that a steel railroad rail supported at both ends would sag one-millionth of an inch when a "fat horsefly" landed on it in the middle.

In addition to using US Conventional units of measure (inch, foot, etc.), industry is gradually converting to metric units of measure (millimeter,

meter, etc.), called the *International System of Units* (abbreviated SI). A micrometer is one-millionth of a meter (0.000001 m).

All of the familiar measuring tools are available with scales graduated in metric units, **Figure 4-1.** An SI Metric (millimeter) rule is compared with conventional fractional and decimal rules in **Figure 4-2.** Metric-based measuring tools should offer no problems for the user. As a matter of fact, they are often *easier* to read than inch-based measuring tools.

Although you will measure in very tiny units when you go to work in industry, you must first learn to read a rule to 1/64″ and 0.5 mm. Then, you can progress through 1/1000″ (0.001″) and 1/100 mm (0.01 mm) by learning to use micrometer and Vernier-type measuring tools. Finally, you can progress to 1/10,000″ (0.0001″) and 1/500 mm (0.002 mm) by using the Vernier scale on some micrometers.

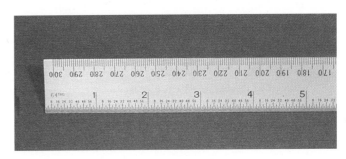

Figure 4-1. *This rule can be used to make measurements in both US Conventional and SI Metric units.*

4.1 THE RULE

The *steel rule,* often incorrectly called a *scale,* is the simplest of the measuring tools found in the shop. **Figure 4-2** shows the three basic types of rule graduations. A few of the many rule styles are shown in **Figure 4-3.**

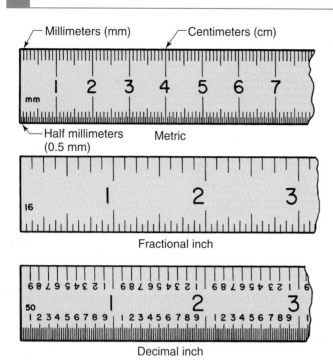

Figure 4-2. *Compare the metric (millimeter-graduated) rule with the more familiar rules graduated in fractional and decimal inch units.*

4.1.1 Reading the Rule (US Conventional)

A careful study of the enlarged rule section will show the different fractional divisions of the inch from 1/8 to 1/64, **Figure 4-4.** The lines representing the divisions are called *graduations.* On many rules, every fourth graduation is numbered on the 1/32 edge, and every eighth graduation on the 1/64 edge.

To become familiar with the rule, begin by measuring objects on the 1/8 and 1/16 scales. Once you become comfortable with these scales, begin using the 1/32 and 1/64 scales. Practice until you can quickly and accurately read measurements. Some rules are graduated in 10ths, 20ths, 50ths, and 100ths. Additional practice will be necessary to read these rules.

Fractional measurements are *always* reduced to the lowest terms. A measurement of 14/16" is reduced to 7/8", 2/8" becomes 1/4", and so on.

4.1.2 Reading the Rule (Metric)

Most metric rules are divided into millimeter or one-half millimeter graduations. They are

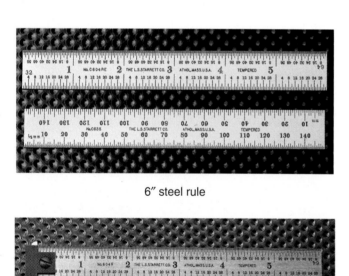

6" steel rule

Rule with adjustable hook

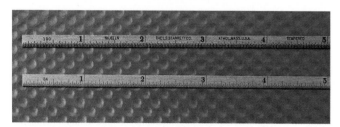

Narrow rule

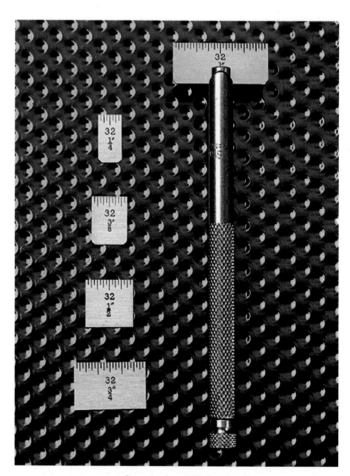

Small rules with holder

Figure 4-3. *Many different types of rules are used to make measuring quicker and more accurate. (L. S. Starrett Co.)*

numbered every 10 mm. See **Figure 4-5.** The measurement is determined by counting the number of millimeters.

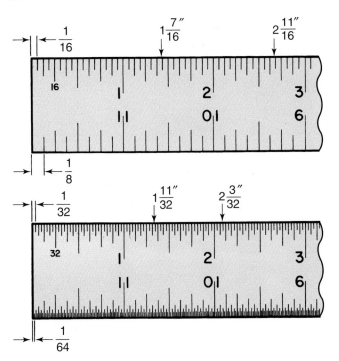

Figure 4-4. *These are the fractional graduations found on a rule. Measurements are taken by counting the number of graduations.*

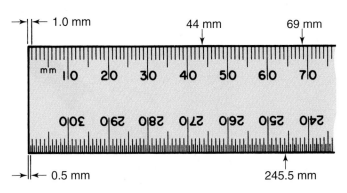

Figure 4-5. *Most metric rules are graduated in millimeters and half-millimeters. They are available in a variety of sizes.*

4.1.3 Care of the Rule

The steel rule is precision-made and, like all tools, its accuracy depends upon the care it receives. Here are a few suggestions:

- Use the rule for measurements only. Do not adjust screws or open paint cans with it. Be careful not to bend your rule.
- Keep the rule clear of moving machinery. Never use it to clean metal chips as they form on the cutting tool. This is extremely dangerous and will ruin the rule.
- Avoid laying other tools on the rule.

- Wipe steel rules with an oily cloth before storing. This will prevent rust. If the rule is to be stored for a prolonged period, coat it with wax or rust preventative.
- Clean the rule with steel wool to keep the graduations legible.
- Make measurements and tool settings from the 1″ line (10 mm line on a metric rule) or other major graduations, rather than from the end of the rule.
- Store rules separately. Do not throw them in a drawer with other tools.
- Use the rule with care to protect the ends from nicks and wear.
- Use the correct rule for the job being done.

4.2 THE MICROMETER CALIPER

A Frenchman, Jean Palmer, devised and patented a measuring tool that made use of a screw thread, making it possible to read measurements quickly and accurately without calculations. It incorporated a series of engraved lines on the sleeve and around the thimble. The device, called *Systeme Palmer,* is the basis for the modern micrometer caliper, **Figure 4-6.**

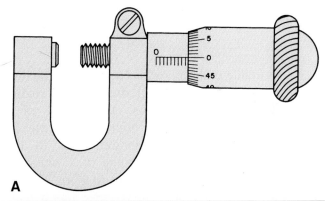

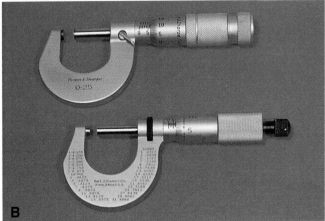

Figure 4-6. *The micrometer caliper, past and present. A—A drawing of the* Systeme Palmer *measuring device. B—These modern micrometer calipers operate on the same principle as the original 1848 invention.*

The *micrometer caliper,* also known as a "mike," is a precision tool capable of measuring to 0.001" or 0.01 mm. When fitted with a Vernier scale, it will read to 0.0001" or 0.002 mm.

4.2.1 Types of Micrometers

Micrometers are produced in a wide variety of models. Digital display is included in many micrometers, making measuring easier. Some of the most popular models are the following:

- An *outside micrometer* measures external diameters and thickness, **Figure 4-7.**

Figure 4-7. *This digital outside micrometer can be used to measure in both US Conventional and SI Metric units. (Mitutoyo/MTI Corp.)*

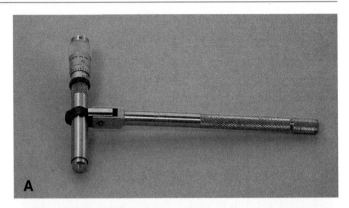

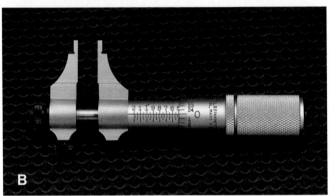

Figure 4-8. *Inside micrometers. A—A conventional inside micrometer. B—The caliper jaws on this inside micrometer allow quick and accurate measurements. The divisions on the sleeve are numbered in the reverse order of a conventional outside micrometer. (L. S. Starrett Co.)*

- An *inside micrometer* has many uses, including measuring internal diameters of cylinders, rings, and slots. The range of a *conventional inside micrometer* can be extended by fitting longer rods to the micrometer head. The range of a *jaw-type inside micrometer* is limited to 1" or 25 mm. The jaw-type inside micrometer has a scale graduated from right to left. See **Figure 4-8.**
- A *micrometer depth gage* measures the depths of holes, slots, and projections. See **Figure 4-9.** The measuring range can be increased by changing to longer spindles. Measurements are read from right to left.
- A *screw-thread micrometer* has a pointed spindle and a double-V anvil shaped to contact the screw thread, **Figure 4-10.** It measures the pitch diameter of the thread, which equals the outside (major) diameter of the thread minus the depth of one thread. Since each thread micrometer is designed to measure only a limited number of threads per inch, a set of thread micrometers is necessary to measure a full range of thread pitches.

Figure 4-9. *A standard micrometer depth gage.*

- A *chamfer micrometer* will accurately measure countersunk holes and other chamfer-type measurements. With fastener

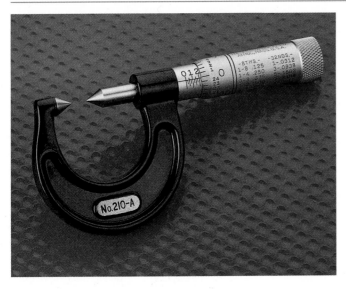

Figure 4-10. *This screw thread micrometer can measure threads as wide as 7/8". (L. S. Starrett Co.)*

tolerances so critical on some aerospace and other applications, it is important that countersunk holes and tapers on fasteners meet specifications. A chamfer micrometer makes it possible to check these critical areas.

Special micrometers are available for other applications. These micrometers are devised to handle nonstandard measurement tasks.

4.2.2 Reading an Inch-Based Micrometer

A micrometer uses a very precisely made screw thread that rotates in a fixed nut. The screw thread is ground on the *spindle* and is attached to the *thimble*. The spindle advances or recedes from the *anvil* as the thimble is rotated. See **Figure 4-11**. The threaded section has 40 threads per inch; therefore, each revolution of the thimble moves the spindle 1/40" (0.025").

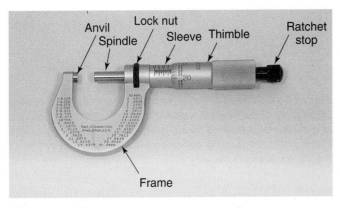

Figure 4-11. *Basic parts of a micrometer caliper.*

The line engraved lengthwise on the sleeve is divided into 40 equal parts per inch (corresponding to the number of threads per inch on the spindle). Each vertical line equals 1/40", or 0.025". Every fourth division is numbered, representing 0.100", 0.200", etc.

The beveled edge of the thimble is divided into 25 equal parts around its circumference. Each division equals 1/1000" (0.001"). On some micrometers, every division is numbered, while every fifth division is numbered on others.

The micrometer is read by recording the highest number on the sleeve (1 = 0.100, 2 = 0.200, etc.). To this number, add the number of vertical lines visible between the number and thimble edge (1 = 0.025, 2 = 0.050, etc.). To this total, add the number of thousandths indicated by the line that coincides with the horizontal sleeve line.

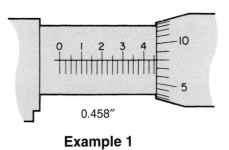

0.458"

Example 1

Add the readings from the sleeve and the thimble:

4 large graduations:	4 × 0.100	= 0.400
2 small graduations:	2 × 0.025	= 0.050
8 thimble graduations:	8 × 0.001	= 0.008
Total mike reading		= 0.458"

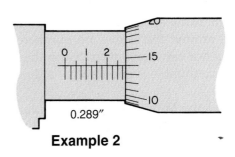

0.289"

Example 2

Add the readings from the sleeve and the thimble:

2 large graduations:	2 × 0.100	= 0.200
3 small graduations:	3 × 0.025	= 0.075
14 thimble graduations:	14 × 0.001	= 0.014
Total mike reading		= 0.289"

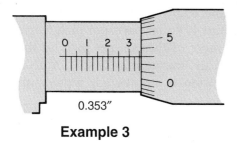

Example 3

Add the readings from the sleeve and the thimble:

3 large graduations:	$3 \times 0.100 = 0.300$
2 small graduations:	$2 \times 0.025 = 0.050$
3 thimble graduations:	$3 \times 0.001 = \underline{0.003}$
Total mike reading	$= 0.353''$

4.2.3 Reading a Vernier Micrometer

On occasion, it is necessary to measure more precisely than 0.001". A Vernier micrometer caliper is used in these situations. This micrometer has a third scale around the sleeve that will furnish the 1/10,000" (0.0001") reading. See **Figure 4-12.**

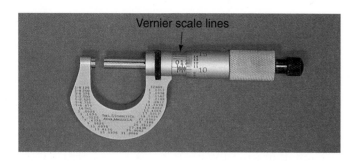

Figure 4-12. A Vernier micrometer caliper includes a Vernier scale on the sleeve.

The Vernier scale has 11 parallel lines that occupy the same space as 10 lines on the thimble. The lines around the sleeve are numbered 1 to 10. The difference between the spaces on the sleeve and those on the thimble is one-tenth of a thousandth of an inch.

To read the Vernier scale, first obtain the thousandths reading, then observe which of the lines on the Vernier scale coincides (lines up) with a line on the thimble. Only one of them can line up. If the line is 1, add 0.0001 to the reading; if line 2, add 0.0002 to the reading, etc. See **Figure 4-13.**

4.2.4 Reading a Metric-Based Micrometer

The metric-based micrometer is read as shown in **Figure 4-14.** If you are able to read the conven-

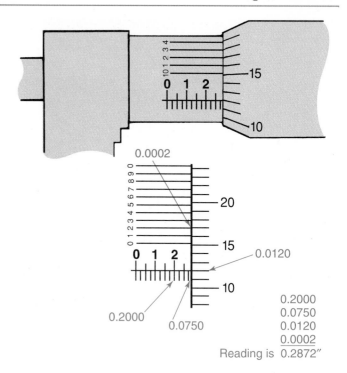

	0.2000
	0.0750
	0.0120
	0.0002
Reading is	0.2872"

Figure 4-13. How to read a Vernier micrometer caliper. Add the total reading in thousandths, then observe which of the lines on the Vernier scale coincides with a line on the thimble. In this case, it is the second line, so 0.0002 is added to the reading.

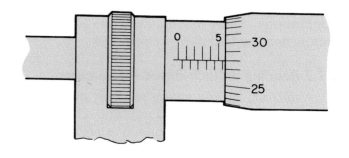

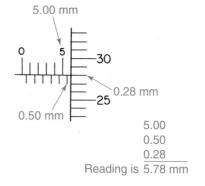

	5.00
	0.50
	0.28
Reading is	5.78 mm

Figure 4-14. To read a metric micrometer, add the total reading in millimeters visible on the sleeve to the reading of hundredths of a millimeter, indicated by the graduation on the thimble. Note that the thimble reading coincides with the longitudinal line on the micrometer sleeve.

tional inch-based micrometer, reading the metric-based tool will offer no difficulties.

4.2.5 Reading a Metric Vernier Micrometer

Metric Vernier micrometers are read in the same way as standard metric micrometers. However, using the Vernier scale on the sleeve, an additional reading of two-thousandths of a millimeter can be obtained, **Figure 4-15.**

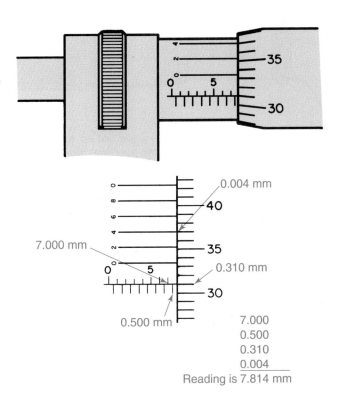

0.004 mm

7.000 mm

0.310 mm

0.500 mm

7.000	
0.500	
0.310	
0.004	

Reading is 7.814 mm

Figure 4-15. *Reading a metric-based Vernier micrometer caliper. To the regular reading in hundredths of a millimeter (0.01), add the reading from the Vernier scale that coincides with a line on the thimble. Each line on the Vernier scale is equal to two thousandths of a millimeter (0.002 mm).*

4.2.6 Using the Micrometer

The proper way to hold a micrometer when making a measurement is shown in **Figure 4-16.** The work is placed into position, and the thimble rotated until the part is clamped *lightly* between the anvil and spindle. Guard against excessive pressure, which will cause an erroneous reading. Some micrometers have features to help regulate pressure:

- A *ratchet stop* is used to rotate the spindle. When the pressure reaches a predetermined amount, the ratchet stop slips and prevents further spindle turning. Uniform contact pressure with the work is ensured, even if different people use the same micrometer. Refer again to **Figure 4-11.**
- A *friction thimble* may be built into the upper section of the thimble. This produces the same

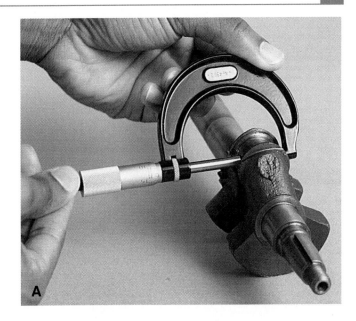

A

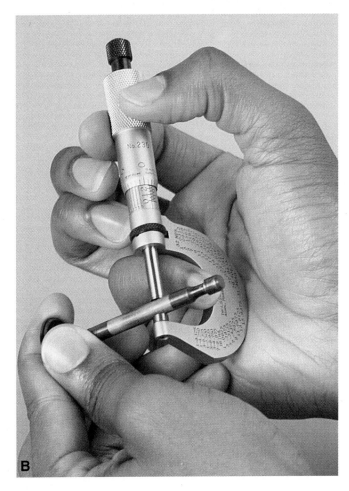

B

Figure 4-16. *Proper technique of handling a micrometer. A—Use very light pressure when turning the thimble. B—When the piece being measured must also be held, position the micrometer as shown, with a finger in the micrometer frame.*

results as the ratchet stop but permits one-handed use of the micrometer.

• A *lock nut* is used when several identical parts are to be gaged. Refer again to **Figure 4-11.** The nut locks the spindle into place. Gaging parts with a micrometer locked at the proper setting is an easy way to determine whether the pieces are sized correctly.

4.2.7 Reading an Inside Micrometer

To get a correct reading with an inside micrometer, it is important that the tool be held square across the diameter of the work. It must be positioned so that it will measure across the diameter on the exact center, **Figure 4-17.**

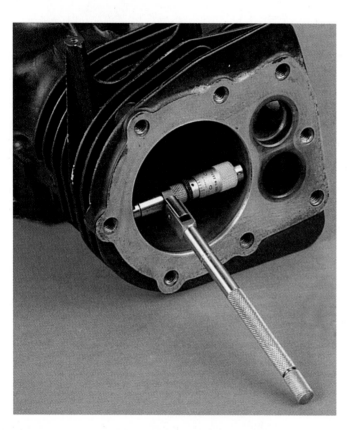

Figure 4-17. *Using an inside micrometer. Extension rods can be added to increase the tool's measuring range.*

Measurement is made by holding one end of the tool in place and then "feeling" for the maximum possible setting by moving the other end from left to right, and then in and out of the opening. The measurement is made when no left or right movement is felt, and a slight drag is noticeable on the in-and-out swing. It may be necessary to take several readings and average them.

4.2.8 Reading a Micrometer Depth Gage

Be sure to read a micrometer depth gage correctly. The graduations on this measuring tool are in

reverse order of the graduations on an outside micrometer. See **Figure 4-18.** The graduations under the thimble must be read, rather than those that are exposed.

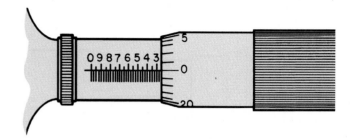

Figure 4-18. *A micrometer depth gage. When making measurements with a depth gage, remember that the graduations are in reverse order. This gage indicates a depth of 0.250.*

4.2.9 Care of a Micrometer

Micrometers are precision instruments and must be handled with care. The following techniques are recommended:

• Place the micrometer on the work carefully so the faces of the anvil and spindle will not be damaged. The same applies when removing the tool after a measurement has been made.

• Keep the micrometer clean. Wipe it with a slightly oiled cloth to prevent rust and tarnish. A drop of light oil on the screw thread will keep the tool operating smoothly.

• Avoid "springing" a micrometer by applying too much pressure when you are making a measurement.

• Clean the anvil and spindle faces before use. This can be done with a soft cloth or by *lightly* closing the jaws on a clean piece of paper and drawing the paper out.

• Check for accuracy by closing the spindle gently on the anvil and note whether the zero line on the thimble coincides with the zero on the sleeve. If they are not aligned, follow the manufacturer's recommended adjustments.

• Avoid placing a micrometer where it may fall on the floor or have other tools placed on it.

• If the micrometer must be opened or closed a considerable distance, do not "twirl" the frame; gently roll the thimble with your palm. See **Figure 4-19.**

• Never attempt to make a micrometer reading until a machine has come to a complete stop.

• Clean and oil the tool if it is to be stored for some time. If possible, place the micrometer in a small box for protection.

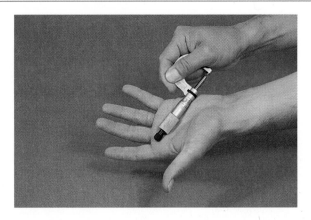

Figure 4-19. *Micrometers must be treated carefully. Roll the micrometer thimble on the palm of your hand if the instrument must be opened or closed a considerable distance.*

4.3 VERNIER MEASURING TOOLS

The Vernier principle of measuring was named for its inventor, Pierre Vernier, a French mathematician. The *Vernier caliper* can make accurate measurements to 1/1000″ (0.001″) and 1/50 mm (0.02 mm). See **Figure 4-20.**

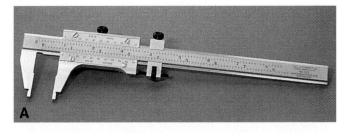

Figure 4-20. *Vernier calipers can be used to make very accurate measurements. A—Standard Vernier caliper. B—Modern digital calipers are easier to read than mechanical instruments.*

The design of the tool permits measurements to be made over a large range of sizes. It is manufactured as a standard item in 6″, 12″, 24″, 36″, and 48″ lengths. SI Metric Vernier calipers are available in 150 mm, 300 mm, and 600 mm lengths. The 6″, 12″, 150 mm, and 300 mm sizes are most commonly used. Unlike the micrometer caliper, the Vernier caliper can be used for both inside and outside measurements, **Figure 4-21.**

Figure 4-21. *Vernier calipers can be used to make both internal and external measurements. (L. S. Starrett Co.)*

The following measuring instruments may include a Vernier scale:

- Height and depth gages are used for layout work and to inspect the locations of features. See **Figure 4-22.**
- Gear tooth calipers are used to measure gear teeth and threading tools, **Figure 4-23.**
- Universal Vernier bevel protractors are used for the layout and inspection of angles, **Figure 4-24.**

Vernier measuring tools, with the exception of the Vernier bevel protractor, consist of a graduated beam with fixed jaw or base and a Vernier slide assembly. The Vernier slide assembly is composed of a movable jaw or scribe, Vernier plate, and clamping screws. The slide moves as a unit along the beam.

Unlike other Vernier measuring tools, the beam of the Vernier caliper is graduated on both sides. One side is for making outside measurements, the other for inside measurements. Many of the newer Vernier measuring tools are graduated to make both inch and millimeter measurements.

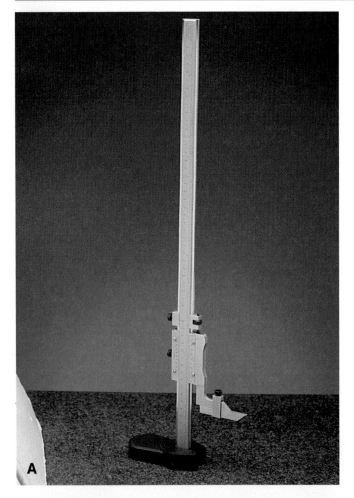

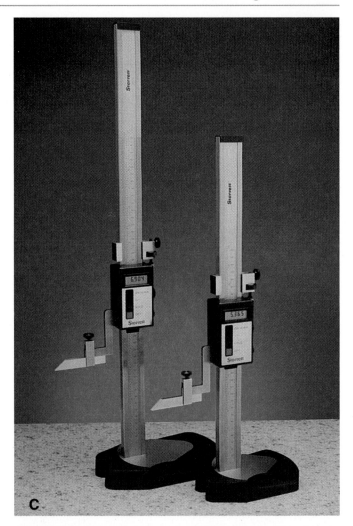

Figure 4-22. *Many instruments are equipped with a Vernier scale. A—Height gage. B—Depth gage. C—The digital readout on this type of height gage serves the same function as a standard Vernier scale. (L. S. Starrett Co.)*

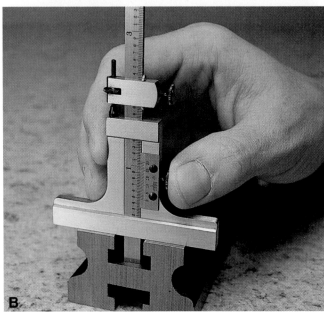

4.3.1 Reading an Inch-Based Vernier Scale

These measuring tools are available with either 25-division or 50-division Vernier plates. Both plates can be read to 0.001″.

On measuring tools using the *25-division Vernier plate*, every inch section on the beam is graduated

Figure 4-23. *Gear tooth Vernier calipers are used to measure gear teeth, form tools, and threaded tools. (L. S. Starrett Co.)*

Figure 4-24. A universal Vernier bevel protractor is used to accurately measure angles. (L. S. Starrett Co.)

into 40 equal parts. Each graduation is 1/40″ (0.025″). Every fourth division, representing 0.100″, is numbered.

There are 25 divisions on the Vernier plate. Every fifth line is numbered: 5, 10, 15, 20, and 25. The 25 divisions occupy the same space as 24 divisions on the beam. This slight difference, equal to 0.001 (1/1000″) per division, is the basis of the Vernier principle of measuring.

To read a 25-division Vernier plate measuring tool, note how many inches (1, 2, 3, etc.), tenths (0.100, 0.200, etc.), and fortieths (0.025, 0.050, or 0.075) there are between the "0" on the Vernier scale and the "0" line on the beam, then add them. Then count the number of graduations (each graduation equals 0.001″) that lie between the "0" line on the Vernier plate and the line that coincides (corresponds exactly) with a line on the beam. Only one line will coincide. Add this to the above total for the reading.

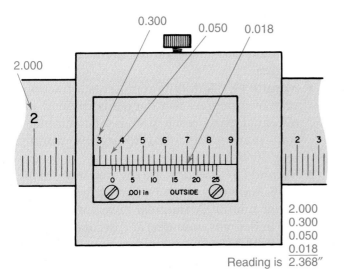

Reading is 2.368″

The "0" line on the Vernier plate is:

Past the 2:	2 × 1	= 2.000
Past the 3:	3 × 0.100	= 0.300
Plus 2 graduations:	2 × 0.025	= 0.050
Plus 18 Vernier scale graduations:	18 × 0.001	= 0.018
Total reading		= 2.368″

On the *50-division Vernier plate*, every second graduation between the inch lines is numbered, and equals 0.100″. The unnumbered graduations equal 0.050″.

The Vernier plate is graduated into 50 parts, each representing 0.001″. Every fifth line is numbered: 5, 10, 15 . . . 40, 45, and 50.

To read a 50-division Vernier measuring tool, first count how many inches, tenths (0.100), and twentieths (0.050) there are between the "0" line on the beam, and the "0" line on the Vernier plate. Then add them. Then count the number of 0.001 graduations on the Vernier plate from its "0" line to the line that coincides with a line on the beam. Add this to the above total.

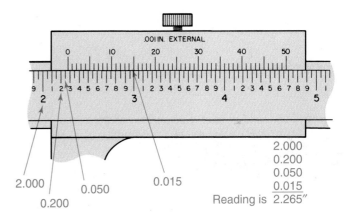

Reading is 2.265″

The "0" line on the Vernier plate is:

Past the 2:	2 × 1.000	= 2.000
Past the 2:	2 × 0.100	= 0.200
Plus one graduation:	1 × 0.050	= 0.050
Plus 15 Vernier scale graduations:	15 × 0.001	= 0.015
Total reading		= 2.265″

4.3.2 Reading a Metric-Based Vernier Scale

The principles used in reading metric Vernier measuring tools are the same as those used for US Conventional measure. However, the readings on the Vernier scale are obtained in 0.02 mm precision. A 25-division Vernier scale is illustrated in **Figure 4-25,** while a 50-division scale is described in **Figure 4-26.**

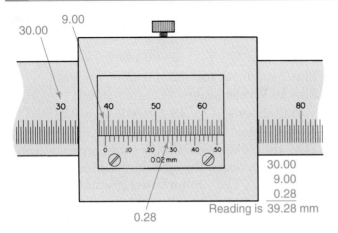

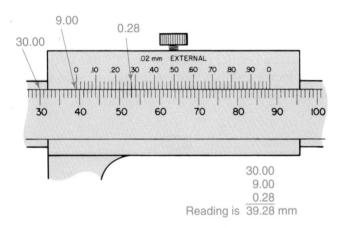

Figure 4-25. How to read a 25-division metric-based Vernier scale. Readings on the scale are obtained in units of two hundredths of a millimeter (0.02 mm).

Figure 4-26. How to read a 50-division metric-based Vernier scale. Each division equals two hundredths of a millimeter (0.02 mm).

Figure 4-27. Dial calipers provide direct readings of measurements. (L.S. Starrett Co.)

4.3.3 Using the Vernier Caliper

As with any precision tool, a Vernier caliper must not be forced on the work. Slide the Vernier assembly until the jaws nearly contact the section being measured. Lock the clamping screw. Make the tool adjustment with the fine adjusting nut. The jaws must contact the work firmly, but not tightly.

Lock the slide on the beam. Carefully remove the tool from the work and make your reading. For precise layout work, divider and trammel point settings are located on the outside measuring scale and on the slide assembly.

Dial calipers. These direct-reading instruments resemble Vernier calipers. They can be used to make outside, inside, and depth measurements (with the addition of a depth attachment). A lock permits the tool to be employed for repetitive measurements. See **Figure 4-27.**

The beam is graduated into 0.10″ increments. The caliper dial is graduated into 100 divisions. The reading is made by combining the division on the beam and the dial reading.

The dial hand makes one full revolution for each 0.10″ movement. Each dial graduation, therefore, represents 1/100 of 0.10″, or 0.001″. On the metric version, each dial graduation represents 0.02 mm.

4.3.4 Universal Vernier Bevel Protractor

A quick review of the circles, angles, and units of measurement associated with them will help in understanding how to read a universal Vernier bevel protractor.

- **Degree (°)**—Regardless of its size, a circle contains 360°. Angles are also measured by degrees.
- **Minute (′)**—A minute represents a fractional part of a degree. If a degree is divided into 60 equal parts, each part is one minute. A foot mark (′) is used to signify minutes (e.g. 30°15′).
- **Second (″)**—Minutes are divided into smaller units known as seconds. There are 60 seconds in one minute. An angular measurement written in degrees, minutes, and seconds appears as 36°18′22″. This would read "36 degrees, 18 minutes, and 22 seconds."

A *universal bevel protractor* has several parts: a dial, a base or stock, and a sliding blade. The dial is graduated into degrees, and the blade can be extended in either direction and set at any angle to the stock. The blade can be locked against the dial by tightening the blade clamp nut. The blade and dial can be rotated as a unit to any desired position, and locked by tightening the dial clamp nut.

The protractor dial is graduated into 360° and reads from 0° to 90° and then back down to 0°. Every ten degree division is numbered, and every five degrees is indicated by a fine line longer than those on either side. The Vernier scale is divided into twelve equal parts on each side of the "0." Every third graduation is numbered (0, 15, 30, 45, 60), representing minutes. Each division equals five minutes. Since each degree is divided into 60 minutes, one division is equal to 5/60 of a degree.

To read the protractor, note the number of degrees that can be read up to the "0" on the Vernier plate. To this, add the number of minutes indicated by the line beyond the "0" on the Vernier plate that aligns exactly with a line on the dial.

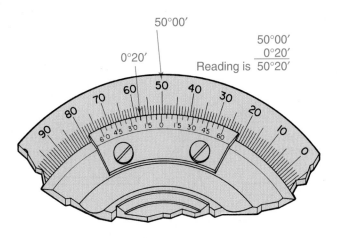

In this example the "0" is past the 50° mark, and the Vernier scale aligns at the 20' mark. Therefore, the measurement is 50°20'.

4.3.5 Care of Vernier Tools

Reasonable care in handling these expensive tools will maintain their accuracy.

- Wipe the instrument with a soft, lint-free cloth before using. This will prevent dirt and grit from being ground in, which could eventually affect the accuracy of the tool.
- Wipe the tool with a lightly oiled, soft cloth after use and before storage.
- Store the tool in its case.

- Never force the tool when you are making measurements.
- Use a magnifying glass or a jeweler's loupe to make Vernier readings. Hold the tool so the light is reflected on the scale.
- Handle the tool as little as possible. Sweat and body acids cause rusting and staining.
- Periodically check for accuracy. Use a measuring standard, Jo-block, or ground parallel. Return the tool to the manufacturer for adjustments and repairs.
- Lay Vernier height gages on their side when not in use. Then there will be no danger that they will be knocked over and damaged.

4.4 GAGES

It is impractical to check every dimension on every manufactured part with conventional measuring tools. Specialized tools, such as plug gages, ring gages, and optical gages are used instead. These gaging devices can quickly determine whether the dimensions of a manufactured part are within specified limits or tolerances.

Measuring requires the skillful use of precision measuring tools to determine the exact geometric size of the piece. *Gaging* involves checking parts with various gages. Gaging simply shows whether the piece is made within the specified tolerances.

When great numbers of an item with several critical dimensions are manufactured, it might not be possible to check each piece. It then becomes necessary to decide how many randomly selected pieces must be checked to ensure satisfactory quality and adherence to specifications. This technique is called *statistical quality control.*

Always handle gages carefully. If dropped or mishandled, the accuracy of the device could be affected. Gages provide a method of checking your work and are very important tools.

4.4.1 Plug Gage

Plug gages are used to check whether hole diameters are within specified tolerances. The *double-end cylindrical plug gage* has two gaging members known as *go* and *no-go* plugs, **Figure 4-28.** The *go plug* should enter the hole with little or no interference. The *no-go plug* should not fit.

The go plug is longer than the no-go plug. A *progressive plug gage*, or *step plug gage*, has the go and no-go plugs on the same end. This gage is able to check the dimensions in one motion. See **Figure 4-29.**

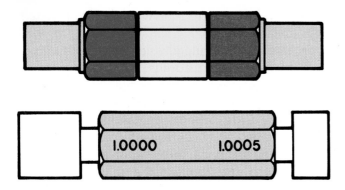

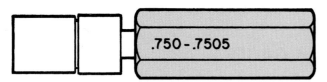

Figure 4-28. A double end cylindrical plug gage.

Figure 4-29. A step plug gage can check for oversize and undersize in a single test.

4.4.2 Ring Gage

External diameters are checked with ring gages. The go and no-go ring gages are separate units, and can be distinguished from each other by a groove cut on the knurled outer surface of the no-go gage. Refer to **Figure 4-30.**

On ring gages, the gage tolerance is the reverse of plug gages. The opening of the go gage is larger than the opening for the no-go gage.

Figure 4-30. Ring gages. The larger sizes are cut away to reduce weight. (Standard Tool Co.)

4.4.3 Snap Gage

A *snap gage* serves the same purpose as a ring gage. Snap gages are designed to check internal diameters, external diameters, or both. There are three general types:

- An *adjustable snap gage* can be adjusted through a range of sizes. See **Figure 4-31.**
- A *nonadjustable snap gage* is made for one specific size. See **Figure 4-32.**

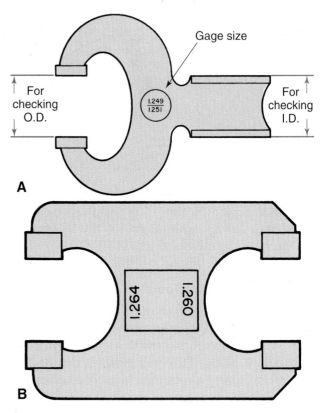

Figure 4-31. An adjustable snap gage. (Taft-Pierce Co.)

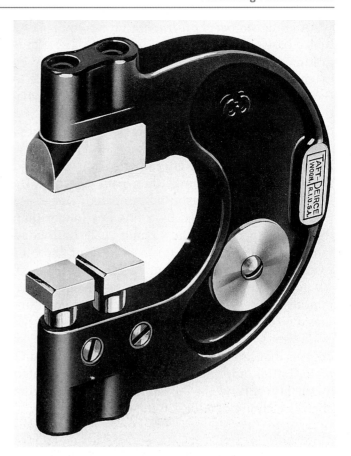

Figure 4-32. Diagram of a nonadjustable snap gage. A—A combination internal-external nonadjustable snap gage. B—An external nonadjustable snap gage.

- A *dial indicator snap gage* measures the amount of variation in the part measurement. The dial face has a double row of graduations reading in opposite directions from zero. Minus graduations are red and plus graduations are black. Both adjustable and nonadjustable indicating snap gages are available. See **Figure 4-33.**

On snap gages, the anvils should be narrower than the work being measured. This will avoid uneven wear on the measuring surfaces.

Figure 4-33. A dial indicator snap gage. (L.S. Starrett Co.)

4.4.4 Thread Gages

Several types of gages are used to check screw thread fits and tolerances. These gages are similar to the gages already discussed:
- Thread plug gage.
- Thread ring gage.
- Thread roll snap gage.

These gages are illustrated in **Figure 4-34.**

4.4.5 Gage Blocks

Gage blocks, commonly known as *Jo-blocks or Johansson blocks,* are precise steel measuring standards. Gage blocks can be purchased in various sets ranging from a few commonly used block sizes to more complete sets. See **Figure 4-35.**

Gage blocks are used to verify the accuracy of master gages. They are also used as working gages and for setting up machining work requiring great accuracy. The Federal Accuracy Grades for gage blocks are shown in **Figure 4-36.**

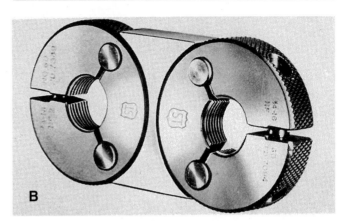

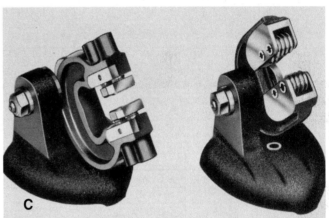

Figure 4-34. Thread gages. A—Thread plug gage. B—Thread ring gage. C—Go/no-go thread snap gage. (Standard Tool Co. and Taft-Pierce Co.)

When working with gage blocks, keep the following tips in mind:
- Improper handling can cause temperature changes in the block, resulting in measurement errors. For the most accurate results, blocks should be used in a temperature-controlled room. Handle the blocks as little as possible. When you must handle the blocks, use the tips of your fingers, as shown in **Figure 4-37A.**
- When wringing gage blocks together to build up to desired size, wipe the blocks and then carefully slide them together. They should adhere to each other strongly. Separate the blocks when you are finished. Leaving gage blocks together for extended periods may cause the contacting surfaces to corrode. See **Figure 4-37B.**

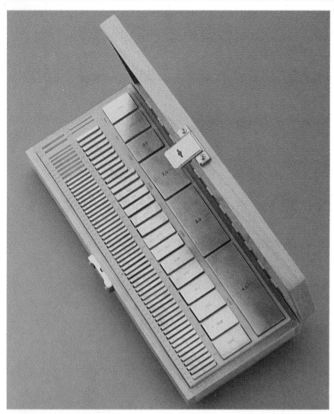

Figure 4-35. A typical set of gage blocks. (Federal Products Co.)

Federal Accuracy Grades			
		Tolerance	
Accuracy grade	Former designation	US Conventional system (inch)	Metric system (millimeter)
0.5	AAA	±.000001″	±.00003 mm
1	AA	±.000002″	±.00005 mm
2	A +	+.000004″ −.000002″	+.0001 mm −.00005 mm
3	A&B	+.000006″ −.000002″	+.00015 mm −.00005 mm

Reference temperature: 68°F (20°C)
One inch = 25.4 millimeters exactly

Figure 4-36. Federal Accuracy Grades for gage blocks.

• Wipe gage blocks with a soft cloth or chamois treated with oil. Be sure the oil is one recommended by the gage manufacturer. See **Figure 4-37C.**

4.5 DIAL INDICATORS

Industry is constantly searching for ways to reduce costs without sacrificing quality. Inspection has always been a costly part of manufacturing. To speed up this phase of production without sacrificing accuracy, dial indicators and electronic gages are receiving increased attention.

Dial indicators are designed with shockproof movements and have jeweled bearings (similar to fine watches). There are two types of indicators: balanced and continuous. *Balanced indicators* can take measurements on either side of a zero line. *Continuous indicators* read from "0" in a clockwise direction. See **Figure 4-38.**

Dial faces are available in a wide range of graduations. They usually read in the following increments:

• 1/1000″ (0.001″)
• 1/100 mm (0.01 mm)
• 1/10,000″ (0.0001″)
• 2/1000 mm (0.002 mm)

Much use is made of dial indicators for centering and aligning work on machine tools, checking for eccentricity, and visual inspection of work. Dial indicators must be mounted to rigid holding devices, **Figure 4-39.**

A *digital electronic indicator,* **Figure 4-40,** features direct digital readouts and a traditional graduated dial for fast, accurate reading. These indicators are available as both self-contained and remote readout units.

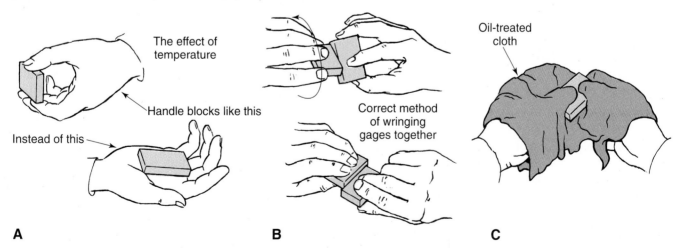

Figure 4-37. Proper care of gage blocks. A—Handling gage blocks. B—Wipe blocks and slide them together. Do not leave blocks together for extended periods. C—Wipe blocks with a soft cloth before storing. (Webber Gage Div., L.S. Starrett Co.)

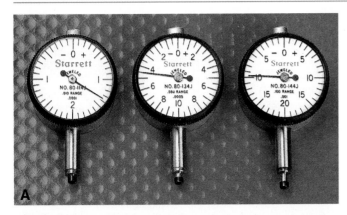

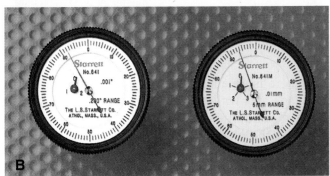

Figure 4-38. *The two basic varieties of dial indicators. A—Balanced indicators. B—Continuous indicators. (L. S. Starrett Co.)*

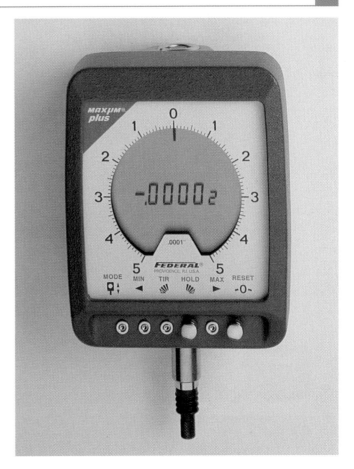

Figure 4-40. *This digital electronic indicator has numeric read-outs and a conventional graduated dial. (Federal Products Company)*

Figure 4-39. *Mounting this dial indicator on a magnetic base permits it to be attached to any ferrous metal surface. A push-button releases the magnet.*

4.5.1 How to Use a Dial Indicator

The hand on the dial is actuated by a sliding plunger. Place the plunger lightly against the work until the hand moves. The dial face is turned until the "0" line coincides with the hand. As the work touching the plunger is slowly moved, the indicator hand will measure movement.

The dial indicator can show the difference between the high and low points, or the total run-out of the piece in a lathe. When machining, adjustments are made until there is little or no indicator movement.

4.6 OTHER GAGING TOOLS

Industry makes wide use of other types of gaging tools. Most of these tools are used for special purposes and are not usually found in a school shop. However, since you might need to use them in industry, it is important to learn about such tools.

4.6.1 Air Gage

An *air gage* uses air pressure to measure hole sizes and hard-to-reach shaft diameters, **Figure 4-41.** This type of gage is especially helpful when measuring deep internal bores. The basic operation of an air gage is illustrated in **Figure 4-42.**

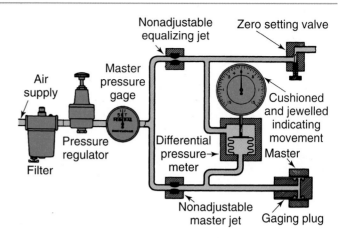

Figure 4-42. This diagram illustrates the operation of an air gage.

There is no actual contact between the measuring gage and wall of the bore being measured. The bore measurement depends on the air leakage between the plug and the hole wall. (The larger the bore diameter, the greater the leakage.) Pressure builds up and the measurement of the back pressure gives an accurate measurement of the hole size.

Change in pressure (air leakage) is measured by a dial indicator, a cork floating on the air stream, or by a *manometer* (U-shaped tube in which the height of fluid in the tube indicates pressure).

4.6.2 Electronic Gage

An *electronic gage,* **Figure 4-43,** is another type of gaging tool used to make extremely precise measurements. Electronic gages are comparison gages: they compare the size of the work to a reference size. Some are calibrated by means of master gage blocks and others use replaceable gaging probes. These instruments measure in both US Conventional and SI Metric units.

4.6.3 Laser Gaging

A *laser* is a device that produces a very narrow beam of extremely intense light. Lasers are used in communication, medical, and industrial applications. Laser is an acronym for *light amplification by stimulated emission of radiation.*

The laser is another area of technology that has moved from the laboratory into the shop. When employed for inspection purposes, it can check the accuracy of critical areas in machined parts quickly and accurately. Refer to **Figure 4-44.**

4.6.4 Optical Comparator

The *optical comparator* uses magnification as a means for inspecting parts, **Figure 4-45.** An

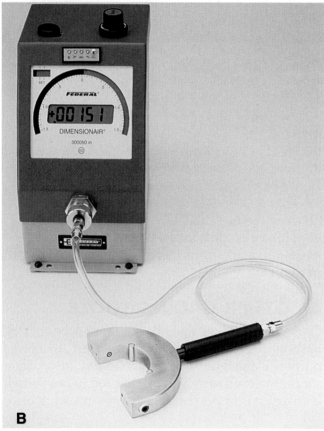

Figure 4-41. Digital air gages are available with either US Conventional or SI Metric readouts. They can check either inside or outside diameters. A—An air gage set up to inspect an internal dimension. The master ring shown with the gage is used to set zero on the readout. B—This gage has an air fork, which is used to check hard-to-reach diameters, such as crankshaft journals. (Federal Products Company)

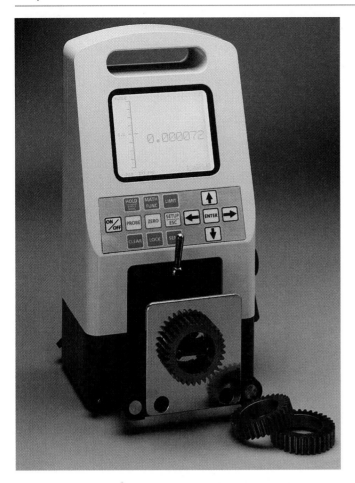

Figure 4-43. *This electronic bore gaging system can deliver electronic resolution as fine as 0.00001″ (0.0002 mm). Using replaceable gaging probes, the self-contained unit measures diameters ranging from 0.370″ to 2.900″. It also measures in millimeters. It can be linked to a computer for statistical process control (SPC) data collection. (Sunnen Products Company)*

Figure 4-44. *This laser is being used to inspect a part from a car's automatic transmission. Manually, one person could inspect no more than four units an hour. The laser can inspect over 120 parts an hour. (Ford Motor Co.)*

Figure 4-45. *This 50-power optical comparator permits a fast check of the tooth formation on a tap.*

enlarged image of the part is projected upon a screen for inspection. The part image is superimposed upon an enlarged, accurate drawing of the correct shape and size. The comparison is made visually. Variations as small as 0.0005″ (0.012 mm) can be noted by a skilled operator.

4.6.5 Optical Flats

Optical flats are precise measuring instruments that use light waves as a measuring standard, **Figure 4-46.** The flats are made of quartz and have one face ground and polished to optical flatness. When this face is placed on a machined surface and a special light passed through it, light bands appear on the surface, **Figure 4-47.** The shape of these bands indicate to the inspector the accuracy of the part. See **Figure 4-48.**

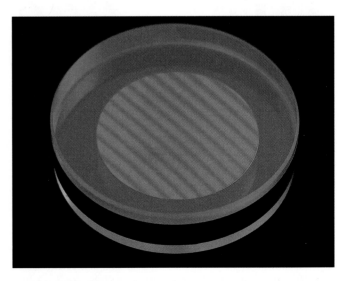

Figure 4-46. *Optical flats are used for precision flatness, parallelism, size, and surface variations. (L. S. Starrett Co.)*

Interference bands
indicate difference
in size between
ball bearing and Jo-block

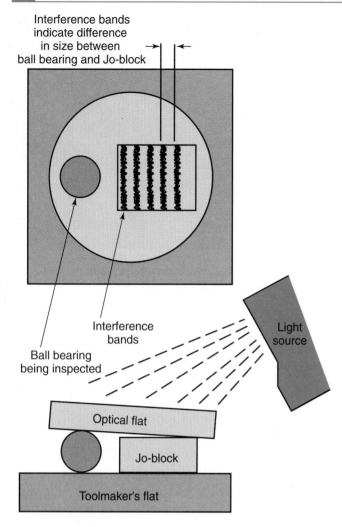

Interference
bands

Ball bearing
being inspected

Figure 4-47. *Optical flat set-up. Optical flat is placed on top of the work and light is positioned above the flat.*

4.6.6 Thickness (Feeler) Gage

Thickness gages are pieces or leaves of metal manufactured to precise thickness, **Figure 4-49.** Thickness gages are made of tempered steel and are usually 1/2″ (12.7 mm) wide.

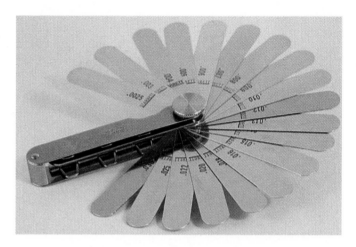

Figure 4-49. *Thickness or feeler gages.*

Thickness gages are ideal for measuring narrow slots, setting small gaps and clearances, determining fit between mating surfaces, and for checking flatness of parts in straightening operations. See **Figure 4-50.**

4.6.7 Screw Pitch Gage

Screw pitch gages are used to determine the pitch or number of threads per inch on a screw, **Figure 4-51.** Each blade is stamped with the pitch or

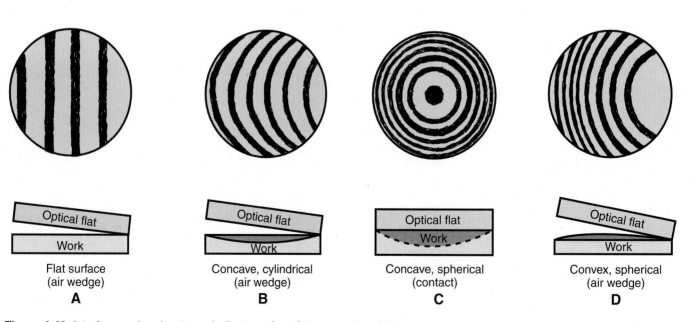

| Flat surface (air wedge) | Concave, cylindrical (air wedge) | Concave, spherical (contact) | Convex, spherical (air wedge) |
| A | B | C | D |

Figure 4-48. *Interference band patterns indicate surface flatness and variations.*

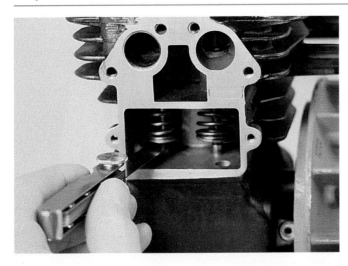

Figure 4-50. A thickness gage is used to check part clearance.

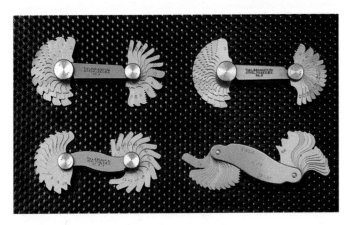

Figure 4-51. Screw pitch gages are made for both inch-based and metric threads. (L. S. Starrett Co.)

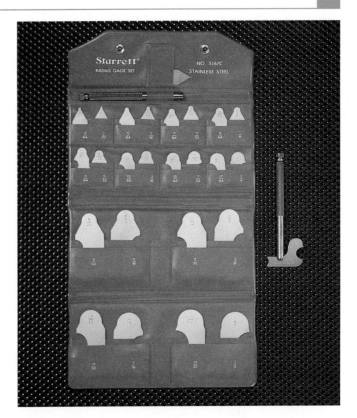

Figure 4-52. A set of radius and fillet gages. (L. S. Starrett Co.)

number of threads per inch. Screw pitch gages are available in US Conventional and SI Metric thread sizes.

4.6.8 Fillet and Radius Gage

The thin steel blades of a *fillet and radius gage*, **Figure 4-52**, are used to check concave and convex radii on corners or against shoulders. The gage is used for layout work and inspection, and as a template when grinding form cutting tools. See **Figure 4-53**. The gages increase in radius in 1/64" (0.5 mm) increments.

4.6.9 Drill Rod

Drill rods are steel rods manufactured to close tolerances to twist drill diameters. They are used to inspect hole alignment, location, and diameter. Drill rods are available in both US Conventional and SI Metric sizes.

4.7 HELPER MEASURING TOOLS

Some measuring tools are not direct reading and require the help of a rule, micrometer, or Vernier caliper to determine the size of the measurement taken. These are called *helper measuring tools.*

4.7.1 Calipers

External or internal measurements of 1/64" (0.4 mm) can be made with calipers, **Figure 4-54**. A caliper does not have a dial or scale that shows a measurement; the distance between points must be measured with a steel rule.

Round stock is measured by setting the caliper square with the work and moving the caliper legs down on the stock. Adjust the tool until the caliper point bears lightly on the center line of the stock. Caliper weight should cause the caliper to slip over the diameter. Hold the caliper next to the rule to make the reading, **Figure 4-55**.

An *inside caliper* is used to make internal measurements where 1/64" (0.4 mm) accuracy is acceptable. Hole diameter can be measured by setting the caliper to approximate size, and inserting the legs into the opening. Hold one leg firmly against the hole wall, and adjust the thumbscrew until the other leg lightly touches the wall exactly opposite the first

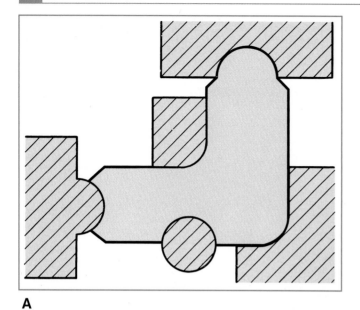

A

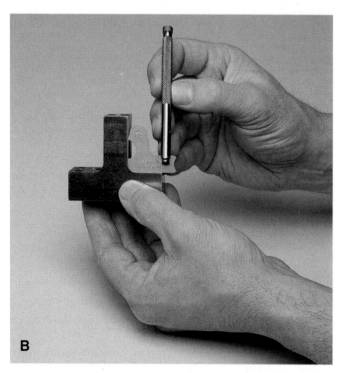

B

Figure 4-53. *Using a radius gage. A—Various ways a radius gage can be used. B—Using a radius gage holder.* *(L. S. Starrett Co.)*

leg. The legs should drag slightly when moved in and out, or from side to side.

Considerable skill is required to make accurate measurements with a caliper. See **Figure 4-56.** Much depends upon the machinist's sense of touch. With practice, measurements with accuracy of 0.003″ (0.07 mm) can be made. However, a micrometer or Vernier caliper is preferred and must be utilized when greater accuracy is required.

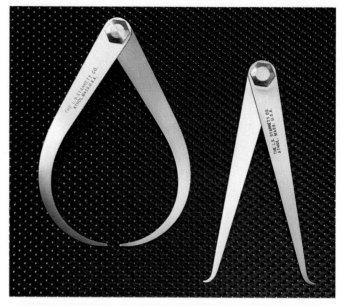

Figure 4-54. *Inside and outside calipers. (L. S. Starrett Co.)*

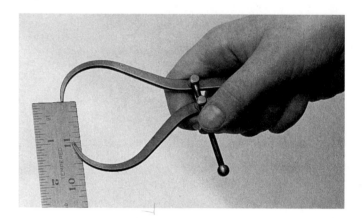

Figure 4-55. *The outside caliper is read with a steel rule.*

4.7.2 Telescoping Gage

A *telescoping gage* is intended for use with a micrometer to determine internal dimensions, **Figure 4-57.** Sets of telescoping gages with varying ranges are available, **Figure 4-58.**

To use a telescoping gage, compress the contact legs. The legs telescope within one another under spring tension. Insert the gage into the hole and allow the legs to expand, **Figure 4-59.** After the proper fitting is obtained, lock the contacts into position. Remove the gage from the hole and make your reading with a micrometer, **Figure 4-60.**

4.7.3 Small Hole Gage

A *small hole gage* is used to measure openings that are too small for a telescoping gage, **Figure 4-61.** The contacts are designed to allow accurate

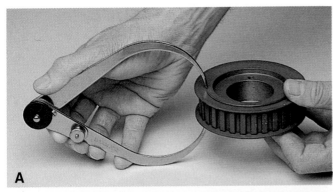

Figure 4-56. Using outside and inside calipers.
(L. S. Starrett Co.)

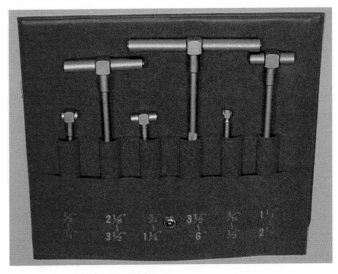

Figure 4-58. A typical set of telescoping gages.

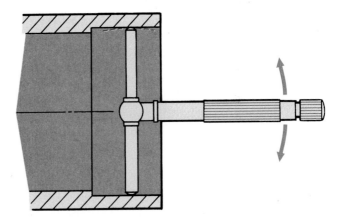

Figure 4-59. Positioning a telescoping gage to measure an inside diameter.

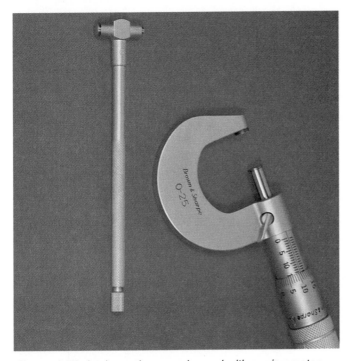

Figure 4-57. A telescoping gage is used with a micrometer.

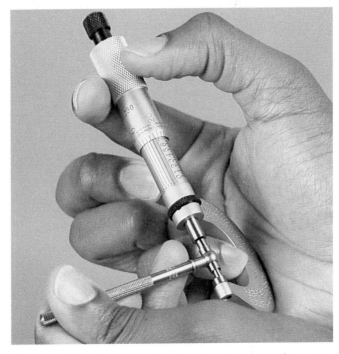

Figure 4-60. After removing the locked telescoping gage, measure it with a micrometer.

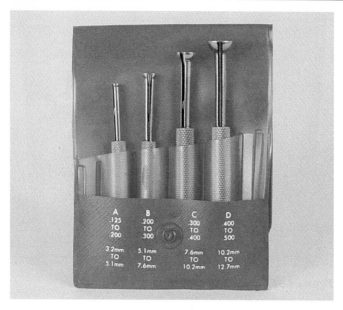

Figure 4-61. Small hole gages are used to measure the diameter of holes that are too small for telescoping gages.

measurement of shallow grooves, and small diameter holes. They are adjusted to size by the knurled knob at the end of the handle. Measurement is made over the contacts with a micrometer, **Figure 4-62.**

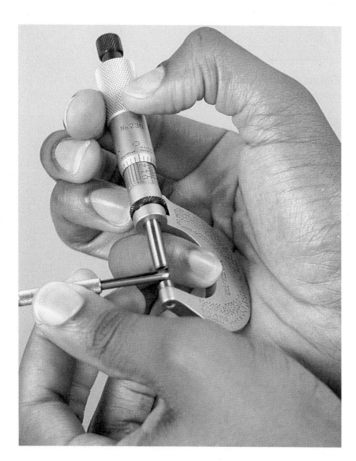

Figure 4-62. The correct way to measure a small hole gage with a micrometer.

TEST YOUR KNOWLEDGE

Please do not write in this text. Write your answers on a separate sheet of paper.

1. Make readings from the rules.

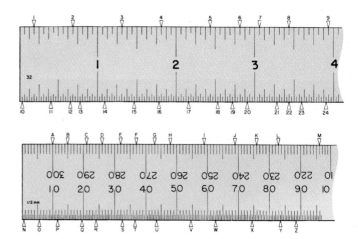

2. Make readings from the Vernier scales shown below.

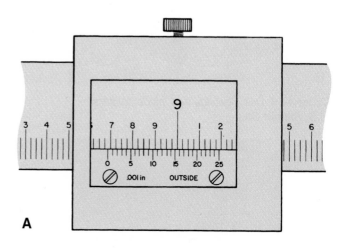

A

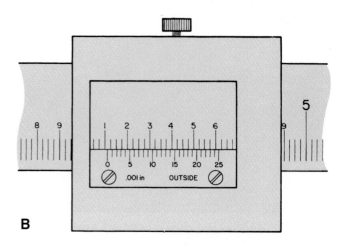

B

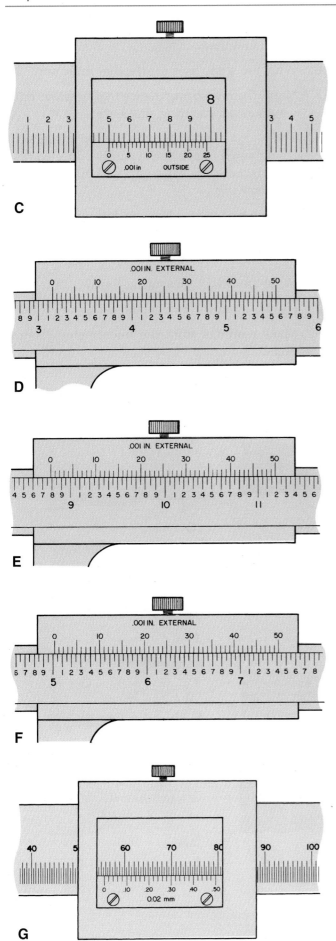

C

D

E

F

G

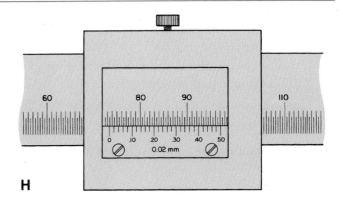

H

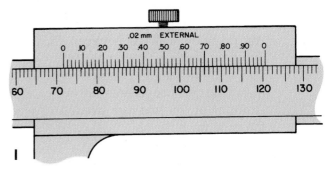

I

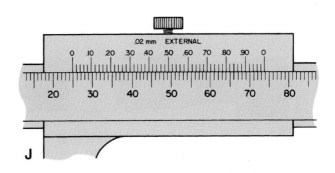

J

Answer the following questions as they pertain to measurement.

3. The micrometer is nicknamed _____.

4. One-millionth part of a standard inch is known as a _____.

5. One-millionth part of a meter is known as a _____.

6. A micrometer is capable of measuring accurately to the _____ and _____ part of standard inch and (in metric versions) to _____ and _____ millimeters.

7. The Vernier caliper has several advantages over the micrometer. List two of them.

8. A Vernier caliper can measure to the _____ part of the inch and (in the metric version) to _____ millimeters.

9. List six precautions that must be observed when using a micrometer or Vernier caliper.

10. The Vernier-type tool for measuring angles is called a _____.

11. How does a double-end cylindrical plug gage differ from a step plug gage?

12. A ring gage is used to check whether _____ are within the specified _____ range.

13. Gage blocks are often referred to as _____ blocks.

14. An air gage employs air pressure to measure deep internal openings and hard-to-reach shaft diameters. It operates on the principle of:
    a. Air pressure leakage between the plug and hole walls.
    b. The amount of air pressure needed to insert the tool properly in the hole.
    c. Amount of air pressure needed to eject the gage from the hole.
    d. All of the above.
    e. None of the above.

15. The dial indicator is available in two basic types. List them.

16. What are some uses for the dial indicator?

17. Name the measuring device that employs light waves as a measuring standard.

18. The _____ is used for production inspection. An enlarged image of the part is projected on a screen where it is superimposed upon an accurate drawing.

19. The pitch of a thread can be determined with a _____.

20. Of what use are fillet and radius gages?

21. What are helper measuring tools?

22. How is a telescoping gage used?

23. Make readings from the micrometer illustrations.

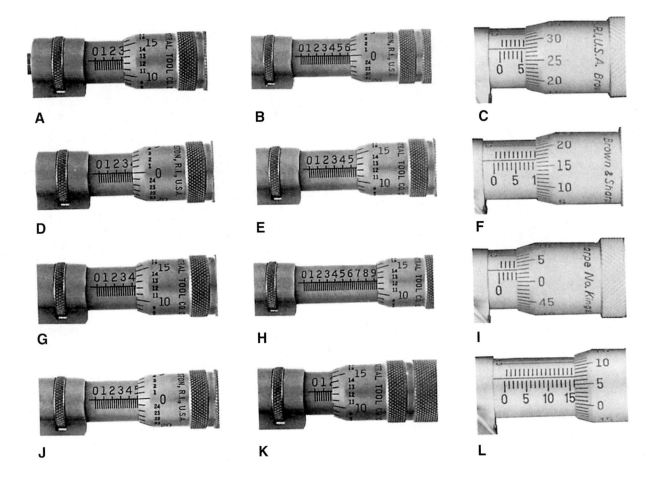

A

B

C

D

E

F

G

H

I

J

K

L

Chapter 5

Layout Work

LEARNING OBJECTIVES

After studying this chapter, you will be able to:
- ○ Explain why layouts are needed.
- ○ Identify common layout tools.
- ○ Use layout tools safely.
- ○ Make basic layouts.
- ○ List safety rules for layout work.

IMPORTANT TERMS

divider	*scriber*
hardened steel square	*straightedge*
layout dye	*surface gage*
plain protractor	*surface plates*
reference line	*V-blocks*

Laying out is the term that describes the locating and marking of lines, circles, arcs, and points for drilling holes or making cuts. These lines and reference points on the metal show the machinist where to machine.

The tools for this work are known as *layout tools.* Many common hand tools fall into this category. The accuracy of the job will depend upon the proper and careful application of these tools. See **Figure 5-1.**

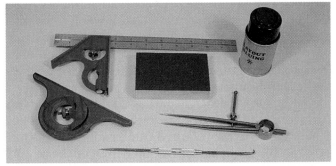

Figure 5-1. A few of the tools needed to make a simple layout.

5.1 MAKING LINES ON METAL

The shiny finish of most metals make it difficult to distinguish layout lines. For this reason, a coating must be placed on the metal before layout.

5.1.1 Layout Dye

There are many coatings used to make layout lines stand out better. Of these coatings, *layout dye* is probably the easiest to use. When applied to the metal, this blue-colored fluid offers an excellent contrast between the metal and the layout lines. All dirt, grease, and oil must be removed before applying the dye. If these substances are present on the surface, the dye will not adhere properly.

Chalk will also work on hot finished steel as a layout background. A pencil should not be used because it marks too wide and rubs off.

5.1.2 Scriber

An accurate layout requires fine lines that must be scribed (scratched) into the metal. A scriber will produce these lines, **Figure 5-2.** The point is made of hardened steel, and is kept needle-sharp by frequent honing on a fine oilstone. Many styles of scribers are available.

Never carry a scriber in your pocket. It can puncture the skin easily.

5.1.3 Divider

The scriber is used to draw straight lines. A *divider* is used to draw circles and arcs, **Figure 5-3.** It is essential that both legs of the tool be equal in length and kept pointed. Measured distances can be laid out with a divider, **Figure 5-4.** To set the tool to the correct distance, set one point on the inch or centimeter mark of a steel rule, and open the divider until the other leg is set to the proper measurement, **Figure 5-5.**

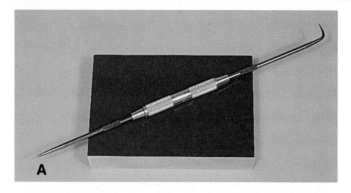

Figure 5-2. Scribers are used to mark parts during layout. A—The long bent point on this scriber can reach through holes. B—This pocket scriber has a removable point. The point can be reversed when the scriber is not being used, protecting the tip and making the work area safer.

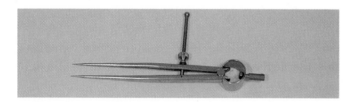

Figure 5-3. A divider is used to mark lines, arcs, and circles.

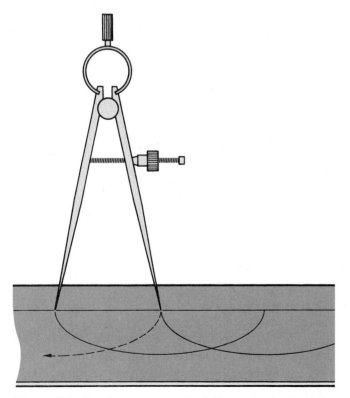

Figure 5-4. Equal spaces can be laid out by "walking" the divider.

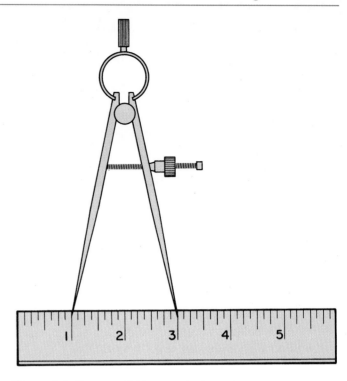

Figure 5-5. To set a divider to a desired dimension, place it on a rule as shown.

Circles and arcs that are too large to be made with a divider are drawn with a *trammel*, **Figure 5-6.** This consists of a long thin rod, called a *beam*, on which two sliding heads with scriber points are mounted. One head is equipped with an adjusting screw. *Extension rods* can be added to the beam to increase the capacity of the tool.

The *hermaphrodite caliper* is a layout tool with one leg that is shaped like a caliper and the other pointed like a divider, **Figure 5-7.** Lines parallel to the edge of the material, either straight or curved, can be drawn with the tool, **Figure 5-8.** It can also be used to locate the center of irregularly shaped stock.

5.1.4 Surface Gage

A *surface gage* has many uses, but is most frequently employed for layout work, **Figure 5-9.** It consists of a *base, spindle,* and *scriber.* An adjusting screw is fitted for making fine adjustments. The scriber is mounted in a way that allows it to be pivoted into any position. A surface gage can be utilized to scribe lines at a given height and parallel to the surface, **Figure 5-10.** A V-slot in the base permits the tool to be also employed on a curved surface.

To check whether a part is parallel to a given surface, fit the surface gage with a dial indicator. Set the indicator to the required dimension with the aid of gage blocks. The tool is then moved back and forth along the work, **Figure 5-11.**

A height gage can be used in the same manner as a surface gage. See **Figure 5-12.**

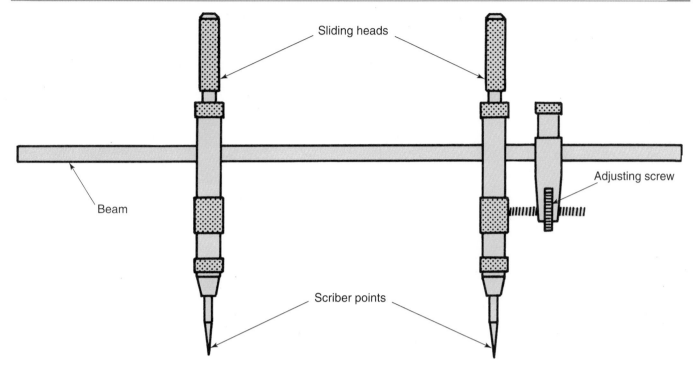

Figure 5-6. *Large circles and arcs are drawn with a trammel.*

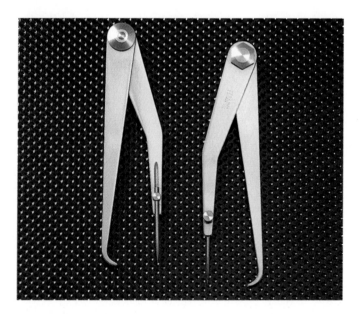

Figure 5-7. *A hermaphrodite caliper has a blunt end for the sliding surface and a point for scribing. (L.S. Starrett Co.)*

5.1.5 Surface Plate

Every linear measurement depends upon an accurate reference surface. **Surface plates** provide a reference surface (plane) for layout and inspection.

Surface plates can be purchased in sizes up to 72″ by 144″ (1800 mm by 3600 mm) and in various grades. Surface plate grade differences are given in degrees of flatness:

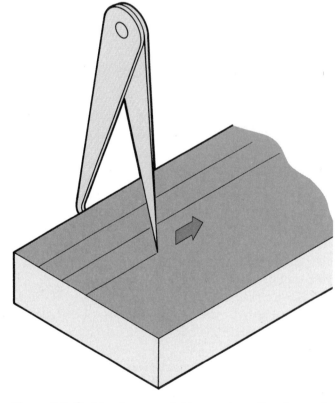

Figure 5-8. *Scribing lines parallel to an edge with a hermaphrodite caliper.*

- Grade AA for laboratories.
- Grade A for inspection.
- Grade B for too room and layout applications.

Figure 5-9. This small surface gage is designed for light work. (L. S. Starrett Co.)

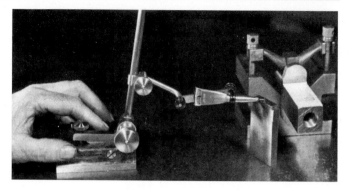

Figure 5-11. A machinist is using gage blocks to set the indicator, which is mounted on the surface gage. (Lufkin Rule Co.)

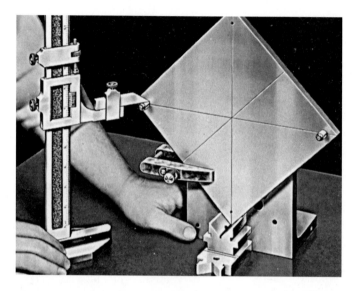

Figure 5-12. A height gage can be used to scribe lines on a workpiece. Here, a machinist is using a Vernier height gage. Note that a V-block and angle plate support the work during layout. (L. S. Starrett Co.)

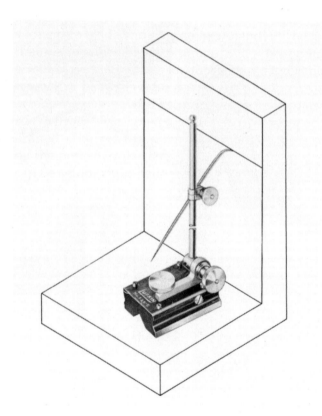

Figure 5-10. Carefully slide the surface gage to scribe lines parallel to the base. Handle the gauge carefully because sharp points can cause injury.

Most surface plates made today are granite but some are semisteel, **Figure 5-13.** Granite is more stable. Semisteel surface plates are more affected by temperature changes.

Surface plates are used primarily for layout and inspection work. They should never be used for any job that could mar or nick the surface.

When square reference surfaces are needed, a *right angle plate* is used, **Figure 5-14.** The plates can be placed in any position with the work clamped to the face for layout, measurement, or inspection.

An accurate surface parallel to the surface plate can be obtained using *box parallels*, **Figure 5-15.** All surfaces are precision-ground to close tolerances.

5.1.6 V-Blocks

V-blocks support round work for layout and inspection, **Figure 5-16.** They are furnished in matched pairs with surfaces that are ground square to close tolerances. Ribs are cast into the body for weight reduction. The ribs also can be used as clamping surfaces.

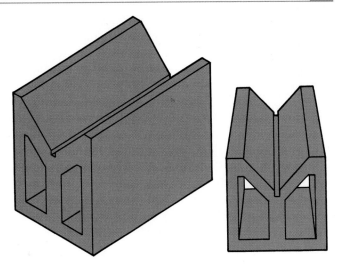

Figure 5-16. V-blocks can be used to hold round stock for layout and measurement work.

5.1.7 Straightedge

Long flat surfaces are checked for accuracy with a *straightedge*, **Figure 5-17.** This tool is also used for laying out long straight lines. Straightedges can be made from steel or granite, with steel being more common.

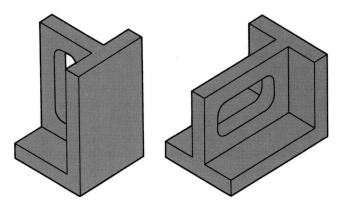

Figure 5-13. Surface plates are available in various grades and materials. A—A granite surface plate. B—A semisteel surface plate. (L. S. Starrett Co. and Challenge Machinery Co.)

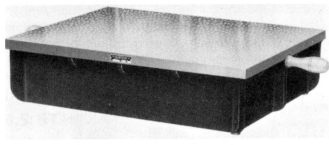

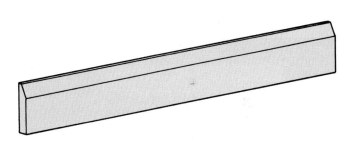

Figure 5-17. Steel straightedges are very common. Straightedges are also available in granite.

Figure 5-14. Right angle plates are often used to check perpendicular surfaces.

5.2 SQUARES

The *square* is employed to check 90° (square) angles. The tool is also used for laying out lines that must be at right angles to a given edge or parallel to another edge. Some simple machine setups can be made quickly and easily with a square.

Many different types of squares are available. The following are a few of the most common:

- *Hardened steel square*—This square is recommended when extreme accuracy is required. See **Figure 5-18.** The square has true right angles, both inside and outside. It is accurately ground and lapped for straightness and parallelism. The tool comes in sizes up to 36″ (910 mm).

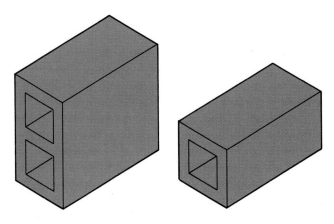

Figure 5-15. Box parallels are available in a number of sizes.

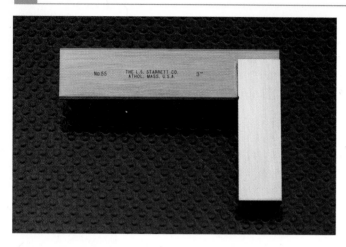

Figure 5-18. *The hardened steel square is handy during layout work. (L. S. Starrett Co.)*

• *Double square*—This is more practical than the steel square for many jobs because the sliding blade is adjustable and interchangeable with other blades, **Figure 5-19.** The tool should not be used where great precision is required. The bevel blade has one angle for checking *octagons* (45° angles), and another for checking *hexagons* (60° angles).

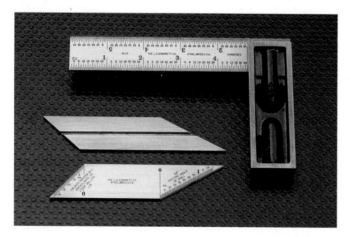

Figure 5-19. *This double square has a graduated blade, beveled blade, and a drill grinding blade. (L. S. Starrett Co.)*

A drill grinding blade is also available for the double square. One end is beveled to 59° for drill grinding and the other end is beveled at 41° for checking the cutting angles of machine screw countersinks. Both ends are graduated for measuring the length of the cutting lips, to ensure that the cutting tools are sharpened on center.

• *Combination set*—This tool consists of a hardened blade, square head, center head, and bevel protractor. The blade fits all three heads, **Figure 5-20.** Combination sets are adaptable to a large variety of operations, making them especially valuable in the shop. The square head, which has one 45° edge, makes it possible for the tool to serve as a miter square. By projecting the blade the desired distance below the edge, it also serves as a depth gage, **Figure 5-21.** The spirit level, fitted in one edge, allows it to be used as a simple level.

With the rule properly inserted, the *center head* can be used to quickly find the center of round stock. This is illustrated in **Figure 5-22.** The *protractor head* can be rotated through 180° and is graduated accordingly. The head can be locked with a locking nut, making it possible to accurately determine and scribe angles, **Figure 5-23.** The head also has a level

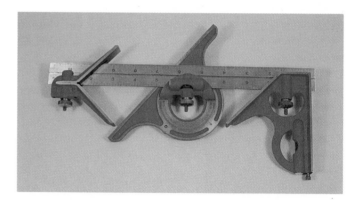

Figure 5-20. *A combination set will perform various layout tasks.*

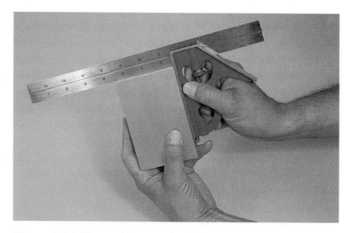

Figure 5-21. *The combination set can be used to check squareness and to measure like a depth gage.*

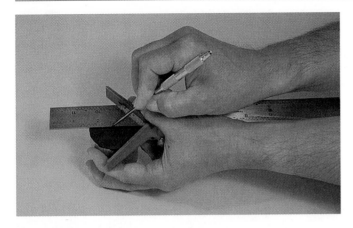

Figure 5-22. *Using a center head and rule to locate the center of a piece of round stock.*

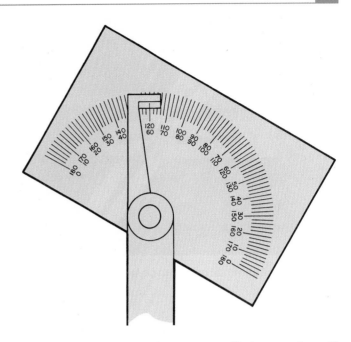

Figure 5-24. *A plain steel protractor will show angles with moderate precision.*

Figure 5-23. *Angular settings on layouts can be made with the protractor head and rule of the combination set.*

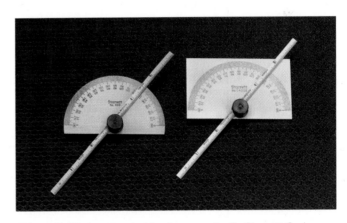

Figure 5-25. *A protractor depth gage. (L. S. Starrett Co.)*

built in, making it possible to use it as a level for positioning angles for inspection, layout, or machining.

Handle a square with care. The blade is mounted solidly, but if the tool is dropped, the blade can be "sprung," ruining the square.

5.3 MEASURING ANGLES

In addition to the protractor head of the combination set, other angle measuring tools are employed in layout work. The accuracy required by a job will determine which tool must be used.

When angles do not need to be laid out or checked to extreme accuracy, a *plain protractor* can be used, **Figure 5-24.** The head is graduated from 0° to 180° in both directions for easy reading.

A *protractor depth gage* is suitable for checking angles and measuring slot depths, **Figure 5-25.**

A *universal bevel* is useful for checking, laying out, and transferring angles, **Figure 5-26.** Both blade and stock are slotted, making it possible to adjust the blade into any desired position. A thumbscrew locks it tightly in place.

When a job requires extreme accuracy, the machinist uses a *Vernier protractor*, **Figure 5-27.** With this tool, angles of 1/12 of a degree (5 minutes) can be accurately measured.

5.4 SIMPLE LAYOUT STEPS

Each layout job requires planning before the operation can be started. **Figure 5-28** shows a typical job. Use the following planning procedure:

1. Carefully study the drawings.

2. Cut stock to size and remove all burrs and sharp edges.

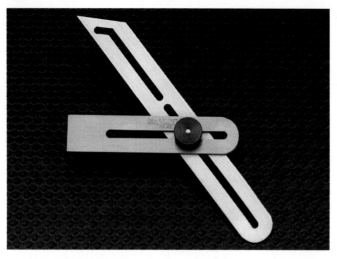

Figure 5-26. *A universal bevel can be locked at various angles. (L. S. Starrett Co.)*

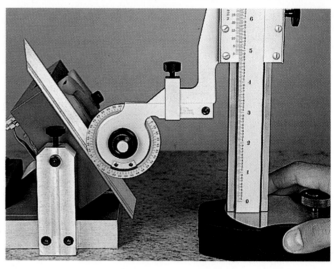

Figure 5-27. *Precise angular measurements are made with a Vernier protractor. In this view, the protractor is mounted on a height gage. (L. S. Starrett Co.)*

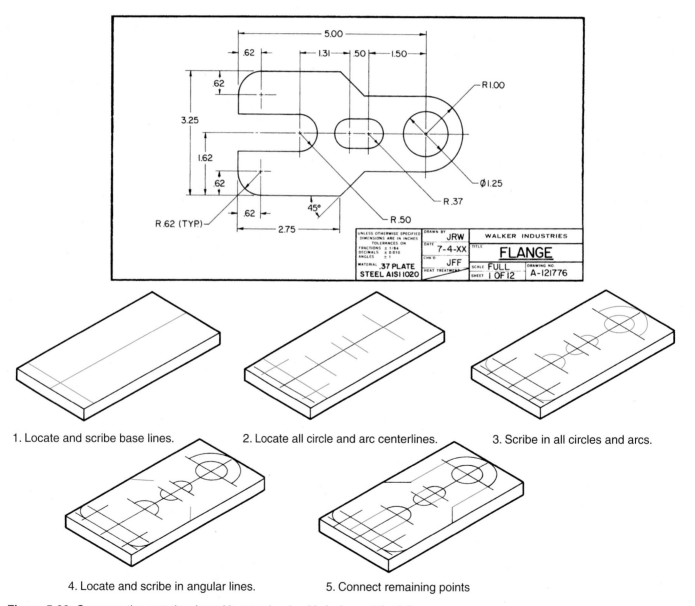

1. Locate and scribe base lines.

2. Locate all circle and arc centerlines.

3. Scribe in all circles and arcs.

4. Locate and scribe in angular lines.

5. Connect remaining points

Figure 5-28. *Compare the part drawing with steps involved in laying out the job.*

3. Clean all dirt, grease and oil from the work surface. Apply layout dye.

4. Locate and scribe a *reference line* (*base line*). You will make all measurements from this line. If the material has one true edge, it can be used in place of the base line.

5. Locate and center points of all circles and arcs.

6. Use a *prick punch* to mark the point where centerlines intersect. The sharp point (30° to 60°) of this punch makes it easy to locate the position. After the prick punch mark has been checked, it is enlarged slightly with a *center punch*, **Figure 5-29.**

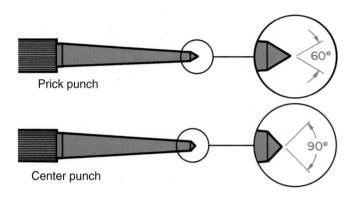

Figure 5-29. *A prick punch has a more sharply angled point than a center punch. The prick punch is used to mark a location. After the prick punch mark is checked, it is enlarged with a center punch.*

7. Scribe in all circles and arcs with a divider or trammel.

8. If angular lines are necessary, scribe them using the proper layout tools. You can also locate the correct points by measuring and connecting them using a rule or straightedge and a scribe.

9. Scribe in all other internal openings.

10. Lines should be clean and sharp. Any double or sloppy line work should be removed by cleaning it off with a solvent. Then apply another coat of dye before scribing the line again.

5.5 LAYOUT SAFETY

- Never carry an open scriber, divider, trammel, or hermaphrodite caliper in your pocket.
- Always cover sharp points with a cork when the tool is not being used.
- Wear goggles when grinding scriber points.
- Get help when you must move heavy items, such as angle plates or V-blocks.
- Remove all burrs and sharp edges from stock before starting layout work.

TEST YOUR KNOWLEDGE

Please do not write in the text. Write your answers on a separate sheet of paper.

1. What is used to make layout lines easier to see?

2. Why are layout lines used?

3. Straight layout lines are drawn with a _____.

4. Circles and arcs are drawn on work with a _____.

5. Large circles and arcs are drawn with a _____.

6. What is wrong with using a pencil to make layout lines on metal?

7. A _____ is the flat granite or steel surface used for layout and inspection work.

8. What layout operations can be performed with a combination set?

9. Round stock is usually supported on _____ for layout and inspection.

10. Long flat surfaces can be checked for trueness with a _____.

11. The center of round stock can be found quickly with the _____ and rule of a combination set.

12. Angular lines that must be very accurate should be laid out with a _____.

13. The _____ punch has a sharper point than the _____ punch.

14. List three safety precautions that you should observe when doing layout work.

A large tap and die set like this one is found in many shops. It includes a complete set of taps and dies, in US Conventional and metric sizes, along with tap wrenches and die stocks. Note that tables matching die and tap sizes to drill sizes are embossed on the inside of the cover for ready reference.

Chapter 6

Hand Tools

LEARNING OBJECTIVES

After studying this chapter, you will be able to:
○ Identify the most commonly used machine shop hand tools.
○ Select the proper hand tool for the job.
○ Maintain hand tools properly.
○ Explain how to use hand tools safely.

IMPORTANT TERMS

abrasive
American National Thread
  System
blind hole
classes of fits
foot-pounds

newton-meters
number sizes
safe edges
torque
Unified System

Selecting and using hand tools correctly will help you do a job safely, with a minimum expenditure of time. When a hand tool is used *incorrectly*, it can be damaged; more importantly, you or someone else may be injured. It is to your advantage to learn to work properly with hand tools.

6.1 CLAMPING DEVICES

Clamping devices are employed to hold and/or position material while it is being worked on. Several types of clamping devices are used in machining.

6.1.1 Vises

The *machinist's vise*, or *bench vise*, is used for many holding tasks. It should be mounted on the bench edge far enough out to permit clamping long work in a vertical position. A vise may be a solid base type or may have a swivel base, which allows the vise to be rotated. See **Figure 6-1**.

A

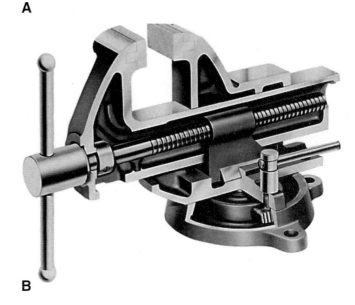

B

Figure 6-1. Machinist's (bench) vise. A—Solid base type. (Wilton Tool Mfg., Inc.) B— Cutaway of swivel base vise. The base is made in two parts so that the body can be rotated to any desired position. (Columbian Vise and Mfg. Co.)

Small precision parts may be held in a small bench vise or *toolmaker's vise*, **Figure 6-2**. This type vise can be rotated and tilted to any desired position. Vise *size* is determined by the width of the jaws, **Figure 6-3**.

Figure 6-2. A small vise used by a toolmaker. It can be rotated and pivoted to desired working position. (Wilton Tool Mfg., Inc.)

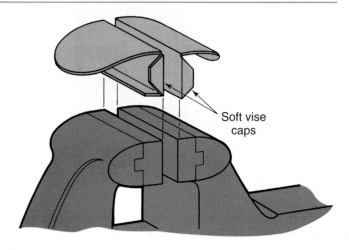

Figure 6-4. Caps made of copper, lead, or aluminum are slipped over hardened vise jaws to protect work from becoming marred or damaged by jaw serrations.

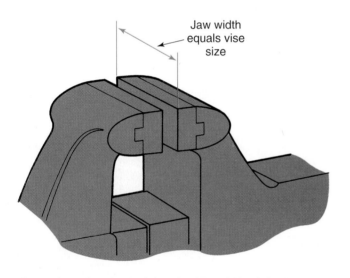

Figure 6-3. Vise size is determined by width of vise jaws.

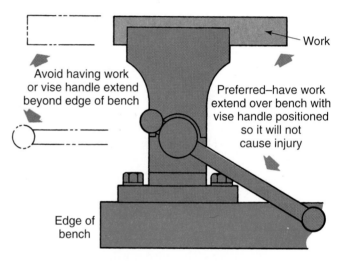

Figure 6-5. To prevent injury, avoid letting the vise handle or work project into the aisle.

A vise's clamping action is obtained from a heavy screw turned by a handle. The handle is long enough to apply ample pressure for any work that will fit the vise. Under *no circumstances* should the vise handle be hammered tight, nor should additional pressure be applied using a length of pipe on the handle for leverage.

Vise jaws are hardened. To clamp work that would be damaged or marred by the jaw serrations, the jaws should be covered with soft copper, brass, or aluminum caps, **Figure 6-4.**

When clamping a job in a vise, do not allow the vise handle or work to project into the aisle, **Figure 6-5**.

6.1.2 Clamps

The *C-clamp* and the *parallel clamp* hold parts together while they are worked on. The C-clamp, **Figure 6-6,** is made in many sizes. Jaw opening determines clamp size.

A parallel clamp is ideal for holding small work. For maximum clamping action, the jaw faces must be parallel. See **Figure 6-7.** Placing strips of paper the width of the clamp jaw between the work and the jaws will improve clamping action.

6.2 PLIERS

Combination pliers, also called *slip-joint pliers,* are widely used for holding tasks. See **Figure 6-8.**

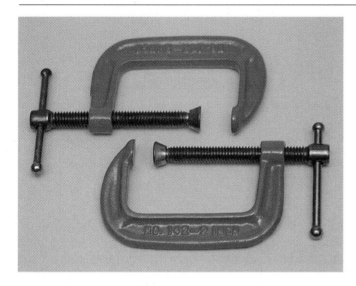

Figure 6-6. *C-clamps are available in a range of sizes.*

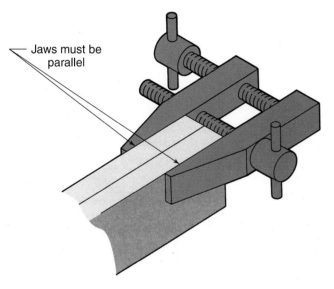

Jaws must be parallel

Figure 6-7. *For maximum clamping action with a parallel clamp, adjust the jaws until they are parallel.*

Figure 6-8. *Combination or slip-joint pliers are used for holding tasks.*

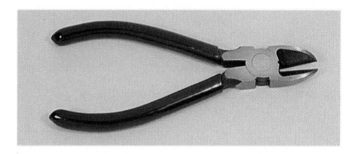

Figure 6-9. *Diagonal pliers will cut flush with a surface.*

Figure 6-10. *Side-cutting pliers have square jaws for holding and cutting tasks.*

The slip-joint permits the pliers to be opened wider at the hinge pin to grip larger size work. They are made in 5″, 6″, 8″, and 10″ sizes. The *pliers size* indicates the overall length of the tool.

Some combination pliers are made with cutting edges for clipping wire and small metal sections to needed lengths. The better grade pliers are of forged construction.

Diagonal pliers are another widely used tool for light cutting tasks, **Figure 6-9.** The cutting edges are at an angle to permit the pliers to cut *flush* (even) with the work surface. Diagonal pliers are made in 4″, 5″, 6″, and 7″ lengths.

Side-cutting pliers are capable of cutting heavier wire and pins, **Figure 6-10.** Some of these pliers have a wire stripping groove and insulated handles. They are made in 6″, 7″, and 8″ lengths.

Round-nose pliers, **Figure 6-11,** are helpful when forming wire and light metal. Their jaws are smooth and will not mar the metal being grasped. Round-nose pliers are available in 4″, 4 1/2″, 5″, and 6″ sizes.

Needle-nose pliers, are available in both straight and curved-nose types. They are handy for holding small work and when work space is limited. They will reach into cramped places. See **Figure 6-12.**

Tongue and groove pliers have aligned teeth for flexibility in gripping different size work, **Figure 6-13.** The size of the jaw opening can be adjusted easily. Tongue and groove pliers are made in many different sizes. The 6″ size usually has five adjustments, while the larger 16″ size has eleven adjustments.

Adjustable clamping pliers are a relatively new addition to the pliers family. On these pliers, the jaw

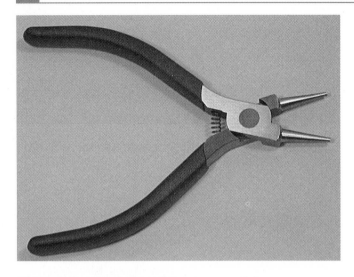

Figure 6-11. Round-nose pliers have smooth jaws.

A

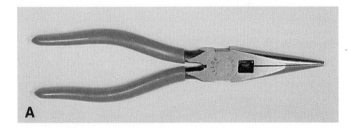

B

Figure 6-12. Needle-nosed pliers. A—Straight pliers can be used to grasp smaller hard-to-reach objects. B—Curved pliers are helpful when working in areas with limited space.

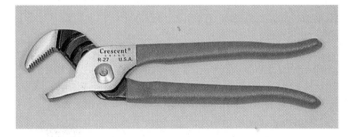

Figure 6-13. Tongue and groove pliers allow the jaws to expand to hold large objects.

opening can be adjusted through a range of sizes by using a threaded mechanism on one handle. See **Figure 6-14**. After adjustment, a squeeze of the hand can lock the jaws onto the work with more than a

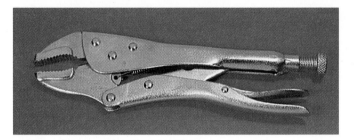

Figure 6-14. Adjustable clamping pliers can be locked on work of different sizes.

ton of pressure. Jaw pressure can be relieved by using the quick release on the handle. These pliers are made in many sizes with straight, curved, or long-nose jaws. They are known by several names, including locking pliers, Vise Grip® pliers, and Tag-L-Lock® pliers. The newest type of adjustable pliers, called Robo-Grip® pliers, permits one-handed jaw-size adjustment by merely squeezing the handles. See **Figure 6-15**. This type of adjustable pliers does not have a locking feature, however.

Figure 6-15. The newest type of adjustable pliers offers one-handed operation.

6.2.1 Care of Pliers

Like many tools, pliers will give long, useful service if a few simple precautions are taken:

- *Never* use pliers as a substitute for a wrench.
- Do not try to cut metal sizes that are too large, or work that has been heat-treated. Pliers with cutters will deform or break if used in this way. Breakage will also occur if additional leverage is applied to the handles.
- Occasionally clean and oil pliers to keep them in good working condition
- Store pliers in a clean, dry place. Avoid throwing them in a drawer or toolbox with other tools.
- Use pliers that are large enough for the job.

6.3 WRENCHES

Wrenches comprise a family of tools designed for use in assembling and disassembling many types of threaded fasteners. They are available in a vast number of types and sizes. Only the most commonly used wrenches will be covered.

6.3.1 Torque-Limiting Wrenches

Torque is the amount of turning or twisting force applied to a threaded fastener or part. It is measured in force units of *foot-pounds* (ft.-lbs.) or the SI Metric equivalent, *newton-meters* (N·m). Torque is the product of the force applied times the length of the lever arm. See **Figure 6-16**.

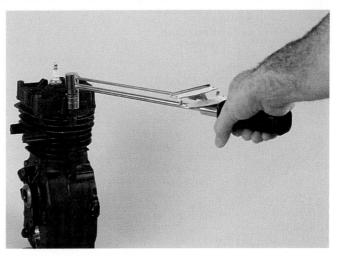

Figure 6-17. *Torque-limiting wrenches are used when fasteners must be tightened to within certain limits to prevent undue stresses and strains from developing in the part.*

There are many types of torque-limiting wrenches, **Figure 6-18**. It is possible to obtain torque wrenches that are direct reading, or that feature a sensory signal (clicking sound or momentary release) when a preset torque is reached.

The right and wrong methods of gripping the wrench handle are shown in **Figure 6-19**. You should *never* lengthen the handle for additional leverage. These tools are designed to take a specific maximum force load. Any force over this amount will destroy the accuracy of the wrench.

Torque-limiting wrenches will provide accurate measurements whether they are pushed or pulled. However, to prevent hand injury, the preferred method is to *pull* on the wrench handle.

6.3.2 Adjustable Wrenches

The term "adjustable wrench" is a *misnomer* (a name not properly applied). Other wrenches, such as the "monkey wrench" and pipe wrench, are also adjustable. However, the wrench that is somewhat like an open-end wrench, but with an adjustable jaw, is commonly referred to as an *adjustable wrench*, **Figure 6-20**.

As the name implies, the wrench can be adjusted to fit a range of bolt-head and nut sizes. Although it is convenient at times, the adjustable wrench is *not* intended to take the place of open-end, box, and socket wrenches.

Three important points must be remembered when using the adjustable wrench:

- The wrench should be placed on the bolt head or nut so that the movable jaw *faces the direction* the fastener is to be rotated, **Figure 6-21**.

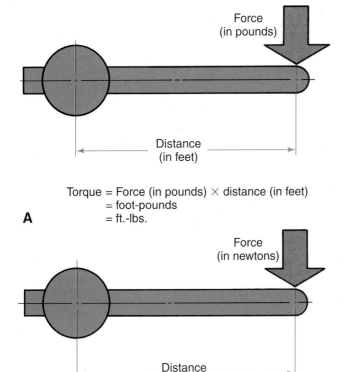

Torque = Force (in pounds) × distance (in feet)
= foot-pounds
A = ft.-lbs.

Torque = Force (in newtons) × distance (in meters)
= newton-meters
B = N•m

Figure 6-16. *Torque measurement. A—In the US Conventional system, torque is measured in foot-pounds. B—Torque values in SI Metric are given in newton-meters.*

A *torque-limiting wrench* allows you to measure the tightening of a threaded fastener in foot-pounds or newton-meters. This provides maximum holding power, without danger of the fastener or part failing or causing the work to warp or spring out of shape. See **Figure 6-17**.

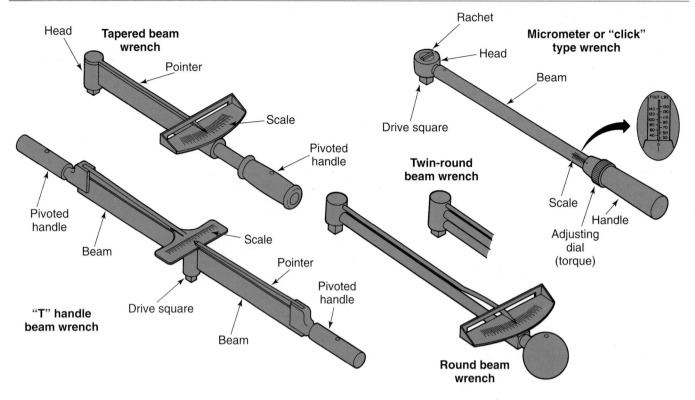

Figure 6-18. Several types of torque-limiting wrenches.

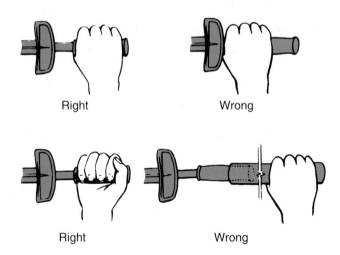

Figure 6-19. The right and wrong ways to apply pressure to a torque-limiting wrench handle.

- Adjust the thumbscrew so the jaws fit the bolt head or nut *snugly*, **Figure 6-22**.
- *Do not* place an extension on the wrench handle for additional leverage. Never hammer on the handle to loosen a stubborn fastener. Use the smallest wrench that will fit the fastener on which you are working. This will minimize the possibility of twisting off the fastener.

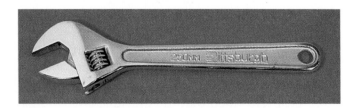

Figure 6-20. An adjustable wrench is handy when a full wrench set is not available.

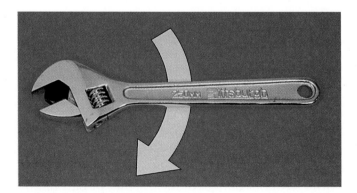

Figure 6-21. The movable jaw of the wrench should always face the direction of rotation.

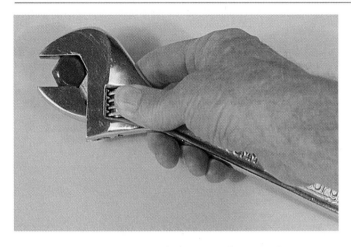

Figure 6-22. A wrench must fit the nut or bolt snugly.

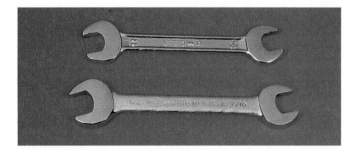

Figure 6-24. An open-end wrench is acceptable when the torque applied is low.

It is dangerous to *push* on, rather than *pull*, any wrench. If the fastener fails or loosens unexpectedly, you will almost always strike and injure your knuckles on the work. This operation is commonly known as "knuckle dusting."

The *pipe wrench* is designed to grip round stock, **Figure 6-23.** However, the jaws always leave marks on the work. Do not use a pipe wrench on bolt heads or nuts unless they cannot be turned with another type of wrench. For instance, you might need a pipe wrench to remove a bolt if the corners of its head have been rounded.

when higher torque must be applied than is possible with an open-end wrench. See **Figure 6-25.** A properly fitted box wrench will not normally slip. It is preferred for many jobs. Box wrenches are available in the same sizes as open-end wrenches and with straight and offset handles.

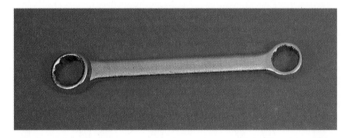

Figure 6-25. A box wrench can handle more torque than an open-end wrench.

Figure 6-23. The pipe wrench has jaws that will grasp round objects.

6.3.5 Combination Open-end and Box Wrenches

A *combination open-end and box wrench* has an open-end wrench at one end of the handle and a box wrench at the other end. These wrenches are made in standard and metric sizes, **Figure 6-26.**

Figure 6-26. Combination wrenches are handy because of two end configurations.

6.3.3 Open-end Wrenches

Open-end wrenches are usually double ended, with two different size openings, **Figure 6-24.** They are made about 0.005″ (0.13 mm) oversize to permit them to easily slip on bolt heads and nuts of the specified wrench size. Openings are at an angle to the wrench body, so that the wrench can be applied in close quarters. Standard and metric open-end wrenches are available. Because of the open end, they can be used only when applied torque is low.

6.3.4 Box Wrenches

The body or jaw of the *box wrench* completely surrounds the bolt head or nut, so it can be used

6.3.6 Socket Wrenches

Socket wrenches are box-like and are made with a tool head-socket (opening) that fits many types of handles (either solid bar or ratchet type). A typical socket wrench set contains various handles and a wide range of socket sizes, **Figure 6-27.** Many sets include both standard and metric sockets. Various types of socket openings are shown in **Figure 6-28.**

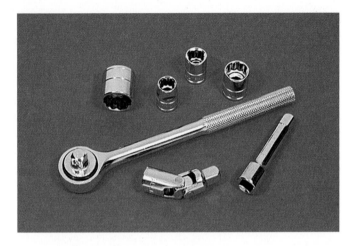

Figure 6-27. A typical socket wrench and sockets. The wrench has a right- and left-hand ratchet mechanism.

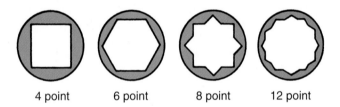

| 4 point | 6 point | 8 point | 12 point |

Figure 6-28. Types of socket openings available. The 12-point socket can be used with both square and hex head fasteners.

6.3.7 Spanner Wrenches

Spanner wrenches are special wrenches with drive lugs, and are designed to turn flush- and recessed-type threaded fittings. The fittings have slots or holes to receive the wrench end. They are usually furnished with machine tools and attachments. See **Figure 6-29.**

A *hook spanner* is equipped with a single lug that is placed in a slot or notch cut in the fitting. An *end spanner* has lugs on both faces of the wrench for better access to the fitting. The lugs fit notches or slots machined into the face of the fitting. On *pin spanner wrenches,* the lugs are replaced with pins that fit into holes on the fitting, rather than into notches.

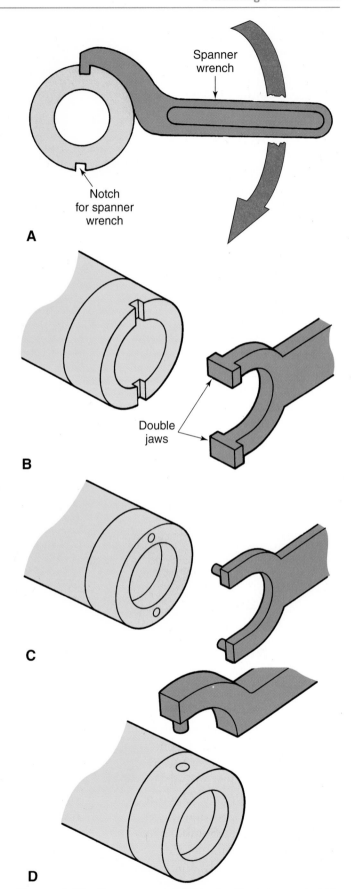

Figure 6-29. Spanner wrenches. A—Hook-type spanner wrench. Some can be adjusted to fit different size fasteners. B—End spanner wrench. C and D—Two types of pin spanner wrenches.

6.3.8 Allen Wrenches

The wrench that is used with socket-headed fasteners is commonly known as an *Allen wrench,* **Figure 6-30**. It is manufactured in many sizes to fit fasteners of various standard and metric dimensions.

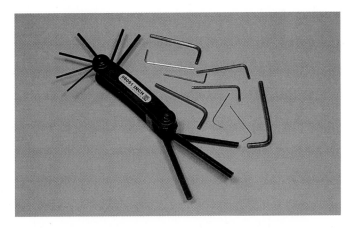

Figure 6-30. Allen wrenches are used with socket-headed fasteners. They are made in both inch and metric sizes.

6.3.9 Wrench Safety

- Always *pull* on a wrench; never *push*. You have more control over the tool and there is less chance of injury.
- Select a wrench that fits properly. A loose-fitting wrench, or one with worn jaws, may slip and cause injury. It can also round off and ruin the bolt or nut on which it is being used.
- Never hammer on a wrench to loosen a stubborn fastener.
- Lengthening a wrench handle for additional leverage is a dangerous practice. Use a larger wrench.
- Before using a wrench, clean any grease or oil off the handle and the floor in the work area. This will reduce the possibility of your hands or feet slipping.
- Never try to use a wrench on moving machinery.

6.4 SCREWDRIVERS

Screwdrivers are manufactured with many different tip shapes, **Figure 6-31**. Each shape has been designed for a particular type of fastener. The standard and Phillips type screwdrivers are familiar to all shop workers. The other shapes may not be as well-known.

The *Phillips screwdriver* has an +-shaped tip for use with Phillips recessed head screws. Four sizes (#1, #2, #3, and #4) will handle the full range of this type fastener. They are manufactured in the same general styles as the standard screwdriver.

The *Pozidriv® screwdriver* tip is similar in appearance to the Phillips tip but has a slightly different shape. This type has been designed for Posidriv screws used extensively in the aircraft, automotive, electronic, and appliance industries.

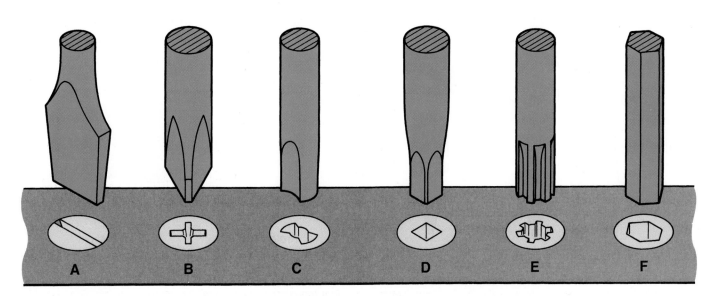

Figure 6-31. Types of screwdriver tips. A—Standard. B—Phillips. C—Clutch. D—Square. E—Torx. F—Hex.

The tip of this screwdriver has a black oxide finish to distinguish it from the Phillips tool. Using a Phillips tip will damage the opening in the head of the Pozidriv screw.

Clutch head, Robertson, Torx®, and *hex screwdrivers* are used for special industrial and security applications.

A *standard screwdriver* has a flattened wedge-shaped tip that fits into the slot in a screw head. This tool is made in 3″ to 12″ lengths. The shank diameter and the width and thickness of the tip are proportional with the length. Screwdriver *length* is measured from blade tip to the bottom of the handle. The blade is heat-treated to provide the necessary hardness and toughness to withstand the twisting pressures.

A few of the standard screwdriver types are shown in **Figure 6-32**. The *double-end offset screwdriver* can be used where there is not enough space for a conventional straight shank tool. The conventional *straight shank screwdriver* is widely used for a variety of work. The *electrician's screwdriver* has a long thin blade and an insulated handle.. The long thin blade will reach into tight areas. A *heavy-duty screwdriver* has a thick, square shank that permits a wrench to be applied for driving or removing large or stubborn screws. The *stubby* or *close quarters screwdriver* is designed for use where work space is limited. The *ratchet screwdriver* moves the screw on the power stroke, but not on the return stroke. It can be set for right-hand or left-hand operation.

6.4.1 Using a Screwdriver

Always select the correct size screwdriver for the screw being driven, **Figure 6-33**. A poor fit will damage the screw slot and often will damage the tool's tip. Damaged screw heads are dangerous, and are often difficult to drive or remove. They should be replaced.

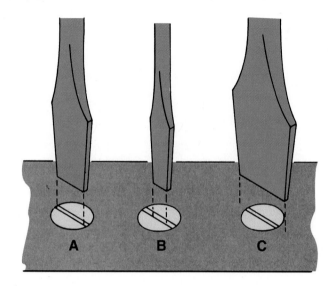

Figure 6-33. Use the correct screwdriver tip for the job. Tip A is the correct width. Tip B is too narrow and will damage the screw head. Tip C is too wide and will damage the work.

When driving or removing a screw, hold the screwdriver square with the fastener. Guide the tip with your free hand.

A worn screwdriver tip, such as the one shown at right in **Figure 6-34**, must be reground. A fine grinding wheel and light pressure is required. *Avoid overheating the tip during the grinding operation. It will destroy the tool.* Check the tip during the grinding operation by fitting it to a screw slot. A properly ground tip will fit snugly and hold the head firmly in the slot.

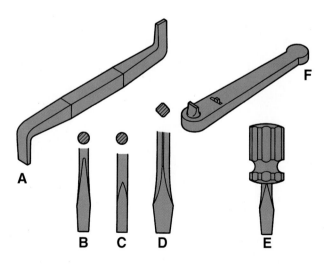

Figure 6-32. Styles and types of standard screwdrivers. A—Double end offset. B—Conventional straight shank. C—Electrician's. D—Heavy-duty. E—Stubby or close quarter. F—Ratchet type offset.

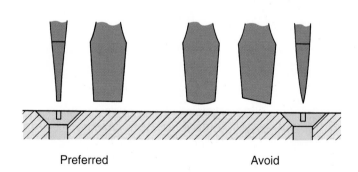

Preferred   Avoid

Figure 6-34. Tips on the right are to be avoided. They are worn or improperly sharpened. The tip at left is ground correctly. Note that the sides are concave; this holds tip in slot when pressure is applied.

6.4.2 Screwdriver Safety

- A screwdriver is not a substitute for a chisel, nor is it made to be hammered on or used as a pry bar.
- Wear safety goggles when regrinding screwdriver tips.
- Screws with burred heads are dangerous. They should be replaced or the burrs removed with a file or abrasive cloth.
- Always turn electric power off before working on electrical equipment. The screwdriver should have an insulated handle specifically designed for electrical work.
- Avoid carrying a screwdriver in your pocket. It is a dangerous practice that can cause injury to you or to someone else. It can also damage your clothing.

6.5 STRIKING TOOLS

The machinist's ball-peen hammer, **Figure 6-35,** is the most commonly used shop hammer. It has a hardened striking face and is used for all general purpose work that requires a hammer.

Figure 6-35. *The ball-peen hammer is most common type in a machine shop.*

Ball-peen hammer *sizes* are classified according to the weight of the head, without the handle. They are available in weights of 2, 4, 8, and 12 ounces, and 1, 1 1/2, 2, and 3 pounds.

A **soft-face hammer** or **mallet** permits heavy blows to be struck without damaging the part or surface. A steel face hammer would damage or mar the work surface. Soft-face hammers are especially

useful for setting work tightly on parallels (steel bars) when mounting material in a vise.

Soft-face hammers are made of many different materials: copper, brass, lead, rawhide, and plastic. See **Figure 6-36.**

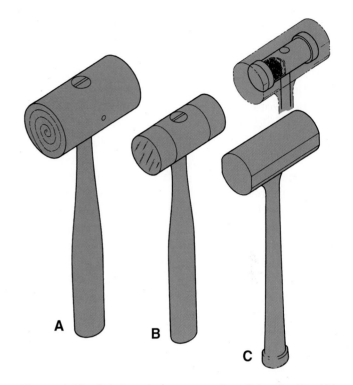

Figure 6-36. *Soft-face hammers and mallets. A—Rawhide mallet. B—Plastic-face hammer. C—Dead blow hammer. The head of the dead blow hammer contains tiny steel shot encased in plastic. This provides a hammer that has striking power but will not rebound (bounce back) as will other mallets and soft-face hammers.*

6.5.1 Striking Tool Safety

- Never strike two hammers together. The faces are very hard; the blow might cause a chip to break off and fly out at high speed.
- Do not use a hammer unless the head is on tightly and the handle is in good condition.
- Do not "choke up" too far on the handle when striking a blow, or you may injure your knuckles.
- Strike each blow squarely, or the hammer may glance off of the work and injure you or someone working nearby.
- Place a hammer on the bench carefully. A falling hammer can cause a painful foot injury, or damage precision tools on the bench.

6.6 CHISELS

Not all cutting in metalworking is done by machine. Chisels are one of several basic hand tools, such as hacksaws and files, that are considered cutting implements. These tools, when in good condition, sharp, and properly handled, are safe to use.

The *chisel* is used mostly to cut cold metal, hence the term "cold" chisel. The four chisels illustrated in **Figure 6-37** are the most common types. The general term *cold chisel* is used when referring to these chisels. Other chisels in this category are variations or combinations of these chisels.

The work to be cut will determine how the chisel should be sharpened, **Figure 6-38.** A chisel with a slightly curved cutting edge will work better when cutting on a flat plate. The curved edge will help prevent the chisel from cutting unwanted grooves in the surrounding metal, as when shearing rivet heads. If the chisel is to be used to shear metal held in a vise, the cutting edge should be straight.

The chisel is frequently employed to chip surplus metal from castings. Chipping is started by holding the chisel at an angle, as shown in **Figure 6-39.** The angle must be great enough to cause the cutting edge to enter the metal.

After the chisel cut has been started and the proper depth reached, the chisel angle can be decreased enough to keep the cutting action at the proper depth. Cut depth can be reduced by decreasing the chisel angle. However, if the cutting

angle is decreased too much, the chisel will ride on the heel of the cutting edge and lift out of the cut.

When shearing metal in a vise, position it so the layout line is just below the vise jaws. This will leave sufficient metal to finish by filing or grinding. When cutting, it is usually best to hold the metal in a vise *without* using jaw caps. This provides a better shearing action between the vise jaws and chisel. Advance the chisel after each blow so the cutting is done by the center of the cutting edge.

A chisel is an ideal tool for removing rivets. The head can be sheared off and the rivet punched out. A variation of the conventional cold chisel for removing rivet heads is called a *"rivet buster,"* **Figure 6-40.**

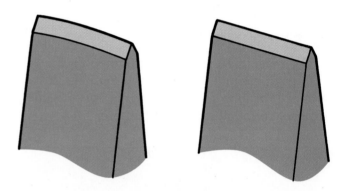

Figure 6-38. *The work to be done determines how a chisel should be sharpened. Left—Slightly rounded edge for cutting on flat plate. Right—Straight edge for shearing.*

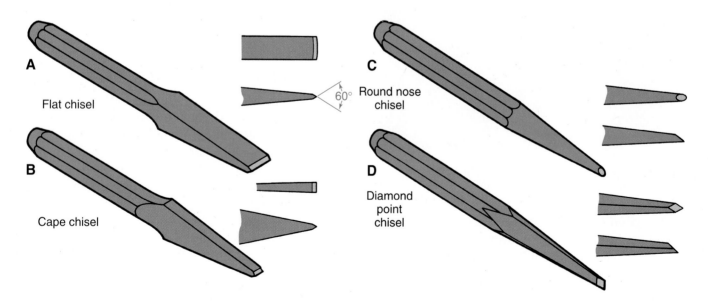

Figure 6-37. *Cold chisels. A—Flat chisel is used for general cutting and chipping work. B—Cape chisel has a narrower cutting edge than the flat chisel and is used to cut grooves. C—Round nose chisel can cut radii and round grooves. D—Diamond point chisel is principally used for squaring corners.*

If the head is so large that it cannot be removed in one piece, make a saw cut almost through the head, then use the chisel to cut away half the head at a time. **Figure 6-41B** shows how this is done. Rivets also can be removed by drilling and using the narrow cape chisel, **Figure 6-41C.**

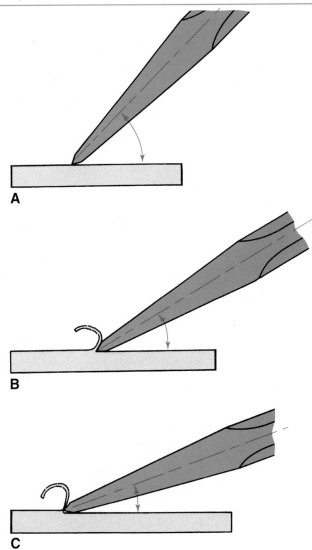

Figure 6-39. Proper chisel angles for various cutting operations. A—Starting the cut. B—Maintaining cut at desired depth. C—Reducing the cutting angle too much will cause chisel to lift out of cut.

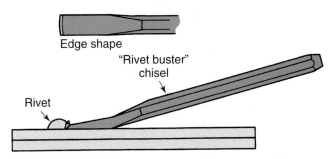

Figure 6-40. This variation of the flat chisel is often referred to as a "rivet buster." Upper drawing shows how it is sharpened.

When there is not enough room to swing a hammer with sufficient force to cut a rivet, an alternate procedure can be used, **Figure 6-41A**. Drill a hole, about the size of the rivet body, almost through the head. The head can then be removed easily with the chisel.

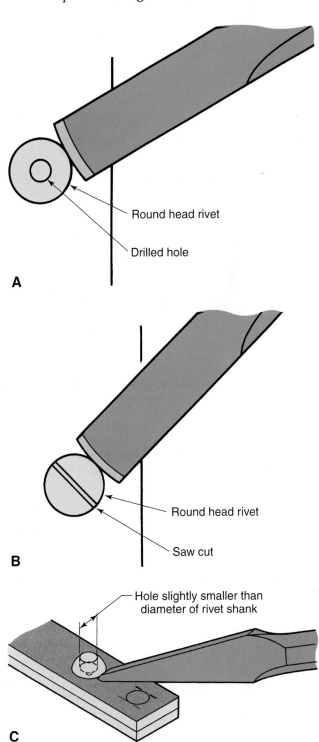

Figure 6-41. Alternate methods for removing rivet heads. A—When there is not enough room to swing hammer with sufficient force. B—When rivet head is too large to be removed as one piece. C—A cape chisel may also be used to remove rivets.

6.6.1 Chisel Safety

- Flying chips are dangerous! When cutting metal with a chisel, wear safety goggles and erect a shield around the work. These steps will protect you and people working nearby.
- Hold a chisel so that if you miss it with the hammer, you will not strike and injure your hand. If one is available, use a chisel holder.
- A **mushroomed chisel head** is extremely dangerous, since jagged metal can be knocked or chipped off and cause serious injury. Remove this hazardous condition by grinding away the excess metal. See **Figure 6-42**.
- Edges on metal cut with a chisel are sharp and can cause bad cuts. Remove them by grinding or filing.

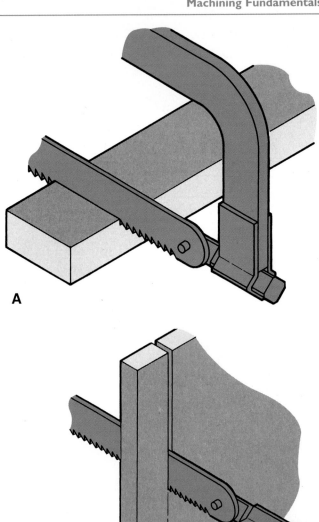

A

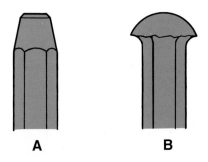

A   **B**

Figure 6-42. Chisel safety. A—Chisel head ground to a safe condition. B—A dangerous mushroomed head.

6.7 HACKSAW

The typical *hacksaw* is composed of a frame with a handle and a replaceable blade, **Figure 6-43**. Almost all hacksaws made today are adjustable to accommodate several different blade lengths. They are also made so the blade can be installed in either a vertical or horizontal position, **Figure 6-44**.

When placing a blade in the saw frame, make sure the frame is adjusted for the blade length being inserted. There should be sufficient adjustment remaining to permit tightening the blade until it "pings" when snapped with your finger. Frequently,

B

Figure 6-44. Blade positions. A—The blade is set to cut in a conventional vertical position. B—The blade has been pivoted 90º to cut a long narrow strip of stock.

a new blade must be retightened after a few strokes because it will stretch slightly from the heat produced while cutting.

The hacksaw blade must be positioned with the teeth pointing *away* from the handle, **Figure 6-45**. This will make it cut on the forward (push) stroke.

6.7.1 Holding Work for Sawing

The work must be held securely, with the point to be cut as close to the vise as practical. This helps to eliminates "chatter" and vibration that will dull the saw teeth.

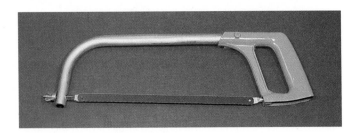

Figure 6-43. A typical hacksaw.

Figure 6-45. A hacksaw blade must be inserted with the teeth pointing away from the handle. This positions it to cut on the forward stroke of the hacksaw.

Figure 6-46 shows some methods preferred for holding work that is irregular shape. The work is clamped so the cut is started on a flat side rather than on a corner or edge. This lessens the possibility of ruining the teeth or breaking the blade.

6.7.2 Starting a Cut

When starting a cut to a marked line, it is best to notch the work with a file, **Figure 6-47.** You can also use the thumb of your left hand to guide the blade until it starts the cut, **Figure 6-48.** Work carefully to avoid injury. Some hacksaw blades are manufactured with very fine teeth at the *front* to make starting a cut easier. Use enough pressure so the blade will begin to cut immediately.

6.7.3 Making the Cut

Grasp the hacksaw firmly by the handle and front of the frame. Apply enough pressure on the forward stroke to make the teeth cut. Insufficient pressure will permit the teeth to slide over the

material, dulling the teeth. Also, lift the saw slightly on the return stroke.

Cut the *full length* of the blade and make about 40 to 50 strokes per minute. More strokes per minute may generate enough heat to draw the blade

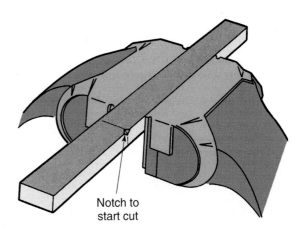

Notch to start cut

Figure 6-47. Using a file to notch or nick the edge of a piece to be cut permits easier starting of the hacksaw cut.

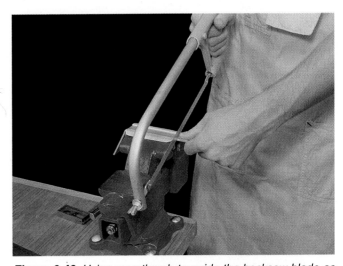

Figure 6-48. Using your thumb to guide the hacksaw blade as the cut is started.

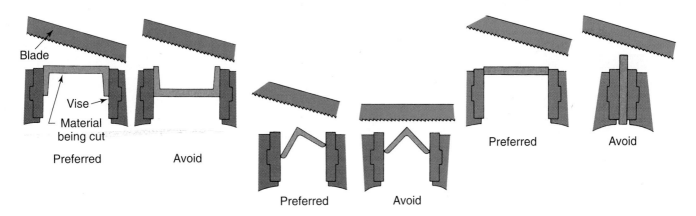

Blade

Vise

Material being cut

Preferred   Avoid

Preferred   Avoid

Preferred   Avoid

Figure 6-46. Preferred methods of holding irregular stock for sawing.

temper and dull the teeth. Keep the blade moving in a straight line. Avoid any twisting or binding, which can bend or break the blade.

Dulling or breaking a hacksaw blade. If you start a cut with an old blade and the blade breaks or dulls, do *not* continue in the same cut with a new blade. As a blade become dull, the *kerf* (slot made by blade) becomes narrower. If you try to continue the cut in the same slot, the new blade will usually bind and be ruined in the first few strokes. If possible, rotate the work and start a new cut on the other side.

6.7.4 Finishing a Cut

When the blade has cut almost through the material, saw carefully. Support the stock being cut off with your free hand to prevent it from dropping when the cut is completed.

6.7.5 Saw Blades

All hacksaw blades are heat-treated to provide the hardness and toughness needed to cut metal.

The shape and kind of material to be cut has an important bearing on blade choice, in terms of the number of teeth per inch, **Figure 6-49.**

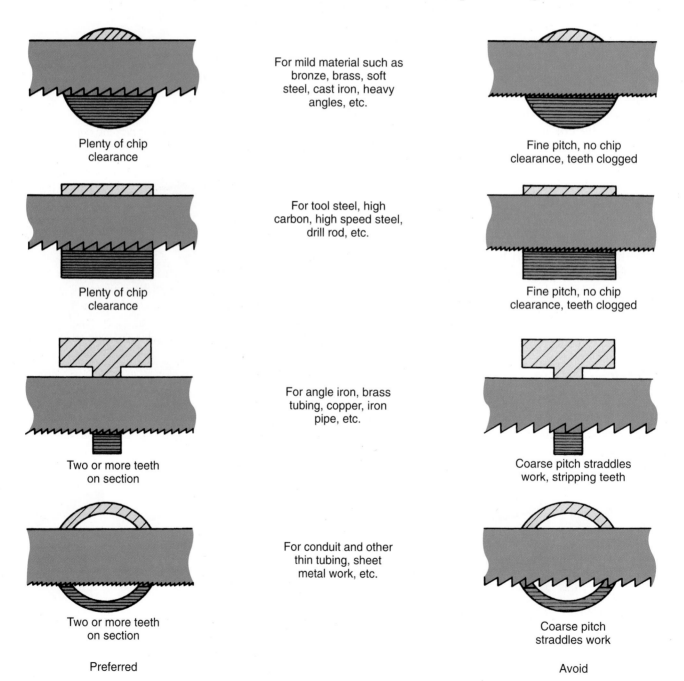

Figure 6–49. The proper hacksaw blade should be used for each job to assure long blade life and rapid cutting action. Study recommendations.

The *flexible back blade* has only the teeth hardened. The *all-hard blade* is hardened throughout. The hardness is reduced near the end holes, however, to reduce the possibility of breakage at these points.

Flexible back blades are best for sawing soft materials or materials with thin cross sections. An all-hard blade is best for cutting hard metals. It does not buckle when heavy pressure is applied.

Two or *three teeth* should be cutting at all times; otherwise, the teeth will straddle the section being cut and snap off when cutting pressure is applied.

The *set* of the blade provides the necessary clearance, and prevents the blade from binding in the cut. A blade may have one of three sets: *undulating, raker,* or *alternate.* These are shown in **Figure 6-50.**

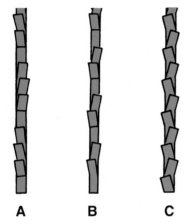

A   B   C

Figure 6-50. Types of sets in hacksaw teeth. A—Undulating. B—Raker. C—Alternate.

6.7.6 Unusual Cutting Situations

Cutting soft metal tubing can be a problem. The blade may bind and tear the tubing or the tubing may flatten. This can be eliminated by inserting a wood dowel of the proper size into the tubing. Then cut through both tubing and dowel. See **Figure 6-51.**

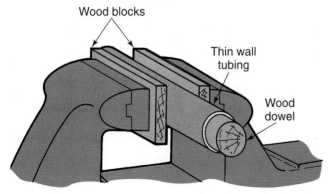

Wood blocks

Thin wall tubing

Wood dowel

Figure 6-51. A snug-fitting dowel slid into thin-wall tubing will make cutting the tubing easier. If tubing is to be held in a vise for cutting, place soft wood blocks between the vise jaws and the work to prevent marring the exterior surface of the tubing.

Cutting a long narrow strip from thin metal can be done by setting the blade at right angles to the frame. Make the cut in the usual way, as in **Figure 6-52.** Strips of any width, up to the capacity of the saw frame, can be made in this manner.

Thin metal can be cut more easily and precisely by putting it between two pieces of wood, and cutting through both wood and metal, **Figure 6-53.**

6.7.7 Hacksaw Safety

- Never test the sharpness of a blade by running your fingers across its teeth.
- Store saws in a way that will prevent accidentally grasping the teeth when you pick up a saw.
- Burrs formed on the cut edge of metal are sharp and can cause a serious cut. Do not brush away chips with your hand; use a brush.
- Always wear safety goggles while using a hacksaw. All-hard blades can shatter and produce flying chips.
- Be sure the hacksaw blade is properly tensioned. If it should break while you are on the cutting stroke, your hand may strike the work, causing a painful injury.

6.8 FILES

A *file* is used for hand smoothing and shaping operations. The modern file is made from high-grade carbon steel and is heat-treated to provide the necessary hardness and toughness.

In manufacturing a file, the first production step is to cut the blank to approximate shape and size. See **Figure 6-54.** The tang and point are formed next. Then, the blank is annealed and straightened. The point and tang are trimmed after the sides and faces have been ground and the teeth cut. After another straightening, the file is heat-treated, cleaned, and oiled. Tests are made continually to assure a quality tool.

6.8.1 File Classifications

Files are classified by their shape. The shape is the general outline and cross section. The outline is either tapered or blunt, **Figure 6-55A.**

Files are also classified according to the cut of the teeth: *single-cut, double-cut, rasp,* and *curved tooth,* **Figure 6-55B,** and to the coarseness of the teeth: *rough, coarse, bastard, second-cut, smooth,* and *dead smooth.*

6.8.2 File Care

A file should *never* be used without a handle. It is too easy to drive the unprotected tang into your hand.

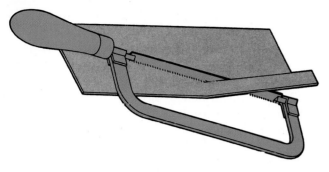

Figure 6-52. *The hacksaw blade can be pivoted to a horizontal position for cutting long narrow strips. Best results can be obtained if strip is bent up slightly, as shown, during the sawing operation.*

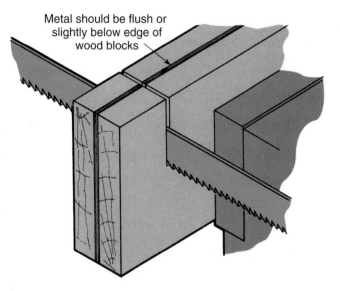

Metal should be flush or slightly below edge of wood blocks

Figure 6-53. *Sandwiching thin metal between two pieces of wood will make sawing easier and more precise.*

A   B   C   D   E   F   G

Figure 6-54. *Steps in manufacturing a file. A—Steel bar cut to correct length. B—Bar forged to shape. C—Blank after it has been annealed. D—Annealed blank straightened and ground smooth. E—Teeth cut on blank. F—Blank trimmed and coated for heat treatment. G—Completed file cleaned and inspected. (Nicholson File Co.)*

Fit a handle to the file by drilling a hole in the handle. The hole should equal in diameter the width of the tang at its midpoint, **Figure 6-56**. Mate the file and handle by inserting the tang in the hole, then sharply striking the handle on a solid surface.

Store files so that they are always separated, **Figure 6-57;** *never* throw files together in a drawer or store them in a damp place.

Clean files frequently with a **file card** or **brush, Figure 6-58.** Some soft metals cause **pinning,** a condition in which the teeth become loaded or clogged with material the file has removed. Pinning will cause gouging and scratching of the work surface. The particles can be removed from the file with a pick or scorer. A file card combines the card, brush, and pick for file cleaning.

6.8.3 Selecting a File

The number of different kinds, shapes, and cuts of files that are manufactured is almost unlimited, **Figure 6-59**. For this reason, only the general classification of files will be covered.

Files have three distinct characteristics: *length,* kind or *type,* and *cut.* The file length is always measured from the heel to the point, **Figure 6-60**. The tang is not included in the measurement.

The file type refers to its *shape,* such as flat, mill, half-round, or square. The file cut indicates the relative *coarseness* of the teeth.

Single-cut files are usually used to produce a smooth surface finish. They require only light pressure to cut.

Double-cut files remove metal much faster than single-cut files. They require heavier pressure and they produce a rougher surface finish.

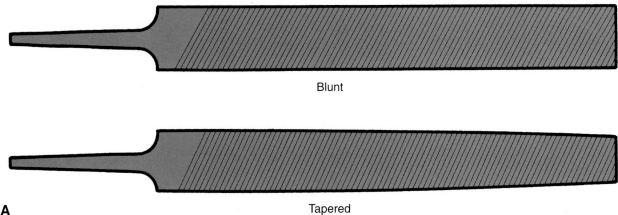

Blunt

A

Tapered

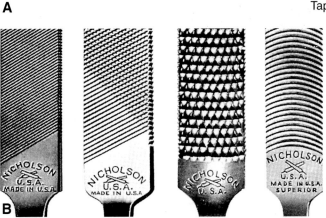

B

Figure 6-55. File classifications. A—Blunt and tapered files. B—Single-cut, double-cut, rasp, and curved-tooth files.

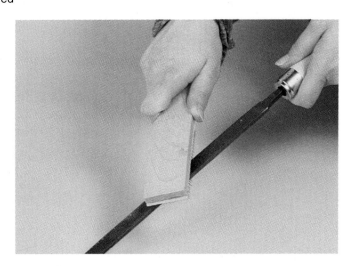

Figure 6-58. Using file card to clean a file.

Figure 6-56. File handle hole should be equal in diameter to width of file tang at the point indicated.

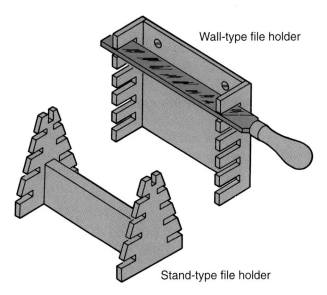

Wall-type file holder

Stand-type file holder

Figure 6-57. Storing files properly, in holders like these, will greatly extend their useful life.

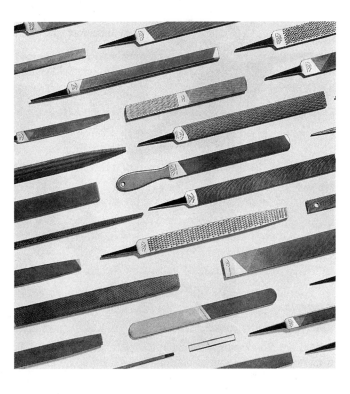

Figure 6-59 A few of the many hundred different kinds of files. (Nicholson File Co.)

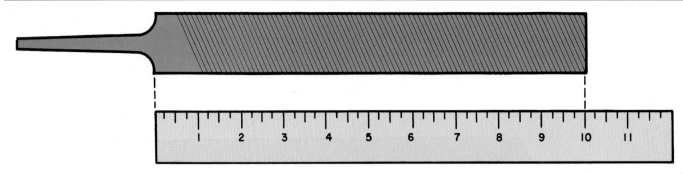

Figure 6-60. How a file is measured.(Nicholson File Co.)

Rasps are best for working wood or other soft materials where a large amount of stock must be removed in a hurry.

A *curved-tooth file* is used to file flat surfaces of aluminum and sheet steel.

Some files have *safe edges.* This denotes that the file has one or both edges without teeth, **Figure 6-61.** This permits filing corners without danger to the portion of the work that is not to be filed.

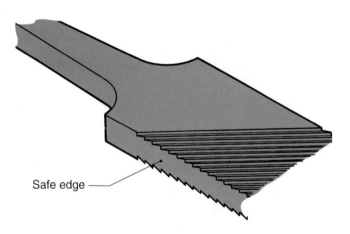

Safe edge

Figure 6-61. The safe edge of a file does not have teeth.

In selecting the file, many factors must be considered if maximum cutting efficiency is to be attained:
- The nature of the work (flat, concave, convex, notched, etc.).
- Kind of material.
- Amount of material to be removed.
- Surface finish and accuracy demanded.

Of the many file shapes available, the most commonly used are flat, pillar, square, 3-square, knife, half-round, crossing, and round, **Figure 6-62.** Each shape is available in many sizes and degrees of coarseness: rough, coarse, bastard, second-cut, smooth, and dead smooth, **Figure 6-63.** Note that a small (4″) rough cut file may be as fine as a large (16″) second-cut file.

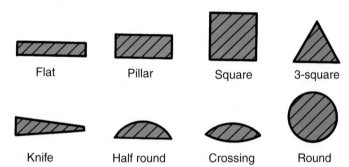

| Flat | Pillar | Square | 3-square |

| Knife | Half round | Crossing | Round |

Figure 6-62. Cross-sectional views of the most widely used file shapes.

6.8.4 Types of Files

The wide variety of files can be divided into five general groups:

The *machinists' file* is used whenever metal must be removed rapidly, and the finish is of secondary importance. It is made in a large range of shapes and sizes, and is double-cut.

The *mill file* is a single-cut and tapers for the last third of its length away from the tang. It is suitable for general filing when a smooth finish is required.

Figure 6-63. Range in coarseness of a typical machinist flat bastard file. File sizes range from 4″ (100 mm) to 16″ (400 mm). (Nicholson File Co.)

A mill file works well for draw filing, lathe work, and working on brass and bronze.

Swiss pattern and *jewelers' files* are manufactured in more than a hundred different shapes. They are used primarily by tool-and-die makers, jewelers, and others who do precision filing.

The *rasp* has teeth that are individually formed and disconnected from each other. It is used for relatively soft materials (plastic for example) when large quantities of the material must be removed.

The *special purpose files* group includes those specifically designed to cut one type of metal or for one kind of operation. An example is the long-angle lathe file, **Figure 6-64**, which does an efficient filing job on the lathe.

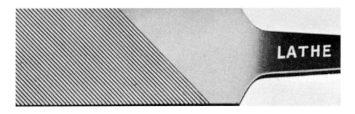

Figure 6-64. *The long-angle lathe file is an example of a special-purpose file.*

6.8.5 Using a File

Efficient filing requires that the work be held solidly. Where practical, hold the work at about elbow height for general filing, **Figure 6-65**. If large quantities of metal must be removed by heavy filing, mount the work slightly lower.

Straight or *cross filing* consists of pushing the file lengthwise across the work, either straight ahead or at a slight angle. Grasp the file as shown in **Figure 6-66A**. Heavy-duty filing requires heavy pressure and can best be done if the file is held as in **Figure 6-66B**.

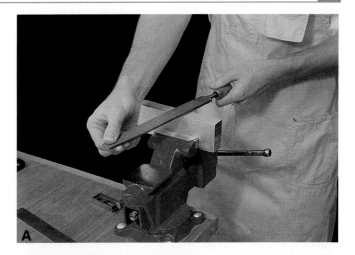

Figure 6-66. *Holding a file. A—The proper way to hold a file for straight or cross filing. B—Holding method used to apply the additional pressure required when a considerable quantity of metal must be removed.*

A file can be ruined by using either *too much* pressure or *too little* pressure on the cutting stroke. Apply just enough pressure to permit the file to cut on the entire forward stroke. Too little pressure allows the file to slide over the work. This will dull the file. Too much pressure "overloads" the file, causing the teeth to clog and chip.

Lift the file from the work on the reverse stroke, except when filing soft metal. The pressure on the return stroke when filing soft metal should be no more than the weight of the file itself

Draw filing, when properly done, will produce a finer finish than straight filing. Hold the file as shown in **Figure 6-67**. Do not use a *short-angle* file for draw filing The short-angle file can cause scoring or scratching, instead of the desired shaving and shearing action, as the file is pushed and pulled across the work. Use a double-cut file to "rough down" the surface, then a single-cut file to produce the final finish.

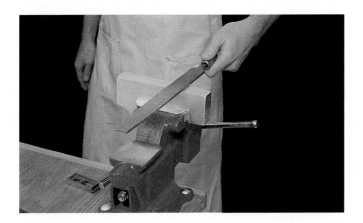

Figure 6-65. *Mount work at elbow height for general filing.*

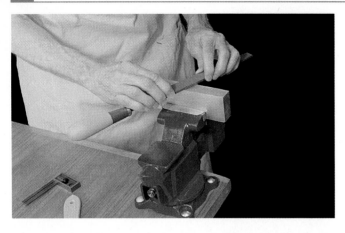

Figure 6-67. Draw filing, when done properly, improves the surface finish.

6.8.6 File Safety

- Never use a file without a handle. Painful injuries may result!
- Clean files with a file card, not your hand. The chips can penetrate your skin and result in a painful infection.
- Do not clean a file by slapping it on the bench, since it may shatter.
- Files are very brittle. Never use one for prying tasks.
- Use a piece of cloth, not your bare hand, to clean the surface being filed. Sharp burrs are formed in filing and can cause serious cuts.
- Never hammer on or with a file. It can shatter, causing chips to fly in all directions.

6.9 REAMERS

A drill does not produce a smooth or accurate enough hole for a precision fit. *Reaming* is the operation that will produce smoothness and accuracy. Ordinarily, *hand reaming* is used only for final sizing of a hole.

6.9.1 Hand Reamer

A *hand reamer* has a square shank end so that it can be held in a tap wrench. See **Figure 6-68A**. Reamers may be made of high-speed steel or carbon steel, and are available in sizes from 1/8" to 1 1/2" (3.175 mm to 38.1 mm). The cutting end is ground with a slight taper to provide easy starting in the hole.

Straight-fluted reamers are suitable for most work. However, when reaming a hole with a keyway or other interruption, it is better to have a *spiral-fluted reamer*.

When preparing a piece to be reamed by hand, 0.005" to 0.010" (0.15 mm to 0.25 mm) of stock should be left in the hole for removal by the reaming tool.

An *expansion hand reamer* is used when a hole must be cut a few thousandths inch over nominal size for fitting purposes, **Figure 6-68B**. Slots are cut into the center of the tool. The center opening is machined on a slight taper. The reamer is expanded by tightening a taper screw into this opening. The amount of expansion is limited; the reamer may be broken if expanded too much.

Because of the danger of producing *oversize holes*, do not use an expansion reamer instead of a solid reamer unless absolutely necessary.

The *adjustable hand reamer* is threaded its entire length and is fitted with tapered slots to receive the adjustable blades, **Figure 6-68C**. The blades are tapered along one edge to correspond with the taper slots in the reamer body, so that the cutting edges of the blades remain parallel.

Reamer diameter is set by loosening one adjusting nut and tightening the other. The blades can be moved in either direction. This type of reamer is manufactured in sizes ranging from 3/8" to 3 1/2" (9.5 mm to 85 mm). Each reamer has sufficient adjustment to increase its diameter to the next larger reamer size.

The *taper reamer* will finish a tapered hole accurately and with a smooth finish for taper pins, **Figure 6-68D**. Because of their long cutting edges, taper reamers are somewhat difficult to operate.

To provide for easier removal of surplus metal, a *roughing reamer* is first rotated into the hole. This reamer is slightly smaller (0.010" or 0.25 mm) than the finish reamer. Left-hand spiral grooves are ground along the cutting edges to break up chips.

6.9.2 Using a Hand Reamer

A two-handle tap wrench is commonly used with a reamer, because it permits an even application of pressure. It is virtually impossible to secure a satisfactory hole using an adjustable wrench to turn the reamer.

To start reaming, rotate the tool slowly to allow it to align with the hole. It is desirable to check, at several points around the reamer's circumference, whether it has started square, **Figure 6-69**.

Feed should be steady and rapid. Keep the reamer cutting, or it will start to "chatter," producing a series of tool marks in the surface of the hole. This could also cause the hole to be out-of-round.

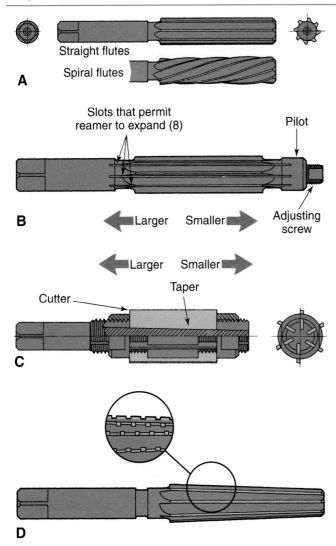

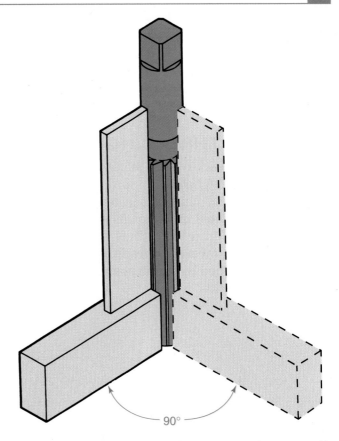

Figure 6-69. Always make sure that the reamer is square with the work.

Figure 6-68. Hand reamers. A—Straight flute and spiral flute solid reamers are used for different applications. B—The expansion hand reamer and how its size is adjusted. C—The adjustable hand reamer can be set for odd sizes. D—A taper hand reamer. Enlarged section shows how the cutting edges are notched on a roughing taper reamer.

Turning pressure is applied evenly with both hands, *always* in a clockwise direction, **Figure 6-70**. Never turn a reamer in a *counterclockwise* direction, since this will dull the cutting edges.

Feed the reamer deeply enough into the hole to take care of the starting taper. The choice of cutting fluid to be applied will depend upon the metal being reamed.

6.9.3 Reaming Safety

- To prevent injury, remove all burrs from holes.
- Never use your hands to remove chips and cutting fluid from the reamer. Use a piece of cotton waste.

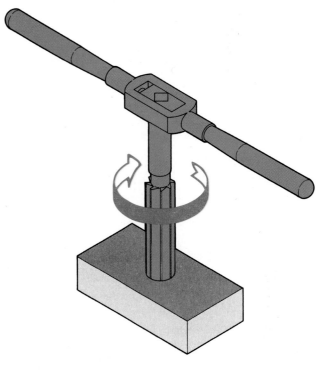

Figure 6-70. Always turn a hand reamer in a clockwise direction.

- Store reamers carefully so they do not touch one another. Never store reamers loose or throw them into a drawer with other tools.
- Clamp work solidly before starting to ream.
- Do not use compressed air to remove chips and cutting fluid or to clean a reamed hole.

6.10 HAND THREADING

Threaded sections have many applications in our everyday life. A *thread* is a spiral or helical ridge found on nuts and bolts. When required on a job, threads are indicated on the plans and drawings in a special way, **Figure 6-71**. They are specified by diameter and number of threads per inch. Metric threads are specified by diameter and thread pitch is given in millimeters.

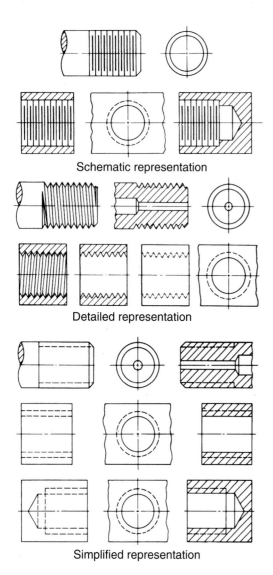

Schematic representation

Detailed representation

Simplified representation

Figure 6-71. *Methods used to visually depict threads on drawings. Only one type will be used on a given drawing.*

The *American National Thread System* was adopted in 1911. It is the common thread form used in the United States and is characterized by the 60° angle formed by the sides of the thread.

The *National Coarse (NC) Thread* is for general purpose work; the *National Fine (NF) Thread* is for precision assemblies. These are the most widely used thread groups in the American National series. The NF group has more threads per inch for a given diameter than the NC group.

A considerable amount of confusion resulted during World War II from the many different forms and kinds of threads used by the Allied nations. As a result, the powers that make up NATO (the North Atlantic Treaty Organization) adopted a standard thread form. It is referred to as the *Unified System,* **Figure 6-72**. It is very similar to the American National Thread System. It differs only in the thread shape. The thread root is rounded and the crest may be flat or rounded. The threads are identified by UNF and UNC (Unified National Fine and Unified National Coarse), **Figure 6-73**. Fasteners using this thread series are interchangeable with fasteners using the American National thread.

In addition to those just described, there are several other thread groups. Included are the *Unified National Extra Fine* (UNEF), *Unified National 8 Series*, and *Unified National 12 Series*. The 8 Series has 8 threads per inch and is used on diameters ranging from 1″ to 6″ in 1/8″ and 1/4″ increments. The 12 Series has 12 threads per inch and is used on diameters that range from 1″ to 6″.

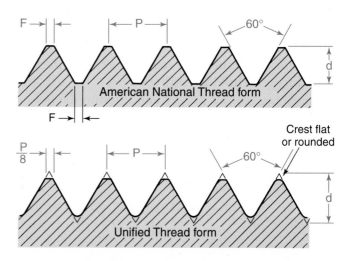

Figure 6-72. *Drawings illustrate similarities and differences of the American National Thread form and the Unified Thread*

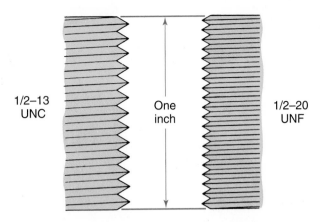

Figure 6-73. Comparison of Unified Coarse (UNC), and Unified Fine (UNF) threads. Both have same geometric shape.

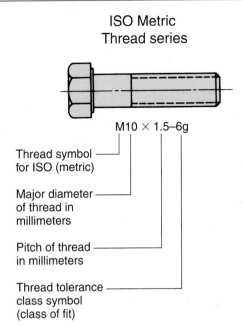

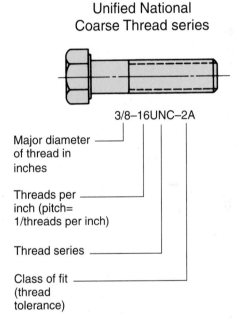

Figure 6-74. How thread size is noted and what each term means.

Metric unit threads have the same shape as the Unified Thread, but are specified in a different manner, **Figure 6-74.** Metric threads and Unified National Series threads are not interchangeable. See **Figure 6-75.**

6.10.1 Thread Size

Threads of the Unified National system smaller than 1/4" diameter are not measured as *fractional* sizes. They are given by **number sizes** that range from #0 (approximately 1/16" or 0.060" diameter) to #12 (just under 1/4" or 0.216" diameter). Both UNC and UNF series are available.

Care must be taken so the number denoting the thread diameter and the number of threads are not mistaken for a fraction. For example: a #8-32 UNC thread would be a thread that is a #8 (0.164") diameter and 32 threads per inch, *not* an 8/32 (1/4") diameter fastener with a UNC series thread.

6.10.2 Cutting Threads

Because thread dimensions have been standardized, the use of **taps** to cut *internal* threads, and dies to cut *external* threads have become universal practice whenever threads are to be cut by hand. See **Figure 6-76.**

6.10.3 Internal Threads

Internal threads are made with a tap, **Figure 6-77.** Taps are made of carbon steel or high-speed steel (HSS) and are carefully heat-treated for long life. Taps are quite brittle and are easily broken if not handled properly.

To meet demands for varying degrees of thread accuracy, it became necessary for industry to adopt standard working tolerances for threads. Working tolerances for threads have been divided into *classes of fits,* which are indicated by the last number on the thread description (1/2-1 3 UNC-2).

Fits for inch-based threads are:

Class 1 – Loose fit.
Class 2 – Free fit.
Class 3 – Medium fit.
Class 4 – Close fit.

ISO Metric
Thread series

Unified National
Coarse Thread series

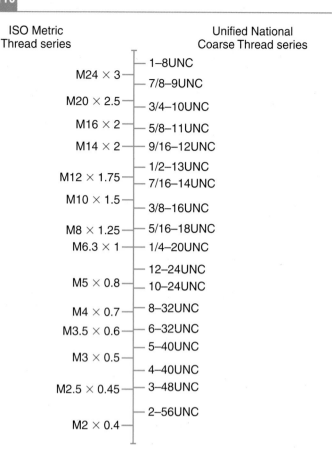

- 1–8UNC
M24 × 3 ┤
  - 7/8–9UNC
M20 × 2.5 ┤
  - 3/4–10UNC
M16 × 2 ┤
  - 5/8–11UNC
M14 × 2 ┤
  - 9/16–12UNC
  - 1/2–13UNC
M12 × 1.75 ┤
  - 7/16–14UNC
M10 × 1.5 ┤
  - 3/8–16UNC
M8 × 1.25 ┤
  - 5/16–18UNC
M6.3 × 1 ┤
  - 1/4–20UNC
  - 12–24UNC
M5 × 0.8 ┤
  - 10–24UNC
M4 × 0.7 ┤
  - 8–32UNC
M3.5 × 0.6 ┤
  - 6–32UNC
  - 5–40UNC
M3 × 0.5 ┤
  - 4–40UNC
M2.5 × 0.45 ┤
  - 3–48UNC
  - 2–56UNC
M2 × 0.4 ┤

Figure 6-75. While ISO Metric threads may appear to be similar in diameter to the Unified National Thread series, the two are not interchangeable.

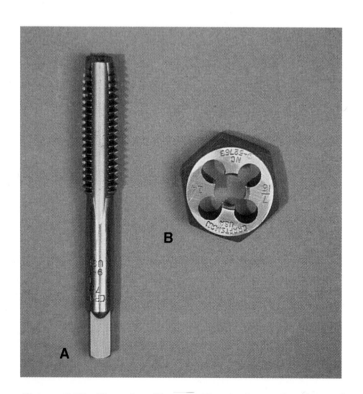

Figure 6-76. Thread cutting. A—Tap is for cutting internal threads. B—Die is for cutting external threads.

Figure 6-77. A machinist is cutting internal threads with a tap.

Under revised ISO standards, there are two classes of thread tolerances for external threads: 6g for *general-purpose threads* and 5g6g for *close tolerance threads*. There is only one tolerance class for internal threads, 6H. A *lower case* letter indicates the tolerance on a bolt and a *capital* letter is used for the nut.

6.10.4 Taps

Standard hand taps are made in sets of three. They are known as taper, plug, and bottoming taps. See **Figure 6-78.**

Threads are started with a *taper tap.* It is tapered back from the end 6 to 10 threads before full thread diameter is reached.

The *plug tap* is used after the taper tap has cut threads as far into the hole as possible. It tapers back 3 or 4 threads before full thread diameter is reached.

Figure 6-78. *Standard hand taps. A—Taper for starting thread. B—Plug tap for continuing thread after taper tap has cut as far into hole as it can. C—Bottoming for continuing threads to bottom of a blind hole. (Threadwell Manufacturing Co.)*

Threads are cut to the bottom of a **blind hole** (one that does not go through the part) with a **bottoming tap.** This tap tapers back 1 or 20 threads before full thread diameter is reached.It is necessary to use the full set of taps only when a blind hole is to be tapped, **Figure 6-79.**

Another tap used in the shop is a pipe tap. A **pipe tap** cuts a tapered thread, so there is a

Figure 6-79. *Cutaway of metal block illustrates three types of threaded holes. Left—Open or through hole. Center—Blind hole that has been drilled deeper than desired threads. Right—Blind hole with threads tapped to bottom.*

"wedging" action set up to make a leak-tight joint. The fraction that indicates pipe tap size may be confusing at first because it indicates pipe size and not the thread diameter. See **Figure 6-80.**

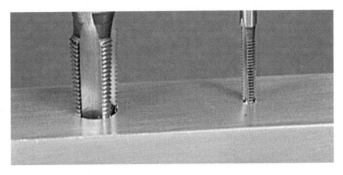

Figure 6-80. *There is considerable difference between a 1/8" standard thread tap and a 1/8" pipe thread tap.*

A **pipe thread** is indicated by the abbreviation **NPT** (National Pipe Thread). To obtain the wedging action needed for leakproof joints, the threads taper 3/4" per foot of length.

6.10.5 Tap Drill

The drill used to make the hole prior to tapping is called a **tap drill.** Theoretically, it should be equal in diameter to the minor diameter of the screw that will be fitted into the tapped hole. See **Figure 6-81**.

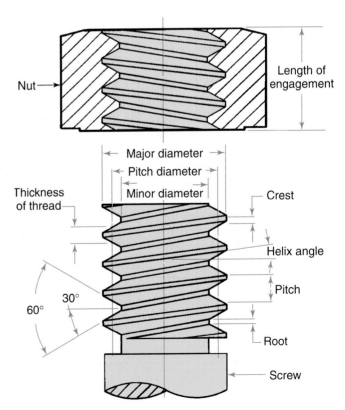

Figure 6-81. *Nomenclature of a fastener thread.*

This situation would cause the tap to cut a full thread, however. The pressure required to rotate the tap would be so great that tap breakage could occur. Full-depth threads are not necessary because three-quarter-depth threads are strong enough that the fastener usually breaks before the threads strip.

Drill sizes can be secured from a tap drill chart, **Figures 6-82** and **6-83.**

6.10.6 Tap Wrenches

Two types of tap wrenches are available, **Figure 6-84.** The type to be employed will depend upon tap size. A *T-handle tap wrench* should be used with all small taps. It allows a more sensitive "feel" when tapping. The *hand tap wrench* is best suited for large taps where more leverage is required.

When tapping by hand, the chief requirement is to make sure that the tap is started straight, and remains square during the entire tapping operation, **Figure 6-85.** The tap must be *backed off* (reversed in rotation) for one-half of a turn every one or two cutting turns. This will break the chips free and allow them to drop through the tap flutes. Backing off prevents chips from jamming the tap and damaging the threads.

Size	Threads per inch	Major dia.	Minor dia.	Pitch dia.	Tap drill 75% thread	Decimal equiv-alent	Clearance drill	Decimal equiv-alent
2	56	.0860	.0628	.0744	50	.0700	42	.0935
	64	.0860	.0657	.0759	50	.0700	42	.0935
3	48	.099	.0719	.0855	47	.0785	36	.1065
	56	.099	.0758	.0874	45	.0820	36	.1065
4	40	.112	.0795	.0958	43	.0890	31	.1200
	48	.112	.0849	.0985	42	.0935	31	.1200
6	32	.138	.0974	.1177	36	.1065	26	.1470
	40	.138	.1055	.1218	33	.1130	26	.1470
8	32	.164	.1234	.1437	29	.1360	17	.1730
	36	.164	.1279	.1460	29	.1360	17	.1730
10	24	.190	.1359	.1629	25	.1495	8	.1990
	32	.190	.1494	.1697	21	.1590	8	.1990
12	24	.216	.1619	.1889	16	.1770	1	.2280
	28	.216	.1696	.1928	14	.1820	2	.2210
1/4	20	.250	.1850	.2175	7	.2010	G	.2610
	28	.250	.2036	.2268	3	.2130	G	.2610
5/16	18	.3125	.2403	.2764	F	.2570	21/64	.3281
	24	.3125	.2584	.2854	I	.2720	21/64	.3281
3/8	16	.3750	.2938	.3344	5/16	.3125	25/64	.3906
	24	.3750	.3209	.3479	Q	.3320	25/64	.3906
7/16	14	.4375	.3447	.3911	U	.3680	15/32	.4687
	20	.4375	.3725	.4050	25/64	.3906	29/64	.4531
1/2	13	.5000	.4001	.4500	27/64	.4219	17/32	.5312
	20	.5000	.4350	.4675	29/64	.4531	33/64	.5156
9/16	12	.5625	.4542	.5084	31/64	.4844	19/32	.5937
	18	.5625	.4903	.5264	33/64	.5156	37/64	.5781
5/8	11	.6250	.5069	.5660	17/32	.5312	21/32	.6562
	18	.6250	.5528	.5889	37/64	.5781	41/64	.6406
3/4	10	.7500	.6201	.6850	21/32	.6562	25/32	.7812
	16	.7500	.6688	.7094	11/16	.6875	49/64	.7656
7/8	9	.8750	.7307	.8028	49/64	.7656	29/32	.9062
	14	.8750	.7822	.8286	13/16	.8125	57/64	.8906
1	8	1.0000	.8376	.9188	7/8	.8750	1- 1/32	1.0312
	14	1.0000	.9072	.9536	15/16	.9375	1- 1/64	1.0156
1-1/8	7	1.1250	.9394	1.0322	63/64	.9844	1- 5/32	1.1562
	12	1.1250	1.0167	1.0709	1- 3/64	1.0469	1- 5/32	1.1562
1-1/4	7	1.2500	1.0644	1.1572	1- 7/64	1.1094	1- 9/32	1.2812
	12	1.2500	1.1417	1.1959	1-11/64	1.1719	1- 9/32	1.2812
1-1/2	6	1.5000	1.2835	1.3917	1-11/32	1.3437	1-17/32	1.5312
	12	1.5000	1.3917	1.4459	1-27/64	1.4219	1-17/32	1.5312

National Coarse and National Fine threads and tap drills

Figure 6-82. Thread and tap drill chart for Unified National threads.

| Nominal size | Internal thread minor diameter | | Tap drill diameter |
	Max.	Min.	
M1.6×0.35	1.321	1.221	1.25
M2×0.4	1.679	1.567	1.6
M2.5×0.45	2.138	2.013	2.05
M3×0.5	2.599	2.459	2.5
M3.5×0.6	3.010	2.850	2.9
M4×0.7	3.422	3.242	3.3
M5×0.8	4.334	4.134	4.2
M6.3×1	5.553	5.217	5.3
M8×1.25	6.912	6.647	6.8
M10×1.5	8.676	8.376	8.5
M12×1.75	10.441	10.106	10.2
M14×2	12.210	11.835	12.0

| Nominal size | Internal thread minor diameter | | Tap drill diameter |
	Max.	Min.	
M16×2	14.210	13.885	14.0
M20×2.5	17.744	17.294	17.5
M24×3	21.252	20.752	21.0
M30×3.5	26.771	26.211	26.5
M36×4	32.370	31.670	32.0
M42×4.5	37.799	37.129	37.5
M48×5	43.297	42.587	43.0
M56×5.5	50.796	50.046	50.5
M64×6	58.305	57.505	58.0
M72×6	66.305	65.505	66.0
M80×6	74.305	73.505	74.0
M90×6	84.305	83.505	84.0
M100×6	94.305	93.505	94.0

Figure 6-83. Thread and tap drill chart for metric threads.

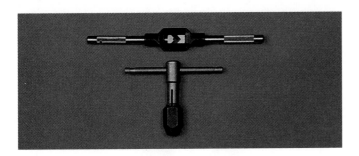

Figure 6-84. Tap wrenches Top—Hand tap wrench. Bottom—T-handle tap wrench.

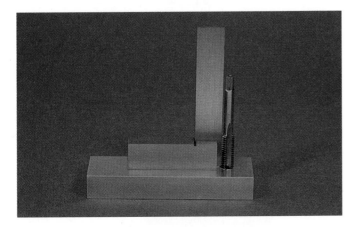

Figure 6-85. A tap must be started squarely with the hole. A quick way to check this is to use a machinist's square.

When tapping a blind hole, some machinists place a dab of grease, or a piece of grease pencil or wax crayon in the hole. As the tap cuts the threads, the grease is forced up and out of the hole, carrying the chips along.

Use a cutting fluid designed for the particular metal you are tapping.

6.10.7 Power Tapper

Tapping can also be done using an *articulating arm power tapper*, **Figure 6-86**, which can be fitted with an auto reverse, automatic tap lubrication, and a digital depth readout. Holes can be tapped quickly with little chance of tap breakage.

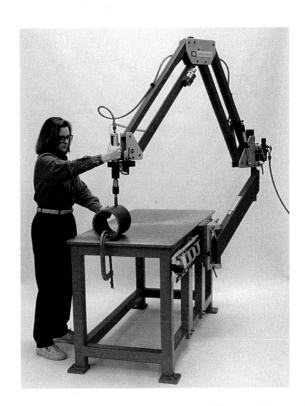

Figure 6-86. Articulated arm power tapper. The machine can be fitted with an auto reverse system (the tap is run out of the hole automatically when the threads are at the proper depth), automatic lubrication of the tap, and a digital depth readout. (Procyon Machine Tools)

6.10.8 Care in Tapping

Considerable care must be exercised when tapping:

- Use the correct size tap drill. Secure this information from a tap drill chart.
- Use a sharp tap and apply sufficient quantities of cutting fluid. With some cutting fluids, the area is flooded with fluid; with others, a few drops are sufficient. Read the container label.
- Start the taper tap square with the work.
- Do not force the tap to cut. Remove the chips using a piece of cloth or cotton waste, not your fingers.
- Avoid running a tap to the bottom of a blind hole and continuing to apply pressure. Do not allow the hole to fill with chips and jam the tap. Either condition can cause the tap to break (especially if the tap is small).
- Remove burrs on the tapped hole with a smooth file.

6.10.9 Dealing with Broken Taps

Taps sometimes break off in a hole. Several tools and techniques have been developed for removing them without damaging the threads already cut. Remember that these methods do not always work and the part may have to be discarded.

Frequently, a tap will *shatter* in the hole. It may then be possible to remove the fragments with a pointed tool such as a scribe.

Broken carbon steel taps can sometimes be removed from steel if the work can be heated to annealing temperature. The tap can then be drilled out. This *cannot* be done with high-speed steel taps. If the HSS tap is large enough, it can be ground out with a hand grinder.

A *tap extractor* can sometimes be used to remove a broken tap. See **Figure 6-87**. Penetrating oil should be applied and allowed to "soak in" for a short time before the fingers of the tap extractor are

fitted into the flutes of the broken tap. The collar on the extractor is slipped down flush with the work surface. A tap wrench is fitted on the extractor. The tap extractor is then carefully twisted back and forth to loosen the tap segments. After the broken parts have been loosened, it is a simple matter to remove them.

In some shops, a *tap disintegrator* is used to remove broken taps. This device makes use of an electric arc to cause the tap to disintegrate. If used properly, it will break up the tap without affecting the metal surrounding the broken tool.

6.10.10 External Threads

External threads are cut with a die, **Figure 6-88**. *Solid dies* are not adjustable and for that reason are

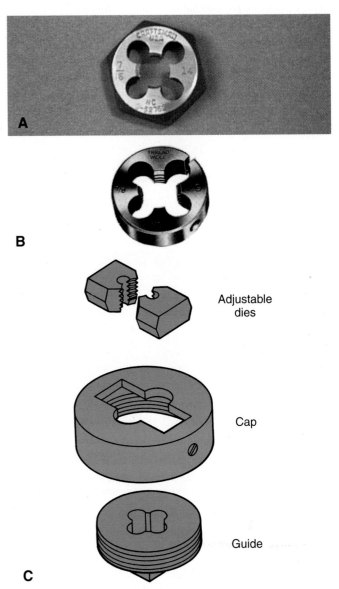

Figure 6-88. Types of dies. A—Solid dies used to cut external threads by hand. They cannot be adjusted. B—A small screw on one side of the split permits small changes in size of an adjustable die. C—Construction of a multi-part adjustable die.

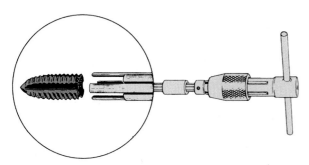

Figure 6-87. Tap extractor will help remove a broken tap. Close-up shows fingers of extractor and how they fit into flutes of broken tap. It does not always work.

not often used. The *adjustable die,* and the *two-part adjustable die,* are preferred. The two-part die has a wide range of adjustment and is fitted with guides to keep it true and square on the work. Dies are available for cutting most standard threads.

6.10.11 Die Stocks

A *die stock* holds the die and provides leverage for turning the die on the work. See **Figure 6-89.**

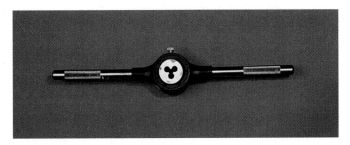

Figure 6–89. The die is held in a die stock.

When cutting external threads, it is necessary to remember:

- Material diameter is the *same size* as the desired thread diameter. That is, 1/2-13 UNC threads are cut on a 1/2" diameter shaft. What diameter shaft would be needed to cut 1/2-20 UNF threads?
- Mount work solidly in a vise.
- Set the die to the proper size. Make trial cuts on a piece of scrap until the proper adjustment is found.
- Grind a small chamfer on the shaft end, as shown in **Figure 6-90.** This permits a die to start easily.

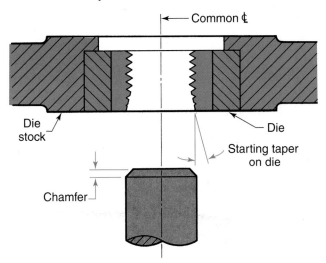

Figure 6-90. A die will start more easily if a small chamfer is cut or ground on the end of the shaft to be threaded. Section through die and die stock shows proper way to start threads.

- Start the cut with the tapered end of the die.
- Back off the die every one or two turns to break the chips.
- Use cutting oil. Place a paper towel down over the work to absorb excess cutting oil. The towel will also prevent the oil from getting on the floor.
- Remove any burrs from the finished thread with a fine cut file.

6.10.12 Threading to a Shoulder

When a thread must be cut by hand to a shoulder, start and run the threads as far as possible in the usual manner, **Figure 6-91.** Remove the die. Turn it over with the taper up. Run the threads down again to the shoulder. *Never try this operation without first starting the threads in the usual manner.*

6.10.13 Problems in Cutting External Threads

The most common problem encountered when cutting external threads with a die is *ragged threads.* They are caused by:

- Applying little or no cutting oil.
- Dull die cutters.
- Stock too large for the threads being cut.
- Die not started square.
- One set of cutters upside down when using a two part die.

6.10.14 Hand Threading Safety

- If a tap or threaded piece must be cleaned of chips with compressed air, protect your eyes from flying chips by wearing goggles. Take care not to endanger persons working in the area near you!
- Chips produced by hand threading are sharp. Use a brush or piece of cloth, *not* your hand, to remove them!
- Newly-cut external threads are very sharp. Again, use a brush or cloth to clean them.
- Wash your hands after using cutting fluids or oils! Some cause skin rash. This can develop into a serious skin disorder if the oils are left on the hands for an extended period.
- Have cuts treated by a qualified person. Infections can occur when cuts and other injuries are not properly treated.
- Be sure the die is clamped firmly in the die stock. If not, it can fall from the holder and cause injury.
- Broken taps have very sharp edges and are very dangerous. Handle them as you would broken glass!

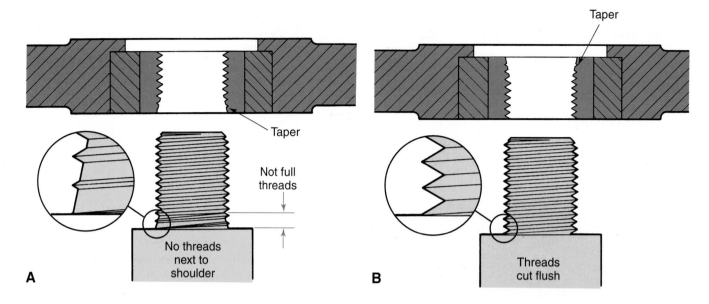

Figure 6-91. *How to cut threads to a shoulder. After die has been run down as far as possible, the die is reversed. When rotated down the shaft, it will cut threads almost flush with shoulder. A—Running die down normally. B—Reversing die to cut flush.*

6.11 HAND POLISHING

An *abrasive* is commonly thought of as any hard substance that will wear away another material. The substance, grain size, backing material, and the manner in which the substance is bonded to the backing material determines the performance and efficiency of an abrasive.

6.11.1 Abrasive Materials

Emery is a natural abrasive. It is black in color and cuts slowly, with a tendency to polish.

Aluminum oxide has replaced emery as an abrasive when large quantities of metal must be removed. It is a *synthetic* (manufactured) material that works best on high-carbon and alloy steels. Aluminum oxide that is designed for use on metal is manufactured with a grain shape that is not as sharp as that made for woodworking.

Silicon carbide is the hardest and sharpest of the synthetic abrasives. Silicon carbide is greenish-black in color. It is superior to aluminum oxide in its ability to cut fast under light pressure. It is ideal for "sanding" metals like cast iron, bronze, and aluminum.

Crocus may be synthetic or natural iron oxide. It is bright red in color, very soft, and is used for cleaning and polishing when a minimum of stock is to be removed.

Diamonds are the hardest *natural* substance known. However, they can also be manufactured. Synthetic diamonds have no value as gems; they are used almost exclusively by industry for polishing and grinding. *Diamond dust polishing compound* is made by crushing synthetic diamonds. It is the only abrasive hard enough to polish the newer heat-treated, exotic alloy steels used by industry.

The table in **Figure 6-92** shows a comparison of abrasive grain size and indicates how the various abrasives are graded.

Technical grades		Simplified grades	Other grades
Mesh	Aluminum oxide Silicon carbide	Emery	Emery polishing
600			4/0
			3/0
500			2/0
400	10/0		0
360			
320	9/0		1/2
280	8/0		
240	7/0		1 G
220	6/0		2
180	5/0		3
150	4/0	Fine	
120	3/0		
100	2/0	Medium	
80	0	coarse	
60	1/2		
50	1	Extra coarse	
40	1 1/2		
36	2		
30	2 1/2		
24	3		
20	3 1/2		
16	4		
12	4 1/2		

Figure 6-92. *A comparative grading chart for abrasives. (Coated Abrasive Manufacturers Institute)*

6.11.2 Coated Abrasives

A *coated abrasive* is cloth or paper with abrasive grains bonded to one surface. Because of its flexibility, cloth is used as a backing material for abrasives found in metalworking or machining. It is available in 9″ × 11″ (210 mm × 280 mm) sheets. It is also available in rolls starting at 1/2″ (12.5 mm) in width, and is called *abrasive cloth.*

6.11.3 Using Abrasive Cloth

Abrasive cloth is quite expensive. Use only what you need. Tear the correct amount from the sheet or roll. Do not discard abrasive cloth unless it is completely worthless. Used cloth is excellent for polishing.

If a job has been filed properly, only a fine-grain cloth will be needed to polish the surface. However, if scratches are deep, start the polishing operation by using a coarse-grain cloth first. Change to a medium-grain cloth next, and finally a fine-grain abrasive. A few drops of oil will speed the operation. For a high polish, leave the oil on the surface after the scratches have been removed. Reverse the cloth and rub the smooth backing over the work.

Abrasive cloth must be properly supported to work efficiently. To get such support, wrap the cloth around a block of wood or file, **Figure 6-93.** Apply pressure and rub the abrasive back and forth in a straight line, parallel to the long side of the work.

Avoid using abrasives on machined surfaces.

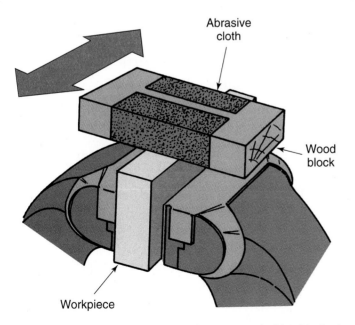

Figure 6-93. *Abrasive cloth should be supported with a block of wood or a file. Do not support it with your fingers alone.*

6.11.4 Abrasives Safety

- Avoid rubbing your fingers or hand over polished surfaces or surfaces to be polished. Burrs on the edges of the metal can cause painful cuts.
- Wash your hands thoroughly after polishing operations.
- Treat all cuts immediately, no matter how small!
- Place all oily rags in a closed metal container. Never put them in your apron/shop coat or in a locker!
- Wipe up any oil dropped on the floor during polishing operations.
- If a lathe is used for polishing operations, make sure the machine is protected from the abrasive grains that fall from the polishing cloth. Stop the machine when inspecting your work.

TEST YOUR KNOWLEDGE

Please do not write in the text. Write your answers on a separate sheet of paper.

1. List two variations of the machinist's vise.

2. How is vise size determined?

3. Work held in a vise can be protected from damage by the jaw serrations if _____ are placed over the jaws.

4. To prevent injuries, what should be avoided when mounting work in a vise?

5. Work is often held together with a _____ and/or _____ while being machined or worked on.

6. How do combination pliers have an advantage over many other types of pliers?

7. Why are the cutting edges on diagonal pliers set at an angle?

8. List three ways of extending the working life of pliers.

9. What are adjustable clamping pliers?

10. Of what use are torque-limiting wrenches?

11. Do torque-limiting wrenches give a more accurate reading when they are pushed or when they are pulled?

12. Several different wrenches can be classified as adjustable wrenches. Name three.

13. List three points that should be observed when using an adjustable wrench.

14. Round work can be gripped with a _____ wrench. Its main disadvantage is that the jaws will probably _____ the work.

15. Describe socket wrenches.

16. What wrenches are employed to turn flush and recessed types of threaded fasteners? The fasteners have slots or holes to receive the wrench lugs.

17. Rather than lengthen the wrench handle for additional leverage, it is better to use a _____ wrench.

18. List five safety precautions that should be observed when using a wrench.

19. What is the difference between a standard screwdriver tip and a Phillips screwdriver tip?

Match each phrase with the correct screwdriver name listed below.

20. _____ Pozidriv®.

21. _____ Standard.

22. _____ Electrician.

23. _____ Heavy-duty.

24. _____ Stubby.

25. _____ Ratchet.

    a.   Has a flattened wedge-shaped tip.
    b.   Moves the fastener on the power stroke, but not on the return stroke.
    c.   Has a square shank to permit additional force to be applied with a wrench.
    d.   Useful when handling small screws.
    e.   Tip is similar to that of a Phillips head screwdriver.
    f.   Has an insulated handle.
    g.   Is short and is used when space is limited.

26. List three safety precautions that should be observed when using a screwdriver.

27. How is the size of a ball-peen hammer determined?

28. Why are soft-face hammers and mallets used in place of a ball-peen hammer?

29. List three safety precautions that should be observed when using striking tools.

30. There are few things more dangerous than a chisel with a head that has become _____ from use. This danger can be removed by _____ .

31. The chisel is an ideal tool for _____ .

32. List the four general types of cold chisels.

33. The standard hacksaw is designed to accommodate _____ .

34. A hacksaw cuts best at about _____ to _____ strokes per minute.

35. Why should the work be mounted solidly and close to the vise before cutting with a hacksaw?

36. If a blade breaks or dulls before completing a cut, you should not continue in the same cut with a new blade. Why?

37. The number of teeth per inch on a hacksaw blade has an important bearing on the shape and kind of metal being cut. At least _____ or _____ should be cutting at all times, otherwise _____.

38. What is the best way to hold thin metal for hacksawing?

39. What is the best way to hold thin wall tubing for hacksawing?

40. Files are cleaned with a _____, never with _____.

41. Files are classified according to the cut of their teeth. List the four cuts.

42. What are the most commonly used file shapes?

43. List three safety precautions that should be observed when files are used.

44. When is reaming done?

45. How much stock should be left in a hole for hand reaming?

46. A _____ is used to cut internal threads. External threads are cut with a _____ .

47. The hole to be tapped must be:
    a.   The same diameter as the desired thread.
    b.   A few thousandths larger than the desired thread.
    c.   A few thousandths (0.003"–0.004") smaller than the threads.
    d.   All of the above.
    e.   None of the above.

48. The drill used to make the hole prior to threading is called a _____.

49. How does the UNC thread series differ from the UNF thread series?

50. List the correct sequence taps should be used to form threads the full depth of a blind hole.

51. Should a shaft be larger or smaller than the finished size if external threads are to be cut on it?

52. Taps are turned in with a _____ . A _____ is used with dies.

53. What is an abrasive?

54. _____ surfaces are never polished with an abrasive.

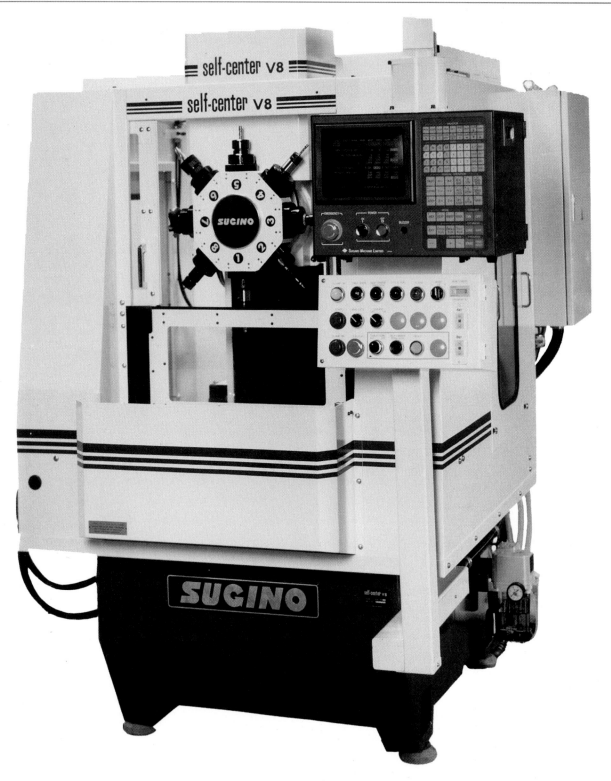

Since many products make use of threaded fasteners for assembly, fast and very accurate drillling and tapping of holes in components is necessary. For larger volumes of parts, automated equipment like this CNC drill and tap center are often used. Note that tools are mounted on an 8-spindle turret for quick tool changes. The workpiece is held stationary on a fixed position worktable, with all movement in the X, Y, and Z axes made by the traveling column holding the tool turret. (Sugino Corp.)

Chapter 7

Fasteners

LEARNING OBJECTIVES

After studying this chapter, you will be able to:
○ Identify several types of fasteners.
○ Explain why inch-based fasteners are not interchangeable with metric-based fasteners.
○ Describe how some fasteners are used.
○ Select the proper fastening technique for a specific job.
○ Describe chemical fastening techniques.

IMPORTANT TERMS

adhesives
assembly
cyanoacrylate quick
  setting adhesives
fastener
keyway
machine bolts
permanent assemblies
setscrews
threaded fasteners
washers

A *fastener* is any device used to hold two objects or parts together. This definition would include bolts, nuts, screws, pins, keys, rivets, and even chemical bonding agents or adhesives. The most common types of fasteners will be explained and illustrated in this chapter.

It is critical to choose the proper fasteners for each job, **Figure 7-1**. A poorly selected fastener can greatly reduce the safety and dependability originally designed into a product. Choosing improper fasteners could increase assembly costs and result in an inferior or faulty product. To improve quality, several different fastening techniques are often employed in the same or related assemblies. For example, one auto manufacturer uses more than 11,000 kinds and sizes of fasteners.

7.1 THREADED FASTENERS

Threaded fasteners make use of the *wedging action* of the screw thread to clamp parts together.

Figure 7-1. Complex products, such as this large earthmover and the flatbed truck being used to transport it, require the use of many types and sizes of fasteners. Reliability of the product, and the safety of persons using it, can be greatly affected if improper fasteners are selected in the design or assembly phases.

To achieve maximum strength, a threaded fastener should screw into its mating part at least a distance equal to one and one-half times the thread diameter. See **Figure 7-2**.

Threaded fasteners vary in cost from thousands of dollars for special bolts that attach the wings to the fuselage of large aircraft, to a fraction of a penny for small machine screws. See **Figure 7-3**.

Most threaded fasteners are available in metric sizes. Many American manufacturers now use metric-sized fasteners in their products, which has led

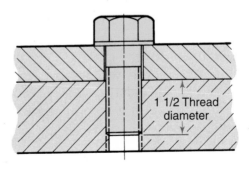

1 1/2 Thread diameter

Figure 7-2. For maximum strength, a threaded fastener must screw into the mating part a distance equal to 1 1/2 times the diameter of the thread.

Figure 7-3. *There is a wide range of threaded fasteners available, from tiny machine screws used in precision instruments to large bolts used in building construction. This 1" anchor bolt is being used to mount a steel column on a concrete foundation pier.*

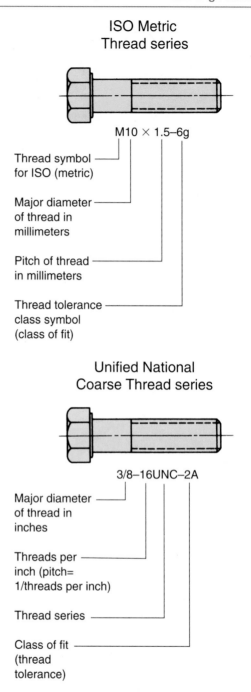

Figure 7-4. *Metric threads have the same basic profile (shape) as the Unified Thread series; however, the Unified and Metric threads are* not *interchangeable.*

to some problems. Metric threads and the common unified (inch-based) threads have the same **basic profile** (shape), but are *not* interchangeable. See **Figure 7-4**.

Until a complete changeover to metric sizes is made, and products already made with unified threads wear out or are discarded, some easy-to-use method must be devised to distinguish between metric threaded fasteners and inch-based fasteners. While no foolproof method has yet been contrived, **Figures 7-5** and **7-6** illustrate two possible solutions.

7.1.1 Machine Screws

Machine screws are widely used in general assembly work. They have slotted or recessed heads, and are made in a number of head styles, **Figure 7-7**.

Machine screws are available in body diameters ranging from #0000 (0.021") to 3/4" (0.750") and in lengths from 1/8" (0.125") to 3". Metric sizes are also manufactured. Nuts, in either square or hexagonal shapes, are purchased separately.

7.1.2 Machine Bolts

Machine bolts are employed to assemble parts that do not require close tolerances. They are manufactured with square and hexagonal heads, in diameters ranging from 1/2" to 3". The nuts are similar in shape to the bolt head. They are usually furnished with the machine bolts. Tightening the nut produces a clamping action to hold parts together, **Figure 7-8**.

On larger metric bolts thread diameter
is often stamped on the bolt head

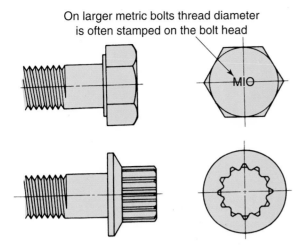

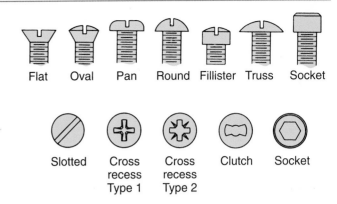

Figure 7-7. A sampling of the many types of machine screws available.

Figure 7-5. Metric fasteners are manufactured in the same variety of head shapes as inch-based fasteners. However, there is a problem in finding an easy way to distinguish between the two fastener types. Top—Some larger size hex-head metric fasteners have the size stamped on the head. Bottom—A twelve-spline flange head is under consideration for use on eight sizes of metric fasteners: 5, 6.3, 8, 10,13, 14, 16, and 20 mm.

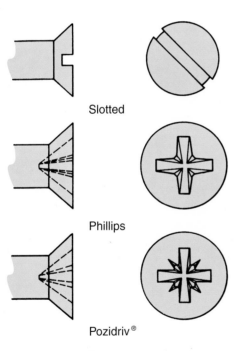

Slotted

Phillips

Pozidriv®

Figure 7-6. The Pozidriv® cross-recess head has been suggested as a means of identifying metric screws. It would replace the Phillips cross-recess head, which would be used only with inch-based screws.

7.1.3 Cap Screws

Cap screws are found in assemblies requiring a higher quality and a more finished appearance, **Figure 7-9**. Instead of tightening a nut to develop clamping action, as with the machine bolt, the cap screw passes through a clearance hole in one of the pieces and screws into a threaded hole in the other

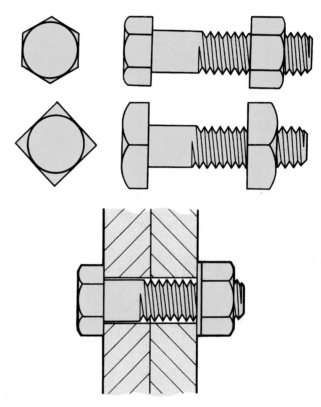

Figure 7-8. Machine bolts use a nut to produce clamping force.

part. Clamping action is accomplished by tightening the bolt into the threaded part.

Cap screws are held to much closer tolerances in their manufacture than machine screws. They are provided with a machined or semifinished bearing surface under the head. Some type of cap screws are heat-treated.

The application determines the strength of the cap screw to be used. Required strength is indicated on the print. Since all steel hex head cap screws are similar in appearance, a series of markings on the bolt head indicate strength capabilities. See **Figure 7-10**. The stronger the cap screw, the more expensive it is.

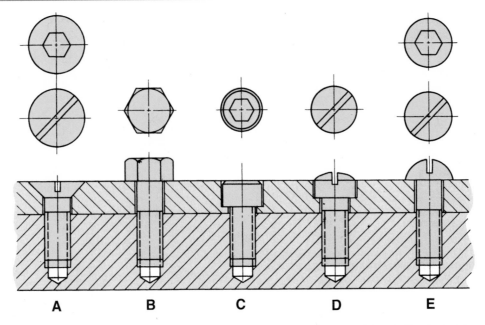

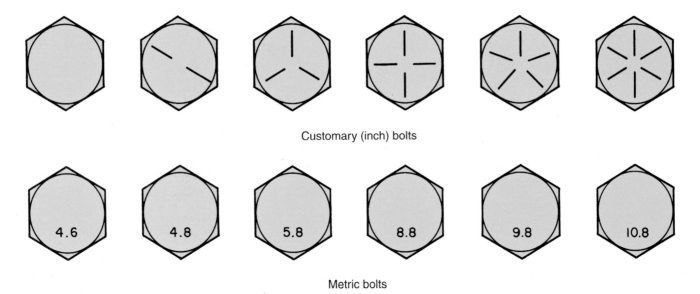

Figure 7-9. Cap screws are manufactured in various types. A—Flat head. B—Hexagonal head. C—Socket head. D—Fillister head. E—Button or round head.

Customary (inch) bolts

| 4.6 | 4.8 | 5.8 | 8.8 | 9.8 | 10.8 |

Metric bolts

Figure 7-10. Identification marks (inch size) and class numbers (metric size) are used to indicate the relative strength of hex head cap screws. As identification marks increase in number, or class numbers become larger, increasing strength is indicated.

Cap screws are stocked in coarse and fine thread series and in diameters from 1/4″ to 2″. Lengths from 3/8″ to 10″ are available. Metric sizes can also be supplied.

7.1.4 Setscrews

Setscrews are semipermanent fasteners that are used for such applications as preventing pulleys from slipping on shafts, holding collars in place on assemblies, and positioning shafts on assemblies. See **Figure 7-11**. Setscrews are usually made of

Figure 7-11. A typical setscrew application, to hold a gear onto a shaft.

heat-treated steel. They are classified in two ways, by head style, and by point style, **Figure 7-12**.

The *thumbscrew* is a variation of the setscrew that can be turned by hand. It is typically used in place of a setscrew for assemblies that require rapid or frequent disassembly, **Figure 7-13**. Thumbscrews are available with points similar to those on setscrews.

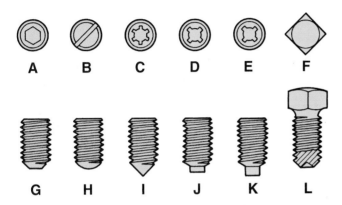

Figure 7-12. *Setscrew head and point designs. A—Socket head. B—Slotted head. C, D, and E—Fluted head. F—Square head. G—Flat point. H—Oval point. I—Cone point. J—Half dog point. K—Full dog point. L—Cut point.*

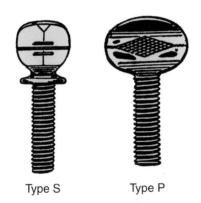

Type S   Type P

Figure 7-13. *Thumb screws can be removed or installed by hand. (Parker-Kalon)*

7.1.5 Stud Bolts

Stud bolts are headless bolts that are threaded for their entire length, or (more commonly) on both ends, **Figure 7-14**. One end is designed for semipermanent installation in a tapped hole; the other end is threaded for standard nut assembly to clamp the pieces together.

7.1.6 Removing Broken or Sheared Bolts

Bolts that have broken or sheared off can be hard to remove without proper tools. The *drill-out power extractor*, **Figure 7-15**, is available

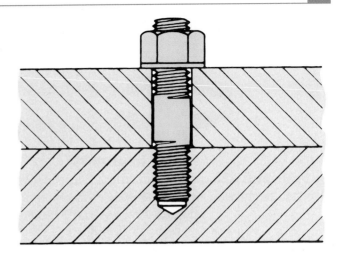

Figure 7-14. *One end of stud bolt usually threads into the part, while other end accepts a nut.*

Figure 7-15. *The Drill-Out Power Extractor™ combines a drill with an adjustable extractor collar. The drill makes the required-size hole, then the power drill is reversed. This causes the extractor to grip the broken bolt and torque it out. (Alden Corp.)*

in several sizes. It is used with a 3/8 in. capacity variable speed/reversing power drill. The built-in drill cuts the proper size hole for the extractor unit to fit. After the hole is drilled and the extractor is placed into position, drill speed is reduced and reversed. This will remove the broken bolt.

The type of extractor shown in **Figure 7-16** is also available in several sizes. A chart furnished with the extractor indicates drill size to be used. After the hole has been drilled, the extractor is inserted and turned counterclockwise with an appropriate size tap wrench.

Extractors of the type shown in **Figure 7-17** are designed to remove sheared machine screws. The blade is made of heat-treated tool steel. A hole is drilled in the screw, then the tapering blade is

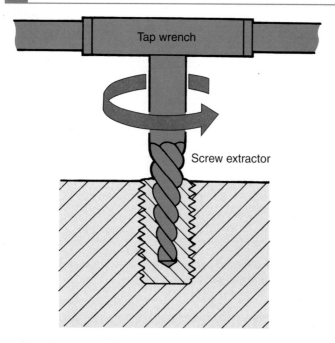

Figure 7-16. *Spiral-flute broken bolt extractor. A hole of proper size is drilled, a tap wrench applied, and the broken bolt is turned out.*

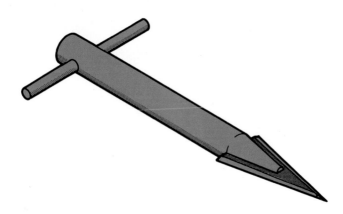

Figure 7-17. *An extractor designed to remove machine screws. A hole is drilled in the broken-off screw. The extractor blade is tapped into place and carefully turned to remove broken screw.*

lightly driven into the hole. The extractor is turned counterclockwise to remove the screw.

Treatment with penetrating oil will often make it easier to remove stubborn sheared bolts and machine screws.

7.1.7 Nuts

For most threaded fasteners, **nuts** with hexagonal or square shapes are used with bolts having the same-shape head. They are available in various degrees of finish. Nuts are usually manufactured of the same materials as their mating bolts.

A **regular nut** is **unfinished** (not machined), except on the threads, **Figure 7-18**.

A **regular semifinished nut** is machined on the bearing face to provide a truer surface for the washer, **Figure 7-19**.

A **heavy semifinished nut** is identical in finish to the regular semifinished nut. However, the body is thicker for additional strength, **Figure 7-20**.

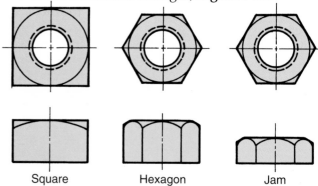

| Square | Hexagon | Jam |

Figure 7-18. *Only the threads of regular nuts are machined.*

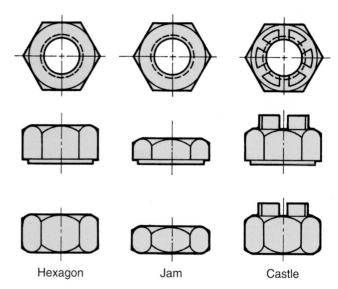

| Hexagon | Jam | Castle |

Figure 7-19. *Regular semifinished nuts have a machined bearing surface.*

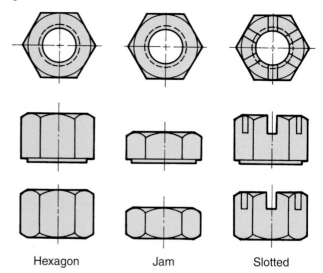

| Hexagon | Jam | Slotted |

Figure 7-20. *Heavy semifinished nuts are thicker than regular nuts.*

The *jam nut* is thinner than the standard nut. It is frequently used to lock a full-size nut in place.

Castellated and *slotted nuts* have slots across the flats so they can be locked in place with a cotter pin or safety wire. A hole is drilled in the bolt or stud, and a cotter pin or wire is inserted through the slot and hole to prevent the nut from turning loose.

These types of locking nuts are being replaced on many applications by *self-locking nuts*, **Figure 7-21**. Self-locking nuts are slightly deformed to produce a friction fit, or have a nylon insert, so they cannot vibrate loose. No hole through the bolt is required when self-locking nuts are employed in an assembly.

In critical assemblies, use a *new* self-locking nut to replace one that has been removed for any reason. The used nut may not have adequate locking action remaining and may loosen in service.

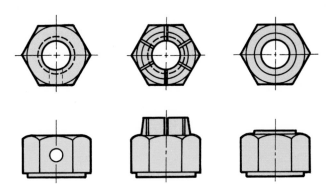

Figure 7-21. Self-locking nuts.

Acorn nuts are used when appearance is of primary importance, or where projecting threads must be protected. They are available in high or low crown styles. See **Figure 7-22**.

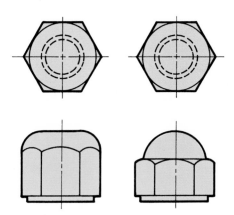

Figure 7-22. Acorn or cap nuts look good and will protect threads.

The *wing nut* is found where frequent adjustment or frequent removal is necessary. It can be loosened and tightened rapidly without the need of a wrench. Refer to **Figure 7-23**.

Figure 7-23. Wing nut looks as if it has "wings." Like the thumbscrew, it is turned by hand. (Parker-Kalon)

7.1.8 Inserts

An *insert* is a special form of nut or internal thread. Inserts are designed to provide higher strength threads in soft metals and plastics. The types shown in **Figure 7-24** are frequently used to replace damaged or stripped threads. The threaded hole is drilled and tapped. The insert is then screwed into the hole. Its internal thread is standard size and form. For optimum results, inserts must be installed according to the manufacturer's instructions.

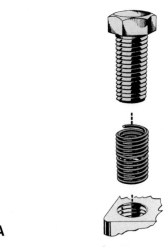

A

B

Figure 7-24. Thread repair inserts. A—An insert is frequently used to replace damaged or stripped threads in a part. (Heli-Coil Corp.) B—These keylocking inserts can be easily installed or removed without special tools. They are available in both carbon steel and stainless steel. (Jergens, Inc.)

7.1.9 Washers

Washers provide an increased bearing surface for bolt heads and nuts, distributing the load over a larger area. They also prevent surface marring. The *standard washer* is produced in light, medium, heavy-duty, and extra heavy-duty series. See **Figure 7-25**.

Figure 7-27. *Lock washer and screw units and lock washer and nut units are frequently used to simplify assembly. (Shakeproof Div., Illinois Tool Works, Inc.)*

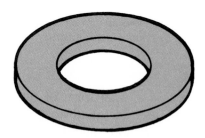

Figure 7-25 . *The standard flat washer provides a bearing surface for a fastener.*

7.1.10 Lock Washers

A *lock washer* will prevent a bolt or nut from loosening under vibration. The *split-ring lock washer* is rapidly being replaced by the *tooth-type lock washer,* which has greater holding power on most applications. See **Figure 7-26**.

Preassembled lock washer and screw nuts and *lock washer and nut units* have a washer mounted on the nut. They are employed to lower assembly time and reduce waste in the mass-assembly market, **Figure 7-27**.

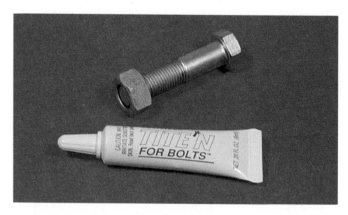

Figure 7-28. *There are many liquid thread locks available. They prevent a bolt, nut, or screw from vibrating loose, but allow easy removal should disassembly be required.*

7.1.11 Liquid Thread Lock

Nuts, bolts, and machine screws can be prevented from loosening due to vibration though use of a liquid thread lock, **Figure 7-28**. Although the thread lock material will prevent fasteners from vibrating loose, it allows easy removal of the fastener should disassembly be necessary. When using a liquid thread lock, follow the manufacturer's recommendations for maximum effectiveness.

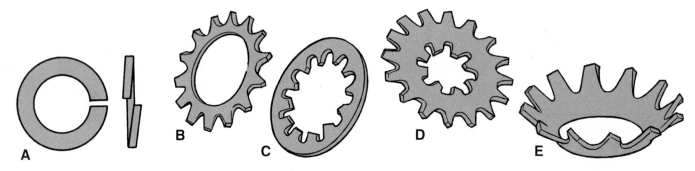

Figure 7-26. *Lock washer variations. A—Split-ring type. B—External type. This type should be used whenever possible, because it provides greatest resistance to turning. C—Internal type. It is used with small head screws and to hide teeth, either for appearance or to prevent snagging. D—Internal-external type. It is employed when mounting holes are oversize. E—Countersunk type. It is used with flat or oval head screws.*

7.1.12 Thread-Forming Screws

Thread-forming screws produce a thread in the part as they are driven, **Figure 7-29**. This feature eliminates a costly tapping operation. A variation of the thread-forming screw eliminates expensive hole-making (drilling or punching) and aligning operations because the screw drills its own hole as it is driven into place. See **Figure 7-30**.

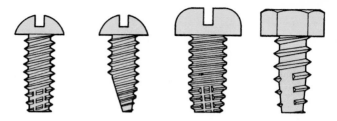

Figure 7-31. *Variations among thread cutting screws.*

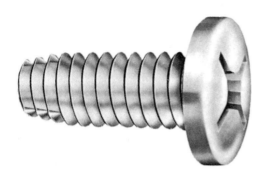

Figure 7-29. *One type of thread-forming screw.*

Figure 7-30. *A self-drilling screw. This type is also known as TEKS®. (USM Corp., Fastener Group)*

7.1.13 Thread Cutting Screws

Thread cutting screws differ from thread forming screws because they actually cut threads into the material when driven. Refer to **Figure 7-31**. Thread cutting screws are hardened. They are employed to join heavy-gage sheet metal, and to thread into nonferrous metal assemblies.

7.1.14 Drive Screws

Drive screws are simply hammered into a drilled or punched hole of the proper size. A permanent assembly results, **Figure 7-32**.

7.2 NONTHREADED FASTENING DEVICES

Nonthreaded fasteners comprise a large group of mechanical holding devices. These include dowel pins, cotter pins, retaining rings, rivets, and keys. Each has its advantages.

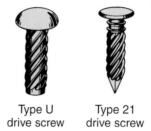

Type U drive screw   Type 21 drive screw

Figure 7-32. *Drive screws are hammered or forced into place in a presized hole. (Parker-Kalon)*

7.2.1 Dowel Pins

Dowel pins are made of heat-treated alloy steel and are found in assemblies where parts must be accurately positioned and held in absolute relation to one another. See **Figures 7-33** and **7-34**. They assure perfect alignment and facilitate quicker disassembly and reassembly of parts in exact relationship to each other. They are fitted into reamed holes and are available in diameters from 1/16″ to 1″. They are also available in metric sizes.

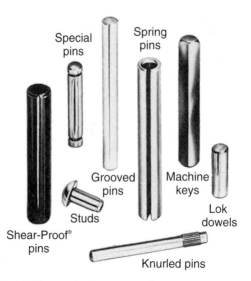

Figure 7-33. *Types of dowel pins. They are made in a wide range of sizes and types. (Driv-lok Inc.)*

Figure 7-34. A taper pin is often used to lock a handle to its shaft.

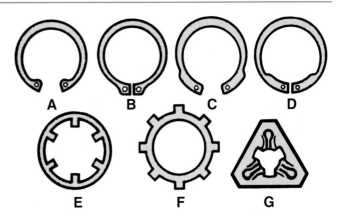

Figure 7-36. Retaining rings. A—Basic internal ring. B—Basic external ring. C—Inverted internal ring. D—Inverted external ring. E—External self-locking ring. F—Internal self-locking ring. G—Triangular self-locking ring.

Dowel pins are normally 0.0002″ (0.005 mm) oversize (identified by a plain steel finish) but are available in 0.001″ (0.025 mm) oversize (identified by a black finish) for repairs.

Taper pins are made with a uniform taper of 1/4″ per foot in lengths up to 6″, with diameters as small as 5/32″ at the large end.

7.2.2 Cotter Pins

A *cotter pin* is fitted into a hole drilled crosswise through a shaft, **Figure 7-35**. The pin prevents parts from slipping or rotating off. Other types of retaining devices are replacing the cotter pin.

7.2.3 Retaining Rings

The *retaining ring*, **Figure 7-36**, has been developed for both internal and external applications. Retaining rings reduce both cost and weight of the product on which they are employed. While most retaining rings must be seated in grooves, **Figure 7-37**, a *self-locking* type does not require this special recess. Special pliers are needed for rapid installation and removal of the retaining rings, **Figure 7-38**.

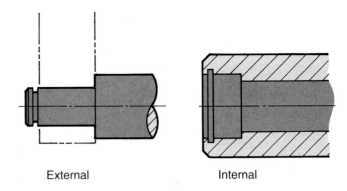

External   Internal

Figure 7-37. Grooves are machined into parts to receive retaining rings. They eliminate many other expensive machining operations.

7.2.4 Rivets

Permanent assemblies (those that do not have to be taken apart) can be made with rivets, **Figure 7-39**. Solid rivets can be *set*, or deformed to become larger on one end, by hand or machine methods.

Blind rivets are mechanical fasteners that have been developed for applications where the joint is accessible from only one side. They require special tools for installation, **Figure 7-40**. Blind rivet types are shown in **Figure 7-41**.

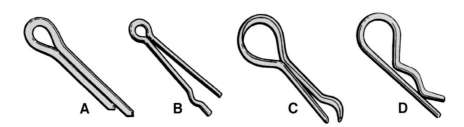

Figure 7-35. Types of cotter pins. A—Standard. B—Humped. C—Clinch. D—Hitch.

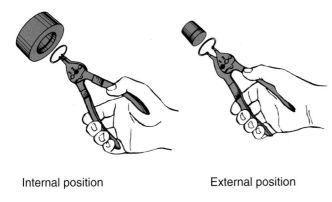

Internal position   External position

Figure 7-38. Special pliers are used to install some types of retaining rings. (Waldes Kohinoor Inc.)

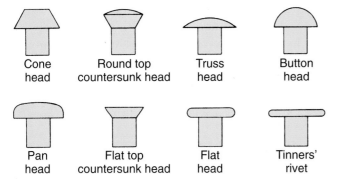

Cone head   Round top countersunk head   Truss head   Button head

Pan head   Flat top countersunk head   Flat head   Tinners' rivet

Figure 7-39. Rivet head styles.

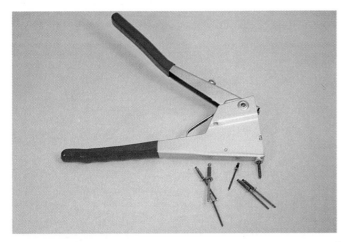

Figure 7-40. A pliers-type rivet gun is used to insert one type of blind rivet.

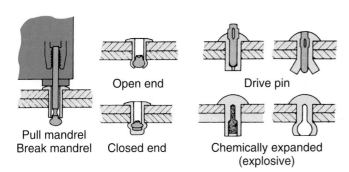

Pull mandrel Break mandrel   Open end   Drive pin

Closed end   Chemically expanded (explosive)

Figure 7-41. Types of blind rivets.

7.2.5 Keys

A *key* is a small piece of metal that prevents a gear or pulley from rotating on its shaft. One-half of the key fits into a *keyseat* on the shaft while the other half of the key fits into a *keyway* in the hub of the gear or pulley, **Figure 7-42**. Commonly used keys are shown in **Figure 7-43**.

A *square key* is usually one-fourth the shaft diameter. It may be slightly tapered on the top to make it easier to install.

The *Pratt & Whitney key* is similar to the square key, but is rounded at both ends. It fits into a keyseat of the same shape.

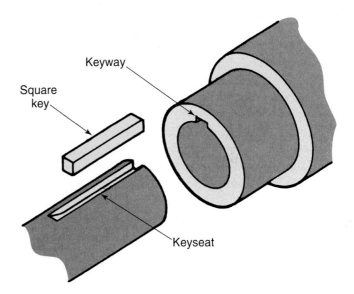

Figure 7-42. A square key is used to prevent a pulley or gear from rotating on its shaft.

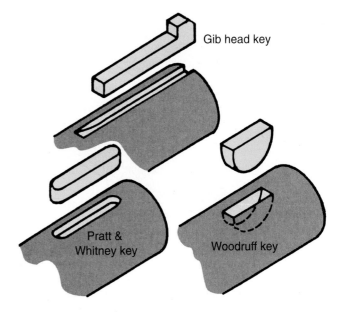

Figure 7-43. Three types of keys.

The *gib head key* is interchangeable with the square key. The head design permits easier removal from the assembly.

A *Woodruff key* is semicircular and fits into a keyseat of the same shape. The top of the key fits into the keyway of the mating part.

7.3 ADHESIVES

Adhesives provide one of the newer ways to join metals and to keep threaded fasteners from vibrating loose. In some applications, the resulting joints are stronger than the metal itself. Adhesive bonded joints do not require costly and time-consuming operations such as drilling, countersinking, riveting, etc.

The major drawback to the use of adhesives is heat. While some adhesives retain their strength at temperatures as high as 700°F (371°C), most should not be used for assemblies that will be exposed to temperatures above 150–200°F (66–93°C).

Adhesives for locking threaded fasteners in place are made in a number of chemical formulations. The desired permanence of the threaded joint will determine the type of adhesive to be employed.

Adhesive-bonded assemblies offer many advantages over other fastening techniques:

- The load is distributed evenly over the entire joined area, **Figure 7-44**.
- There is continuous contact between the mating surfaces, **Figure 7-45**.

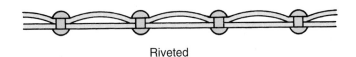

Adhesive-bonded

Riveted

Figure 7-45. On parts joined with an adhesive, the mating surfaces are in continuous contact.

- Full strength of the mating parts is maintained, since holes do not have to be made to insert fasteners. The extreme heat required for joining methods like welding is not necessary with adhesive bonding. This means there is no danger of the work becoming distorted or having its heat treatment affected.
- Smooth surfaces result from adhesive bonding—there are no external projections (as with rivets or bolts), and the surface is not marred by the heat and pressure necessary to join pieces with spot welds.

Many commercial adhesives are sold in small quantities. They are suitable for use in training areas and the home, **Figure 7-46**.

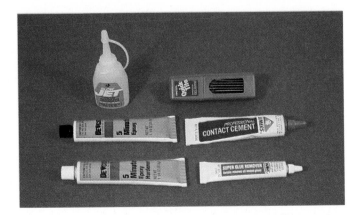

Figure 7-46. Adhesives for joining metal to metal and metal to other materials are available in good hardware stores. They are similar to those found in industry.

Adhesives are available in liquid, paste, or solid form. Many can be applied directly from the container. Others must be mixed with a catalyst or hardener. A few pressure-sensitive adhesives are manufactured in sheet form.

One type of adhesive has found growing use in machining technology to make temporary bonds. *Cyanoacrylate quick setting adhesives* (known by such trade names as Eastman 910™,

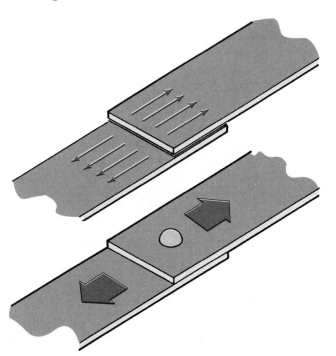

Figure 7-44. When an adhesive is used to join metal, the load is distributed evenly over the entire joint. A rivet or conventional threaded fastener localizes the load in a small area.

Super Glue™, and Crazy Glue™) are used to hold matching metal sections together while they are being machined. Round stock too small for existing collets can be glued into larger stock for turning, milling, or grinding. Some fragile parts have been glued to backup blocks for machining, and parts like the gun sight casting shown in **Figure 7-47** have been glued to a fixture for such operations as machining the sighting and bottom grooves.

After machining, the parts can be removed from the holding device by an application of heat (175°F or 79°C maximum). Very small parts can be removed by applying a cyanoacrylate debonder.

For successful use of cyanoacrylate adhesives, the part and mounting surface must be prepared according to the adhesive manufacturer's directions.

When using cyanoacrylate adhesives, always wear approved eye protection and keep fingers away from your eyes and mouth. Since this adhesive can instantly bond fingers to each other or to other surfaces, always have a debonder available for immediate use. Should you get adhesive in your eyes, see a physician immediately.

7.3.1 Using Adhesives

Most adhesives require following a five-step process to produce solidly bonded joints:

1. Surface preparation is critical! All adhesives require clean surfaces to produce full-strength bonds. Preparation may range from simply wiping surfaces with a solvent to performing multistage cleaning and chemical treatment.

2. Adhesive preparation must be done properly. Mixing, delivery to the work area, setting up equipment, etc. must all be done according to the manufacturer's directions.

3. Adhesive application may be done by brushing, rolling, spraying, dipping, or methods designed for a specific assembly. See **Figure 7-48**.

4. Assembly involves positioning of materials to be joined. This often requires the use of jigs or fixtures for alignment.

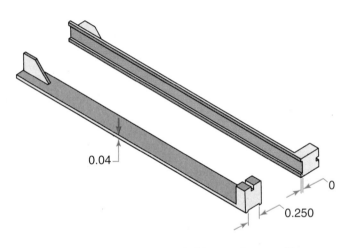

Figure 7-47. This gunsight was held on a fixture with cyanoacrylate adhesive to allow machining. The thin base section of the part made it difficult and expensive to mount to a fixture by other means.

 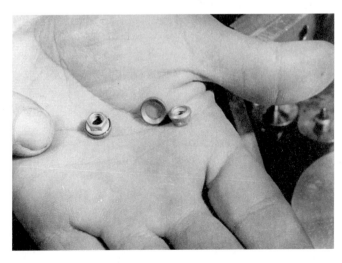

Figure 7-48. Self-aligning nuts used in aircraft are assembled with a cyanoacrylate adhesive. The two part fasteners are assembled automatically and drop onto an indexing table. Setup produces 3600 assemblies an hour. Left—A needle-tip applicator applies a precise amount of adhesive before parts are brought together. Right—Parts after and before assembly. (Loctite Corp.)

5. Bond development is the process of evaporation of solvents and curing of the adhesive. It may involve application of pressure and/or heat. See **Figure 7-49**.

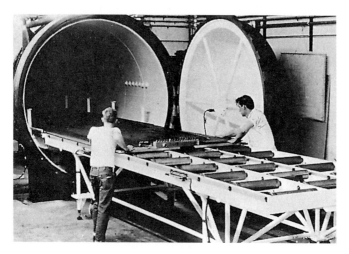

Figure 7-49. *This autoclave is used to bond large assemblies, with steam applying heat and pressure simultaneously. The assembly to be bonded is wrapped in a rubber blanket to protect the bond from moisture. (3M Co.)*

7.4 FASTENER SAFETY

- Wear approved eye protection when drilling, punching, or countersinking openings for fasteners.
- Never use your hands to remove metal chips from holes drilled for fasteners! Do not use your fingers to check whether burrs have been removed. Burrs around the opening can cause nasty cuts. Use a brush.
- If you use compressed air to clean drilled or tapped holes, wear approved eye protection. Make sure that there is no danger of flying chips injuring nearby workers.
- Do not apply adhesives near areas where there are open flames. Solvents used in some adhesives are highly flammable and/or toxic. Apply them only in well-ventilated areas, and wear a suitable respirator.
- The chemicals in adhesives for joining metals and other materials can cause severe skin irritation. To be safe, wear disposable plastic gloves when preparing or applying all types of adhesives.

- Carefully follow all instructions on the adhesive container when mixing and using adhesives. Only mix the amount you will need. Promptly remove any adhesive from your skin by washing in water.
- Cyanoacrylate adhesives cure in 5 to 15 seconds. Do not allow any of this adhesive to get onto your fingers, since it will bond skin together. Unless a suitable solvent is available, surgery might be needed to separate the joined fingers.

TEST YOUR KNOWLEDGE

Please do not write in the text. Write your answers on a separate sheet of paper.

1. For maximum strength, a threaded fastener should screw into its mating part a distance equal to _____ times the diameter of the thread.

2. There are many ways of joining material. List four types of threaded fasteners. Describe how each is used.

3. _____ screws are used for general assembly work.

4. How is the strength of hex-head cap screws indicated?

5. When removing stubborn sheared bolts, what can be done to make their removal easier?

6. To prevent a pulley from slipping on a shaft, a _____ is often employed.

7. The _____ bolt is threaded at both ends.

8. _____ or _____ are employed when the parts are to be joined permanently.

9. Why are lock washers used?

10. While most _____ must be seated in grooves, a self-locking type does not require the special recess.

11. When is a jam nut employed?

12. The shape of the _____ nut permits it to be loosened and tightened without a wrench.

Match each word in the left column with the most correct sentence in the right column. Place the appropriate letter in the blank.

13. _____ Rivet.

14. _____ Jam nut.

15. _____ Drive screw.

16. _____ Thread-cutting screw.

17. _____ Acorn nut.

18 _____ Dowel pin.

19. _____ Blind rivet.

20 _____ Keyway.

21. _____ Keyseat.

22. _____ Key.

a. Developed for use in confined area, where a joint is only accessible from one side.

b. Used where parts must be aligned accurately and held in absolute relation with one another.

c. Prevents a pulley or gear from slipping on a shaft.

d. Protects projecting threads.

e. Is hammered into a drilled or punched hole.

f. Used to make permanent assemblies.

g. Slot cut in gear or pulley to receive "c."

h. Locks a regular nut in place.

i. Eliminate costly tapping operations.

j. Slot cut in shaft to receive "c."

23. List the steps, in their proper sequence, that must be used to join metals with adhesives.

24. List at least five safety precautions that must be observed when using fasteners.

The Bell 609 is the latest design for transporting people in and out of inner city locations. After a helicopter-like take-off, the blades transition into a horizontal position and the 609 flies like a conventional aircraft. The blades are returned to an upright position for a vertical landing. In addition to the use of conventional-type fasteners, high-strength adhesives are used extensively in its manufacture.(Bell Helicopter/ A Subsidiary of TEXTRON, Inc.)

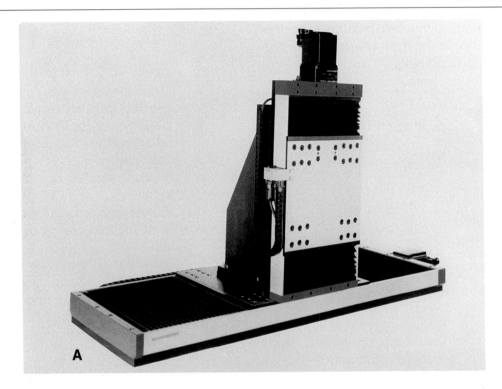

Positioning tables are used with fixtures and jigs to rapidly and accurately move the work-piece into the proper relationship with a cutting tool. A—This linear table moves in both the horizontal and vertical axes, using precision-ground ballscrews and guide rails to achieve positioning as accurate as ± 10 microns per 300 mm. (Schneeberger, Inc.) B—Rotary tables can be mounted vertically or horizontally, depending upon the application. They offer precise and repeatable indexing of the workpiece through a full 360° rotation. (Yukiwa Seiko USA, Inc.)

Chapter 8

Jigs and Fixtures

LEARNING OBJECTIVES

After studying this chapter, you will be able to:
- ○ Explain why jigs and fixtures are used.
- ○ Describe a jig.
- ○ Describe a fixture.
- ○ Elaborate on the classifications of jigs and fixtures.

IMPORTANT TERMS

box jig
bushings
closed jig
drill template
fixture

fixture holding devices
jig
open jig
plate jig
slip bushings

Jigs and fixtures are devices that are used extensively in production machine shops to hold work while machining operations are performed. They position the work and guide the cutting tool or tools so that all of the parts produced are uniform and within specifications. When large numbers of identical and interchangeable parts must be produced, the use of jigs and fixtures helps to reduce manufacturing costs. The use of jigs and fixtures is often justified when limited production is required, because they allow relatively unskilled workers to operate the machines.

Jigs and fixtures are also employed in assembly work for operations, such as welding or riveting. They position and hold work while standardized parts are being fabricated.

8.1 JIGS

A *jig* is a device that holds a workpiece in place and guides the cutting tool during a machining operation such as drilling, reaming, or tapping. Hardened steel *bushings* are used to guide the drill or cutting tool, **Figure 8-1**.

Figure 8-1. A lift-type drill jig. Left—Jig is open to receive the part shown in foreground. Right—Jig is closed, with the part in place, ready for drilling. (Ex-Cell-O Corp.)

The jig is seldom mounted solidly to the drill press table. For safety, however, it is usually nested between guide bars that *are* mounted solidly to the table, **Figure 8-2**.

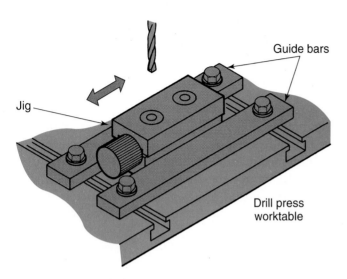

Figure 8-2. *A drill jig is nested between guide bars to prevent dangerous and undesirable "merry-go-round" rotation.*

8.1.1 Jig types

Drill jigs fall into two general types: open jigs and box (closed) jigs.

The *drill template* or *plate jig* is the simplest form of the *open jig*. It consists of a plate with holes to guide the drill. The jig fits over the work, **Figure 8-3**.

In a more elaborate form of open-type drill jig, **Figure 8-4**, clamps are used to hold the work in place. Drill jigs may be fitted with a base plate to provide clearance for the drill as it breaks through the work. Such a base plate is used on the circular jig shown in **Figure 8-5**. This jig is a variation of the plate jig.

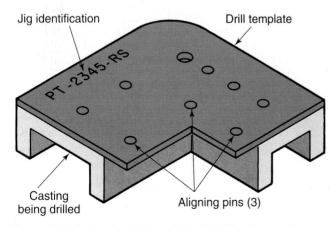

Figure 8-3. *A simple drill template. Identification numbers on jigs and fixtures allow these devices to be located easily when stored away between uses.*

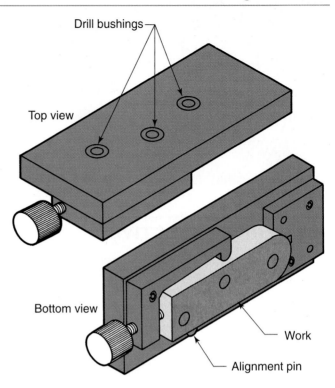

Figure 8-4. *This open-type jig has a clamp to hold the work in position for drilling. A V-notch at one end and an alignment pin at the other end position the work properly in the jig.*

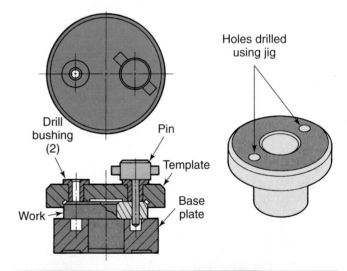

Figure 8-5. *With this circular type drill jig, a pin is placed in first hole after it is drilled. This holds the workpiece in position when drilling the second hole. Note the base plate that provides clearance for the drill as it breaks through the workpiece.*

The *box jig* or *closed jig* encloses the work, **Figure 8-6**. This type is more costly to make than an open jig, but is often used when holes must be drilled in several directions. **Figure 8-7** illustrates a box jig in its simplest form. The work is fitted into the jig through a hinged or swinging cover. The clamps that hold the work in place are permanently mounted to the jig. A more complex type of box jig is shown in **Figure 8-8**.

When several different operations must be performed on a job, a combination of open and box jigs is often used. *Slip bushings* are utilized to guide the drills. They are then removed for subsequent operations such as reaming, tapping, countersinking, counterboring, or spot facing.

8.2 FIXTURES

A *fixture* is employed to position and hold a workpiece while machining operations are performed on it, **Figure 8-9**. Unlike a jig, a fixture does *not* guide the cutting tool.

Fixtures fall into many classifications. The class is determined by the type of machine tool on which the fixture is used, such as a machining center,

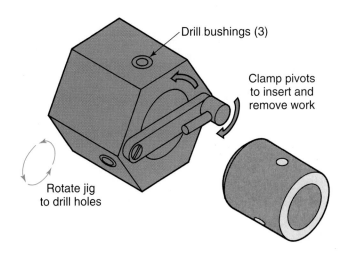

Figure 8-7. *A light closed (box) jig used to drill three equally spaced holes in a base end cap. Since only a limited production run was required, it was not necessary to construct a more elaborate jig.*

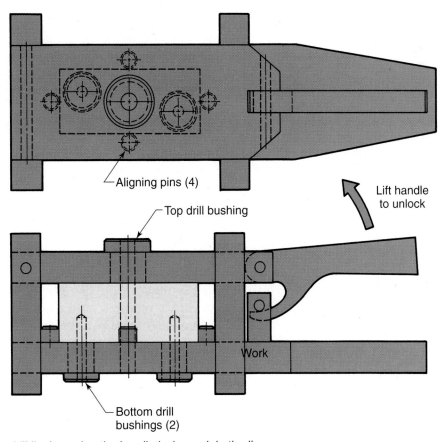

Figure 8-6. *A box type drill jig. Lowering the handle locks work in the jig.*

Figure 8-8. To hold a drill head casting in place for machining, a fairly complex jig is used. (Clausing Industrial, Inc.)

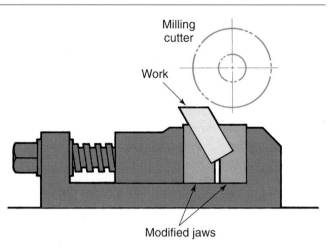

Figure 8-10. This simple fixture consists of vise jaws that have been modified to position a workpiece so that an angular cut can be made on it.

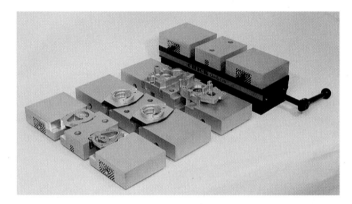

Figure 8-9. Machining centers often require special fixture holding devices. The pockets to hold the work are machined directly into the body of the jaws. Three examples are shown in the foreground. The jaws, which snap on and off the workholding system, are shown in the background. New setups can be made in a very short period of time. (Chick Machine Tool, Inc.)

Figure 8-11. Many fixtures are used in the assembly of an aircraft fuselage. In fixture construction, extreme accuracy is critical, requiring use of lasers to assure precise alignment. (The Boeing Co.)

milling machine (vertical or horizontal), lathe, band saw, or grinder. Fixture designs range from simple vise jaw modifications, **Figure 8-10**, to very large and complex devices used by the aerospace industry, **Figure 8-11**.

8.3 JIG AND FIXTURE CONSTRUCTION

Jigs and fixtures are devices that are designed for *specific* jobs. Their complexity is determined by the number of pieces to be produced, the degree of accuracy required, and the kind of machining operations that must be performed.

The body of a jig or fixture may be built-up, welded, or cast. Commercial components are available in a wide range of sizes, types, and shapes. See **Figure 8-12**. Special *fixture holding devices* have been developed for machining centers and other CNC machine tools that permit multiple setups. See **Figures 8-13** and **8-14**.

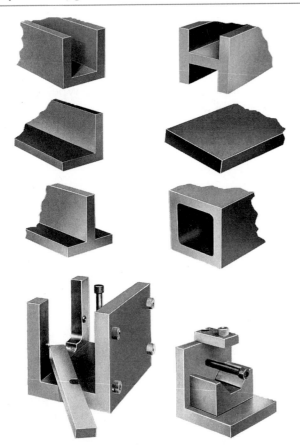

Figure 8-12. Standard cast iron shapes are machined parallel and square to save time and money in both designing and building jigs and fixtures. Sections of different shapes can be bolted together to form complex holding devices. Two completed units are shown at bottom of illustration. (Ex-Cell-O Corp.)

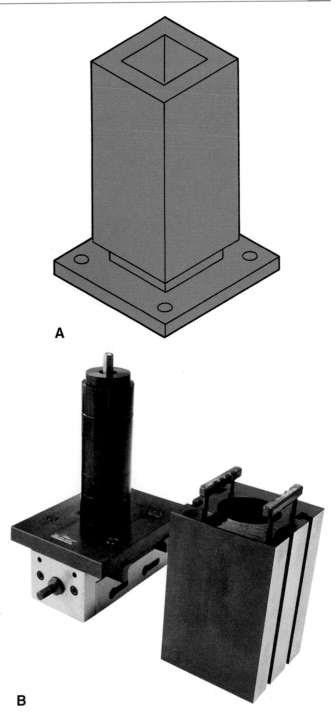

A

B

Figure 8-14. "Tombstones" are a common fixture-holding device used with machining centers and other CNC machine tools. They are made from heavy castings and are precisely machined. A—A typical tombstone intended for mounting on a machine's worktable. Fixtures are mounted directly onto the tombstones. The tombstone may be blank, as shown, have drilled and tapped mounting holes, or use T-slots for mounting fixtures. B—A modular tombstone system consisting of a receiver module and mating tombstone modules. An expansion mechanism on the receiver's centerpost allows the tombstone module to be rotated, then locked in position. The assembly at the bottom of the receiver module is a dovetail adapter that allows the assembly to be locked in a standard machine vise. The receiver module can also be bolted directly to the worktable. (Interlen Products Corporation)

Figure 8-13. Special fixture holding devices for machining centers and other CNC machine tools. Workholding pockets are cut directly into the jaw blocks. Pivoting the vertical setup will bring the next set of workpieces into position. (Chick Machine Tool, Inc.)

TEST YOUR KNOWLEDGE

Please do not write in the text. Write your answers on a separate sheet of paper.

1. Jigs and fixtures are devices used in _____ to _____ while machining operations are performed.

2. Why are jigs and fixtures used?

3. What is a jig?

4. Jigs fall into two general types. List and briefly describe each type.

5. A combination of the two jig types listed in question 4 is often used when _____.

6. What is a fixture?

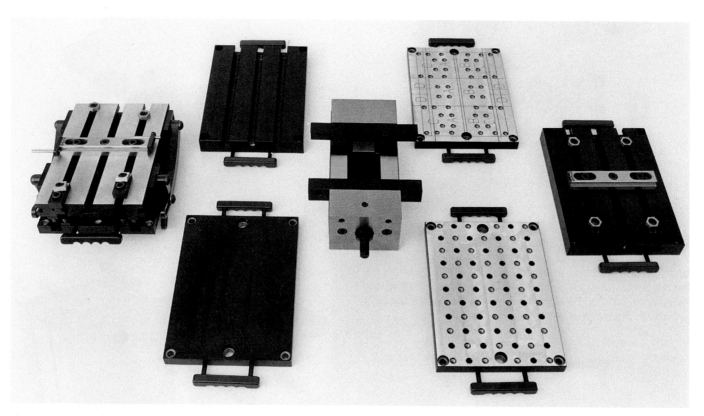

Special quick-change pallets have been developed to lock into a receiver that mounts on a standard machine vise. Shown are some of the interchangeable pallets, which come in a variety of T-slot and grid-hole patterns for mounting different workpieces and fixtures. (Interlen Products corporation)

Chapter 9

Cutting Fluids

LEARNING OBJECTIVES

After studying this chapter, you will be able to:
○ Understand why cutting fluids are necessary.
○ List the types of cutting fluids.
○ Describe each type of cutting fluid.
○ Discuss how cutting fluids should be applied.

IMPORTANT TERMS

chemical cutting fluids
contaminants
cutting fluids
emulsifiable oils
gaseous fluid
lubricating

mineral oils
misting
noncorrosive
semichemical cutting
   fluids

Cutting fluids are required to do many things simultaneously. These functions include:
- Cooling the work and cutting tool, **Figure 9-1**.
- Improving surface finish quality.

- *Lubricating* to reduce friction and cutting forces, thereby extending tool life.
- Minimizing material buildup on tool cutting edges.
- Protecting machined surface against corrosion.
- Flushing away chips, **Figure 9-2**.

In addition, cutting fluids must comply with all federal, state, and local regulations for human safety, air and water pollution, waste disposal, and shipping restrictions.

9.1 TYPES OF CUTTING FLUIDS

Cutting fluids fall into four basic types:
- Mineral oils.
- Emulsifiable oils.
- Chemical and semichemical fluids.
- Gaseous fluids.

Figure 9-1. For maximum results, coolant should flood the area being machined and cutting tool to provide the most efficient removal of the heat generated. (Kesel/JRM International, Inc.)

Figure 9-2. An additional function of cutting fluids is to flush away chips from the area where cutting is taking place. (EROWA Technology, Inc.)

9.1.1 Mineral Cutting Oils

Cutting oils made from mineral oil may be used straight or combined with additives. Straight *mineral oils* are best suited for light-duty (low speed, light feed) operations where high levels of cooling and lubrication are not required.

They are *noncorrosive* and are usually used with high machinability metals, such as aluminum, magnesium, brass, and free-cutting steel.

Mineral oils are often combined with animal and vegetable oils and contain sulfur, chlorine, and/or phosphorus. Their use is limited by high cost, operator health problems, and danger from smoke and fire. Mineral oils also stain some metals. They have a tendency to become rancid, so the tank containing them must be cleaned periodically and the fluid replaced.

When working in situations where cutting fluid mists or vapors are present, always wear an approved respirator. A simple dust mask is not sufficient protection.

9.1.2 Emulsifiable Oils

Emulsifiable oils are also known as *soluble oils*. They are composed of oil droplets that are suspended in water by blending the oil with emulsifying agents and other materials. Emulsifiable oils range in appearance from milky to translucent. They are available in many variations for metal removal applications that generate considerable heat.

Emulsifiable oils offer a number of advantages over straight cutting oils. They provide increased cooling capacity in some applications. They are cleaner to work with than other cutting fluids, and provide cooler and cleaner parts for the machinist to handle. These oils reduce the *misting* and fogging that are health hazards for machine operators. Because they are diluted with water, they offer increased economy and present no fire hazard.

Emulsifiable cutting oils can be used in most light- and moderate-duty machining operations. For economy and best machining results, these oils must be mixed according to the manufacturer's recommendations. These take into account the material being machined and the machining operation performed. Fluid maintenance must be performed on a routine basis to control rancidity.

Water-based cutting fluids must never be used when machining magnesium.

9.1.3 Chemical and Semichemical Cutting Fluids

Chemical cutting fluids generally contain no oil. They have various rates of dilution depending upon use. A wetting agent is often added to provide moderate lubricating qualities.

Semichemical cutting fluids may have a small amount of mineral oil added to improve the fluid's lubricating qualities. Semichemical cutting fluids incorporate the best qualities of both chemical and emulsifiable cutting fluids.

Chemical and semichemical cutting fluids offer the following advantages:
- Fluids dissipate heat rapidly.
- They are clean to use.
- After machining, residue is easy to remove.
- The fluids are easy to mix and do not become rancid.

Their disadvantages are:
- Some formulas have minimal lubricating qualities.
- Fluids may cause skin irritation in some workers.
- When they become contaminated with other oils, disposal can be a problem.

9.1.4 Gaseous Fluids

Compressed air is the most commonly used *gaseous fluid* coolant. It cools by forced convection.

In addition to cooling the workpiece and tool, compressed air also blows chips away at high velocity. Workers in surrounding areas must be shielded from the flying chips.

9.2 APPLICATION OF CUTTING FLUIDS

Machining and grinding applications require a continuous flooding of fluid around the cutting tool and work to provide efficient removal of the heat generated, **Figure 9-3.** Coolant nozzles must be positioned carefully so that, in addition to cooling the work area, the cutting fluid will also carry the chips away. In some machining operations, a conveyor system, **Figure 9-4**, is used to remove chips and cutting fluid from the cutting area. The cutting fluid is filtered to remove **contaminants** and returned to the machine's coolant tank for reuse.

Figure 9-3. *Coolant fluid must surround the cutting area for maximum effect. Shown is a modular hose system for applying air and liquids. The units snap together and can be shaped to fit any job. (Lockwood Products, Inc.)*

Figure 9-4. *This powered conveyor system moves chips away from a machine's cutting area. A rotating drum filter is used to separate chips and metal particles as fine as 50 microns from the cutting fluid. The filtered fluid is recycled to the machine at rates of up to 30 gallons per minute. (Jorgensen Conveyors, Inc.)*

9.2.1 Evaluation of Cutting Fluids

It is not possible in this text's limited space to cover all cutting fluids, nor does space permit recommending specific cutting fluids for every machining operation. This information can be obtained from data published by cutting fluid manufacturers. Recommendations for cutting fluid use are included in the chapters of this text dealing with each type of machine tool. In general, however, cutting fluids (*gaseous fluids* excepted) are compatible with HSS (high-speed steel) and carbide tooling. See **Figure 9-5.** Since carbide tooling operates at higher

cutting speeds and generates higher cutting temperatures, cutting fluids that have high cooling rates should be used in such applications.

Machining with ceramic tooling is usually accomplished without the use of cutting fluids, **Figure 9-6**.

Figure 9-5. *Carbide cutting inserts are widely used on machine tools, but require coolants that can remove heat at higher rates. This tungsten carbide insert is coated with a thin film of diamond material for longer life when cutting highly abrasive materials. (Kennametal, Inc.)*

Figure 9-6. *This ceramic cutting insert is being used to turn a titanium workpiece. Cutting fluids are not necessary with ceramic tooling. (Kennametal, Inc.)*

TEST YOUR KNOWLEDGE

Please do not write in the text. Write your answer on a separate sheet of paper.

1. Cutting fluids must do many things simultaneously. What does this include?

2. List the four basic types of cutting fluids.

3. What type cutting oil is recommended for machining aluminum, magnesium, brass, and free-machining steels?

4. Why does the above type of cutting fluid have limited use?

5. _____ cutting fluids are also known as soluble oils.

6. What advantages do the emulsifiable oil cutting fluids have over the cutting fluids indicated in Question 3?

7. _____ cutting fluids contain no oils.

8. When small amounts of mineral oil are added to the cutting fluid described in Question 7, it is known as _____ cutting fluid.

9. What are the advantages of the cutting fluids indicated in Questions 7 and 8?

10. What is dangerous about using compressed air to cool the area being machined?

Some machine tools are designed with built-in fluid delivery systems that surround the cutting area, flooding it with coolant to remove heat and flush away chips. In other machines. the delivery system consists of separate nozzles and tubing. (Sharnoa Corp.)

Chapter 10

Drills and Drilling Machines

LEARNING OBJECTIVES

After studying this chapter, you will be able to:
- ○ Select and safely use the correct drills and drilling machine for a given job.
- ○ Make safe setups on a drill press.
- ○ Explain the safety rules that pertain to drilling operations.
- ○ List various drill series.
- ○ Sharpen a twist drill.

IMPORTANT TERMS

blind hole
center finder
countersinking
drill point gage
flutes
lip clearance

machine reamer
multiple spindle drilling
  machines
spotfacing
twist drills

10.1 DRILLING MACHINES

A *drilling machine* is a power-driven machine that holds the material and cutting tool and brings them together so a round hole is made in the material. Many different types of drilling machines are used in industry. The type of machine used depends on the operation being performed, size of the workpiece, and the variety of operations required of the machine.

The most common drilling machine is the *drill press*, **Figure 10-1.** A drill press operates by rotating a cutting tool, or drill, against the material with sufficient pressure to cause the drill to penetrate the material. See **Figure 10-2.**

The size of a drill press is determined by the largest diameter circular piece that can be drilled on center, **Figure 10-3.** A 17″ drill press can drill to the center of a 17″ diameter piece. The centerline of the drill is 8 1/2″ from the column.

Figure 10-1. *A light 15″ variable speed drill press. (Wilton Corp.)*

Bench drill presses can be used to drill holes in small workpieces, **Figure 10-4.** These presses do not have as many capabilities as the floor models.

Electric hand drills are used to drill small holes in relatively thin material. They are reasonably priced and convenient to use. See **Figure 10-5.**

A *radial drill press* is designed to handle very large drilling work. The drill head is mounted in a way that allows it to be moved back and forth on an

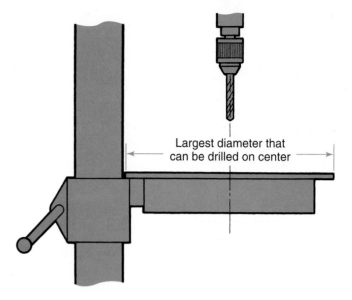

Figure 10-3. *How drill press size is determined .*

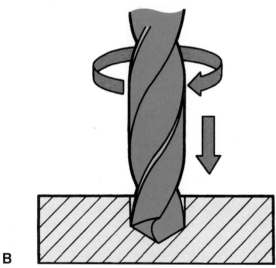

Figure 10-2. *Drilling is the operation most often performed on a drill press. A—Completed hole. B—Both rotating force and a downward pushing force are needed for drilling. (Clausing Industrial, Inc.)*

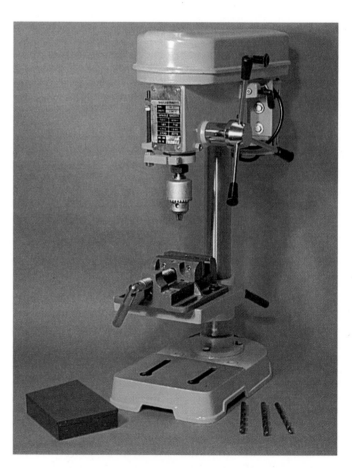

Figure 10-4. *A small bench drill press can be very useful.*

arm that extends from the massive machine column. The arm can be moved up and down and pivoted on the column, **Figure 10-6.**

Often, a large pit is located along one side of the machine to permit the positioning of large, odd-shaped work. The pit is covered when not in use. Holes up to 3 1/2″ (90 mm) in diameter can be drilled with this machine.

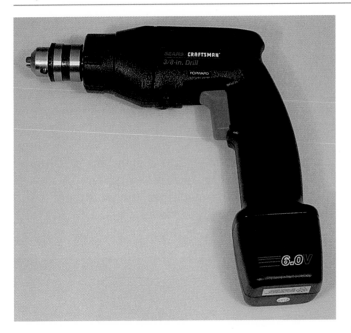

Figure 10-5. *Portable electric drills are manufactured in a wide range of sizes. This model is battery powered.*

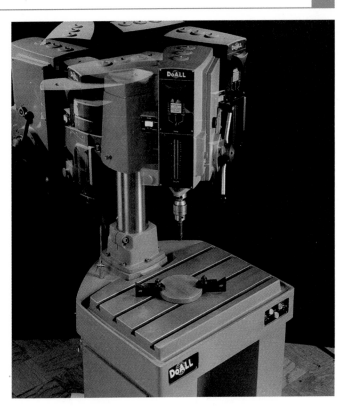

Figure 10-7. *Bench radial drill press. (DoALL Co.)*

Figure 10-6. *This radial drill press can drill large diameter holes in large workpieces. (Sharp Industries, Inc.)*

Smaller **bench radial drill presses** are used to drill smaller holes. See **Figure 10-7.** These units are not as expensive as a full-size radial drill press.

Portable magnetic drills can be used in the shop and on the job site. These machines can be positioned in an upright, horizontal, or vertical position when drilling. See **Figure 10-8.**

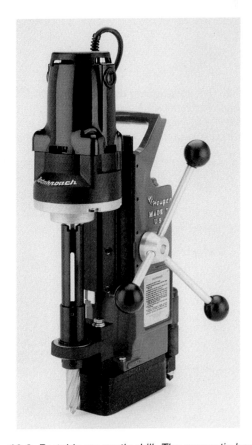

Figure 10-8. *Portable magnetic drill. The magnetic base locks the machine to ferrous metals and enables it to be used in situations where a conventional drill press cannot be employed. (Hougen Manufacturing, Inc.)*

Gang drilling machines consist of several drill assemblies, **Figure 10-9.** The workpiece is moved from one assembly to another. A different operation is performed at each stage.

Multiple spindle drilling machines have several drilling heads. Several operations can be performed without changing drills. See **Figure 10-10.**

Machining centers operate very efficiently and accurately. The center operates under computer numerical control (CNC). These systems are only used in industrial applications. See **Figure 10-11.**

Robotic drilling machines are basically programmable, mobile drilling machines, **Figure 10-12.** The machine is programmed to move along one workpiece or between several workpieces, drilling at specified locations. With a standard drill press, the drill is stationary and the workpiece is moved. With robotic drilling machines, the workpiece is stationary and the machine moves.

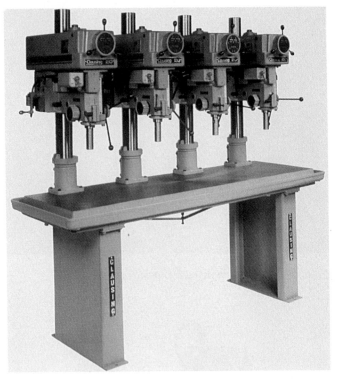

Figure 10-9. *This gang drilling machine has four drills working together. Each machine is fitted with a different cutting tool. The work is held in a drill jig that moves from position to position as each operation is performed. (Clausing Industrial, Inc.)*

Figure 10-11. *Vertical machining center can be programmed to drill holes as part of a machining sequence. No drill jigs are required. Computer numerical control (CNC) is used to set drill position and speed. (Sharnoa Corp.)*

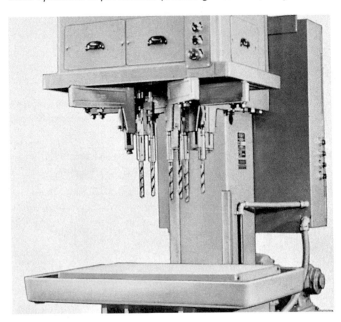

Figure 10-10. *This heavy-duty multiple spindle drill press has drilling heads that can be positioned to meet various drilling requirements. (Deka-Drill, South Bend Lathe)*

Figure 10-12. *This robotic drilling machine moves between workstations on rails. (Northrop-Grumman Corp.)*

10.1.1 Uses of Drilling Machines

Drilling machines are primarily used for cutting round holes. They are also used for many different machining operations, including the following:

- Reaming—An operation performed on an existing hole. The hole is enlarged and finished as the tool removes material from the internal surface of the hole. See **Figure 10-13**.
- Countersinking—Enlarging a hole at the workpiece surface along an angle to allow a screw head to be flush with the surface. See **Figure 10-14**.
- Counterboring—Similar to countersinking, this operation cuts a cylindrical enlargement at the surface of a hole to allow bolt heads to be flush with the surface of the workpiece. See **Figure 10-15**.

- Spotfacing—Putting a smooth finish on a raised area surrounding a hole. See **Figure 10-16**.
- Tapping—This operation cuts screw threads into an existing hole.

10.2 DRILL PRESS SAFETY

- Wear goggles when working on a drill press.
- Remove any jewelry and tuck in loose clothing so they do not become entangled in the rotating drill.

Figure 10-13. *Reaming is being done on this drill press. (Clausing Industrial, Inc.)*

Figure 10-15. *Counterboring is done to prepare a hole to receive a fillister- or socket-head screw. (Clausing Industrial, Inc.)*

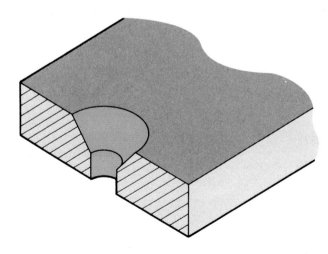

Figure 10-14. *Cross section of a drilled hole that has been countersunk.*

Figure 10-16. *Spotfacing is machining a surface to permit a nut or bolt head to bear uniformly. (Clausing Industrial, Inc.)*

- Check the operation of the machine.
- Be sure all guards are in place.
- Clamp the work solidly. Do not hold work with your hand.
- When removing a drill, place a piece of wood below it. Small drills can be damaged if dropped and larger drills can cause injuries.
- Never attempt to operate a drilling machine while your senses are impaired by medication or other substances.
- Use sharp tools.
- Always remove the key from the chuck before turning on the power.
- Let the drill spindle come to a stop after completing the operation. Do not stop it with your hand.
- Clean chips from the work with a brush, not your hands.
- Keep the work area clear of chips. Place them in an appropriate container. Do not brush them onto the floor.
- Wipe up all cutting fluid that spills on the floor right away.
- Place all oily and dirty waste in a closed container when the job is finished.

10.3 DRILLS

Common drills are known as *twist drills* because most are made by forging or milling rough flutes and then twisting them to a spiral shape. After twisting, the drills are milled and ground to approximate size, **Figure 10-17.** Then, they are heat-treated and ground to exact size.

Most drills are made of *high-speed steel* (HSS) or *carbon steel.* High-speed steel drills can be operated at much higher cutting speeds than carbon steel drills without danger of burning and drill damage.

Figure 10-17. *Close-up of flute milling operation in the manufacture of a large drill. (Chicago-Latrobe)*

Most drills are available with straight or taper shanks and with tungsten carbide tips. Coating drills with titanium nitride greatly increases tool life.

10.3.1 Types of Drills

Industry uses special drills to improve the accuracy of the drilled hole, to speed production, and to improve drilling efficiency.

The *straight-flute gun drill* is designed for ferrous and nonferrous metals, **Figure 10-18.** It is usually fitted with a carbide cutting tip.

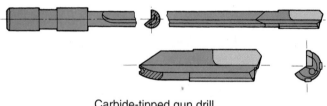

Carbide-tipped gun drill

Figure 10-18. *Gun drill. The tip is shown in larger scale. The light-colored portion is tungsten carbide. The larger area does the cutting; the smaller sections act as wear surfaces.*

An *oil-hole drill* has coolant holes through the body, which permit fluid or air to remove heat from the point, **Figure 10-19.** The pressure of the fluid or air also ejects the chips from the hole while drilling.

Three- and four-flute core drills are used to enlarge core holes in a casting. See **Figure 10-20.**

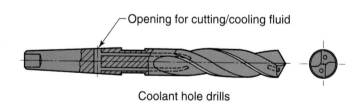

Opening for cutting/cooling fluid

Coolant hole drills

Figure 10-19. *A taper shank twist drill with holes to direct coolant to the cutting edges.*

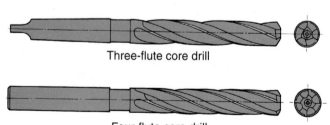

Three-flute core drill

Four-flute core drill

Figure 10-20. *Three- and four-flute core drills.*

Special *step drills* eliminate drilling operations in production work. **Figure 10-21** illustrates this type drill. A *combination drill and reamer* also speeds up production by eliminating one operation, **Figure 10-22.**

Microdrills, **Figure 10-23,** have diameters smaller than 0.0135". They require special drilling equipment.

Half-round straight-flute drills, **Figure 10-24,** are designed for producing holes in brass, copper alloys, and other soft nonferrous materials. Heavy duty carbide-tipped versions are available for drilling hardened steels. These drills are manufactured in fractional and number sizes.

Indexable-insert drills, **Figure 10-25,** are capable of drilling at much higher speeds than high-speed steel drills. The low-cost carbide inserts with multiple cutting edges eliminate costly sharpening. The entire drill does not need to be replaced when the cutting edges are worn—only the insert is changed.

Indexable-insert drills do have limitations. Hole depth is limited to approximately four times the hole diameter. The smallest size available is 5/8" diameter.

Spade drills have replaceable cutting tips that are normally made of tungsten carbide, **Figure 10-26.** These drills are available in sizes from 1" to 5" (25 mm to 125 mm). They are less expensive than twist drills of the same size.

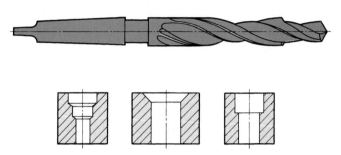

Figure 10-21. *Step drill.*

Figure 10-22. *Combination drill and reamer.*

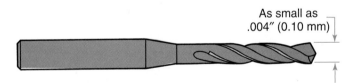

As small as
.004" (0.10 mm)

Figure 10-23. *Microdrills are smaller than the #80 drill (0.0135" diameter).*

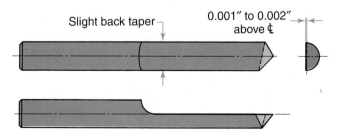

Slight back taper

0.001" to 0.002" above ₡

Figure 10-24. *A half-round drill.*

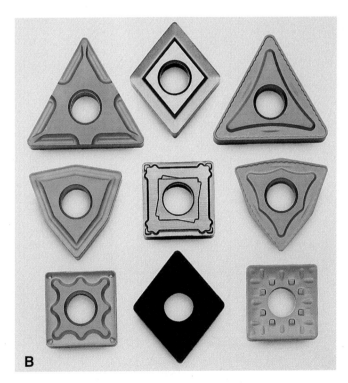

A

B

Figure 10-25. *Indexable-insert drill. A—When the carbide insert becomes worn, it can be replaced with a new insert. B—Different insert shapes are used with different materials and operations.*
(Iscar Metals, Inc. and Hartel Cutting Technologies, Inc.)

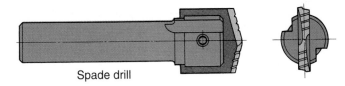

Figure 10-26. *This spade drill has an interchangeable cutting tip.*

10.3.2 Drill Size

Drill sizes are expressed by the following series:

- *Numbers*—#80 to #1 (0.0135" to 0.2280" diameters).
- *Letters*—A to Z (0.234" to 0.413" diameters).
- *Inches and fractions*—1/64" to 3 1/2" diameters.
- *Metric*—0.15 mm to 76.0 mm diameters.

The drill size chart will give an idea of this vast array of drill sizes, **Figure 10-27.**

10.3.3 Drill Measurements

Most drills, with the exception of small drills in the number series, have the diameter stamped on the shank. These figures frequently become obscured, making it necessary to determine the diameter by measuring.

When a micrometer is used for measuring, the measurement is made across the drill margins. However, if the drill is worn, the measurement is made on the shank at the end of the flutes. See **Figure 10-28** for both techniques.

Diameter can also be checked with a ***drill gage***, **Figure 10-29.** Drill gages are made for various drill series; however, 1/2" drills are the largest that can be checked. New drills are checked at the points, worked drills are checked at the end of the flutes.

Always check the drill diameter before using it. Using the wrong size drill can be a very expensive and time-consuming mistake.

Inch	mm	Wire gage	Decimals of an inch
		80	.0135
		79	.0145
1/64			.0156
	.4		.0157
		78	.0160
		77	.0180
	.5		.0197
		76	.0200
		75	.0210
	.55		.0217
		74	.0225
	.6		.0236
		73	.0240
		72	.0250
	.65		.0256
		71	.0260
	.7		.0276
		70	.0280
		69	.0293
	.75		.0295
		68	.0310
1/32			.0313
	.8		.0315
		67	.0320
		66	.0330
	.85		.0335
		65	.0350
	.9		.0354
		64	.0360
		63	.0370
	.95		.0374
		62	.0380
		61	.0390
	1		.0394
		60	.0400
		59	.0410
	1.05		.0413
		58	.0420
		57	.0430
	1.1		.0433
	1.15		.0453
		56	.0465

Inch	mm	Wire gage	Decimals of an inch
3/64			.0469
	1.2		.0472
	1.25		.0492
	1.3		.0512
		55	.0520
	1.35		.0531
		54	.0550
	1.4		.0551
	1.45		.0571
	1.5		.0591
		53	.0595
	1.55		.0610
1/16			.0625
	1.6		.0630
		52	.0635
	1.65		.0650
	1.7		.0669
		51	.0670
	1.75		.0689
		50	.0700
	1.8		.0709
	1.85		.0728
		49	.0730
	1.9		.0748
		48	.0760
	1.95		.0768
5/64			.0781
		47	.0785
	2		.0787
	2.05		.0807
		46	.0810
		45	.0820
	2.1		.0827
	2.15		.0846
		44	.0860
	2.2		.0866
	2.25		.0886
		43	.0890
	2.3		.0906
	2.35		.0925
		42	.0935

Inch	mm	Wire gage	Decimals of an inch
3/32			.0938
	2.4		.0945
		41	.0960
	2.45		.0966
		40	.0980
	2.5		.0984
		39	.0995
		38	.1015
	2.6		.1024
		37	.1040
	2.7		.1063
		36	.1065
	2.75		.1083
7/64			.1094
		35	.1100
	2.8		.1102
		34	.1110
		33	.1130
	2.9		.1142
		32	.1160
	3		.1181
		31	.1200
	3.1		.1220
1/8			.1250
	3.2		.1260
	3.25		.1280
		30	.1285
	3.3		.1299
	3.4		.1339
		29	.1360
	3.5		.1378
		28	.1405
9/64			.1406
	3.6		.1417
		27	.1440
	3.7		.1457
		26	.1470
	3.75		.1476
		25	.1495
	3.8		.1496
		24	.1520
	3.9		.1535
		23	.1540

Inch	mm	Wire gage	Decimals of an inch
5/32			.1563
		22	.1570
	4		.1575
		21	.1590
		20	.1610
	4.1		.1614
	4.2		.1654
		19	.1660
	4.25		.1673
	4.3		.1693
		18	.1695
11/64			.1719
		17	.1730
	4.4		.1732
		16	.1770
	4.5		.1772
		15	.1800
	4.6		.1811
		14	.1820
		13	.1850
	4.7		.1850
	4.75		.1870
3/16			.1875
	4.8		.1890
		12	.1890
		11	.1910
	4.9		.1929
		10	.1935
		9	.1960
	5		.1969
		8	.1990
	5.1		.2008
		7	.2010
13/64			.2031
		6	.2040
	5.2		.2047
		5	.2055
	5.25		.2067
	5.3		.2087
		4	.2090
	5.4		.2126
		3	.2130
	5.5		.2165
7/32			.2188
	5.6		.2205
		2	.2210
	5.7		.2244
	5.75		.2264
		1	.2280
	5.8		.2883

Figure 10-27. *Decimal equivalents of drill sizes.*

Inch	mm	Letter sizes	Decimals of an inch
	5.9		.2323
		A	.2340
15/64			.2344
	6		.2362
		B	.2380
	6.1		.2402
		C	.2420
	6.2		.2441
		D	.2460
	6.25		.2461
	6.3		.2480
1/4		E	.2500
	6.4		.2520
	6.5		.2559
		F	.2570
	6.6		.2598
		G	.2610
	6.7		.2638
17/64			.2656
	6.75		.2657
		H	.2660
	6.8		.2677
	6.9		.2717
		I	.2720
	7		.2756
		J	.2770
	7.1		.2795
		K	.2810
9/32			.2812
	7.2		.2835
	7.25		.2854
	7.3		.2874
		L	.2900
	7.4		.2913
		M	.2950
	7.5		.2953
19/64			.2969
	7.6		.2992
		N	.3020
	7.7		.3031
	7.75		.3051
	7.8		.3071
	7.9		.3110
5/16			.3125
	8		.3150
		O	.3160
	8.1		.3189
	8.2		.3228
		P	.3230
	8.25		.3248
	8.3		.3268

Inch	mm	Letter sizes	Decimals of an inch
21/64			.3281
	8.4		.3307
		Q	.3320
	8.5		.3346
	8.6		.3386
		R	.3390
	8.7		.3425
11/32			.3438
	8.75		.3345
	8.8		.3465
		S	.3480
	8.9		.3504
	9		.3543
		T	.3580
	9.1		.3583
23/64			.3594
	9.2		.3622
	9.25		.3642
	9.3		.3661
		U	.3680
	9.4		.3701
	9.5		.3740
3/8			.3750
		V	.3770
	9.6		.3780
	9.7		.3819
	9.75		.3839
	9.8		.3858
		W	.3860
	9.9		.3898
25/64			.3906
	10		.3937
		X	.3970
		Y	.4040
13/32			.4063
		Z	.4130
	10.5		.4134
27/64			.4219
	11		.4331
7/16			.4375
	11.5		.4528
29/64			.4531
15/32			.4688
	12		.4724
31/64			.4844
	12.5		.4921
1/2			.5000
	13		.5118
33/64			.5156
17/32			.5313
	13.5		.5315

Inch	mm	Decimals of an inch
35/64		.5469
	14	.5512
9/16		.5625
	14.5	.5709
37/64		.5781
	15	.5906
19/32		.5938
39/64		.6094
	15.5	.6102
5/8		.6250
	16	.6299
41/64		.6406
	16.5	.6496
21/32		.6563
	17	.6693
43/64		.6719
11/16		.6875
	17.5	.6890
45/64		.7031
	18	.7087
23/32		.7188
	18.5	.7283
47/64		.7344
	19	.7480
3/4		.7500
49/64		.7656
	19.5	.7677
25/32		.7812
	20	.7874
51/64		.7969
	20.5	.8071
13/16		.8125
	21	.8268
53/64		.8281
27/32		.8438
	21.5	.8465
55/64		.8594
	22	.8661
7/8		.8750
	22.5	.8858
57/64		.8906
	23	.9055
29/32		.9063
59/64		.9219
	23.5	.9252
15/16		.9375
	24	.9449
61/64		.9531
	24.5	.9646
31/32		.9688
	25	.9843
63/64		.9844

Inch	mm	Decimals of an inch
1		1.0000
	25.5	1.0039
1 1/64		1.0156
	26	1.0236
1 1/32		1.0313
	26.5	1.0433
1 3/64		1.0469
1 1/16		1.0625
	27	1.0630
1 5/64		1.0781
	27.5	1.0827
1 3/32		1.0938
	28	1.1024
1 7/64		1.1094
	28.5	1.1220
1 1/8		1.1250
1 9/64		1.1406
	29	1.1417
1 5/32		1.1562
	29.5	1.1614
1 11/64		1.1719
	30	1.1811
1 3/16		1.1875
	30.5	1.2008
1 13/64		1.2031
1 7/32		1.2188
	31	1.2205
1 15/64		1.2344
	31.5	1.2402
1 1/4		1.2500
	32	1.2598
1 17/64		1.2656
	32.5	1.2795
1 9/32		1.2813
1 19/64		1.2969
	33	1.2992
1 5/16		1.3125
	33.5	1.3189
1 21/64		1.3281
	34	1.3386
1 11/32		1.3438
	34.5	1.3583
1 23/64		1.3594
1 3/8		1.3750
	35	1.3780
1 25/64		1.3906
	35.5	1.3976
1 13/32		1.4063
	36	1.4173
1 27/64		1.4219
	36.5	1.4370

Figure 10-27. (continued)

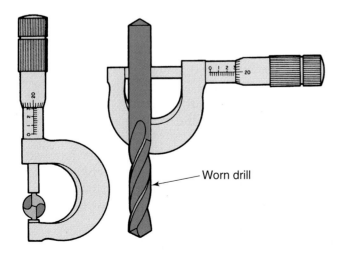

Figure 10-28. Measuring drill size with a micrometer.

Worn drill

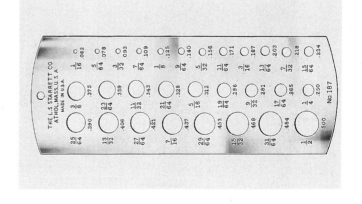

Figure 10-29. This drill gage is used to measure fractional size drills. Similar gages are available for measuring letter, number, and millimeter size drills.

10.3.4 Parts of a Drill

The twist drill is an efficient cutting tool. It is composed of three principal parts: point, shank, and body, **Figure 10-30.**

The *point* is the cone-shaped end that does the cutting. The point consists of the following components:

- *Dead center* refers to the sharp edge at the extreme tip of the drill. This should always be in the *exact center* of the drill axis.
- The *lips* are the cutting edges of the drill.
- The *heel* is the portion of the point back from the lips.
- *Lip clearance* is the amount by which the surface of the point is relieved back from the lips.

10.3.5 Shank

The *shank* is the portion of the drill that mounts into the chuck or spindle. Twist drills are made with shanks that are either straight or tapered, **Figure 10-31.** Straight shank drills are used with a chuck. Taper shank drills have self-holding tapers (No. 1 to No. 5 Morse taper) that fit directly into the drill press spindle.

A *tang* is on the taper shank; it fits into a slot in the spindle, sleeve, or socket, and assists in driving the tool. The tang also provides a means of separating the taper from the holding device.

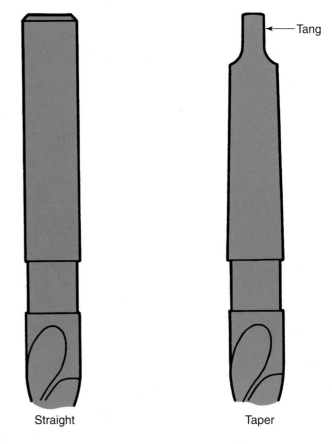

Figure 10-31. *Types of twist drill shanks.*

10.3.6 Body

The *body* is the portion of the drill between the point and the shank. It consists of the following:

- The *flutes* are two or more spiral grooves that run along the length of the drill body. The flutes serve four purposes:
  - Help form the cutting edges of the drill point.
  - Curl the chip tightly for easier removal.
  - Form channels through which the chips can escape as the hole is drilled.
  - Allow coolant and lubricant to flow down to the cutting edges.
- The *margin* is the narrow strip extending back along the entire length of the drill body.

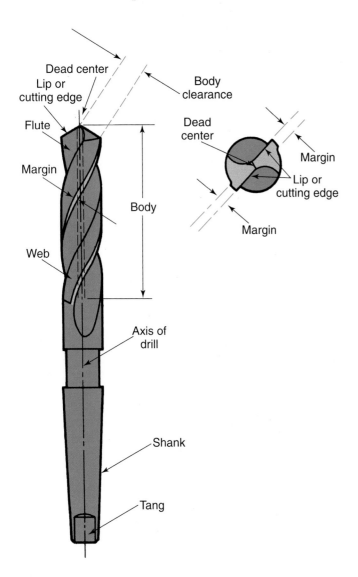

Figure 10-30. *Parts of a twist drill.*

- *Body clearance* refers to the part of the drill body that has been reduced in order to lower friction between the drill and the wall of the hole.
- The *web* is the metal column that separates the flutes. It gradually increases in thickness toward the shank for added strength.

10.4 DRILL-HOLDING DEVICES

A drill is held in the drill press by either of these methods:

- *Chuck:* A movable jaw mechanism for drills with straight shanks. See **Figure 10-32**.
- *Tapered spindle*: A tapered opening for drills with taper shanks, **Figure 10-33**.

Figure 10-32. A drill chuck can be tightened with a key to hold the drill. Keyless chucks are also available. (Yukiwa Seiko USA, Inc.)

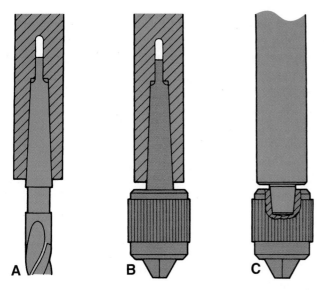

Figure 10-33. Tapered spindles. A—Taper shank drill. B—Drill chuck with a taper shank. C—This is a solid spindle with a short external taper that fits into a drill chuck. The chuck is permanently attached to spindle.

A drill chuck with a taper shank makes it possible to use straight-shank drills when the drill press is fitted with a tapered spindle.

When using a chuck, first insert the drill and tighten the chuck jaws by hand. If the chuck is centered and running true, tighten the chuck with a **chuck key.** Always remove the key from the chuck before turning on the drill press.

Taper shank drills must be wiped clean before inserting the shank into the spindle. Nicks in the shank must be removed with an oilstone; otherwise, the shank will not seat properly. Never attempt to use a taper shank drill mounted in a drill chuck.

Most drill press spindles are made with a No. 2 or No. 3 Morse taper (often indicated as "MT"). A drill with a shank smaller than the spindle taper must be enlarged to fit by using a sleeve, **Figure 10-34**. Drills with shanks larger than the spindle opening can be fit by using a socket, **Figure 10-35**. The taper opening in the socket is larger than the taper on its shank.

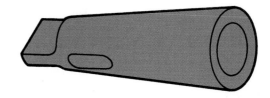

Figure 10-34. A drill sleeve is needed when the drill is too small for the spindle.

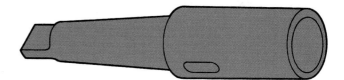

Figure 10-35. A drill socket is used when the spindle is too small for the drill.

A socket should only be used when a larger drill press is not available. It is dangerous to overstress a drill press by using a drill larger than the machine's rated capacity.

Sleeves, sockets, and taper shank drills are separated with a *drift*, **Figure 10-36**. To use a drift, insert it in the slot with the round edge up, **Figure 10-37**. A sharp rap with a lead hammer will cause the parts to separate.

Never use a file tang in place of a drift. It will damage the drill shank and machine spindle. Then

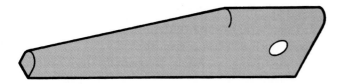

Figure 10-36. *A drift is used to remove taper shank tools from the drill spindle. Never use a file tang.*

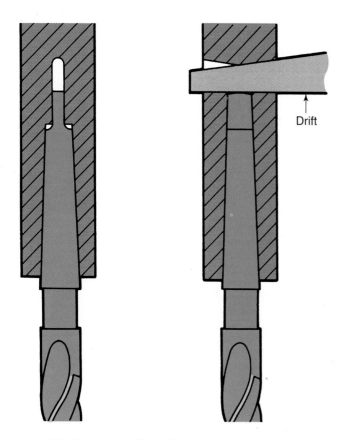

Figure 10-37. *Using a drift. A—The drill is locked in the spindle. B—Using a drift to remove the drill from the spindle.*

other drills will not fit properly. The file could also shatter.

When removing a drill from the spindle, hold the drill to prevent it from falling to the floor. Dropping the drill may damage the drill point. Wrap a piece of clean cloth around the drill to protect your hand from metal chips.

10.5 WORK-HOLDING DEVICES

Work must be mounted solidly on the drilling machine. If work is mounted improperly, it may spring or move, causing drill damage or breakage.

Serious injury can result from work that becomes loose and spins on a drill press. This dangerous situation is nicknamed a "merry-go-round."

10.5.1 Vises

Vises are widely used to hold work, **Figure 10-38.** For best results, the vise must be bolted to the drill table.

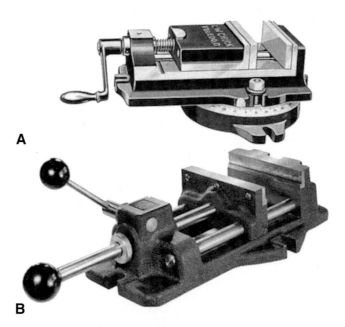

Figure 10-38. *Typical vises used on drill press. A—The swivel base permits the vise body to pivot 180°. B—This quick-acting vise jaw slides to lock parts quickly. (L-W Chuck Co.)*

Parallels are often used to level the work and raise it above the vise base, **Figure 10-39.** This will permit the drill to come through the work and not damage the vise. Parallels can be made from stock steel bars or from special heat-treated steel. Heat-treated parallels are ground to size.

Seat the work on the parallels by tightening the vise and tapping the work with a mallet until the parallels do not move. Loose parallels indicate that the work is not seated properly.

An *angular vise* permits angular drilling without tilting the drill press table. See **Figure 10-40.**

A *cross-slide* permits rapid alignment of the work. Some cross-slides are fitted with a vise, **Figure 10-41.** Others have a series of tapped holes for mounting a vise or another work-holding device.

10.5.2 V-Blocks

V-blocks support round work for drilling, **Figure 10-42.** These blocks are made in many sizes. Some are fitted with clamps to hold the work. Larger sizes must be clamped with the work, **Figure 10-43.**

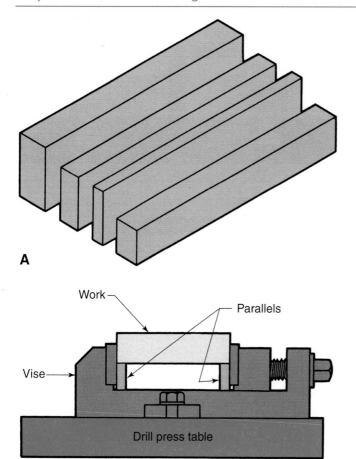

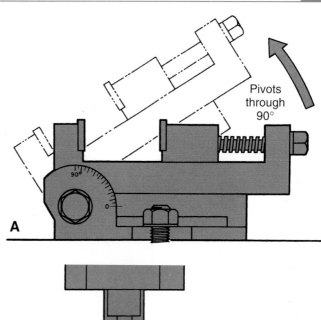

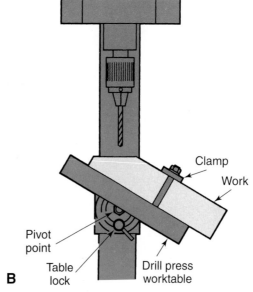

Figure 10-39. *Parallels. A— Steel parallels are available in a large variety of sizes. B—Parallels are often used to raise work above the vise base. This will prevent the drill from cutting into the vise as it goes through the work.*

10.5.3 T-Bolts

T-bolts fit into the drill press table slots and fasten the work or clamping devices to the machine, **Figure 10-44.** A washer should always be used between the nut and the holding device. For convenience, it is desirable to have an assortment of different length T-bolts. To reduce the chance of a setup working loose, place the bolts as close to the work as possible. See **Figure 10-45.**

10.5.4 Strap Clamps

Strap clamps, **Figure 10-46,** make the clamping operation easier. The elongated slot permits some adjustment without removing the washer and nut. Use a strip of copper or aluminum to protect a machined surface that must be clamped.

A *U-strap clamp* is used when the clamp must bridge the work. It can straddle the drill and not interfere with the drilling operation. The small, round section that projects from a *finger clamp* permits the use of small holes or openings in the work for clamping.

Figure 10-40. *Angular vise. A—An angular vise can be adjusted through 90° to permit drilling on an angle without tilting the entire vise or drill table. B—Angular drilling can also be done by tilting the drill table. Be sure the table is locked tightly before starting to drill.*

Figure 10-41. *A cross-slide vise permits rapid alignment of work for drilling.*

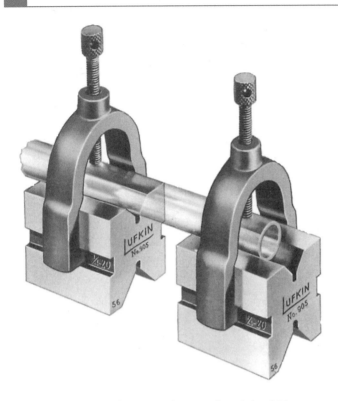

Figure 10-42. V-blocks supporting round work for drilling.

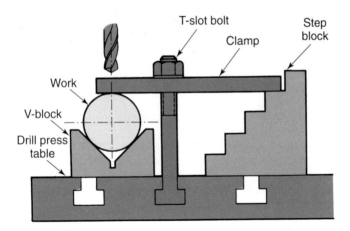

Figure 10-43. One method of clamping large diameter stock for drilling. Always check to be sure that drill will clear the V-block when it comes through work.

Figure 10-44. A few of the many types and sizes of T-bolts available.

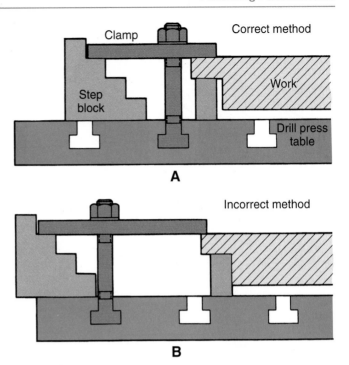

A

B

Figure 10-45. Examples of clamping techniques. A—Correct clamping technique. Note that clamp is parallel to work. Clamp slippage can be reduced by placing a piece of paper between the work and the clamp. B—Incorrect clamping technique. T-bolt is too far from work. This allows the clamp to spring under pressure.

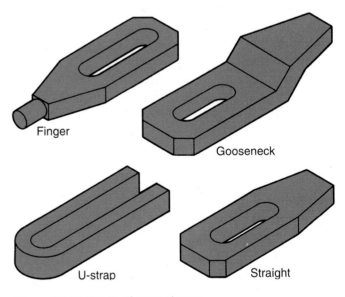

Figure 10-46. Types of strap clamps.

10.5.5 Step Blocks

A *step block* supports the strap clamp opposite the work, **Figure 10-47.** The steps allow the adjustments necessary to keep the strap parallel with the work.

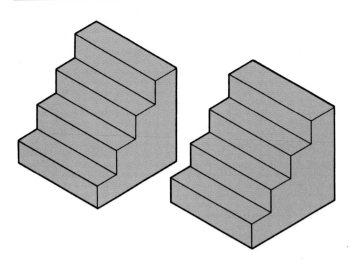

Figure 10-47. *Step blocks are used to support strap clamps.*

10.5.6 Angle Plate

An *angle plate,* **Figure 10-48,** is often used when work must be clamped to a support. The angle plate is then bolted to the machine table, **Figure 10-49.**

10.5.7 Drill Jig

A *drill jig* permits holes to be drilled in a number of identical pieces, **Figure 10-50.** This clamping device supports and locks the work in the proper position. With the use of drill bushings, it guides the drill to the correct location. This makes it unnecessary to lay out each individual piece for drilling.

10.6 CUTTING SPEEDS AND FEEDS

The *cutting speed* is the speed at which the drill rotates. The *feed* is the distance the drill is moved into the work with each revolution. Both are important considerations because they determine the time required to produce the hole.

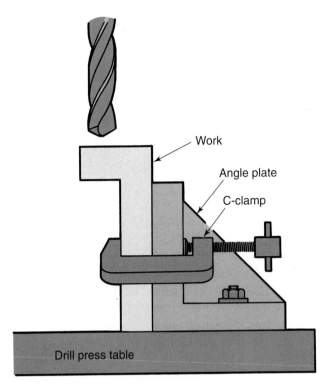

Figure 10-49. *Work must sometimes be mounted against an angle plate for adequate support for drilling.*

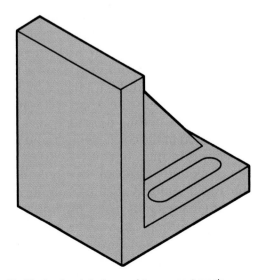

Figure 10-48. *Angle plate is used to support work.*

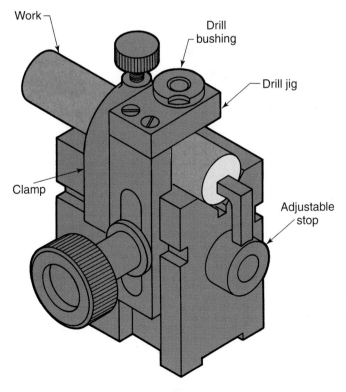

Figure 10-50. *A typical drill jig for holding round stock for drilling through center.*

Drill cutting speed, also known as *peripheral speed*, does not refer to the revolutions per minute (rpm) of the drill, but rather to the distance that the drill cutting edge circumference travels per minute.

10.6.1 Feed

Contrary to popular belief, the spiral shape of a drill flute does not cause the drill to pull itself into the work. Constant pressure must be applied and maintained to advance the drill point at a given rate. This advance is called *feed* and is measured in either decimal fractions of an inch or millimeters.

Because so many variables affect results, there can be no hard and fast rule to determine the exact cutting speed and feed for a given material. For this reason, the *drill speed and feed table* indicates only recommended speeds and feeds, **Figure 10-52.** They are a starting point and can be increased or decreased for optimum cutting.

Feed cannot be controlled accurately on a hand-fed drill press. A machinist must be aware of the cutting characteristics (such as uniform chips) that indicate whether the drill is being fed at the correct rate.

A feed that is too *light* will cause the drill to scrape, "chatter," and dull rapidly. Chipped cutting edges, drill breakage, and drill heating (despite the application of coolant) usually indicate that the rate is too *great*.

10.6.2 Speed Conversion

A problem arises in setting a drill press to the correct speed because its speed is given in revolutions per minute (rpm), while recommended drill cutting speed (CS) is given in feet or meters per minute (fpm or mpm).

The simple formula

$$rpm = \frac{4 \times CS}{D}$$

determines the rpm to operate any diameter drill (D) at any specified speed.

Drill speed problem: At what speed (rpm) must a 1/2″ diameter high-speed steel drill rotate when drilling aluminum?

To solve this problem:

1. Refer to the speed and feed table, **Figure 10-51.** It gives the recommended cutting speed for aluminum (250 fpm).

2. Convert drill diameter (1/2″) to decimal fraction (0.5).

3. Substitute values into the formula

$$rpm = \frac{4 \times CS}{D}$$
$$= \frac{250 \times 4}{0.5}$$
$$= 2000 \text{ rpm}$$

Metric problems are solved in a similar manner using the following formula:

$$rpm = \frac{CS \times 1000}{D \times \pi}$$

Where:
CS = Cutting speed (mpm)
D = Drill diameter (mm)
π = 3 (rounded)

10.6.3 Drill Press Speed Control Mechanisms

With some drill presses, it is possible to set a dial to the desired rpm, **Figure 10-52.** However, on most conventional drilling machines, it is not possible to set the machine at the exact speed desired. The machinist must settle for the available speed nearest the desired speed.

The number of speed settings is limited by the number of pulleys in the drive mechanism, **Figure 10-53.** A decal or an engraved metal chart showing spindle speeds at various settings is attached to many machines. Information on spindle speeds can be found in the operator's manual, or it can be calculated if motor speed and pulley diameters are known.

10.7 CUTTING COMPOUNDS

Drilling at the recommended cutting speeds and feeds generates considerable heat at the cutting point. This heat must be dissipated (carried away) as fast as it is generated, or it will destroy the drill's temper and cause it to dull rapidly.

Cutting compounds are applied to absorb the heat. They cool the cutting tool, serve as a lubricant to reduce friction at the cutting edges, and minimize the tendency for the chips to weld to the lips. Cutting compounds also improve hole finish and aid in the rapid removal of chips from the hole.

There are many kinds of cutting fluids and compounds. Many cutting compounds must be applied liberally. However, some newer compounds should be applied sparingly. Read the instructions on the container for the compound being employed.

Material	Cutting Fluid	Speed feet per minute	Feeds per revolution Over .040 diameter**				
			Under 1/8	1/8 to 1/4	1/4 to 1/2	1/2 to 1	Over 1
Aluminum & aluminum alloys	Sol. oil, ker. & lard oil, lt. oil	200-300	.0015	.003	.006	.010	.012
Aluminum & bronze	Sol. oil, ker. & lard oil, lt. oil	50-100	.0015	.003	.006	.010	.012
Brass, free machining	Dry, sol. oil, ker. & lard oil, lt. min. oil	150-300	.0025	.005	.010	.020	.025
Bronze, common	Dry, sol. oil, lard oil, min. oil	200-250	.0025	.005	.010	.020	.025
Bronze, soft and medium hard	Min. oil with 5%-15% lard oil	70-300	.0025	.005	.010	.020	.025
Bronze, phosphor, 1/2 hard	Dry, sol. oil, lard oil, min. oil	110-180	.0015	.003	.006	.010	.012
Bronze, phosphor, soft	Dry, sol. oil, lard oil, min. oil	200-250	.0025	.005	.010	.020	.025
Cast iron, soft	Dry or airjet	100-150	.0025	.005	.010	.020	.025
Cast iron, medium	Dry or airjet	70-120	.0015	.003	.006	.010	.012
Cast iron, hard	Dry or airjet	30-100	.001	.002	.003	.005	.006
*Cast iron, chilled	Dry or airjet	10-25	.001	.002	.003	.005	.006
*Cast steel	Soluble oil, sulphurized oil. min. oil	30-60	.001	.002	.003	.005	.006
Copper	Dry, soluble oil, lard oil, min. oil	70-300	.001	.002	.003	.005	.006
Magnesium & magnesium alloys	Mineral seal oil	200-400	.0025	.005	.010	.020	.025
Manganese copper, 30% mn.	Soluble oil, sulphurized oil	10-25	.001	.002	.003	.005	.006
Malleable iron	Dry, soluble oil, soda water, min. oil	60-100	.0025	.005	.010	.020	.025
Monel metal	Sol. oil, sulphurized oil, lard oil	30-50	.0015	.003	.006	.010	.012
Nickel, pure	Sulphurized oil	60-100	.001	.002	.003	.005	.006
Nickel, steel 3-1/2%	Sulphurized oil	40-80	.001	.002	.003	.005	.006
Plastics, thermosetting	Dry or airjet	100-300	.0015	.003	.006	.010	.012
Plastics, thermoplastic	Soluble oil, soapy water	100-300	.0015	.003	.006	.010	.012
Rubber, hard	Dry or airjet	100-300	.001	.002	.003	.005	.006
Spring steel	Soluble oil, sulphurized oil	10-25	.001	.002	.003	.005	.006
Stainless steel, free mach'g.	Soluble oil, sulphurized oil	60-100	.0025	.005	.010	.020	.025
Stainless steel, tough mach'g.	Soluble oil, sulphurized oil	20-27	.0025	.005	.010	.020	.025
Steel, free machin'g SAE 1100	Soluble oil, sulphurized oil	70-120	.0015	.003	.006	.010	.012
Steel, SAE-AISI, 1000-1025	Soluble oil, sulphurized oil	60-100	.0015	.003	.006	.010	.012
Steel, .30-.60% carb., SAE 1000-9000 Steel, annealed 150-225 Brinn.	Soluble oil, sulphurized oil	50-70	.0015	.003	.006	.010	.012
Steel, heat treated 225-283 Brinn.	Sulphurized oil	30-60	.0025	.005	.010	.020	.025
Steel, tool hi. car. & hi. speed	Sulphurized oil	25-50	.0025	.005	.010	.020	.025
Titanium	Highly activated sulphurized oil	15-20	.0025	.005	.010	.020	.025
Zinc, alloy	Soluble oil, kerosene & lard oil	200-250	.0015	.003	.006	.010	.012

*Use specially constructed heavy duty drills.
**For drill under .040, feeds should be adjusted to produce chips
and not powder with ability to dispose of same without packing.

A

Drill Speeds in R.P.M.

Diameter of drill	Soft metals 300 fpm	Plastics and hard rubber 200 fpm	Annealed cast from 140 fpm.	Mild steel 100 fpm	Malleable iron 90 fpm	Hard cast iron 80 fpm	Tool or hard steel 60 fpm	Alloy steel cast steel 40 fpm
1/16 (No. 53 to 80)	18320	12217	8554	6111	5500	4889	3667	2445
3/32 (No. 42 to 52)	12212	8142	5702	4071	3666	3258	2442	1649
1/8 (No. 31 to 41)	9160	6112	4278	3056	2750	2445	1833	1222
5/32 (No. 23 to 30)	7328	4888	3420	2444	2198	1954	1465	977
3/16 (No. 13 to 22)	6106	4075	2852	2037	1833	1630	1222	815
7/32 (No. 1 to 12)	5234	3490	2444	1745	1575	1396	1047	698
1/4 (A to E)	4575	3055	2139	1527	1375	1222	917	611
9/32 (G to K)	4071	2712	1900	1356	1222	1084	814	542
5/16 (L, M, N)	3660	2445	1711	1222	1100	978	733	489
11/32 (O to R)	3330	2220	1554	1110	1000	888	666	444
3/8 (S, T, U)	3050	2037	1426	1018	917	815	611	407
13/32 (V to Z)	2818	1878	1316	939	846	752	563	376
7/16	2614	1746	1222	873	786	698	524	349
15/32	2442	1628	1140	814	732	652	488	326
1/2	2287	1528	1070	764	688	611	458	306
9/16	2035	1357	950	678	611	543	407	271
5/8	1830	1222	856	611	550	489	367	244
11/16	1665	1110	777	555	500	444	333	222
3/4	1525	1018	713	509	458	407	306	204

B

Figures are for high-speed drills. The speed of carbon drills should be reduced one-half. Use drill speed nearest to figure given.

Figure 10-51. *A—Drill speed and feed table. B—Drill speeds in rpm. These tables are starting points for drilling different materials. Feeds and speeds should be increased or decreased depending upon the specific metal being drilled and the condition of the drill press. (Chicago-Latrobe)*

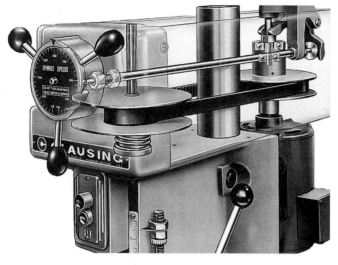

Figure 10-52. This split-pulley speed control mechanism allows speed adjustment by turning the dial in front. (Clausing Industrial, Inc.)

Figure 10-53. With step-pulley speed control, the belt is transferred to different pulley ratios to change drill speed. (Clausing Industrial, Inc.)

Avoid using cutting fluids and compounds when drilling cast iron or other brittle materials. The fluids tend to cause the chips to pack and glaze the opening. Compressed air, used with care, will work when drilling these materials.

Always wash your hands thoroughly with soap and warm water after using cutting compounds and fluids. Before using the cutting compound, check the container to determine what should be done if you get any in your eye.

10.8 SHARPENING DRILLS

A drill becomes dull with use and must be resharpened. Continued use of a dull drill may result in drill breakage or burning. Improper sharpening will cause the same problems.

Remove the entire point if it is badly worn or if the margins are burned, chipped, or worn off near the point. If the drill becomes overheated during grinding, do not plunge it into water. Allow it to cool in still air. The shock of sudden cooling may cause it to crack.

Three factors must be considered when repointing a drill: lip clearance, length and angle of the lips, and proper location of dead center.

Lip clearance. The two cutting edges, or lips, are comparable to chisels, **Figure 10-54.** To cut effectively, the heel (part of the point back of the cutting edge) must be relieved. Without this *lip clearance,* it is impossible for the lips to cut. If there is too much clearance, the cutting edges will be weakened. Too little clearance results in the drill point merely rubbing without penetration into the material.

Gradually increase lip clearance toward the center until the line across dead center stands at an angle of 120° to 135° with the cutting edge. See **Figure 10-55.**

Length and angle of lips. The material being drilled determines the proper point angle, **Figure 10-56.** The angles, in relation to the axis, must be the same (a 59° angle is satisfactory for most metals). If the angles are unequal, only one lip will cut and the hole will be oversized.

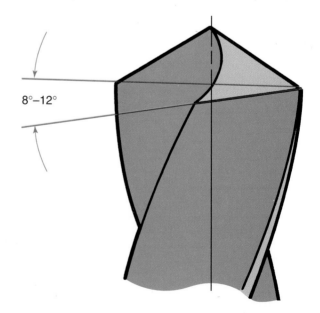

Figure 10-54. Lip clearance of 8° to 12° is satisfactory for most drilling.

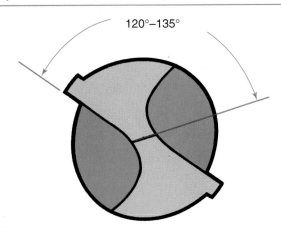

Figure 10-55. *Proper angle of drill dead center.*

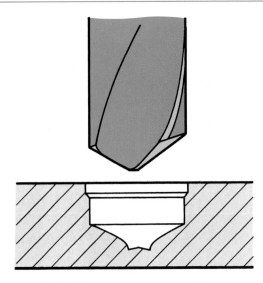

Figure 10-58. *This is the type of hole produced when the drill point has unequal point angles and is sharpened off center.*

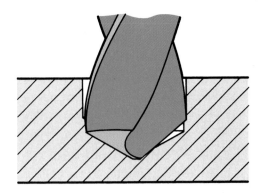

Figure 10-56. *Unequal drill point angles will produce a drilled hole that is oversize.*

The web of a drill increases in thickness toward the shank, **Figure 10-59.** When a drill has been shortened by repeated grindings, the web must be thinned to minimize the pressure required to make the drill penetrate the material. The thinning must be done equally on both sides of the web and care must be taken to ensure that the web is centered.

A *drill point gage* is used to check a drill point while sharpening. Its use is shown in **Figure 10-60.**

10.8.1 Drill Sharpening Procedures

Use a coarse grinding wheel for roughing out the drill point if a large quantity of metal must be removed. Complete the operation on a fine wheel.

Proper location of dead center. The drill dead center must be accurate. Lips of different lengths will result in oversized holes, causing "wobble." This places tremendous pressures on the drill press spindle and bearings. See **Figure 10-57.**

A combination of lip and dead center faults can result in a broken drill. If the drill is very large, permanent damage to the drilling machine can result. The hole produced will be oversized and often out-of-round. Refer to **Figure 10-58.**

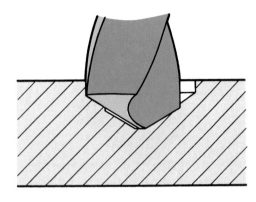

Figure 10-57. *Hole produced when drill is sharpened off center.*

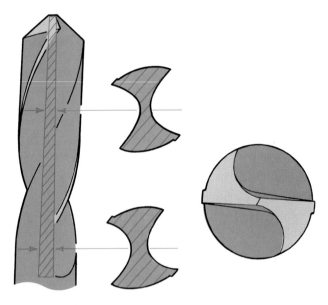

Figure 10-59. *The drill web tapers down toward the tip. The point is sometimes relieved to improve cutting action.*

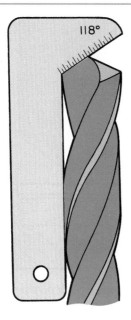

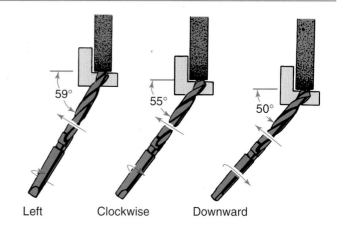

Left   Clockwise   Downward

Figure 10-62. One drill sharpening technique. Hold the point lightly against the rotating wheel and use three motions of the shank: to the left, clockwise rotation, and downward.

Figure 10-60. Using a drill point gage will help ensure proper drill sharpening. For general drilling, an included angle of 118° is used.

Many hand sharpening techniques have been developed. The following technique is suggested:

1. Grasp the drill shank with your right hand and the rest of the drill with your left hand. See **Figure 10-61**.

Figure 10-61. One recommended way to hold a drill when it is being sharpened.

2. Place your left-hand fingers that are supporting the drill on the grinder tool rest. The tool rest should be slightly below center (about 1" down on a 7" diameter wheel, for example).

3. Stand so the centerline of the drill will be at a 59° angle to the centerline of the wheel, **Figure 10-62**. Lightly touch the drill lip to the wheel in a horizontal position.

4. Use your left hand as a pivot point and slowly lower the shank with your right hand. Increase pressure as the heel is reached to ensure proper clearance.

5. Repeat the operation on each lip until the drill is sharpened. Do not quench high-speed steel drills in water to cool them. Allow them to cool in air.

6. Check the drill tip frequently with a drill point gage to ensure a correctly sharpened drill.

Sharpening a drill is not as difficult as it may first appear. However, before attempting to sharpen a drill, secure a properly sharpened drill and run through the motions explained above. When you have acquired sufficient skill, sharpen a dull drill.

To test, drill a hole in soft metal and observe the chip formation. When properly sharpened, chips will come out of the flutes in curled spirals of equal size and length. Tightness of the chip spiral is governed by the rake angle, **Figure 10-63.**

A standard drill point has a tendency to stick when used to drill brass. When brass is drilled, sharpen the drill as shown in **Figure 10-64.**

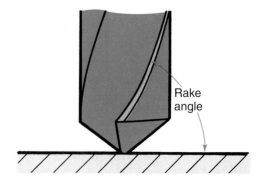

Rake angle

Figure 10-63. Rake angle of the drill.

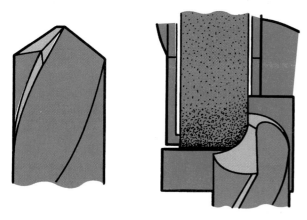

Figure 10-64. Modified rake angle for drilling brass.

10.8.2 Drill Grinding Attachments

A drill sharpening device is shown in **Figure 10-65.** An attachment for conventional grinders is shown in **Figure 10-66.** In the machine shop where a high degree of hole accuracy is required and a large amount of sharpening must be done, these devices are a must.

10.9 DRILLING

Obeying a few simple rules will help you drill accurately. Use the following procedure:

1. Carefully study the drawing to determine hole locations. Lay out the positions and mark the intersecting lines with a prick punch.

2. Secure a drill and check its size.

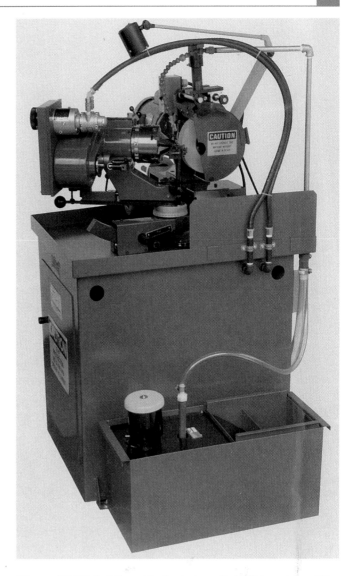

Figure 10-66. A drill and tool grinder with built-in point splitting and web thinning capabilities. The user can choose from a wide variety of drill points. Machines of this type are available in semiautomatic and automatic versions. (Rush Machinery, Inc.)

3. Mount work solidly on the drilling machine. Never hold the work by hand. The workpiece could whip out of your hand and cause serious injuries.

4. Insert a *wiggler* or *center finder* in the drill chuck to position the point to be drilled directly under the chuck or spindle. Turn on the power and center the wiggler point with your fingers. Position the work until the revolving wiggler point does not wiggle when it is lightly dropped into the punched hole location and removed. If there is any point movement, additional alignment is necessary because the work is not positioned properly, **Figure 10-67.**

Figure 10-65. A drill sharpening grinder attachment ensures that cutting edges of the drill will be uniform. (Darex Corp.)

Figure 10-67. *It is difficult to align a workpiece with centerlines by eye. To assist in this job, a center finder, or wiggler, is used.*

5. Remove the wiggler and insert a center drill. Hand tighten the chuck, **Figure 10-68**. Check to be sure the drill runs true. If it does, tighten the chuck with a chuck key. Remember to remove the key before starting the machine.

Figure 10-68. *After alignment with the wiggler, center drilling will ensure that the drill will make the hole in the proper location.*

6. After center drilling, replace the tool with the required drill. Hand tighten it in the chuck. Turn on the machine. If it does not run true, the drill may be bent or may have been placed in the chuck off center. Also check that it will drill to the required depth.

7. Calculate the correct cutting speed and feed if you plan to use a power feed. Adjust the machine to operate as closely as possible to this speed.

8. Turn on the power and apply cutting fluid. Start the cut. Even pressure on the feed handle will keep the drill cutting freely.

9. Watch for the following signs that indicate a poorly cutting drill:
   - A dull drill will squeak and overheat. Chips will be rough and blue, and cause the machine to slow down. Small drills will break.
   - Infrequently, a chip will get under the dead center and act as a bearing, preventing the drill from cutting. Remove it by raising and lowering the drill several times.
   - Chips packed in the flutes will cause the drill to bind and slow the machine or cause the drill to break. Remove the drill from the hole and clean it with a brush that has been dipped in cutting fluid. Do not use cutting fluid when drilling cast iron.

10. Clear chips and apply cutting fluid as needed.

11. The most critical time of the drilling operation occurs when the drill starts to break through the work. Ease up on feed pressure at this point to prevent the drill from "digging in."

12. Remove the drill from the hole and turn off the power. Never try to stop the chuck with your hand. Clear the chips with a brush. Unclamp the work and use a file to remove all burrs.

13. Clean chips and cutting fluid from the machine. Wipe it down with a soft cloth. Return equipment to storage after cleaning.

Observe extreme care in positioning the piece for drilling. A poorly planned setup may permit the drill to cut into the vise or drill table when it breaks through the work.

If a hole must be located precisely, certain additional precautions can be taken to ensure that the hole will be drilled where it is supposed to be drilled. After the center point has been determined, a series of *proof circles* are scribed, **Figure 10-69A**. They will serve as reference points to help check whether the drill remains on center as it starts to penetrate the material.

Even when work is properly centered, the drill may "drift" when starting a hole. Various factors can cause this, such as hard spots in the metal or an improperly sharpened drill. The drill cannot be brought back on center by moving the work, because it will still try to follow the original hole. This condition, **Figure 10-69B,** must be corrected before the full diameter of the drill is reached.

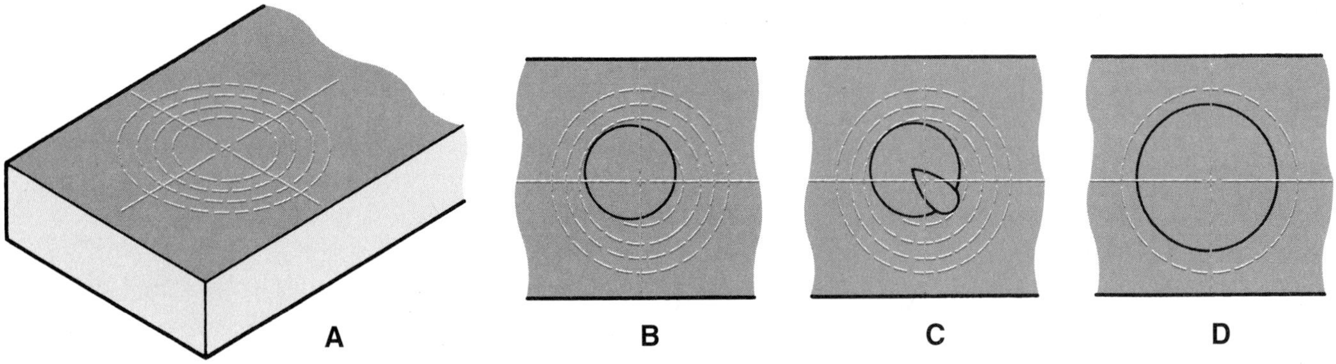

Figure 10-69. How to bring a drill back on center. A—Proof circles. B—Drill has been started off center (exaggerated). C—Groove cut to bring drill back on center. D—Drill back on center. This operation will only work if the drill has not begun to cut to its full diameter.

The drill is brought back on center by using a round-nose cape chisel to cut a groove on the side of the hole where the drill must be drawn, **Figure 10-69C**. This groove will "pull" the drill point to the center. Repeat the operation until the hole is centered in the proof circles, **Figure 10-69D**.

10.9.1 Drilling Larger Holes

Drills larger than 1/2" (12.5 mm) diameter require considerable power and pressure to get started. Even then, they may run off center. The pressure can be greatly reduced and accuracy improved by first drilling a *pilot hole* (*lead hole*) that is smaller in diameter than the final hole. See **Figure 10-70.**

The small pilot hole permits pressure to be exerted directly on the cutting edges of the large drill, causing it to drill faster. The diameter of the pilot hole should be as large as, or slightly larger than, the width of the dead center.

10.9.2 Drilling Round Stock

Holes are more difficult to drill in the curved surface of round stock. Many difficulties can be eliminated by holding the round material in a V-block, **Figure 10-71**. A V-block can be held in a vise or clamped directly to the table.

Use the following procedure to center round stock in a V-block:

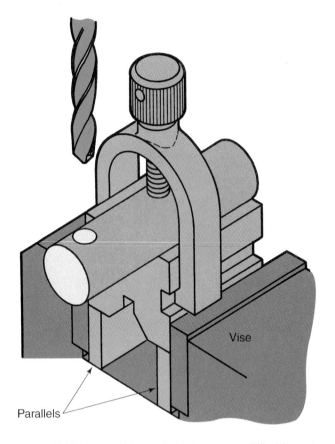

Figure 10-71. The V-block eliminates many difficulties when drilling round stock. Be sure the drill will clear the V-block when it comes through material.

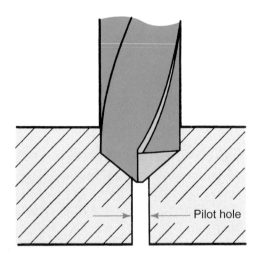

Figure 10-70. A pilot hole makes drilling a large hole much easier.

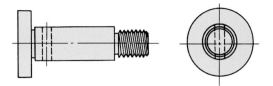

1. Locate the hole position on the stock. Prick punch the intersection of the layout lines. Place the stock in a V-block. If the hole is to go through the piece, make certain that the drill will clear the V-block. Also be sure there is ample clearance between the clamp and drill chuck.

2. To align the hole for drilling through exact center, place the work and V-block on the drill press table or on a surface plate. Rotate the punch mark until it is upright. Place a steel square on the flat surface with the blade against the round stock as shown in **Figure 10-72**. Measure from the square blade to the punch mark, and rotate the stock until the measurement is the same when taken from both sides of the stock.

3. From this point, the drilling sequence is identical to that previously described.

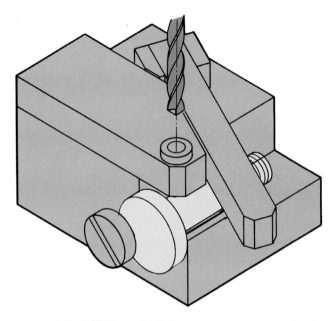

Figure 10-73. This typical drill jig has an arm that lifts to allow easy insertion and removal of the part being drilled.

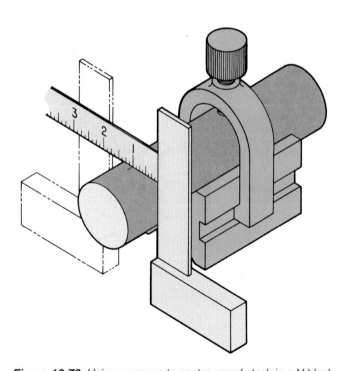

Figure 10-72. Using a square to center round stock in a V-block.

If a large number of identical parts must be drilled, it may be desirable to make a drill jig, **Figure 10-73**. The drill jig automatically positions and centers each piece for drilling.

10.9.3 Blind Holes

A *blind hole* is a hole that is not drilled all the way through the work. *Hole depth* is measured by the distance the *full hole diameter* goes into the work, **Figure 10-74**. Using a drill press fitted with a

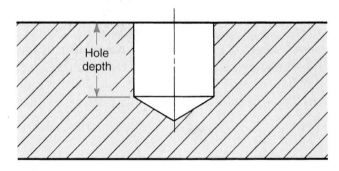

Figure 10-74. Measuring the depth of a blind hole.

depth stop or *depth gage* is the quickest means of achieving proper depth when drilling blind holes, **Figure 10-75**.

10.10 COUNTERSINKING

Countersinking is the operation that cuts a chamfer in a hole to permit a flat-headed fastener to be inserted with the head flush to the surface, **Figures 10-76**.

The tool used to machine countersinks is called a *countersink*, **Figure 10-77**. Countersinks are available with cutting edge angles of 60°, 82°, 90°, 100°,

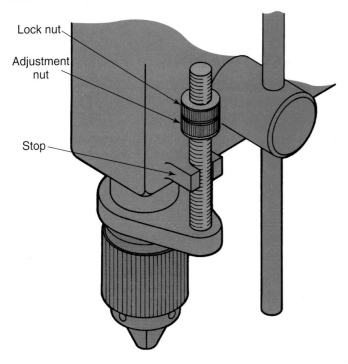

Figure 10-75. *The depth gage attachment provides easy adjustment of how far the drill moves into the work.*

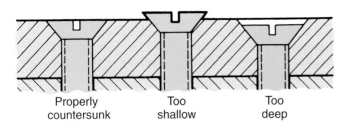

Figure 10-76. *Correctly and incorrectly countersunk holes. The countersink angle must match the fastener head angle.*

Figure 10-77. *Countersinks come in various sizes. (Greenfield Tap & Die)*

110°, and 120° included angles. Countersinks are also used for deburring holes.

Countersinks with indexable carbide inserts, **Figure 10-78,** are available in a number of sizes and point angles. They have two cutting edges per insert and do not require resharpening. Cutting speeds are five to ten times higher than with HSS countersinks.

A single cutting edge countersink, **Figure 10-79,** is free cutting and produces minimum chatter. Chips produced by the cutting edge pass through the hole and are ejected.

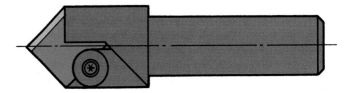

Figure 10-78. *Countersinks with indexing carbide inserts have a life five to ten times longer than similar HSS countersinks.*

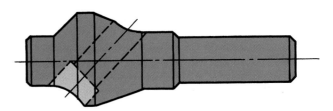

Figure 10-79. *Countersink with a single cutting edge and pilot.*

10.10.1 Using a Countersink

1. The cutting speed should be about one-half that recommended for a similar size drill. This will minimize the probability of chatter.

2. Feed the tool into the work until the chamfer is large enough for the fastener head to be flush.

3. Use the depth stop on the drill press if a number of similar holes must be countersunk.

10.11 COUNTERBORING

The heads of fillister-head and socket-head screws are usually set below the work surface. A *counterbore* is used to enlarge the drilled hole to the proper depth and machine a square shoulder on the bottom to secure maximum clamping action from the fastener, **Figure 10-80.**

The counterbore tool has a guide, called a pilot, which keeps it positioned correctly in the hole. Solid counterbores are available. However, counterbores with interchangeable pilots and cutters are commonly used, **Figure 10-81.** They can be changed easily from one size cutter or pilot to another size. A drop of oil on the pilot will prevent it from binding in the drilled hole.

Counterbores with indexable carbide inserts, **Figure 10-82,** are also available. When the cutting edges become dull, new edges can be indexed into

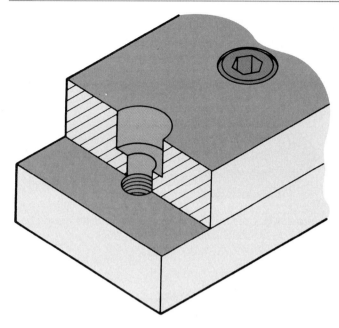

Figure 10-80. *A sectional view of a hole that has been drilled and counterbored to receive a socket-head screw.*

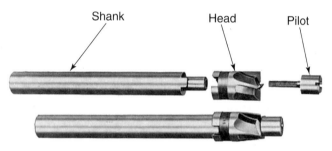

Figure 10-81. *A straight shank interchangeable counterbore. Head and pilot can be quickly changed.*

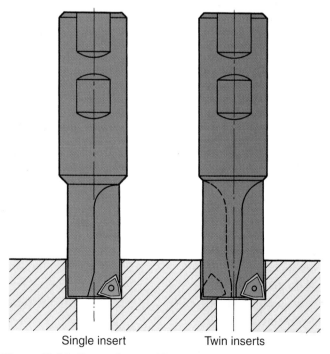

Figure 10-82. *Counterbores with carbide indexable inserts. The inserts are rotated when a cutting edge becomes dull.*

place without affecting opening diameter. Costly sharpening is eliminated.

10.12 SPOTFACING

Spotfacing is the operation during which a circular spot is machined on a rough surface (such as a casting or forging) to provide a bearing surface for the head of bolt, washer, or nut. A counterbore may be used for spotfacing, although a special tool manufactured for inverted spotfacing is available, **Figure 10-83.**

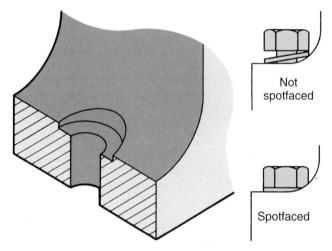

Figure 10-83. *Sectional view of a casting with a mounting hole that has been spotfaced. Profile drawings show the casting before and after spotfacing. The bolt head cannot be drawn down tightly until the mounting hole is spotfaced.*

Special **backspotface** and **backcounterbore** tools are required to perform operations in areas where conventional tools cannot be used. See **Figure 10-84.** The cutting point is lifted up into the workpiece, rather than being pushed down into it. See **Figure 10-85.**

Large diameter openings can be counterbored, spotfaced, or drilled with the APT Multi-Tool™, **Figure 10-86.** A pilot hole is drilled with a conventional twist drill. The required size pilot and blade is inserted and the opening is made. For very large openings, it may be necessary to use a smaller-size blade before machining the specified size. Multi-Tool blades are available in sizes from 1 1/8" diameter to 4" diameter.

10.13 TAPPING

Tapping may be done by hand on a drill press using the following steps:

1. Drill the correct size hole for the tap, **Figure 10-87**.

2. With the work clamped in the machine, insert a small 60° center in the chuck. The center holds the tap vertically.

Figure 10-84. *Special backspotface and backcounterbore tools are used in situations where conventional tools cannot be inserted. Blade setting can be manual or automatic. Standard tools are available for bores and spotfaces of 0.250 and larger. (Parlec, Inc.)*

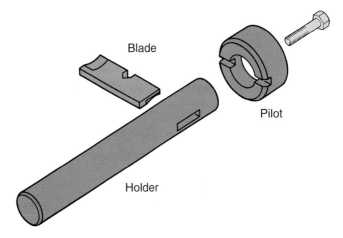

Figure 10-86. *Interchangeable drilling, spotfacing, and counterboring tool. Blades for producing holes up to 4″ diameter are available.*

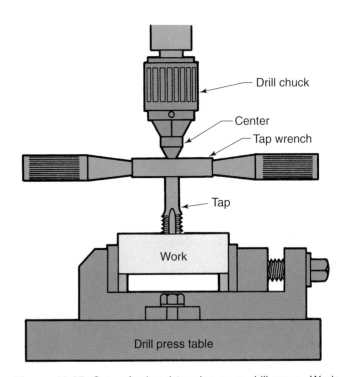

Figure 10-87. *Setup for hand tapping on a drill press. Work must be mounted solidly.*

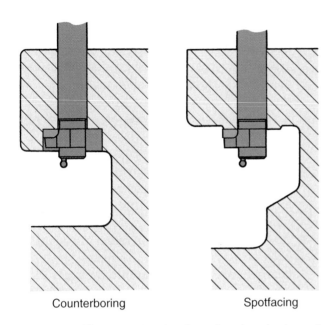

Figure 10-85. *These section drawings show how backcounterboring and backspotfacing tools work.*

3. Place the center point in the tap's center hole.

4. Feed the tap into the work by holding down on the feed handle and turning the tap with a ***tap wrench***.

Never insert a tap into the drill chuck and attempt to use the drill press power to run the tap into the work. The tap will shatter when power is applied. Turn the tap by hand.

Tapping can only be done with power through the use of a *tapping attachment,* **Figure 10-88.** This device fits the standard drill press. It has reducing gears that slows the tap to about one-third of the drill press speed. A table provided with the attachment gives recommended spindle speeds for tapping.

A clutch arrangement drives the tap until it reaches the predetermined depth, at which time the tap stops rotating. Raising the feed handle causes the tap to reverse direction and back out of the hole.

Specially designed *microdrilling machines,* **Figure 10-89,** are required for drilling holes as small as 0.0016" (0.04 mm) with close tolerances. These drills have very accurate spindles and collets to

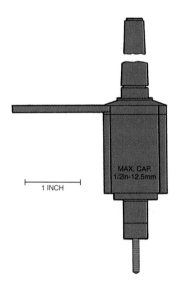

MAX. CAP. 1/2In-12.5mm

1 INCH

Figure 10-88. *Tapping attachment on a drill press.*

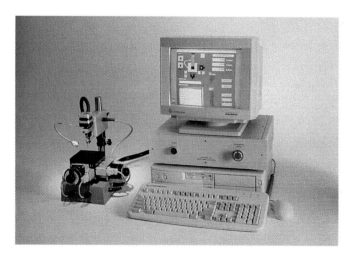

Figure 10-89. *CNC microdrilling machine with a point-to-point programmable control system. The table has 2" × 2" (50 mm × 50 mm) travel with an accuracy of 0.0001" (0.0025 mm). Variable speed control for the drill head motor is controlled directly from the computer. (Minitool Inc.)*

reduce drill flexing and breakage. Many of the microdrilling machines are controlled by computer numerical control (CNC) systems.

Microdrilling uses a "pecking" technique to cut these small diameter holes. In this technique, the drill is repeatedly inserted and removed from the hole. The drills have flutes to pull chips out of the hole, but because they are so small, pecking is necessary for chip removal. The depth of each peck is determined by the drill diameter and the material being drilled.

Small-size holes with other geometric shapes (such as square, rectangular, or hexagonal) are made by Electrical Discharge Machining (EDM). This topic is discussed in Chapter 27, Electromachining Processes.

10.14 REAMING

Reaming produces holes that are extremely accurate in diameter and have an exceptionally fine surface finish. Machine reamers are made in a variety of sizes and styles. They are usually manufactured from high-speed steel. Some are fitted with carbide cutting edges. Descriptions of a few of the more common machine reamers follow. Refer to **Figure 10-90.**

- A *jobber's reamer,* also called a *machine reamer,* is identical to a hand reamer except that a taper shank is available and the tool is designed for machine operation.
- A *chucking reamer* is manufactured with both straight and taper shanks. It is similar to a jobber's reamer but its flutes are shorter and deeper. It is available with straight or spiral flutes.
- A *rose chucking reamer* is designed to cut on its end. The flutes provide chip clearance and are ground to act only as guides. This type of reamer is best used when considerable metal must be removed and the finish is not critical.
- A *shell reamer* is mounted on a special arbor that can be used with several reamer sizes. The arbor can have straight or spiral flutes and is also made in the rose style. The arbor shank may be straight or tapered. A hole in the reamer is tapered to fit the arbor, which is fitted with drive lugs.
- An *expansion chucking reamer* is available with straight flutes and either a straight or taper shank. Slots are cut into the body to permit the reamer to expand when an adjusting screw in the end is tightened.

A regular expansion reamer has several drawbacks. The slots, which are necessary for the reamer

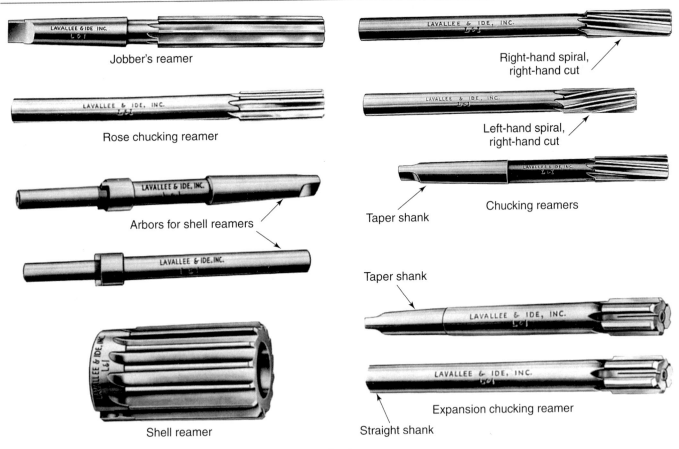

Jobber's reamer

Rose chucking reamer

Arbors for shell reamers

Shell reamer

Right-hand spiral, right-hand cut

Left-hand spiral, right-hand cut

Taper shank

Chucking reamers

Taper shank

Expansion chucking reamer

Straight shank

Figure 10-90. Each type of reamer is suited for a different application.

to expand, reduce tool rigidity. This diminishes accuracy and surface finish. Also, cutting-edge clearance is reduced as the reamer expands, creating a "drag." This often causes the tool to chatter with a resulting decrease in finish quality.

A *solid* expansion reamer provides rigidity and accuracy not possible with conventional expansion reamers. See **Figure 10-91.** To expand this type, a tapered plug is forced into the reamer end. The tool body expands well beyond the tip, and ensures uniform parallel expansion across the full length of the carbide cutting lips. Clearance is automatically provided. The plug can be removed for shimming to a larger size. Once expanded, the reamer diameter cannot be reduced without grinding.

Figure 10-91. This solid expansion reamer has tungsten carbide cutting edges for extended cutting life. (Standard Tool Co.)

10.14.1 Using Machine Reamers

Reamers are expensive precision tools. The quality of the finish and accuracy of the reamed hole will depend on how the tool is used. Obey the following reaming rules:

- Carefully check the reamer diameter before use. If the hole diameter is critical, drill and ream a hole in a piece of similar material to check tool accuracy.
- Use a sharp reamer.
- Mount the reamer solidly.
- Cutting speed for a high-speed steel reamer should be about two-thirds that of a similar size drill.
- Feed should be as high as possible while still providing a good finish and accurate hole size.
- Allow enough material in the drilled hole to permit the reamer to cut rather than burnish (smooth and polish). The following allowances are recommended:
  - Up to 1/4″ (6.3 mm) diameter, allow 0.010″ (0.25 mm).
  - 1/4″ to 1/2″ (6.3 mm to 12.5 mm) diameter allow 0.015″ (0.4 mm).
  - 1/2″ to 1.0″ (12.5 mm to 25.0 mm) diameter allow 0.020″ (0.5 mm).
  - 1.0″ to 1.5″ (25.0 mm to 38.0 mm) diameter allow 0.025″ (0.6 mm).
- Use an ample supply of cutting fluid.

- Remove the reamer from the hole before stopping the machine.
- When not being used, reamers should be stored in separate containers or storage compartments. This will minimize chipping and dulling of the cutting edges.

TEST YOUR KNOWLEDGE

Please do not write in the text. Write your answers on a separate piece of paper.

1. A twist drill works by:
   a. Being forced into material.
   b. Rotating against material and being pulled through by the spiral flutes.
   c. Rotating against material with sufficient pressure to cause penetration.
   d. All of the above.
   e. None of the above.

2. How is drill press size determined?

3. Drills are made from:
   a. High-speed steel.
   b. Carbon steel.
   c. Both of the above.
   d. Neither of the above.

4. Drill sizes are expressed by what four series?

5. What are two techniques used to determine a drill's size?

6. List the two types of drill shanks.

7. _____ shank drills are used with a chuck.

8. _____ shank drills fit directly into the drill press spindle.

9. The spiral grooves that run the length of the drill body are called _____.

10. The spiral grooves in a drill body are used to:
    a. Help form the cutting edge of the drill point.
    b. Curl chips for easier removal.
    c. Form channels through which the chips can escape from the hole.
    d. All of the above.
    e. None of the above.

11. Name the device employed to enlarge a taper shank drill so it will fit the spindle opening.

12. The device used to permit a drill with a taper shank too large to fit the spindle opening is called a(n)_____.

13. What is the name of the tool used to separate a taper shank drill from the above devices?

14. Cutting compounds or fluids are used to:
    a. Cool the drill.
    b. Improve the finish of a drilled hole.
    c. Aid in the removal of chips.
    d. All of the above.
    e. None of the above.

15. List the three factors that must be considered when repointing a drill.

16. What occurs when the cutting lips of a drill are not sharpened to the same lengths?

17. The _____ should be used frequently when sharpening to ensure a correctly sharpened drill.

18. The included angle of a drill point sharpened for general drilling is _____ degrees.

19. What coolant should be used when drilling cast iron?

20. Large drills require a considerable amount of power and pressure to get started. They also have a tendency to drift off center. These conditions can be minimized by first drilling a _____ hole. This hole should be as large as, or slightly larger than, the width of the _____ of the drill point.

21. What is a blind hole?

22. How is the depth of a drilled hole measured?

23. The _____ is almost identical to the hand reamer except that the shank has been designed for machine use.

24. A(n) _____ expansion reamer provides rigidity and accuracy not possible with conventional expansion reamers.

25. How should a reamer be removed from a finished hole?

26. The cutting speed for a high-speed reamer is approximately _____ that for a similar-sized drill.

27. What is the name of the operation employed to cut a chamfer in a hole to receive a flat-head screw?

28. The operation used to prepare a hole for a fillister or socket head screw is called _____.

29. _____ is the operation that machines a circular spot on a rough surface for the head of a bolt or nut.

Chapter 11

Offhand Grinding

LEARNING OBJECTIVES

After studying this chapter, you will be able to:
○ Identify the various types of offhand grinders.
○ Dress and true a grinding wheel.
○ Prepare a grinder for safe operation.
○ Use an offhand grinder safely.
○ List safety rules for offhand grinding.

IMPORTANT TERMS

abrasive belt grinding machines
bench grinder
concentricity
flexible shaft grinders
pedestal grinder
precision microgrinder
reciprocating hand grinder
temper
tool rest
wheel dresser

Grinding is an operation that removes material by rotating an abrasive wheel or belt against the work, **Figure 11-1.** It is used for the following tasks:
- Sharpening tools.
- Removing material too hard to be machined by other techniques.
- Cleaning the parting lines from castings and forgings.
- Finishing and polishing molds used in die casting of metals and injection molding of plastics.

11.1 ABRASIVE BELT GRINDERS

Abrasive belt grinding machines are heavy-duty versions of the belt and disc sanders found in woodworking, **Figure 11-2.** A wide variety of abrasive belts permits these machine tools to be used for grinding to a line, finishing cast and forged parts, deburring, contouring, and sharpening. See **Figure 11-3.**

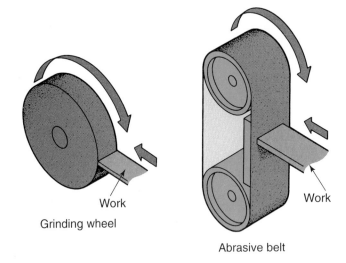

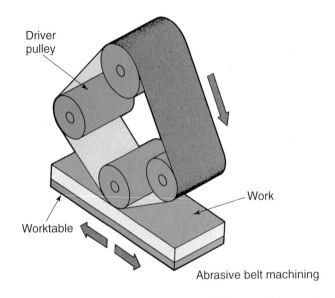

Figure 11-1. Principles of how typical grinding machines work.

11.2 BENCH AND PEDESTAL GRINDERS

The bench grinder and pedestal grinder are the simplest and most widely used grinding machines.

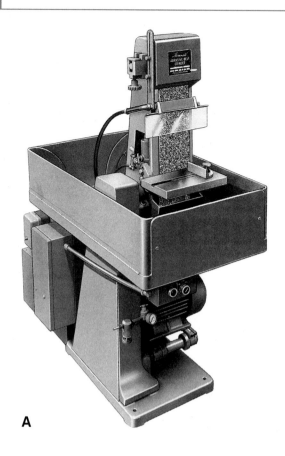

A

Figure 11-3. *Abrasive belt being used to clean up an aluminum casting of a base for an office chair. (American Foundrymen's Society)*

Grinding done on a bench, pedestal, or belt grinder is called *offhand grinding*. This type of work does not require great accuracy. The part is held in your hands and manipulated until it is ground to the desired shape.

The *bench grinder* is one that has been fitted to a bench or table, **Figure 11-4.** The grinding wheels mount directly onto the motor shaft. Normally, one wheel is *coarse,* for roughing, and the other is *fine,* for finish grinding.

Figure 11-2. *Abrasive belt grinding machines. A—Wet-type abrasive belt platen grinder sprays water on the belt to cool the work. It can be quickly changed from vertical to horizontal. B—Flexible-belt grinder adapts to three-dimensional contours. (Hammond Machinery, Inc. and Baldor)*

Figure 11-4. *A bench grinder can be used for many tasks. Never operate a bench or pedestal grinder unless all safety devices are in place and in sound condition. The tool rest must be properly spaced and eye protection must be worn, even though the grinder has eye shields. (Baldor)*

A *pedestal grinder* is usually larger than the bench grinder and is equipped with a pedestal (base) fastened to the floor. See **Figure 11-5.** The *dry-type pedestal grinder* has no provisions for cooling the work during grinding other than a water container. The part is dipped into the water. A *wet-type pedestal grinder,* **Figure 11-6,** has a coolant system built into the grinder. This system keeps the wheels constantly flooded with fluid. The coolant washes away particles of loose abrasive material and metal and cools the work. Cooling prevents localized heat buildup, which can ruin tools and "burn" areas of other types of work.

Wear safety glasses and be sure the grinder eye shield is in place before doing any grinding.

The **tool rest** is provided to support the work being ground. It is recommended that the rest be adjusted to within 1/16″ (1.5 mm) of the wheel, **Figure 11-7.** This will prevent the work from being wedged between the rest and the wheel. After adjusting the rest, turn the wheel by hand to be sure there is sufficient clearance.

Do not make tool rest adjustments while the grinding wheels are revolving.

Figure 11-6. This grinder has a coolant attachment that keeps coolant dripping on the tool being sharpened. (Baldor)

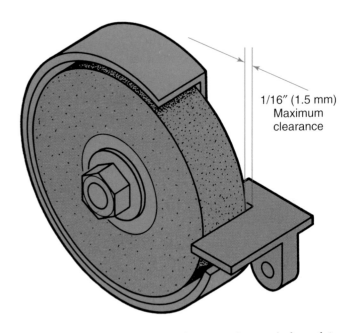

1/16″ (1.5 mm)
Maximum
clearance

Figure 11-7. The tool rest must be spaced properly for safety. Maximum safe clearance is 1/16″ (1.5 mm).

11.3 GRINDING WHEELS

Grinding wheels can be a source of danger and should be examined frequently for *concentricity* (running true), roundness, and cracks. A new wheel can be tested by suspending it on a string or wire, or holding it lightly with one finger, and tapping the side lightly with a metal rod or screwdriver handle. A solid wheel will give off a clear ringing sound. A wheel that does not give off a clear sound should be assumed to have a fault.

Figure 11-5. This pedestal grinder has a roughing wheel on the left and a finishing wheel on the right. (Baldor)

11.3.1 Wheel Dresser

The grinding wheels must run true and be balanced on the shaft. A *wheel dresser*, **Figure 11-8,** is used to true the wheel and remove any glaze that may have formed during grinding operations.

The wheel dresser is supported on the tool rest and is held firmly against the wheel with both hands. It is moved back and forth across the wheel face to remove a thin layer of stone. See **Figure 11-9.**

Figure 11-8. *A mechanical wheel dresser will true stone face. (Hammond Machinery Inc.)*

The appearance of the grinding wheel surface indicates the amount of glaze. A wheel dresser should be used to remove the glaze. **Figure 11-10** shows grinding wheel conditions.

11.3.2 Grinding Rules

To obtain maximum efficiency from a grinder, the following recommendations should be observed:
- Grind using the face of a wheel, not the sides.
- Move the work back and forth across the wheel face. This will wear the wheel evenly and prevent grooves from forming.
- Keep the wheel dressed and the tool rests properly adjusted.
- Soft metals (aluminum, brass, and copper) tend to load (clog) grinding wheels. When possible, these metals should be ground on an abrasive belt grinder.

11.4 ABRASIVE BELT AND GRINDER SAFETY

- Make sure the tool rest is properly adjusted.
- Wear goggles or a face shield when performing grinding operations, even though the machines are fitted with eye shields.
- Never attempt to operate an offhand grinding machine while your senses are impaired by medication or other substances.
- Check the machine thoroughly before using it. Lubricate the machine only as recommended by the manufacturer.

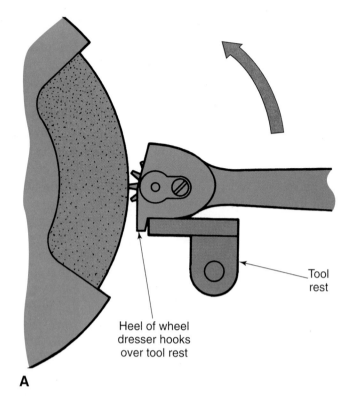

Tool rest

Heel of wheel dresser hooks over tool rest

A

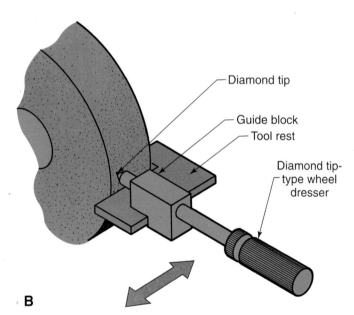

Diamond tip

Guide block

Tool rest

Diamond tip-type wheel dresser

B

Figure 11-9. *The proper way to use a mechanical wheel dresser. A—Move the tool back and forth over the face of the stone. Wear a dust mask, eye protection, and an apron when dressing wheels on grinders. B—Industrial diamonds are also used to dress and true grinding wheels. The guide block is used for grinders with slotted tool rests.*

- Check a grinding wheel for soundness before putting it on the grinder. Destroy wheels that are not sound or that have a worn center hole.
- Do not use a wheel that is glazed or loaded with metal.

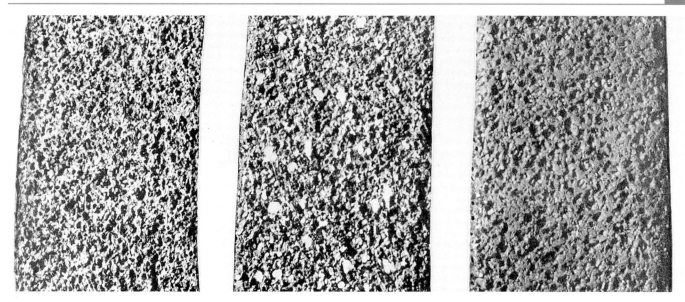

Figure 11-10. Grinding wheels in various conditions. A—Properly dressed. B—Loaded. C—Glazed. (Norton Co.)

- Be sure all wheel guards and safety devices are in place before attempting to use a grinder or abrasive belt machine.
- If the grinding operation is to be performed dry, be sure to hook up all exhaust attachments before starting.
- Stand to one side of the machine during operation. Do not stand directly in front of the wheel.
- Hold small work in a clamp or hand vise. Under no condition should work be held with a cloth.
- Avoid work pressure on the side of the grinding wheel.
- Keep your hands clear of the rotating wheel.
- Never operate a grinding wheel at speeds higher than those recommended by the manufacturer.
- Have injuries caused by rotating grinding wheels treated immediately.
- Allow the wheels or belt to stop completely before attempting to make any machine adjustments.

11.5 USING A DRY-TYPE GRINDER

After examining the grinder and making the necessary adjustments, turn on the machine. Be sure that you wear safety glasses whenever you are in the shop. Stand to one side until the grinder has reached operating speed.

Place the work on the tool rest and slowly push it against the grinding wheel. If too much pressure is applied, the work will begin to "burn" or discolor. Overheating can be minimized by dipping the work

into the water container from time to time. Care must be taken when grinding edge tools because excessive heat will "draw" (remove) the *temper* (hardness) and ruin the tool.

Keep the work moving across the wheel face to prevent the formation of grooves or ridges. Dress and retrue the wheel as neccessary for maximum efficiency.

Pieces of cloth should never be used to hold work while it is being ground. Serious injuries can result if the cloth is pulled into the wheel. Hold the work, especially small lathe cutter bits, in hand vises specially designed for that purpose, **Figure 11-11.**

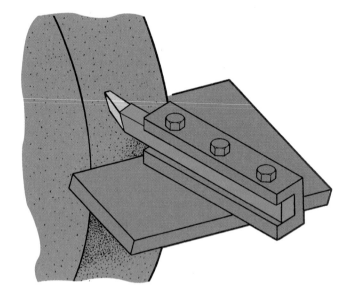

Figure 11-11. A hand vise is used to hold cutter bits while they are sharpened.

11.6 USING A WET-TYPE GRINDER

The wet-type grinder is primarily used to grind carbide-tipped tools. Since a carbide tool is often brazed onto a steel shank, both steel and carbide must be ground away when these tools are sharpened. Aluminum oxide wheels should be used to grind the steel shank and silicon carbide or diamond-impregnated wheels to grind the carbide tip.

Wet-type grinders should normally have a flat face, but a slightly crowned face should be used when grinding carbide-tipped tools, **Figure 11-12.** The crown minimizes the contact between the wheel and the work. This reduces the possibility of the tip being damaged by excessive heat.

The coolant attachment must be adjusted to

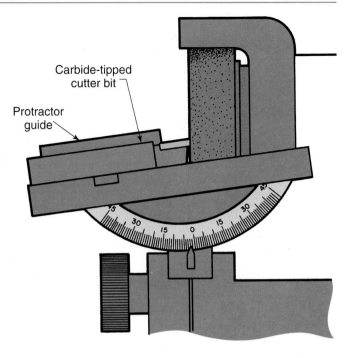

Figure 11-13. The tool rest can be adjusted to any desired angle using a table protractor and protractor guide.

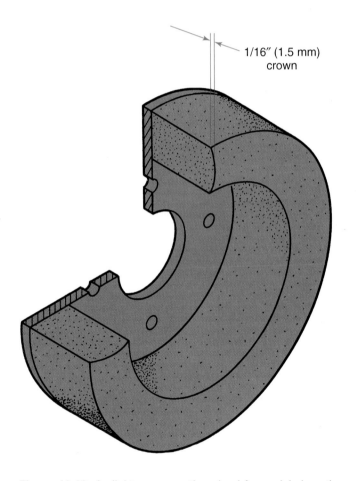

1/16″ (1.5 mm) crown

Figure 11-12. A slight crown on the wheel face minimizes the amount of contact between the work and the wheel. A crown reduces the possibility of heat destroying the carbide tip of a tool.

keep a full flow of liquid directed on the tool at all times. Adjust the tool table rest to obtain the correct clearance angle, **Figure 11-13.** A protractor guide is helpful when compound clearance angles are required.

Use the entire face of the wheel. Keep the tool in continuous motion to minimize wheel wear. Dress and retrue the wheel as needed.

11.7 PORTABLE HAND GRINDERS

Many grinding jobs, from light deburring to die-polishing operations, are done with small portable hand grinders. *Flexible shaft grinders* and *precision microgrinders,* **Figure 11-14,** are used to perform a variety of toolroom and production jobs. They can be powered by electricity or air.

A *reciprocating hand grinder,* **Figure 11-15,** is used to finish dies. Using various attachments, this tool can polish dies to a mirror finish, **Figure 11-16.**

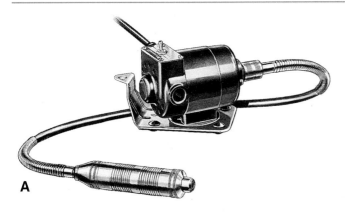

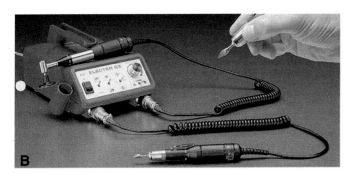

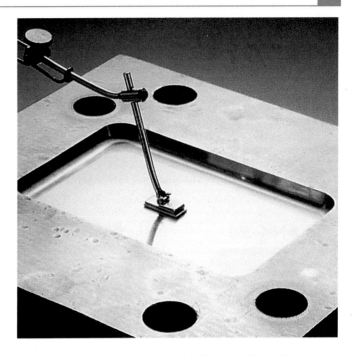

Figure 11-16. A reciprocating finishing grinder being used to polish a die casting mold. The abrasive stone can be replaced with files and hones of various shapes. (NSK America)

Figure 11-14. Portable hand grinders. A—A flexible shaft hand grinding unit is helpful for working in recessed areas on parts. B—A precision electric microgrinder. (Dumore Co. and NSK America)

TEST YOUR KNOWLEDGE

Please do not write in this text. Write your answers on a separate sheet of paper.

1. Describe the grinding operation.

2. How do abrasive belt grinders differ from abrasive wheel grinders?

3. Bench and pedestal grinders are used to do _____ grinding.

4. The grinding technique referred to in the preceding statement is so named because:
   a. It can only do external work.
   b. Work is too hard to be machined by other methods.
   c. Work is manipulated with fingers until desired shape is obtained.
   d. All of the above.
   e. None of the above.

5. Name the two types of pedestal grinders. How do they differ?

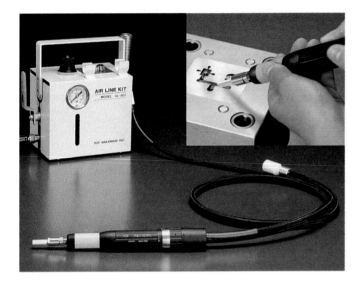

Figure 11-15. A pneumatic reciprocating finishing grinder. (NSK America)

6. The tool rest should be about _____ inches or _____ mm away from the grinding wheel or belt for safety. This prevents the possibility of work being _____ between the tool _____ and _____.

7. How can grinding wheel soundness be checked?

8. Since a grinding wheel cannot be checked each time the grinder is used, it is recommended that the operator:
    a. Not use the grinder.
    b. Check with the instructor whether the wheel is sound.
    c. Stand to one side of the grinder when using the machine.
    d. All of the above.
    e. None of the above.

9. Work will _____ if it is forced against the wheel with too much pressure.

10. Carbide-tipped tools are usually sharpened on a _____ grinder.

11. The face of the wheel on a wet-type grinder is _____ slightly. Why is this done?

12. Never mount a grinding wheel on a grinder without _____.

13. List four safety precautions to be observed when operating a grinder.

Chapter 12

Sawing and Cutoff Machines

IMPORTANT TERMS

all-hard blade	gravity feed
cold circular saw	horizontal band saw
dry abrasive cutting	raker set
flexible-back blades	three-tooth rule
friction saw	wet abrasive cutting

12.1 METAL-CUTTING SAWS

The first step in most machining jobs is to cut the stock to required length. This can be done using power saws, **Figure 12-1**.

Figure 12-1. The first step in most machining jobs is to cut the stock to the desired length. Measure the cutoff length carefully and observe all safety precautions. (DoALL Co.)

There are three principal types of metal-cutting saws, **Figure 12-2**. Reciprocating power saws use a back-and-forth (reciprocating) cutting action. The cutting is done on the backstroke. The blade is similar to that found on a hand hacksaw, only larger and heavier. Band-type power saws have a continuous blade that moves in one direction. Circular-type power saws have a round, flat blade that rotates into the work. A toothed blade, friction blade, or abrasive blade may be used, depending on the material and the operation.

12.2 RECIPROCATING POWER HACKSAW

A *reciprocating power hacksaw*, **Figure 12-3**, uses a blade that moves back and forth across the work. The blade cuts on the backstroke. There are several types of feeds available.

Positive feed produces an exact depth of cut on each stroke. The pressure on the blade varies with the number of teeth in contact with the work.

Definite pressure feed yields a pressure on the blade that is uniform regardless of the number of teeth in contact with the work. The depth of the cut varies with the number of teeth contacting the work. This condition prevails with *gravity feed*.

Feed can be adjusted to meet varying conditions. For best performance, the blade and feed must be selected to permit high-speed cutting and heavy feed pressure with minimum blade bending and breakage.

Standard reciprocating metal cutting saws are available in sizes from 6″ × 6″ (150 mm × 150 mm) to 24″ × 24″ (900 mm × 900 mm). The saws can be fitted with many accessories, including quick-acting vises, power stock feed, power clamping of work, and automatic cycling of the cutting operation. The latter moves the work out the required distance, clamps it, and makes the cut automatically. The cycle is repeated upon completion of the cut.

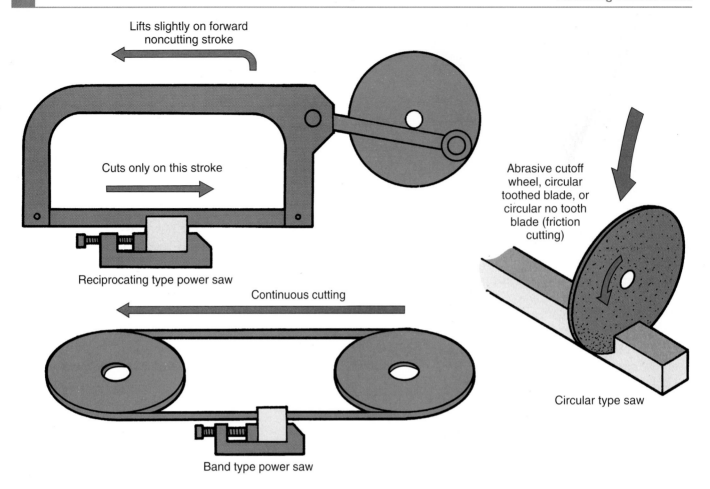

Figure 12-2. *The three principal types of cutoff saws.*

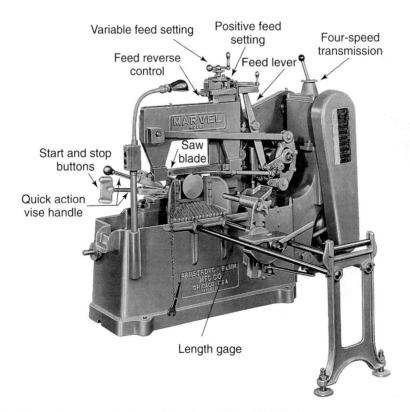

Figure 12-3. *An industrial reciprocating power hacksaw. (Armstrong-Blum Mfg. Co.)*

High-speed cutting requires use of a coolant. Coolant reduces friction, increases blade life, and prevents chip-clogged teeth. Cast iron and some brass alloys, unlike most materials, do not require coolant.

A swivel vise permits angular cuts to be made quickly. See **Figure 12-4.**

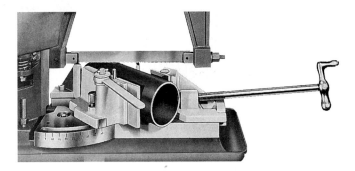

Figure 12-4. A swivel vise permits angular cuts.

12.2.1 Selecting a Power Hacksaw Blade

Proper blade selection is important. Use the *three-tooth rule*—at least three teeth must be in contact with the work. Large sections and soft materials require a coarse-tooth blade. Small or thin work and hard materials require a fine-tooth blade.

For best cutting action, apply heavy feed pressure on hard materials and large work. Use light feed pressure on soft materials and work with small cross sections, **Figure 12-5.**

Blades are made in two principal types: flexible-back and all-hard. The choice depends upon use.

Flexible-back blades should be used where safety requirements demand a shatterproof blade. These blades should also be used for cutting odd-shaped work if there is a possibility of the work coming loose in the vise.

For a majority of cutting jobs, the *all-hard blade* is best for straight, accurate cutting under a variety of conditions.

When starting a cut with an all-hard blade, be sure the blade does not drop on the work when cutting starts. If it falls, the blade could shatter and flying pieces cause injuries.

Blades are also made from tungsten and molybdenum steels, and with tungsten carbide teeth on steel alloy backs. The following "rule-of-thumb" can be followed for selecting the correct blade:

- Use a 4-tooth blade for cutting large sections or readily machined metals.
- Use a 6-tooth blade for cutting harder alloys and miscellaneous cutting.
- Use 10- and 14-tooth blades primarily on light duty machines where work is limited to small sections requiring moderate or light feed pressure.

12.2.2 Mounting a Power Hacksaw Blade

The blade must be mounted to cut on the power (back) stroke. The blade must also lie perfectly flat against the mounting plates, **Figure 12-6.** If long life and accurate cuts are to be achieved, the blade must be properly tensioned.

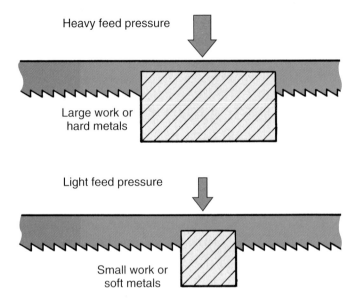

Figure 12-5. Apply heavy feed pressure on hard metals and large work. Use light pressure on soft metals and work with small cross sections.

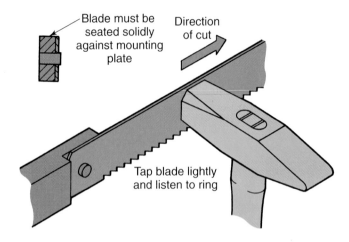

Figure 12-6. The blade must be adjusted to cut on the back stroke. Make sure it is perfectly flat against the mounting plates before tensioning. Tighten the blade until a low musical ring is heard when the blade is tapped with a small hammer. Since blades have a tendency to stretch slightly after making a few cuts, tension should be checked and, if necessary, adjusted.

Many techniques have been developed for properly mounting and tensioning blades. Use a torque wrench and consult the manufacturer's literature. If the information (proper torque for a given blade on a given machine) is not available, the following methods can be used:

- Tighten the blade until a low musical ring is heard when the blade is tapped lightly. A high-pitched tone indicates that the blade is too tight. A dull thud means the blade is too loose.
- The shape of the blade pin hole can serve as an indicator of whether the blade is tensioned properly. When proper tension is achieved, the pin holes will become slightly elongated, **Figure 12-7**.

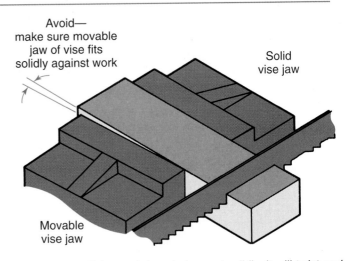

Figure 12-8. *If the work is not clamped solidly, it will twist and the blade will bind and be ruined in the first few seconds of use.*

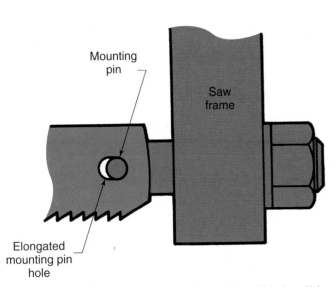

Figure 12-7. *Pin holes on a properly tensioned blade will be slightly elongated, rather than round.*

The blade will become more firmly seated after the first few cuts and will stretch slightly. The blade will require *retensioning* (retightening) before further cutting can be done.

12.2.3 Cutting with a Power Hacksaw

Measure off the distance to be cut. Allow ample material for facing if the work order does not specify the length of cut. Mark the stock and mount the work firmly on the machine, **Figure 12-8**.

If several sections are to be cut, use a *stop gage*, **Figure 12-9**. Apply an ample supply of coolant if the machine has a built-in coolant system.

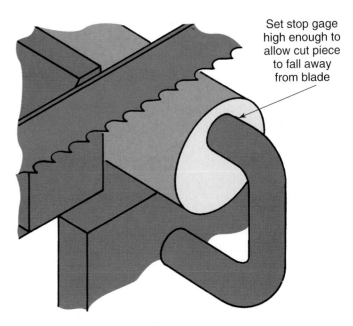

Figure 12-9. *A stop gage is used when several pieces of the same length must be cut. Set it high to permit the work to fall free when completely cut.*

12.3 POWER BAND SAW

The *horizontal band saw*, **Figure 12-10**, is frequently referred to as the cutoff machine. It offers three advantages over the reciprocating hacksaw:

- *Greater precision*—The blade on a band saw can be guided more accurately than the blade on the reciprocating power saw. It is common practice to cut directly "on the line" when band sawing, because finer blades can be used.

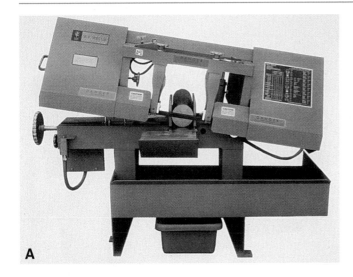

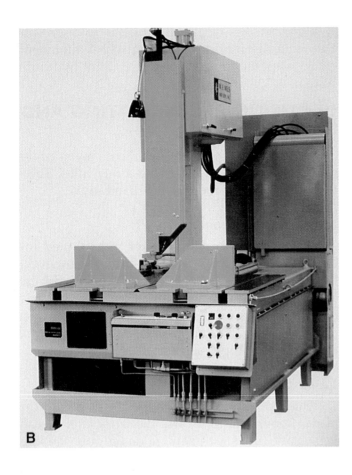

Figure 12-10. Band sawing machines. A—Typical band cutoff saw with built-in coolant system, manually controlled blade tension, and automatic end-of-cut shutoff. B—Band cutoff machines are also available with tilting frame. The frame is capable of tilting up to 45°. (W. F. Wells)

- *Faster speed*—The long, continuous blade moves in only one direction, so cutting is also continuous. The blade can run at much higher speeds because it rapidly dissipates the cutting heat.
- *Less waste*—The small cross section of the band saw blade makes smaller and fewer chips than the thicker reciprocating blade, **Figure 12-11**.

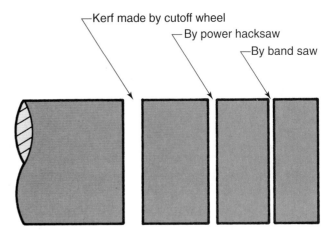

Figure 12-11. Differences in the amount of metal converted to chips (waste) by each cutoff machine.

12.3.1 Selecting a Band Saw Blade

Band saw blades are made with raker teeth or wavy teeth, **Figure 12-12**. Most manufacturers also make variations of these sets. The *raker set* is preferred for general use.

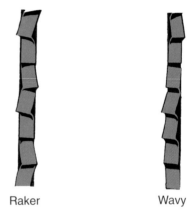

Raker   Wavy

Figure 12-12. Saw blades commonly have raker or wavy teeth. Raker teeth are preferred for general use, cutting large solid sections, and cutting thick plate.

Tooth pattern determines the efficiency of a blade in various materials. The *standard tooth* blade pattern is best suited for cutting most ferrous metals. A *skip tooth* blade pattern is preferred for cutting aluminum, magnesium, copper, and soft brasses. The *hook tooth* blade pattern also is recommended for most nonferrous metallic materials. See **Figure 12-13**.

For best results, consult the blade manufacturer's chart or manual for the proper blade characteristics (set, pattern, and number of teeth per inch) for the particular material being cut.

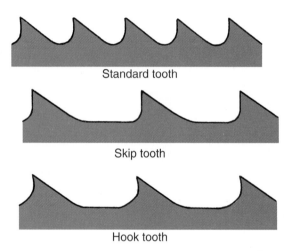

Figure 12-13. *Standard tooth blades, with rounded gullets, are usually best for most ferrous metals, hard bronzes, and hard brasses. Skip tooth blades provide for more chip clearance without weakening the blade body. They are recommended for cutting aluminum, magnesium, copper, and soft brasses. Hook tooth blades offer two advantages over skip tooth blades— easier feeding and less "gumming up."*

12.3.2 Installing a Band Saw Blade

If the saw is to work at top efficiency, the blade must be installed carefully. Wear heavy leather gloves to protect your hands when installing a band saw blade.

Blade guides should be adjusted to provide adequate support, **Figure 12-14**. Proper blade support is required to cut true and square with the holding device.

Follow the manufacturer's instructions for adjusting blade tension. Improper blade tension ruins blades and can cause premature failure of bearings in the drive and idler wheels.

Cutting problems encountered with the band saw are similar to those of the reciprocating hack saw. Most problems are caused by poor machine condition. They can be kept to a minimum if a maintenance program is followed on a regular basis.

Blade guides

Figure 12-14. *Adjust blade guides to provide adequate blade support; otherwise, blade will not cut true. (W.F. Wells)*

This typically includes checking wheel alignment, guide alignment, feed pressure, and hydraulic systems.

12.4 USING RECIPROCATING AND BAND SAWS

Most sawing problems can be prevented by careful planning and observing a few rules. These apply to both reciprocating and band saws.

12.4.1 Blades Breaking

Blades are normally broken when they are dropped on the work. A loose blade or excessive feed can cause the blade to fracture. Loose work can also cause blade damage, as will making a cut on a corner or sharp edge where the three-tooth rule is not observed. Broken blades can normally be avoided with proper machine setup.

12.4.2 Crooked Cutting

This problem is usually the result of a worn blade. Remember to reverse the work after replacing a blade, and start a new cut on the opposite side. See **Figure 12-15**. A loose blade or a blade rubbing on a clamping fixture will cause the same problem. It can also be caused by excessive blade pressure on the work or by worn saw guides.

12.4.3 Blade Pin Holes Breaking Out

This reciprocating blade problem can be caused by dirty mounting plates or too much tension on the blade. Worn mounting plates can cause a blade to twist and strain in such a way that the pin hole will break out.

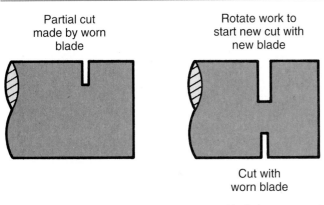

Partial cut made by worn blade

Rotate work to start new cut with new blade

Cut with worn blade

Figure 12-15. Never attempt to start a new blade in a cut made by a worn blade. Reverse the work and start another cut on the opposite side. Cut through to the old cut.

12.4.4 Premature Blade Tooth Wear

When this problem occurs, the teeth become rounded and dull quickly. Insufficient feed pressure (indicated by light, powdery chips) is one of the major causes of this condition. Excessive pressure (indicated by burned chips) causes the same problem.

Insufficient pressure can be corrected by increasing cutting pressure until a full curled chip is produced. If too much pressure is the culprit, reduce feed pressure until a full curled chip is formed.

Lack of coolant or a poorly adjusted machine can also cause rapid wear. Correct by following the manufacturer's recommendations.

12.4.5 Teeth Strip Off

This failure results when the teeth snap off the blade. Starting a cut on a sharp corner is a major cause of this problem. A machine setup with a flat starting surface will greatly reduce tooth stripping. Be sure the work is clamped securely; loose work can also cause the teeth to strip, **Figure 12-16.**

Check the manufacturer's chart to determine the proper blade for the job to be done. A blade with teeth too fine will clog (load) and jam, causing the teeth to shear off. A blade that is too coarse (less than three teeth cutting) will cause the same problem. Make sure the blade is properly mounted and cutting on the power stroke.

12.5 CIRCULAR METAL-CUTTING SAWS

Metal-cutting circular saws are found in many areas of metalworking. Primarily production machines, these saws are divided into three classifications:

- Abrasive cutoff saw.
- Cold circular saw.
- Friction saw.

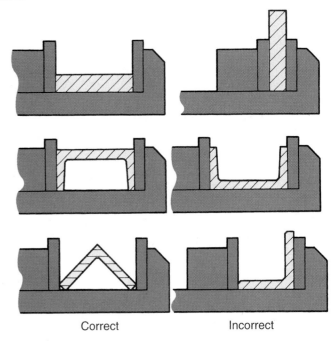

Correct   Incorrect

Figure 12-16. Recommended ways to hold sharp-cornered work for cutting. A carefully planned setup will ensure that at least three teeth will be cutting, greatly extending blade life.

An *abrasive cutoff saw,* **Figure 12-17,** cuts material using a rapidly revolving, thin abrasive wheel. Most materials—glass, ceramics, and metals—can be cut to close tolerances. Hardened steel does not require annealing to be cut. Special heat-resistant abrasive wheels are available for high-speed cutoff of hot stock.

Abrasive cutting falls into two classifications, dry and wet. *Wet abrasive cutting,* while not quite as rapid as dry cutting in some applications, produces a finer surface finish and permits cutting to close tolerances. The cuts are burn-free and have few or no burrs. *Dry abrasive cutting* does not use a coolant and is used for rapid, less-critical cutting.

A *cold circular saw,* **Figure 12-18,** makes use of a circular, toothed blade capable of producing very accurate cuts. Large cold circular saws can sever round metal stock up to 27″ (675 mm) in diameter.

A *friction saw* blade may or may not have teeth. The saw operates at very high speeds (20,000 surface feet per minute or 6000 m per minute) and actually melts its way through the metal.

Teeth are used primarily to carry oxygen to the cutting area. These machines find many applications in steel mills to cut red-hot *billets* (sections of semifinished steel).

12.6 POWER SAW SAFETY

- Never attempt to operate a sawing machine while your senses are impaired by medication or other substances.

Figure 12-17. An industrial abrasive cutoff saw. (Everett Industries, Inc.)

- Get help when you are lifting and cutting heavy material.
- Clean oil, grease, and coolant from the floor around the work area.
- Burrs on cut pieces are sharp. Use special care when handling pieces with burrs.
- Follow the manufacturer's instructions for tensioning a blade. Too much tension can shatter the blade.
- Handle band saw blades with extreme care. They are long and springy and can uncoil suddenly.
- Be sure the work is mounted solidly before starting a cut.
- Be sure all guards are in place before using the saw.
- Always wear a dust mask and full face shield when cutting stock with a dry-type abrasive cutoff saw.
- Avoid standing directly in line with the blade when operating a circular cutoff saw.
- Use a brush to clean chips from the machine. Do not use your hands. Wait for the machine to come to a complete stop before cleaning.

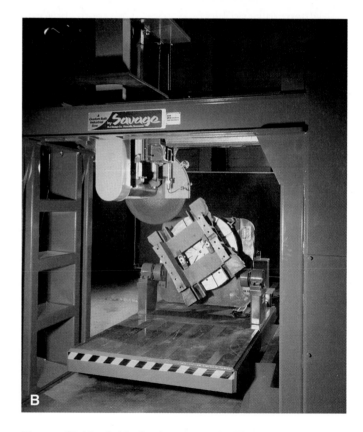

Figure 12-18. Cold circular saws. A—This automated cold circular saw can accept a piece up to 2500 pounds. A laser guide light marks the position of the cut. B—This machine can be fitted with carbide-tipped blades or abrasive disks. (W.J. Savage Co.)

- Keep your hands out of the way of moving parts.
- Stop the machine before making adjustments.
- Have all cuts, bruises, and scratches, even minor ones, treated immediately.

TEST YOUR KNOWLEDGE

Please do not write in the text. Write your answers on a separate sheet of paper.

1. List the three basic types of metal-cutting saws.

2. The _____ type saw has a back-and-forth cutting action. However, it only cuts on the _____ stroke.

3. What is the "three-tooth rule" for sawing?

4. When using a power sawing machine, with which materials should you *not* use coolant?

5. Hacksaw blades are manufactured in two principal types. Name them.

6. The following "rule-of-thumb" should be followed for selecting the correct blade:
   a. _____ teeth per inch for cutting large sections or readily machined materials.
   b. _____ teeth per inch for cutting harder alloys and miscellaneous cutting.
   c. _____ teeth per inch for cutting on the majority of light-duty machines, where work is limited to small sections and moderate to light feed pressures.

7. List three methods used to put proper tension on a power hacksaw blade.

8. When is a stop gage used?

9. What three advantages does the continuous band sawing machine offer over other types of power saws?

10. Band saw blades are made with two types of teeth. Name them.

11. The tooth pattern of a blade determines the efficiency of a blade in various materials.
    a. The _____ tooth is best suited for cutting most ferrous metals.
    b. The _____ tooth pattern is preferred for cutting aluminum, magnesium, copper, and soft brass.
    c. The _____ tooth is also recommended for most nonferrous metallic materials.

12. List the three types of circular metal-cutting saws.

13. List five safety precautions to be observed when operating a power saw.

Grooving and parting are operations frequently performed on the lathe. This grooving tool uses a replaceable insert that clamps into a special thin holder for performing deep face grooving. The insert can be reversed, allowing a quick change to a new cutting edge when one becomes dull or damaged. Inserts for grooves ranging from 2 mm to 6 mm in width are available. (Iscar Metals, Inc.)

Chapter **13**

The Lathe

LEARNING OBJECTIVES

After studying this chapter, you will be able to:
- ○ Describe how a lathe operates.
- ○ Identify the various parts of a lathe.
- ○ Safely set up and operate a lathe using various work-holding devices.
- ○ Sharpen lathe cutting tools.

IMPORTANT TERMS

compound rest
cross-slide
depth of cut
facing
headstock
indexable insert cutting
   tools
plain turning
single-point cutting tool
tailstock
tool post

The *lathe* operates on the principle of the work being rotated against the edge of a cutting tool, **Figure 13-1**. It is one of the oldest and most important machine tools. The cutting tool is controllable and can be moved lengthwise on the lathe bed and across the revolving work at any desired angle. See **Figure 13-2**.

13.1 LATHE SIZE

Lathe size is determined by the *swing* and the *length of the bed*, **Figure 13-3**. The *swing* indicates the largest diameter that can be turned over the *ways* (the flat or V-shaped bearing surface that aligns and guides the movable part of the machine). *Bed length* is the entire length of the ways.

Bed length must not be mistaken for the maximum length of the work that can be turned *between centers*. The longest piece that can be turned is equal to the length of the bed minus the distance taken up by the headstock and tailstock. Refer to measurement B in **Figure 13-3**.

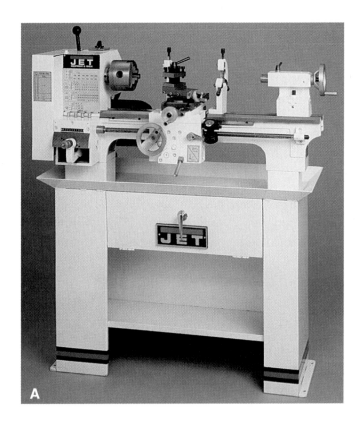

A

B

Figure 13-1. *Metal-cutting lathes. A—The basic lathe. All controls are operated manually. Most machinists begin their training on this type of lathe. (Jet Equipment & Tools) B—Metal-cutting lathes are some of the most versatile machine tools made. There are many variations of the basic lathe for differing applications. This one is CNC controlled. (Harrison/REM Sales Inc.)*

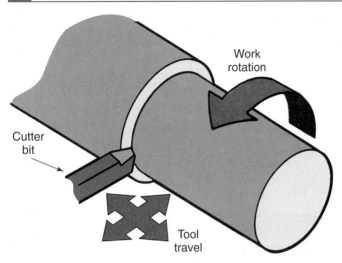

Figure 13-2. Operating principle of the lathe. The cutting tool is fed into the revolving work.

As an example, consider the capacity and clearance of a modern 13″ × 6′ (325 mm × 1800 mm) lathe:

Swing over bed:	13″ (325 mm)
Swing over cross-slide:	8 3/4″ (218 mm)
Bed length:	72″ (1800 mm)
Distance between centers:	50″ (1240 mm)

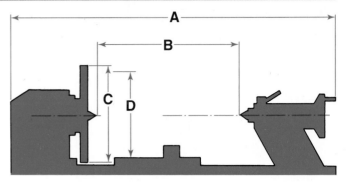

Figure 13-3. Lathe measurements. A—Length of bed. B—Distance between centers. C—Diameter of work that can be turned over the ways. D—Diameter of work that can be turned over the cross-slide.

13.2 MAJOR PARTS OF A LATHE

The chief function of any lathe, no matter how complex it may appear to be, is to rotate the work against a controllable cutting tool. Each of the lathe parts in **Figure 13-4** can be assigned to one of these three categories:

- Driving the lathe.
- Holding and rotating the work.
- Holding, moving, and guiding the cutting tool.

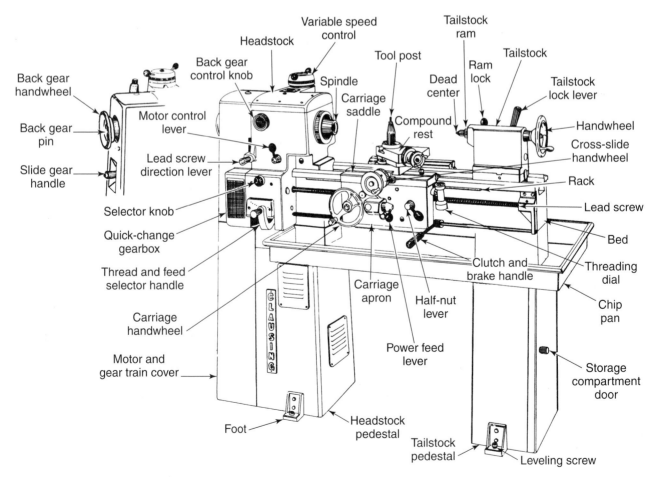

Figure 13-4. The engine lathe and its major parts. (Clausing Industrial, Inc.)

13.2.1 Driving the lathe

Power is transmitted to the drive mechanisms by a belt drive and/or gear train. Spindle speed can be varied by:

- Shifting to a different gear ratio, **Figure 13-5**.
- Adjusting a split pulley to another position, **Figure 13-6.**
- Moving the drive belt to another pulley ratio (seldom used today).
- Controlling the speed hydraulically.

Slower speeds with greater power are obtained on some machines by engaging a *back gear.* See **Figure 13-7.** To avoid damaging the lathe's drive system, *do not engage the back gear while the spindle is rotating.*

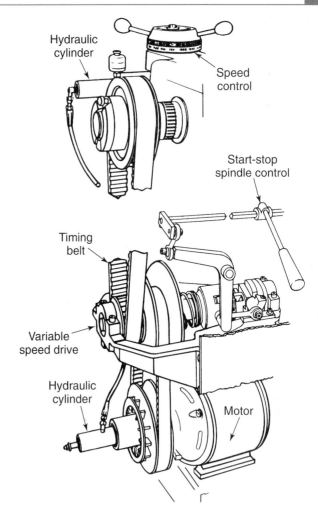

Figure 13-6. This split pulley is hydraulically actuated from the top of the machine by a speed control. A split pulley is used to control spindle speeds on many lathes. (Clausing Industrial, Inc.)

Figure 13-5. Spindle speed control. A—Speed is increased or decreased by shifting to different gear ratios. B—On this machine, spindle speed is controlled by an automatic transmission. Desired speed is dialed in. (LeBlond Makino Machine Tool Co.)

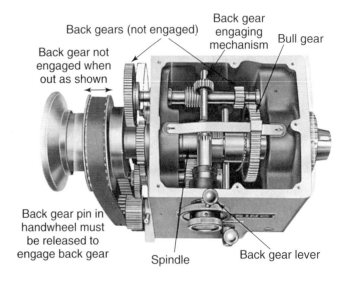

Figure 13-7. The back gear mechanism is clearly visible in this view of a lathe headstock. Direct drive is disengaged before back gear slides into position. (Clausing Industrial, Inc.)

13.2.2 Holding and rotating the work

The *headstock* contains the spindle to which the various work-holding attachments are fitted, **Figure 13-8**. The *spindle* revolves in heavy-duty bearings and is rotated by belts, gears, or a combination of the two. The front of the hollow spindle is tapered internally to receive tools and attachments with taper shanks, **Figure 13-9**. The hole through the spindle permits long stock to be turned without dangerous overhang. It also allows use of a *knock-out bar* to remove taper-shank tools.

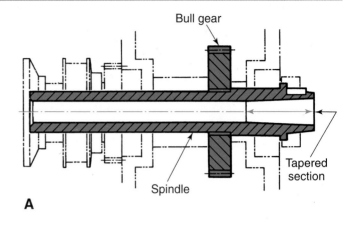

Figure 13-8. *The headstock is the driving end of lathe. (Clausing Industrial, Inc.)*

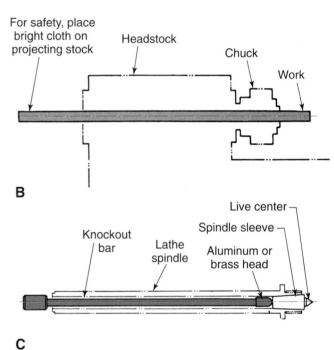

Figure 13-9. *Lathe spindle. A— Hollow construction of the spindle allows long stock to be turned without dangerous overhang. B—To prevent accidents that could cause injury, some sort of flag should be tied to the portion of stock that projects from the rear of the spindle. C—A knockout bar is used to tap tapered shank lathe accessories out of the spindle.*

On the front end, a spindle may be threaded externally or fitted with one of two types of tapered spindle noses to receive work-holding attachments. . See **Figure 13-10**. A *threaded spindle* nose is seldom used on modern lathes. It permits mounting an attachment by screwing it directly on the threads until it seats on the spindle flange.

The *cam-lock spindle* nose has a short taper that fits into a tapered recess on the back of the work-holding attachment. A series of cam locking studs, located on the back of the attachment, are inserted into holes in the spindle nose. The studs are locked by tightening the cams located around the spindle nose.

A *long taper key spindle* has a protruding long taper and key that fits into a corresponding taper and keyway in the back of the work-holding device.

To mount a work-holding device (a chuck or faceplate), the spindle is rotated until the key is on top. See **Figure 13-10C**. The keyway in the back of the work-holding device is slid over the key to support the device until the threaded spindle collar can be engaged with the threaded section of the device, then tightened.

Figure 13-10. *Spindle noses. A—Threaded spindle nose is seldom used today. (South Bend Lathe Corps.) B—A cam-lock spindle nose. (Clausing Industrial, Inc.) C—A long taper key spindle nose. (Clausing Industrial, Inc.)*

Note: Attachment points on the spindle nose and work-holding attachment must be cleaned carefully before mounting the device.

Work is held in the lathe by a chuck, faceplate, or collet, or by mounting it between centers. These attachments will be described in detail later in this chapter.

The outer end of the work is often supported by the lathe's *tailstock*, **Figure 13-11**. The tailstock can be adjusted along the ways to accommodate different lengths of work.

The tailstock is used to mount the "dead" center, or can be fitted with tools for drilling, reaming, and threading. It can also be offset for taper turning.

The tailstock is locked onto the ways by tightening a **clamp bolt nut** or **binding lever**. The tailstock spindle is positioned by rotating the handwheel and can be locked in position by tightening a binding lever.

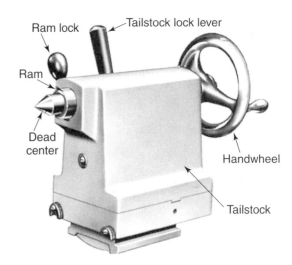

Figure 13-11 *Parts of the tailstock. (Clausing Industrial, Inc.)*

13.2.3 Holding, moving, and guiding the cutting tool

The *bed*, **Figure 13-12**, is the foundation of a lathe. All other parts are fitted to it. Ways are integral with the bed. The V-shaped rails maintain precise alignment of the headstock and tailstock, and guide the travel of the carriage, **Figure 13-13**.

The *carriage* controls and supports the cutting tool, and is composed of a number of parts.

- The *saddle* is fitted to the ways and slides along them.

Figure 13-12. The bed is the foundation of the lathe.

- The *apron* contains a drive mechanism to move the carriage along the ways, using hand or power feed.
- The *cross-slide* permits *transverse* tool movement (movement toward or away from the operator, at a right angle to the axis of the lathe).
- The *compound rest* permits angular tool movement.
- The *tool post* is used to mount the cutting tool.

Power is transmitted to the carriage through the *feed mechanism,* which is located at the left (headstock) end of the lathe, **Figure 13-14.** Power is transmitted through a train of gears to the *quick-change gearbox.* This device, **Figure 13-15,** regulates the amount of tool travel per revolution of the spindle. The gear train also contains gears for reversing tool travel.

The quick change gearbox is located between the spindle and the lead screw. It contains gears of various ratios that make it possible to machine different pitches of screw threads without physically removing and replacing gears. *Longitudinal* (back-and-forth) travel and cross (in-and-out) travel is controlled in the same manner.

Figure 13-13. V-shaped ways guide carriage. The cutting tool is mounted on the carriage.

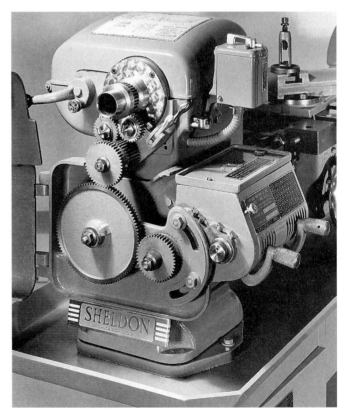

Figure 13-14. A feed mechanism is used to move the carriage along the ways.

Figure 13-15. A quick-change gearbox for cutting both inch and metric size threads. (Clausing Industrial, Inc.)

An *index plate* provides instructions on how to set the lathe shift levers for various threadcutting and feed combinations, **Figure 13-16.** It is located on the face of the gearbox. The *large numbers* on the index plate at left indicate the number of threads that can be cut per inch or pitch of metric threads. The *smaller figures* indicate the carriage longitudinal movement, in thousandths of an inch or in mm for each spindle revolution.

The *lead screw* transmits power to the carriage through a gearing and clutch arrangement in the carriage apron, **Figure 13-17.** *Feed change levers* on the apron control the operation of power longitudinal feed and power cross-feed, **Figure 13-18.**

When the feed change lever is placed in neutral, the *half-nuts* may be engaged for thread cutting. The gear arrangement makes it possible to engage power feed and half-nuts simultaneously. The half-nuts are engaged *only* for thread cutting and are *not* used as an "automatic feed" for regular turning.

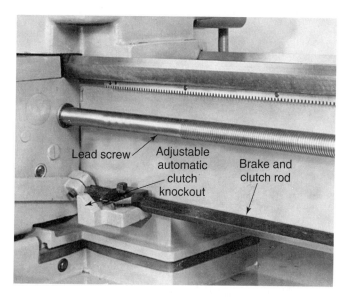

Figure 13-17. The lead screw.

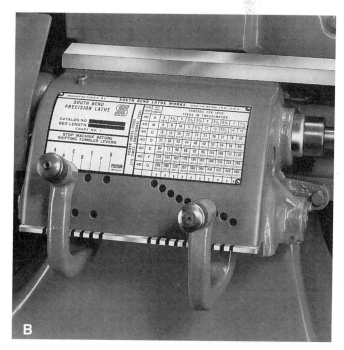

Figure 13-16. Index plates. A—An index plate showing lever positions for inch and metric feeds and threads. (Clausing Industrial, Inc.) B—An index plate for inch feeds and threads. (South Bend Lathe Corp.)

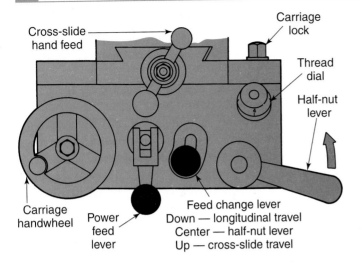

Figure 13-18. Power feed functions are controlled by the feed change levers on the apron. The half-nut lever is engaged only for thread cutting, sometimes called "thread chasing."

Figure 13-19. The carriage should be checked for binding by moving it along the ways. Correct any binding before attempting to use lathe. (Clausing Industrial, Inc.)

13.3 PREPARING LATHE FOR OPERATION

Before an aircraft is permitted to take off, the pilot and crew must go through a checkout procedure to determine whether the engines, controls, and safety features are in first-class operating condition. The same applies to the operation of a machine tool such as a lathe. The operator should inspect the machine for safe and proper operation. The "checkout procedure" for the lathe should include the following actions:

- Clean and lubricate the machine. Use lubricant types and grades specified by the manufacturer. Many recommend a specific lubricating sequence to reduce any possibility of missing a vital lubrication point.
- Be sure all guards are in position and locked in place.
- Turn the spindle over by hand to be sure it is *not* locked nor engaged in back gear (unless you intend to use back gear).
- Move the carriage along the ways, **Figure 13-19**. There should be no binding.
- Check cross-slide movement. If there is too much play, adjust the gibs. See **Figure 13-20**.
- Mount the desired work-holding attachment. Clean the spindle nose with a soft brush. A threaded nose spindle should have a drop of lubricating oil applied before the chuck or faceplate is attached.
- Adjust the drive mechanism for the desired speed and feed.
- If the tailstock is used, check it for proper alignment, **Figure 13-21**.

Figure 13-20. If there is too much play in the cross-slide, adjust gibs according to instructions in manufacturer's handbook. (Clausing Industrial, Inc.)

Figure 13-21. Witness lines on tailstock indicate whether the tailstock is aligned properly with the headstock.

- Clamp the cutter bit into an appropriate tool-holder and mount it in the tool post. Do not permit excessive compound rest overhang, since this often causes tool "chatter" and results in a poorly machined surface, **Figure 13-22**.
- Mount the work. Check for adequate clearance between the work and the various machine parts.

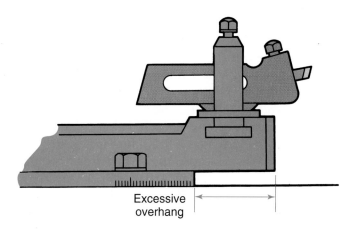

Excessive overhang

Figure 13-22. Excessive overhang of the compound rest usually causes tool "chatter," resulting in a surface that is poorly machined.

In addition to the above procedures, the operator must take some safety precautions. Sleeves should be rolled up and all jewelry removed before beginning to use the lathe.

A lathe board will aid in organizing and holding the tools and measuring instruments needed for the job. See **Figure 13-23**. Loose tools must never be placed on the lathe ways or carriage.

Figure 13-23. A lathe board is easily made in the school. It keeps tools within easy reach and away from chips and turnings.

13.4 CLEANING THE LATHE

To maintain the accuracy built into a lathe, it must be thoroughly cleaned after each work period. Use a 2" paint brush (not a *dust brush*) to remove the accumulated chips.

Lathe chips are sharp; do not remove them with your hands. *Never* use an air hose to remove chips. The flying particles could injure you or others.

Wipe all painted surfaces with a soft cloth. To complete the job, move the tailstock to the extreme right end of the ways. Use a soft cloth to remove any remaining chips, oil, and dirt from the machined surfaces.

To prevent rust until the next time the machine is used, apply a light coating of machine oil to all machined surfaces. The lead screw occasionally needs cleaning. To do so, adjust the screw to rotate at a slow speed, then place a heavy cord around it and start the machine, **Figure 13-24**. With the lead screw revolving, permit the cord to feed along the thread. Hold the cord just tightly enough to remove the accumulated dirt. *Never* wrap the cord around your hand. The cord could catch and cause serious injury.

13.5 LATHE SAFETY

- Do not attempt to operate a lathe until you know the proper procedures and have been checked out on its safe operation by your instructor.

Figure 13-24. This machinist is cleaning lead screw with a piece of cord. The lead screw should rotate at low speed. Do not wrap the cord around your fingers or hands.

- Never attempt to operate a lathe while your senses are impaired by medication or other substances.
- Dress appropriately! Remove any necklaces or other dangling jewelry, wristwatch, or rings. Secure any loose-fitting clothing and roll up long sleeves. Wear an apron or a properly fitted shop coat. Safety glasses are a must!
- Clamp all work solidly. Use the correct size tool and work-holding device for the job. Get help when handling large sections of metal and heavy chucks and attachments.
- Check work frequently when it is being machined between centers. A workpiece expands as it heats up from friction and could damage the tailstock center.
- Be sure all guards are in place before attempting to operate the machine. Never attempt to defeat or bypass a safety switch.
- Turn the faceplate or chuck by hand to be sure there is no binding or danger of the work striking any part of the lathe.
- Keep the machine clear of tools, and always stop the machine before making measurements and adjustments.
- Metal chips are sharp and can cause severe cuts. Do not try to remove them with your hands when they become "stringy" and build up on the tool post. Stop the machine and remove them with pliers, **Figure 13-25**.
- Do not permit small-diameter work to project too far from the chuck without support from the tailstock. Without support, the work will be tapered, or worse, spring up over the cutting tool and/or break. See **Figure 13-26**.

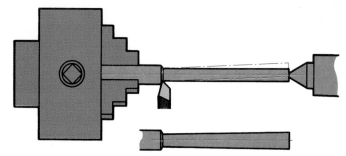

Figure 13-26. If a small-diameter workpiece is not properly supported by a tailstock center, it will spring away from cutting tool and be machined on a slight taper.

- Do not run the cutting tool into the chuck or dog. Check any readjustment of the work or tool to make sure there is ample clearance when the cutter has been moved leftward to the farthest point that will be machined.
- Stop the machine before attempting to wipe down its surface, so the cloth doesn't become caught on rotating parts. When knurling, keep the coolant brush clear of the work.
- Before repositioning or removing work from the lathe, move the cutting tool clear of the work area. This will prevent accidental cuts on your hands and arms from the cutter bit.
- Avoid talking to anyone while running a lathe! Do not permit anyone to fool around with the machine while you are operating it. You are the only one who should turn the machine on or off, or make any adjustments.
- If the lathe has a threaded spindle nose, never attempt to run the chuck on or off the spindle using power. It is also dangerous practice to stop such a lathe by reversing the direction of rotation. The chuck could spin off and cause serious injury to you.
- Before engaging the half-nuts or automatic feed, you should always be aware of the direction of travel and speed of the carriage.
- Always remove the key from the chuck. Make it a habit to never let go of the key until it is out of the chuck and clear of the work area.
- Tools must not be placed on the lathe ways. Use a tool board or place them on the lathe tray, **Figure 13-27**.
- When doing filing on a lathe, make sure the file has a securely fitting handle.
- If any odd sounding noise or vibration develops during lathe operation, stop the machine immediately. If you cannot locate the trouble, get help from your instructor. Do not operate the machine until the trouble has been corrected.

Figure 13-25. To avoid injury, always remove stringy chips with pliers; never with your hands.

Figure 13-27. *Keep tools on a lathe tray; never on the machine ways or in the chip pan.*

- Remove sharp edges and burrs from the workpiece before dismounting it from the machine. Burrs and sharp edges can cause painful cuts.
- Use care when cleaning the lathe. Chips sometimes stick in recesses. Remove them with a paintbrush or wooden stick, not a dust brush. Never clean a machine tool with compressed air.

13.6 CUTTING TOOLS AND TOOL HOLDERS

To operate a lathe efficiently, the machinist must have a thorough knowledge of cutting tools and know how they must be shaped to machine various materials. The *cutting tool* is held in contact with the revolving work to remove material from the work. In most applications, you will be using a *single-point cutting tool* of high-speed steel (HSS).

The square cutter bit body is inserted in a lathe *toolholder*, **Figure 13-28.** Toolholders are made in straight, right-hand, and left-hand models. To tell the difference between right-hand and left-hand toolholders, hold the head of the tool in your hand and note the direction the shank points. The shank of the right-hand holder points to the right, the left-hand toolholders points to the left. A *turret holder* may also be utilized. Turret holders typically have four cutter bits. A bit can be changed by loosening the lock (handle) and pivoting the holder so the new bit is in cutting position, then locking it in place.

13.6.1 High-speed steel cutting tool shapes

Figure 13-29 shows the parts of the cutter bit, and the correct terminology for those parts.

To get best performance, the bit must have a keen, properly shaped cutting edge. The shape

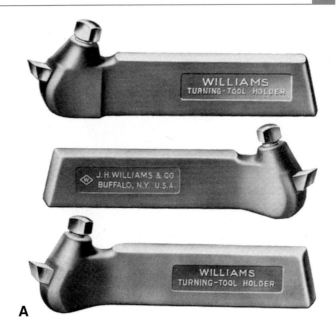

A

B

Figure 13-28. *Toolholders. A—Straight, left-hand, and right-hand toolholders. (J.H.Williams and Co.) B—A turret-type toolholder with four cutter bits. (ENCO Mfg. Co.)*

depends on the type of work, roughing or finishing, and on the metal to be machined. Most cutter bits are ground to cut in one only direction (left or right). The exception is the round-nose tool, which can cut in either direction. Some cutting tools used for general purpose turning are shown in **Figure 13-30**.

13.6.2 Roughing tools

The deep cuts made to remove considerable material from a workpiece are called *roughing cuts*. Roughing tools have a *tool shape* (shape of cutting tip) that consists of a straight cutting edge with a

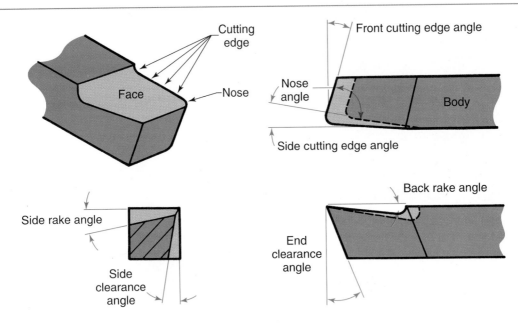

Figure 13-29. Cutter bit nomenclature.

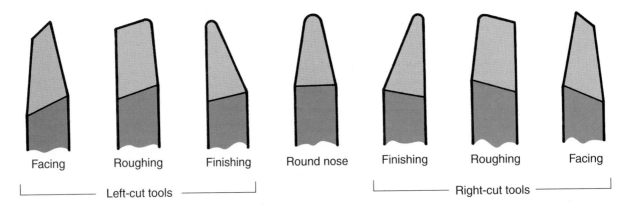

Figure 13-30. Standard HSS cutting tool shapes.

small rounded nose. This shape permits deep cuts at heavy feeds. The slight *side relief* provides ample support to the cutting edge.

The *left-cut* roughing tool cuts most efficiently when it travels from left to right. The *right-cut* roughing tool operates just the opposite, right to left. See **Figure 13-31A**.

13.6.3 Finishing tools

The nose of a *finishing tool* is more rounded than the nose of the roughing tool. See **Figure 13-31B**. If the cutting edge is honed with a fine oil stone after grinding, a finishing tool will produce a smooth finish on the workpiece. A light cut and a fine feed must be used. Like roughing tools, finishing tools are made in left-hand and right-hand models.

13.6.4 Facing tool

The *facing tool* is ground to prevent interference with the tailstock center. The tool point is set at a slight angle to the work face with the point leading slightly. See **Figure 13-32A**.

13.6.5 Round nose tool

A *round nose tool* is designed for lighter turning and is ground flat on the face (without back or side rake) to permit cutting in either direction. See **Figure 13-32B**. A slight variation of the round-nose tool, with a negative rake ground on the face, is excellent for machining brass, **Figure 13-33**.

Machining *aluminum* requires a tool with a considerably different shape from those previously described. As shown in **Figure 13-34**, the tool is set slightly above center to reduce any tendency to

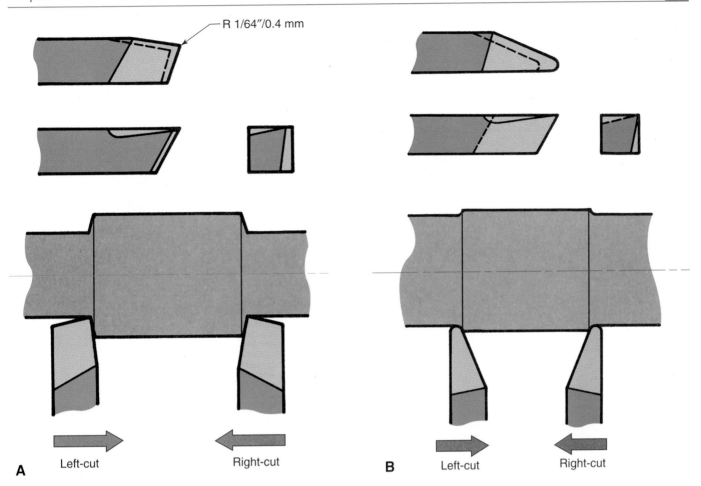

Figure 13-31. Lathe tools. A—The roughing tool is used for rapid material removal. B—The finishing tool will produce a smooth surface.

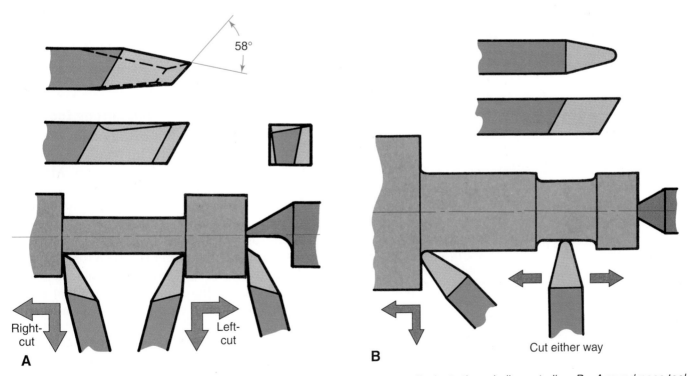

Figure 13-32. Lathe tools. A—A facing tool is used to machine surfaces perpendicular to the spindle centerline. B—A round-nose tool will produce fillets. Its shape permits it to cut either left or right.

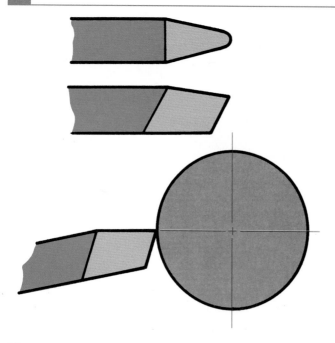

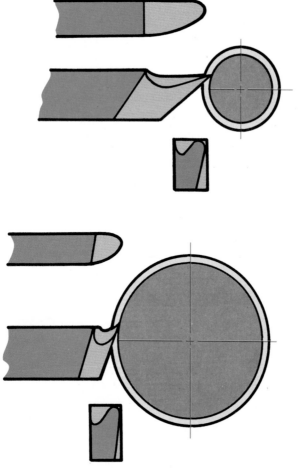

Figure 13-33. Cutter nose shape for machining brass. Note the negative rake.

Figure 13-34. For machining aluminum, cutter bit shapes different from those used for other metals are necessary. The tool is set slightly above center for a smoother cut.

chatter (vibrate rapidly). The tool designs illustrated are typical of cutting tools used to machine aluminum alloys.

13.6.6 Grinding high-speed cutter bits

When first attempting to grind a cutter bit, it may be best if you first practice on square sections of cold finished steel rod. You may also want to use chalk or bluing and draw the desired tool shape on the front portion of the blank, as shown in **Figure 13-35**. The lines will serve as guides for grinding.

Figure 13-36 depicts the recommended grinding sequence for a cutter bit. Side clearance, top clearance, and end relief may be checked with a clearance and cutting angle gage, **Figure 13-37**.

Figure 13-35. This cutter blank has been laid out and marked in preparation for grinding.

13.6.7 Brazed-tip single-point cutting tools

Brazed-tip single-point cutting tools are made by brazing a **carbide** cutting tip onto a shank made from less costly material, **Figure 13-38**. Many tip shapes (tool blanks) are available.

Cutting speeds can be increased by 300% to 400% when using **carbide cutting tools**. Powders of tungsten, carbon, and cobalt are molded into **tool blanks** and heated to extremely high temperatures. The hardness and strength of the blank can be controlled by varying the amount of cobalt that is used to *cement* (bind together) the tungsten and carbon particles.

For best results, these tools should be sharpened on a special **silicon carbide** or **diamond-charged grinding wheel** in which diamond dust particles or chips are embedded. A special type of grinder must

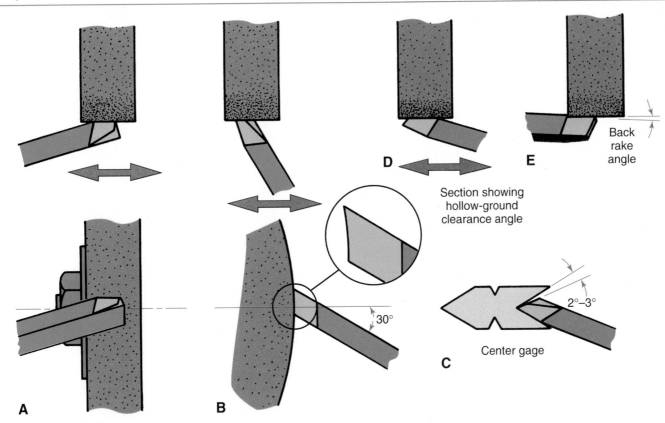

Figure 13-36. Grinding sequence for a cutter bit. A—Two views showing how to position a cutter bit blank on the grinding wheel to shape side clearance angle and side cutting edge angle. B—Shaping end clearance angle and front cutting edge angle. C—Center gage being used to check nose angle. D—Grinding other side clearance angle, when required. E—Grinding back/side rake angles. Accuracy of clearance angles can be checked with cutter bit gage.

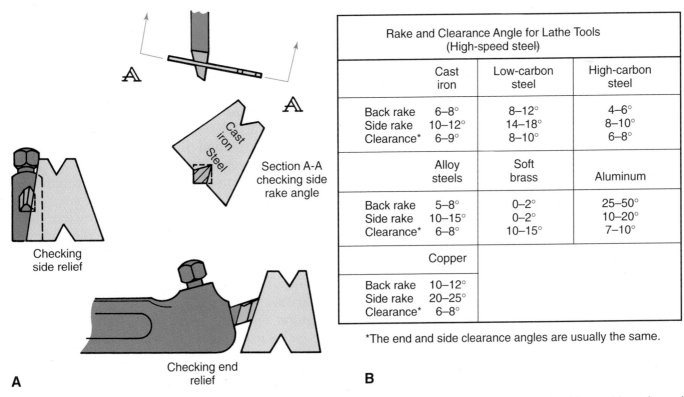

Rake and Clearance Angle for Lathe Tools (High-speed steel)			
	Cast iron	Low-carbon steel	High-carbon steel
Back rake	6–8°	8–12°	4–6°
Side rake	10–12°	14–18°	8–10°
Clearance*	6–9°	8–10°	6–8°
	Alloy steels	Soft brass	Aluminum
Back rake	5–8°	0–2°	25–50°
Side rake	10–15°	0–2°	10–20°
Clearance*	6–8°	10–15°	7–10°
	Copper		
Back rake	10–12°		
Side rake	20–25°		
Clearance*	6–8°		

*The end and side clearance angles are usually the same.

Figure 13-37. Cutter bit gage. A—Bit gage being used to check accuracy after grinding cutter tip. B—This table provides rake and clearance angles for lathe tools to machine different metals.

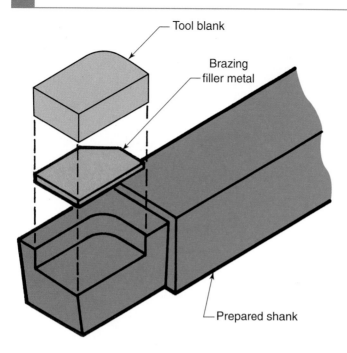

Figure 13-38. *Brazing a carbide tool blank into place on a prepared shank. The brazing must be done properly or the tool blank will not be solidly attached, causing it to wear rapidly. Tungsten carbide tool blanks are available in a wide selection of shapes, sizes, and degrees of hardness.*

be used, **Figure 13-39**. (Also see Chapter 11, Offhand Grinding.)

Cutting tools designed for machining steel are chamfered 0.003″ to 0.002″ (0.050 to 0.075 mm) by honing them lightly with a silicon carbide or diamond hone. If the tools are not honed, the irregular edge produced by grinding will crumble when used. Honing, if done properly, does not interfere with the cutting action.

13.6.8 Carbide tipped straight turning tools

The cutting tools shown in **Figure 13-40** are general purpose tools for facing, turning, and boring. The *square nose* shape permits machining to a square shoulder. Note that the clearance angles of the carbide tools described are not as great as those required for high speed steel cutting tools.

Also shown is a carbide tipped *threading tool* (Style E). This tool has a 60° included angle that conforms with the Unified National 60° included-angle thread. It is used for V-grooving and chamfering.

13.6.9 Indexable insert cutting tools

Brazed-tip cutting tools are being replaced by mechanically clamped *indexable insert cutting tools*, **Figure 13-41**. Indexable insert cutting tools

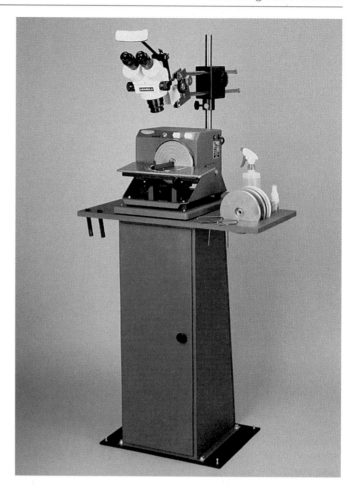

Figure 13-39. *A grinder designed for sharpening carbide, cermet, cubic boron nitride, and polycrystalline diamond (PCD) cutting tools. It uses special diamond charged wheels and is fitted with a microscope inspection system.*

are widely used for turning and milling operations. The inserts are manufactured in a number of shapes and sizes, **Figure 13-42**, for different turning geometries. Six of the most commonly used standard shapes are shown in order of increasing and decreasing strengths. See **Figure 13-43**.

Indexable inserts clamp to special tool holders, **Figure 13-44**. As an edge dulls, the next edge is rotated into position until all edges are dulled. Since it is less costly to replace inserts made from some materials than to resharpen them, they are usually discarded after use.

Inserts are manufactured from a number of materials, with each designed for a different metal requirement. See **Figure 13-45**. Carbide inserts are given increased versatility (higher abrasion resistance, chemical stability, and lubricity) when coated with various combinations of *titanium carbide (TiC)*, *titanium nitride (TiN)*, and *alumina*.

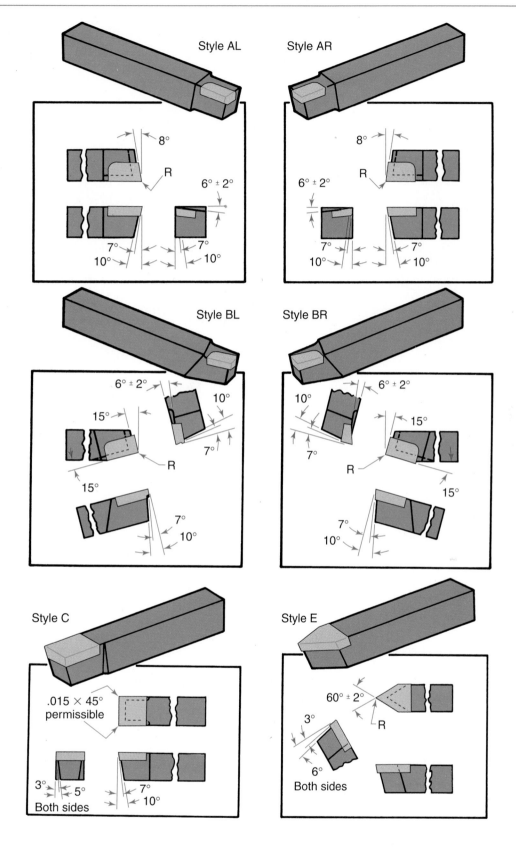

Figure 13-40. *Typical standard cemented-carbide single-point tools. Style E is a carbide tipped threading tool. (Carboloy, Inc.)*

Chipbreakers

When some metals are machined, long continuous chips will be created, unless some method is employed to break the chips into smaller pieces. This is accomplished by a small step or groove, called a *chipbreaker,* that is located on the top of the

Figure 13-41. Indexable insert cutting tools of carbide or sintered oxides (often referred to as cermets) are mechanically clamped into tool holders to perform cutting tasks. This insert is being used to machine stainless steel. (Sandvik Coromant Co.)

Figure 13-44. A selection of typical holders and replaceable carbide insert cutting tools for the lathe. Each insert has three or four cutting tips. The inserts are clamped in place on the holder, and can be indexed (rotated into position) to present a new tip when the one in use becomes dull. (Carboloy, Inc.)

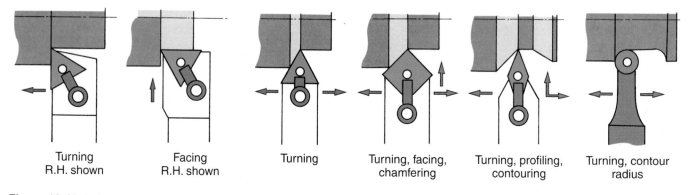

| Turning R.H. shown | Facing R.H. shown | Turning | Turning, facing, chamfering | Turning, profiling, contouring | Turning, contour radius |

Figure 13-42. Indexable inserts are manufactured in a number of different shapes and sizes for different turning operations.

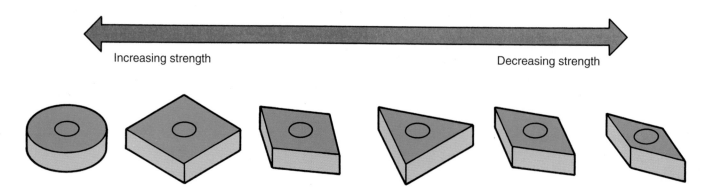

Figure 13-43. Most commonly used indexable insert shapes are shown in order of increasing and decreasing strengths.

Material	Strengths	Weaknesses	Typical Applications
HSS	Superior resistance Versatility	Poor speed capabilities Poor wear resistance	Screw machine and other low-speed operations, interrupted cuts, low-horsepower machining.
Carbide	Most versatile cutting material High shock resistance	Limited speed capabilities	Finishing to heavy roughing of most materials, including irons, steels, exotics, and plastics.
Coated Carbide	High versatility High shock resistance Good performance at moderate speeds	Limited to moderate speeds	Same as carbide, except with higher speed capabilities.
Cermet	High versatility Good performance at moderate speeds	Low shock resistance Limited to moderate speeds	Finishing operations on irons, steels, stainless steels, and aluminum alloys.
Ceramic- Hot/Cold Pressed	High abrasion resistance High-speed capabilities Versatility	Low mechanical shock resistance Low thermal shock resistance	Steel mill-roll resurfacing, finishing operations on cast irons and steels.
Ceramic- Silicon Nitride	High shock resistance Good abrasion resistance	Very limited applications	Roughing and finishing operations on cast irons.
Ceramic- Whisker	High shock resistance High thermal shock resistance	Limited versatility	High-speed roughing and finishing of hardened steels, chilled cast iron, high-nickel superalloys.
Cubic Boron Nitride	High hot hardness High strength High thermal shock resistance	Limited performance on materials below 38Rc Limited applications High cost	Hardened work materials in 45-70 Rockwell C range.
Poly- Crystalline Diamond	High abrasion resistance High-speed capabilities	Limited applications Low mechanical shock resistance	Roughing and finishing operations on abrasive nonferrous or nonmetallic materials.

Figure 13-45. The nine basic categories of cutting tool materials. (Valenite, Inc.)

cutter at the cutting edge. Most inserts manufactured today have molded-in chipbreakers. Other single-point cutting tools must have a chipbreaker ground into the top face of the tool, **Figure 13-46**.

13.6.10 Other types of cutting tools

Diamonds, both natural and manufactured, are employed as single-point cutting tools on materials whose hardness or abrasive qualities make them difficult to machine with other types of cutting

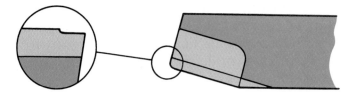

Figure 13-46. Typical chipbreaker on a single-point tool.

tools. These diamonds are known as *industrial diamonds*.

13.7 CUTTING SPEEDS AND FEEDS

The matter of cutting speed and feed is most important, since these factors govern the length of time required to machine the work and the quality of the surface finish.

Cutting speed indicates the distance the work moves past the cutting tool, expressed in *feet per minute (fpm)* or *meters per minute (mpm)*. Measuring is done on the circumference of the work.

To explain this differently: if a lathe were to cut one long chip, the length of the chip cut in one minute (measured in either feet or meters) would be the cutting speed of the lathe. Cutting speed is *not* the *revolutions per minute (rpm)* of the lathe.

Feed is the distance that the cutter moves *lengthwise* along the lathe bed during a single revolution of the work.

There are a number of factors that must be considered when determining the correct cutting speed and amount of feed:

- Material used for the cutting tool.
- Kind of material being machined.
- Desired finish.
- Condition of the lathe.
- Rigidity of the workpiece.
- Kind of coolant being used (if any).
- Shape of the material being machined.
- Depth of cut.

If the machining is done with a cutting speed that is *too slow*, extra time will be needed to complete the job. If speed is *too high*, the cutting tool will dull rapidly and the finish will be substandard.

A *speed and feed chart* takes into consideration the many factors listed earlier. **Figure 13-47** is a chart for use with high-speed steel cutter bits. Cutting speeds and feeds on the chart can be increased by 50% if a coolant is used, and by 300% to 400% if a cemented carbide cutting tool is employed.

13.7.1 Calculating cutting speeds

The cutting speeds shown in **Figure 13-47** should be considered as only a starting point. Depending upon machine condition, they may have to be increased or decreased until optimum cutting conditions are obtained.

Cutting speed (CS), as noted, is given in feet per minute (fpm) or meters per minute (mpm). Speed of the work (spindle speed) is given in revolutions per minute (rpm). Thus, the *peripheral speed* (speed at

Material to be cut	Roughing cut 0.01″ to 0.020″ 0.25 mm to 0.50 mm feed		Finishing cut 0.001″ to 0.010″ 0.025 mm to 0.25 mm feed	
	fpm	mpm	fpm	mpm
Cast iron	70	20	120	36
Steel				
Low carbon	130	40	160	56
Med carbon	90	27	100	30
High carbon	50	15	65	20
Tool steel				
(annealed)	50	15	65	20
Brass—yellow	160	56	220	67
Bronze	90	27	100	30
Aluminum*	600	183	1000	300

The speeds for rough turning are offered as a starting point. It should be all the machine and work will withstand. The finishing feed depends upon the finish quality desired.

* The speeds for turning aluminum will vary greatly according to the alloy being machined. The softer alloys can be turned at speeds upward of 1600 fpm (488 mpm) roughing to 3500 fpm (106 mpm) finishing. High silicon alloys require a lower cutting speed.

Figure 13-47. *Cutting speeds and feeds suggested for turning various metals with high-speed steel tools.*

the circumference or outside edge of the work) must be converted to rpm to determine the required spindle speed. The following formulas are used:

Inch-based:

$$rpm = \frac{CS \times 4}{D}$$

Where rpm = Revolutions per minute.
      CS = Cutting speed recommended for the particular material being machined (steel, aluminum, etc.) in feet per minute.
      D = Diameter of work in inches. Convert all fractions to decimals.

Cutting speed problem: What spindle speed is required to finish-turn 4″ diameter aluminum alloy?

$$rpm = \frac{CS \times 4}{D}$$
$$= \frac{1000 \times 4}{4}$$
$$= 1000 \text{ rpm}$$

      CS = Table recommends a cutting speed of 1000 fpm for finish-turning aluminum alloy.
      D = 4″

Adjust the spindle speed to as close to this speed (1000 rpm) as possible. Increase or decrease speed as needed to obtain desired surface finish.

Metric-based:

$$rpm = \frac{CS \times 1000}{D \times \pi}$$

Where rpm = Revolutions per minute.
       CS = Cutting speed recommended for particular material being machined (steel, aluminum, etc.) in meters per minute (mpm).
        D = Diameter of work in millimeters (mm).
        π = 3 (Since cutting speeds are approximate, π has been rounded off to 3 from 3.1416 to simplify calculation.)

Cutting speed problem: What spindle speed is required to finish-turn 100 mm diameter aluminum alloy?

$$rpm = \frac{CS \times 1000}{D \times 3}$$
$$= \frac{300 \times 1000}{100 \times 3}$$
$$= 1000 \, rpm$$

      CS = Table recommends a cutting speed of 300 m/pm for finish-turning aluminum alloy.
        D = 100 mm
        π = 3

Adjust the spindle to as close to this speed (1000 rpm) as possible. Increase or decrease speed as needed to obtain desired surface finish.

13.7.2 Roughing cuts

Roughing cuts are taken to reduce work diameter to approximate size. The work is left 1/32″ (0.08 mm) oversize for finish turning. Since the finish obtained on the roughing cut is of little importance, use the highest speed and coarsest feed consistent with safety and accuracy.

13.7.3 Finishing cuts

The *finishing cut* brings the work to the required diameter and surface finish. A high-spindle speed, sharp cutting tool, and fine feed are employed.

13.7.4 Depth of cut

The *depth of cut* refers to the distance the cutter is fed into the work surface. The depth of cut, like feed, varies greatly with lathe condition, material hardness, speed, feed, amount of material to be removed, and whether it is to be a roughing or finishing cut.

Depth of the cut can be set accurately with the *micrometer dials* on the cross-slide and compound rest, **Figure 13-48**.

The micrometer dial is usually graduated in 0.001″ or 0.02 mm increments. This means that a movement of *one* graduation feeds the cutting tool into the piece 0.001″ or 0.02 mm. However, material is removed around the periphery (outside edge) of the rotating work at *double* the depth adjustment. For each 0.001″ (0.02 mm) of infeed, for example, the workpiece diameter is reduced by 0.002″ or 0.04 mm. See **Figure 13-49**. This must not be forgotten or *twice* as much material as specified will be removed.

Some lathes, however, have a micrometer dial set up so that the number of graduations the cutter

Figure 13-48. *Combination inch/metric-graduated micrometer dials on the cross-slide and compound rest handwheels of a lathe. (Clausing Industrial, Inc.)*

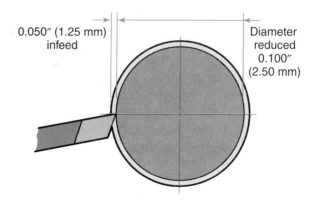

0.050″ (1.25 mm) infeed

Diameter reduced 0.100″ (2.50 mm)

Figure 13-49. *Remember that material is removed from the work on each cut at two times the infeed distance.*

is fed into the work will equal the amount that the work diameter will be reduced. That is, if the cutter is fed in 0.005" (0.10 mm) or 5 graduations, the work diameter will be reduced 0.005" (0.10 mm). Check the lathe you will be using to be sure which system it employs.

A common mistake when using a lathe is to remove *too little* material at *too slow* a speed. Cuts as deep as 0.125" (3 mm) can be handled by light lathes; cuts of 0.250" (6 mm) and deeper can be made by heavier machines without overtaxing the lathe.

13.8 WORK-HOLDING ATTACHMENTS

One of the reasons the lathe is such a versatile machine tool is the great variety of ways that work may be mounted in or on it. The most common way is to mount the work so that it revolves, permitting the cutting tool to move across the work's surface. Large and/or odd-shaped pieces are sometimes mounted on the carriage and machined with a cutting tool that is mounted in the rotating spindle.

Most work is machined while supported by one of the methods shown in **Figure 13-50**:

- Between centers using a faceplate and dog.
- Held in one of the three types of chucks.
  - 3-jaw universal chuck.
  - 4-jaw independent chuck.
  - Jacobs type chuck.
- Held in a collet.
- Bolted to the faceplate.

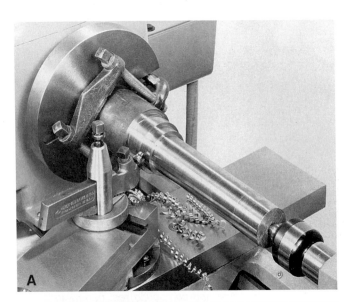

Figure 13-50. Work-holding methods. A—Work being machined between centers. B—Work held in a chuck for machining. (Clausing Industrial, Inc.) C—Work being machined while held in a collet. D—Work bolted to a faceplate for machining.

13.9 TURNING WORK BETWEEN CENTERS

Considerable lathe work is done with the workpiece supported between centers. For this operation, a faceplate, **Figure 13-51,** is attached to the spindle nose. A sleeve and live center are inserted into the spindle opening, **Figure 13-52.**

Either a nonrevolving *dead center* or a heavy duty *ball bearing center* is fitted into the tailstock spindle to support one end of the work. See **Figure 13-53.** The ends of the stock are drilled to fit over the center points.

A lathe dog, **Figure 13-54,** is clamped to one end of the material. Three types of *lathe dogs* are shown in **Figure 13-55:**

- The *bent-tail standard dog* has the setscrew exposed.
- The *bent-tail safety dog* has the setscrew recessed. This type dog is usually preferred over the standard lathe dog.
- The *clamp-type dog* is used for turning square or rectangular work.

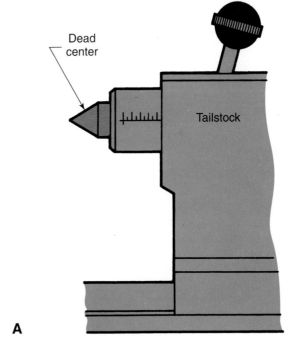

A

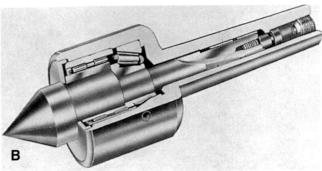

B

Figure 13-53. Tailstock centers. A—A dead center does not rotate. It is fitted into the tailstock spindle. B—Cutaway shows construction of a heavy-duty ball bearing center.

Figure 13-51. Lathe faceplates come in various sizes.

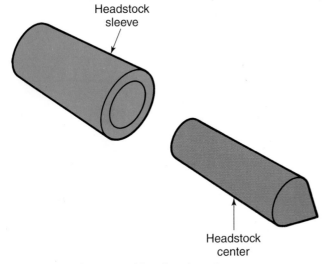

Figure 13-52. Sleeve and headstock center.

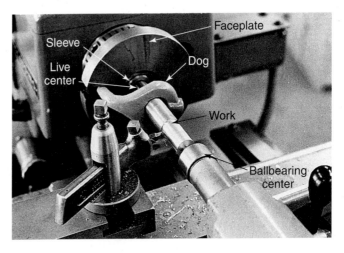

Figure 13-54. Machining work mounted between centers.

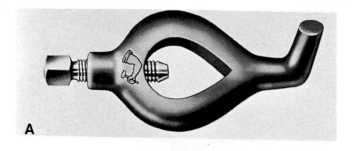

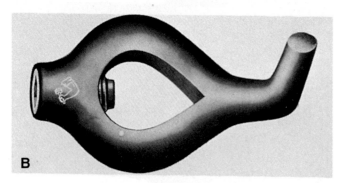

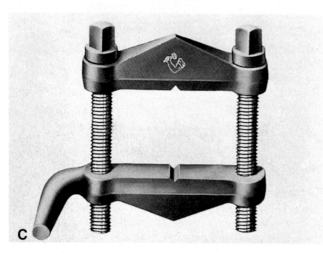

Figure 13-55. *Lathe dogs. A—Bent-tail standard dog. B—Bent-tail safety dog. C—Clamp-type dog. (Armstrong Bros. Tool Co.)*

13.9.1 Drilling center holes

Before work can be mounted between centers, it is necessary to locate and drill center holes in each end of the stock, **Figure 13-56.** Several methods for locating the center of round stock are shown in **Figure 13-57.**

Center holes are usually drilled with a *combination drill and countersink*, **Figure 13-58.** The *drill angle* is identical to that of the center point.

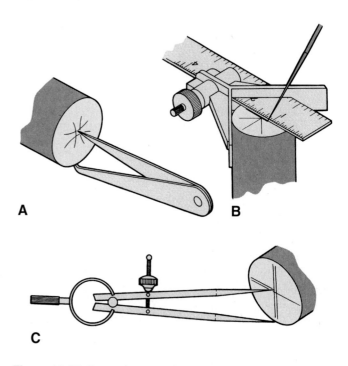

Figure 13-57. *Several ways to locate the center of round stock. A—With a hermaphrodite caliper. B—With center head and rule of combination set (recommended method). C—With dividers.*

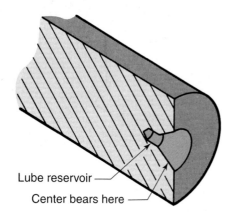

Figure 13-56. *Tail center rides in the drilled and countersunk center hole. A supply of lubricant is placed in reservoir. The lubricant will expand and lubricate the center as metals heats up.*

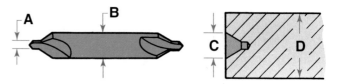

Combination drill and countersink no.	A	B	C	D
1	1/16	13/64	1/8	3/16 to 5/16
2	3/32	3/16	3/16	3/8 to 1
3	1/8	1/4	1/4	1 1/4 to 2
4	5/32	7/16	5/16	2 1/4 to 4

Figure 13-58. *This size chart contains information needed to select correct-size center drill. Combination drill and countersink makes the hole and countersinks it in one operation.*

The *straight drill* provides clearance for the center point and serves as a reservoir for a lubricant. The chart provides information needed to select the correct size center drill.

The center holes can be drilled on a drill press, on the lathe with the work centered in the chuck, or on the lathe with the center drill held in the headstock. See **Figure 13-59**.

The center holes should be drilled deep enough to provide adequate support, **Figure 13-60**.

13.9.2 Checking center alignment

Accurate turning between centers requires centers that run true and are in precise alignment. Since the work must be reversed to machine its entire length, care must be taken to make the live center

A

B

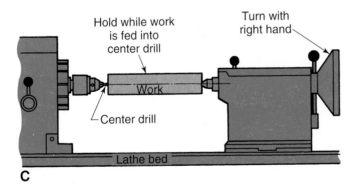

Hold while work is fed into center drill

Turn with right hand

Work

Center drill

Lathe bed

C

Figure 13-59. Drilling center holes. A—Holes can be drilled on a drill press. Mount work in a V-block for support. B—Some work can be held in a lathe chuck for center drilling. C—Center holes can be drilled in large stock by mounting a Jacobs chuck in the headstock. Locate center point of each end and center punch. Support one end on tail center and feed other into center drill mounted in the Jacobs chuck. Repeat operation on second end.

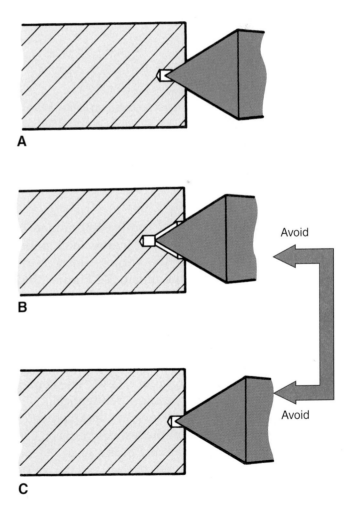

A

B   Avoid

C   Avoid

Figure 13-60. Correctly and incorrectly drilled center holes. A—Properly drilled center hole. B—Hole drilled too deep. C—Hole not drilled deep enough. Does not provide enough support; if used with a dead center, the center point will burn off.

run true. If it does *not*, the diameters will be *eccentric* (not aligned on the same center line), **Figure 13-61**. This can be prevented by truing the live center. If the center is not hardened, a light truing cut can be made. A tool post grinder will be needed to true a hardened center.

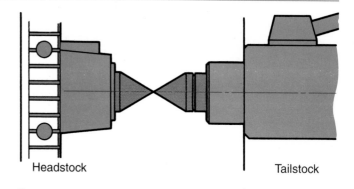

Headstock   Tailstock

A

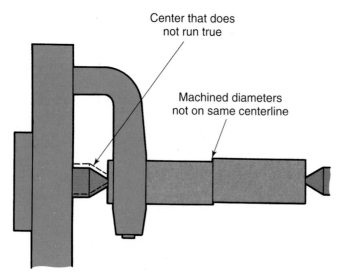

Center that does
not run true

Machined diameters
not on same centerline

Figure 13-61. The workpiece must be reversed in the lathe dog so it can be machined for its entire length. If the live center does not run true, eccentric diameters will result.

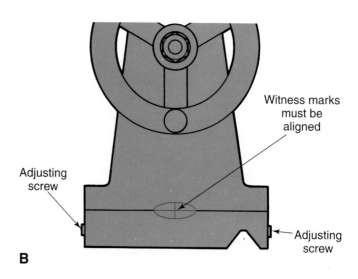

Witness marks
must be
aligned

Adjusting
screw

Adjusting
screw

B

Figure 13-62. Checking center alignment. A—Checking alignment by bringing center points together. View is looking down on top of centers. B—Alignment can be determined by checking witness lines on base of tailstock.

Approximate alignment can be determined by checking centers visually by bringing their points together, or by checking the witness lines on the base of the tailstock for alignment. See **Figure 13-62.**

A more precise method for checking alignment is needed if close tolerance work is to be done. Several such methods are described below:

- Make a light trial cut across a few inches of the material. Check the diameter at each end with a micrometer, **Figure 13-63**. The centers are aligned if the readings are identical.
- Use a steel test bar and dial indicator, **Figure 13-64**. Mount the test bar between centers and position the dial indicator in the tool post and at right angles to the work. Move the indicator contact point against the test bar until a reading is shown. Move the indicator along the test bar. If the readings remain constant, the centers are aligned.

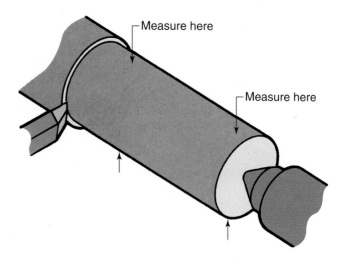

Measure here

Measure here

Figure 13-63. Make a light cut on stock and measure diameter at two points to check alignment. Measurements must be equal.

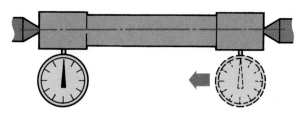

Figure 13-64. Using a test bar and dial indicator to check center alignment.

- Machine a section of scrap, as shown in **Figure 13-65**. Set the cross-feed screw to make a light cut at the right end of the piece. With the same tool setting, move it to the left and continue the cut. Identical micrometer readings indicate center alignment.

Adjusting screws (one on each side at the base of the tailstock) are used to align the centers if checks indicate this is needed. Make adjustments gradually. See **Figure 13-66**.

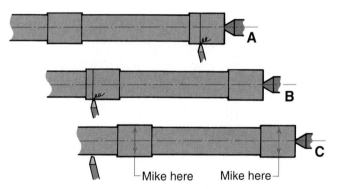

Figure 13-65. Checking for center alignment. A—Machine two shoulders on a test piece. B—Keep same tool setting and make a cut on both shoulders. C—Measure resulting diameters.

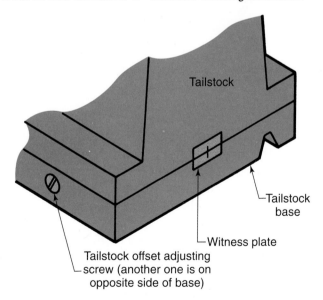

Figure 13-66. Adjusting screws are located on both sides of tailstock base. Their primary use is to set over or shift the tailstock for taper turning.

13.9.3 Mounting work between centers

Clamp a dog to one end of the work. Place a *lubricant* (white lead, graphite and oil, or a commercial center lubricant) in the center hole on the other end. Mount the piece on the centers and adjust the tailstock spindle until the work is snug. If the work is too loose, it will "clatter." If adjusted too tightly, it will score or burn the center point.

Check the adjustment from time to time, since heat generated by the machining process will cause the work to expand. Using a ball bearing center, instead of a dead center, will reduce or eliminate many of the problems involved in working between centers.

Check to see if the dog tail binds on the faceplate slot, **Figure 13-67**. This can cause the work to be pulled off center. When machined, this will produce a surface that is not concentric with the center hole. If binding is occurring, use a different faceplate.

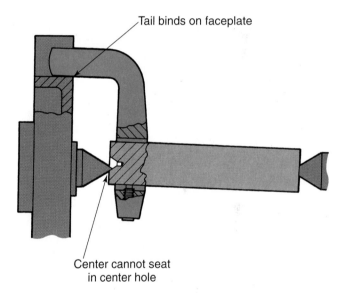

Figure 13-67. Diameter of the turned surface will not be concentric with the center holes if center hole is not seated properly on the live center. A binding lathe dog is a common cause of this problem.

13.9.4 Facing work held between centers

Facing is an operation that machines the end of the work square and reduces it to a specific length. At times, considerable material must be removed. In this situation, it is best to leave the work longer than finished size and drill deeper center holes for better support during the roughing operation.

Face the work to length before starting the finish cut. A *right-cut facing tool* will be needed. The 58° point on this tool provides a slight clearance between the center point and the work face, **Figure 13-68**. Be careful not to damage the cutting tool point by running it into the center. A *half center* makes the operation easier, but is used only for facing— it does not provide an adequate bearing surface for general work and will not hold lubricant.

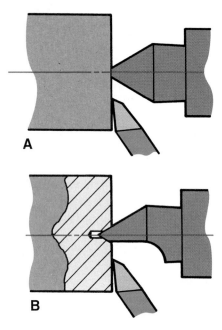

Figure 13-68. *Facing stock. A—Relationship of cutter bit to work face when making a facing cut. B—Using a half center will give more clearance when facing end of stock.*

13.9.5 Facing to length

It is standard machining practice to cut stock slightly longer than needed to permit its ends to be machined square. A steel rule may be employed if the dimension is not critical. For more accuracy, a Vernier caliper or large micrometer may be used. The difference between the rough length and the required length is the amount of material that must be removed.

Set the compound rest at 30°, **Figure 13-69**. Bring the cutting tool up until it just touches the surface to be machined, then lock the carriage. Remove material from each end of the stock until the specified length is attained.

13.9.6 Rough turning between centers

Rough turning is an operation in which excess material is cut away rapidly with little regard for the quality of surface finish. The diameter is reduced to within 1/32″ (0.8 mm) of required size by employing deep cuts and coarse feeds.

Set the compound rest at 30° from a right angle to the work, **Figure 13-70**. This will permit the tool

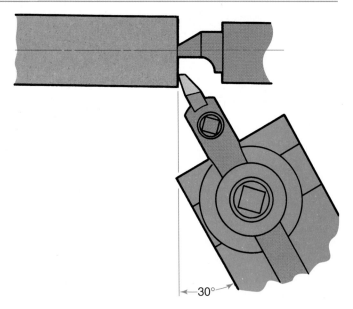

Figure 13-69. *Recommended compound rest setting when facing stock to length.*

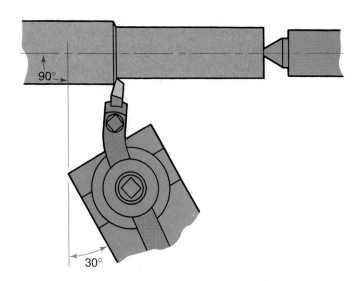

Figure 13-70. *Compound rest setting used for rough turning.*

to cut as close as possible to the left end of the work without the dog striking the compound rest.

CAUTION: Always check the maximum distance that compound rest can be fed toward the dog or chuck without striking them *before* you start the lathe.

Use a left-hand toolholder. Position the tool post as far to left as possible in the compound rest T-slot. Avoid excessive tool overhang, **Figure 13-71**.

Locate the cutting edge of the tool about 1/16″ (1.5 mm) above work center for *each inch* of diameter, **Figure 13-72**. It can be set by comparing it with the tail center point or with an index line scribed on the tailstock ram of some lathes.

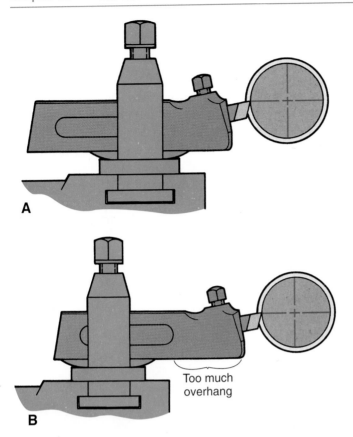

Figure 13-71. Correct and incorrect mounting of tool holder. A—Tool holder and cutter bit in proper position. B—Too much overhang will cause the tool to "chatter" and produce a rough machined surface.

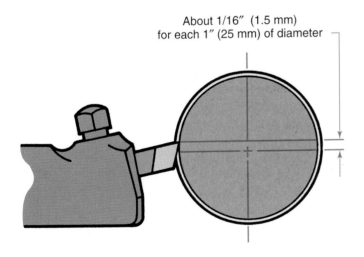

Figure 13-72. Correct location of cutter bit point when rough turning. Lower the point as work diameter decreases in size.

The toolholder must be positioned correctly. If it is not, the heavy side pressure developed during machining will cause it to turn in the tool post, forcing the cutting tool deeper into the work. When the toolholder is correctly positioned, the cutting tool will pivot away from the work. See **Figure 13-73**.

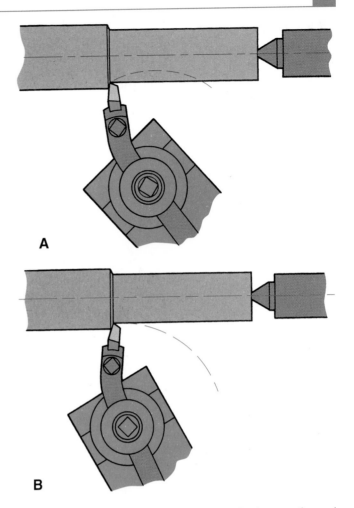

Figure 13-73. Toolholder positioning. A—An incorrectly positioned tool will cut deeper into work if toolholder slips in the tool post. B—A correctly positioned tool will swing clear of work if tool holder slips.

Make a trial cut to true up the stock. Measure the resulting diameter. The difference between the diameter and the required rough diameter is *twice* the distance the tool must be fed into the work. If the piece is greatly oversize, it will be necessary to make two or more cuts to bring it to size.

When depth of cut has been determined, engage the power feed. Observe the condition of the chips. They should be in *small* sections and slightly *blue* in color. Long, stringy chips indicate a cutting tool that is not properly sharpened. Stop the machine and remove stringy chips with pliers. Replace tool with one that is properly sharpened.

After each cut, measure work diameter to prevent excess metal removal. Always stop the machine before making measurements or cleaning out chips.

If a dead center is used in the tailstock, lubricate the center frequently. Stop the machine *immediately* if the center heats up and starts to smoke or "squeal."

13.9.7 Finish turning

After rough turning, the work is still oversize. It must be machined to the specified diameter and to a smooth surface finish by *finish turning,* **Figure 13-74**.

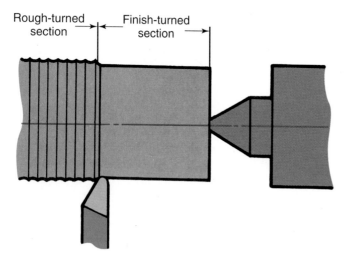

Figure 13-74. *After roughing work to approximate size, turn it to required size with a finishing tool. Cut should be made from right to left.*

Fit a right-cut finishing tool into the toolholder. All rough and finish machining should be done *toward the headstock* (right to left) because the headstock offers a more solid base than the tailstock. Position the tool on center and check for adequate clearance between the compound rest and the revolving lathe dog.

Adjust the lathe for a faster spindle speed and a fine feed. Run the cutting tool into the work until a light cut is being made; then engage the power feed. After a sufficient distance has been machined, disengage the power feed and stop the lathe. Never reverse a lathe; brake it to a stop!

Do not interfere with the cross-slide setting. "Mike" the diameter of the machined area. The difference between the measurement and the specified diameter is the amount of material that must be removed. Move the cutting tool clear of the work and feed it in *one-half* the amount that must be removed. For example, if the diameter is 0.008" (0.20 mm) oversize, tool infeed should be 0.004" (0.10 mm). Make another cut about 1/2" (13 mm) in width at the new depth setting. Measure again to make sure the correct diameter will be machined.

When reversing the work to permit machining its entire length, avoid marring of the finished surface by the lathe dog setscrew. Insert a small piece of soft aluminum or copper sheet between the setscrew and the workpiece.

13.9.8 Turning to a shoulder

Up to this point, only *plain turning* has been described. This is turning in which the entire length of the piece is machined to a specified diameter. However, it is frequently necessary to machine a piece to several *different* diameters.

Locate the points to which the different diameters are to be cut. Scribe lines with a hermaphrodite caliper which has been set to the required length, **Figure 13-75**.

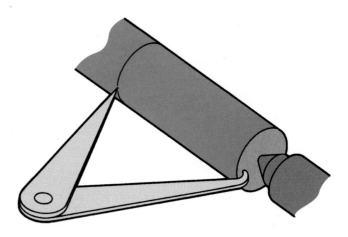

Figure 13-75. *Scribing reference lines on a workpiece with a hermaphrodite caliper.*

Machining is done as previously described, with the exception of cutting the *shoulder,* (the point where the diameters change). **Figure 13-76** shows the four types of shoulders:
- Square.
- Angular.
- Filleted.
- Undercut.

A *right-cut tool* is used to make the square and angular type shoulders. See **Figures 13-77** and **13-78**. For machining a filleted shoulder, a *round nose tool* is ground to the required radius using a fillet or radius gage to check radius accuracy. See **Figure 13-79**.

13.9.9 Grooving or necking operations

It is sometimes necessary to cut a *groove* or neck on a shaft to terminate a thread, or to provide adequate clearance for mating pieces, **Figure 13-80**. As any recess cut into a surface has a tendency to weaken a shaft, it is better to make the groove *round,* rather than square.

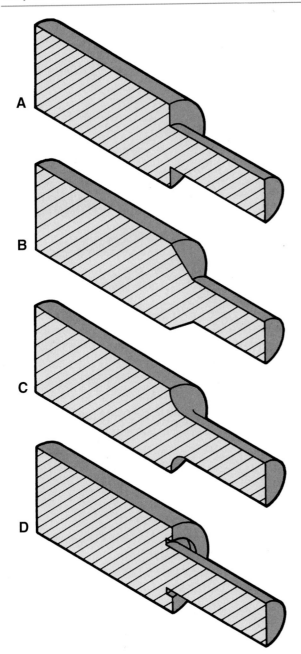

Figure 13-76. *Four types of shoulders. A—Square. B—Angular. C—Filleted. D—Undercut.*

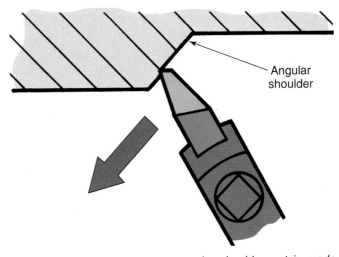

Figure 13-77. *To machine an angular shoulder, cut is made from smaller diameter to larger diameter.*

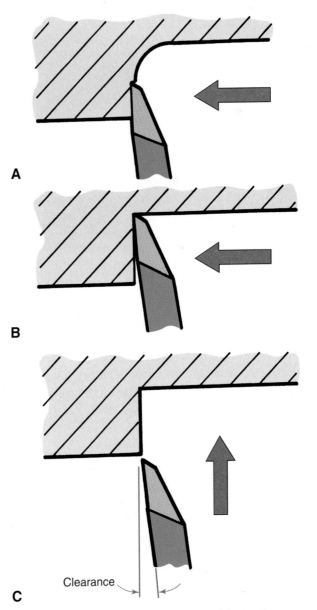

Figure 13-78. *Machining sequence used for cutting a square shoulder. A—First cut. B—Second cut. C—Facing cut.*

The tool is set on center and fed in until it just touches the work surface. Set the cross-feed micrometer dial to zero and feed the tool in the required number of thousandths/millimeters for the specified depth. Square grooves can be machined with a *parting tool.*

13.10 USING LATHE CHUCKS

The chuck is another device for holding work in a lathe. *Chucking* is the most rapid method of mounting work; for that reason, it is widely preferred. Other operations, such as drilling, boring,

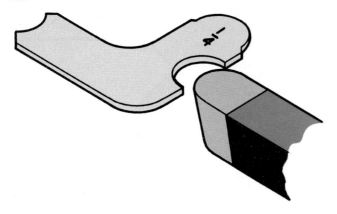

Figure 13-79. *The radius ground on a cutter bit can be checked with a fillet gage.*

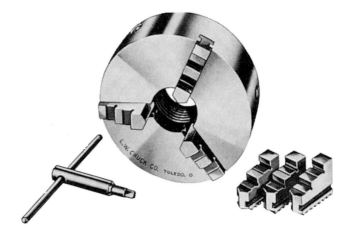

Figure 13-81. *The 3-jaw universal chuck automatically centers round or hexagonal stock. (L-W Chuck Co.)*

Figure 13-80. *A groove or neck can be cut into a shaft with a grooving or parting tool. Shape of the groove may be square, angular (square with sloping sides), or round. (Kaiser Tool Co., Inc.)*

reaming, and internal threading, can be done while the work is held in a chuck. Additional support can be obtained for the piece by supporting the free end with the tailstock center.

The common types of chucks are:
- 3-jaw universal.
- 4-jaw independent.
- Jacobs.
- Draw-in collet.

13.10.1 3-jaw universal chuck

The *3-jaw universal chuck* is designed so that all jaws operate at the same time, **Figure 13-81**. It will automatically center round or hexagonal shaped stock.

Two sets of jaws are supplied with each universal chuck. One set is used to hold large-diameter work; the other set is for small-diameter work, **Figure 13-82**.

The jaws in each set are numbered 1, 2, and 3, as are the slots in which they are fitted. The *jaw number* must correspond with the *slot number* if the work is to be centered. Sets of jaws are made for a specific chuck and are not interchangeable with other chucks. Make sure the chuck and jaws have the same serial number!

13.10.2 Installing chuck jaws

Before installing jaws, clean the jaws, jaw slots, and *scroll* (spiral thread seen in the jaw slots). Turn the scroll until the first thread does not quite show in jaw slot 1. Slide the matching jaw into the slot as far as it will go. Now, turn the scroll until the spiral engages with the first tooth on the bottom of the jaw. Repeat the operation at slots 2 and 3, making sure the proper jaws are inserted.

Remove the chuck key when you finish using it; if left in the chuck, it could become a dangerous missile when the lathe is turned on. Make it a habit to never let go of a lathe chuck key unless you are placing it on the tool tray or lathe board.

Jaws of a universal chuck lose their centering accuracy as the scroll wears. Accuracy is also affected when too much pressure is used to mount the work, or when work is gripped too near the front of the jaws. Avoid gripping work near the front of the jaws. It can fly out and cause injuries.

13.10.3 4-jaw independent chuck

Each of the jaws of *4-jaw independent chuck*, **Figure 13-83**, operates individually, instead of being

3-jaw chuck

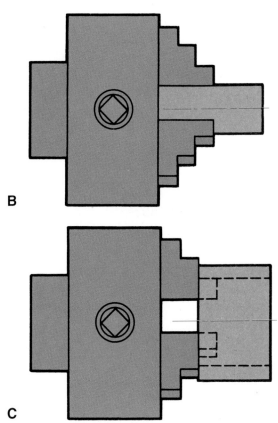

A

Large-diameter
work

B

C

Figure 13-82. Chuck jaws. A—One of the sets of jaws supplied with a 3-jaw universal chuck is used to mount large-diameter work. B—Holding work using the set of jaws supplied for smaller-size workpieces. C—Another method of mounting work in the chuck.

coupled with the other jaws (as in the 3-jaw universal chuck). This permits square, rectangular, and odd-shaped work to be centered. Unlike those of the 3-jaw chuck, the jaws of a 4-jaw chuck can be removed from their slots and reversed.

Figure 13-83. A 4-jaw independent chuck. The jaws on this type chuck are reversible. (L-W Chuck Co.)

This reversing feature permits the jaws to be used to hold large-diameter work in one position and smaller-diameter work when reversed, **Figure 13-84**.

Unlike the 3-jaw chuck, the 4-jaw type is not self-centering. The most accurate way to center round work in this type chuck is to use a dial indicator. The piece is first centered approximately, using the concentric rings on the chuck face as a guide. A dial indicator is then mounted in the tool post, **Figure 13-85**. The jaws are adjusted until the indicator needle does not *fluctuate* (move back and forth) when the work is rotated by hand. After the piece has been centered, all jaws must be tightened securely.

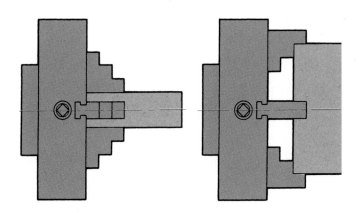

Figure 13-84. Reversing feature of jaws in a 4-jaw independent chuck makes it possible to turn work having extreme differences in diameter without difficulty.

Another centering method uses chalk, **Figure 13-86**. Rotate the work while bringing the chalk into contact with it. Slightly loosen the jaws opposite the chalk mark. Then tighten the jaws on the side where the chalk mark appears. Continue this operation until the work is centered. If the work is oversize enough, a cutting tool may be used instead of chalk.

Figure 13-85. Centering work in a 4-jaw chuck using a dial indicator. The machine shown is a small modelmaker's lathe.

Figure 13-86. Using chalk to center work in a 4-jaw chuck. The chalk mark indicates the "high point." Gradual adjustments, rather than large movements, are most effective in centering work.

Avoid trying to center stock in one or two adjustments, but rather, work in *increments* (very small steps). When making the final small adjustment, it may be necessary to loosen the jaw on the low side and retighten it, after which the high side is given a final tightening. This last method for making final adjustment applies, in particular, when centering work with a dial indicator.

13.10.4 Jacobs chuck

When turning small-diameter work, such as screws or pins, the *Jacobs chuck* can be utilized. This chuck, **Figure 13-87,** is better suited for such work than the larger universal or independent chuck.

A *standard Jacobs chuck* is normally fitted in the tailstock for drilling. However, it also can be mounted by fitting it in a sleeve and then placing the unit in the headstock spindle. Wipe the chuck shank, sleeve, and spindle hole with a clean soft cloth before they are fitted together.

A *headstock spindle Jacobs chuck* is similar to the standard Jacobs chuck, but is designed to fit directly onto a threaded spindle nose, **Figure 13-88**. The chuck has the advantage of not interfering with the compound rest, making it possible to work very close to the chuck.

13.10.5 Draw-in collet chuck

The *draw-in collet chuck* is a work-holding device for securing work small enough to pass through the lathe spindle, **Figure 13-89**.

Figure 13-87. Turning small-diameter work in a Jacobs chuck.

Figure 13-88. *Jacobs chuck that can be mounted on a lathe with a threaded spindle nose. (Jacobs Mfg. Co.)*

Figure 13-89. *A draw-in type collet chuck.*

Collets are accurately made sleeves with one end threaded and the other split into three even sections. The slots are cut slightly more than half the length of the collet and permit the jaws to spring in and clamp the work, **Figure 13-90.**

Figure 13-90. *A variety of collets is necessary to clamp different stock sizes. (Maswerks, Inc.)*

The standard collet has a circular hole for round stock, but collets for holding square, hexagonal, and octagonal material are available.

The chief advantages of collets are their ability to center work automatically and to maintain accuracy over long periods of hard usage. They have the disadvantage of being expensive, since a separate collet is needed for each different size or stock shape.

A collet chuck using steel segments bonded to rubber is also available. An advantage of this chuck is that each collet has a range of 0.100″ (2.5 mm), rather than being a single size, like steel collets. However, these collets are available only for round work.

13.10.6 Mounting and removing chucks

If a chuck is not installed on the spindle nose correctly, its accuracy will be affected. To install a chuck, remove the center and sleeve, if they are in place. Hold the center and sleeve with one hand and tap them loose with a knockout bar. Carefully wipe the spindle end clean of chips and dirt. Apply a few drops of spindle oil. Clean the portion of the chuck that fits on the spindle.

On a chuck that is fitted to a threaded spindle nose, clean the threads with a *spring cleaner*. See **Figure 13-91**.

With the tapered key spindle nose, rotate the spindle until the key is in the up position. Slide the chuck into place and tighten the threaded ring. Pins on the cam-lock spindle are fitted into place and locked.

Fitting a chuck onto a threaded spindle nose requires a different technique. Hold the chuck against the spindle nose with the right hand and turn the spindle with the left hand. Screw the chuck on until it fits firmly against the shoulder.

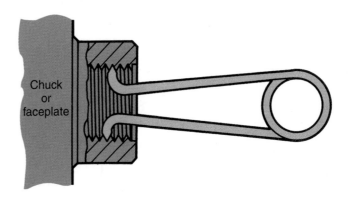

Figure 13-91. *A spring cleaner is used to clean the threads in a chuck before mounting it on a threaded spindle nose.*

To avoid possible injury, do not spin the chuck on rapidly or use power. Release belt tension if possible, to eliminate any chance of power being transferred to the spindle.

During installation, place a board on the ways under the chuck to protect your hands and prevent damage to the machine ways if the chuck is dropped.

13.10.7 Removing a chuck from threaded spindle

There are several accepted methods of removing chucks from a threaded headstock spindle. The first step in any method, regardless of the type of spindle nose, is to place a wooden cradle across the ways beneath the chuck for support. Then use one of the techniques shown in **Figure 13-92.**
* Lock the spindle in back gear and use a chuck key to apply leverage.
* Place a suitable size adjustable wrench on one jaw and apply pressure to the wrench.
* If neither of the preceding methods works, place a block of wood between the rear lathe ways and one of the chuck jaws. Engage the back gear and give the drive pulley a quick rearward turn.

13.10.8 Removing a chuck from other spindle noses

Little difficulty should be encountered when removing a chuck from tapered and cam lock spindle noses. For tapered spindle noses, first lock the spindle in back gear, then place the appropriate spanner wrench in the locking ring. Give it a tap or two with a leather or plastic mallet. Turn the ring until the chuck is released.

Place a wooden cradle under the chuck before attempting to remove it from the spindle. Removal will be easier and hand injuries will be avoided.

13.11 FACING STOCK HELD IN A CHUCK

A round nose cutting tool, held in a straight toolholder, is used to face stock held in a chuck. The compound rest is pivoted 30° to the right. The toolholder is set to less than 90° to face the work, and the cutter bit is exactly on center. The carriage is then moved into position and locked to the way. See **Figure 13-93A.**

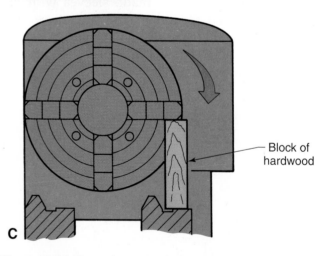

Block of hardwood

Figure 13-92. Removing a chuck. A— Using a chuck wrench to loosen the chuck. Note the wooden cradle placed under the chuck for support. B—Fitting an adjustable wrench to one of the jaws may help you loosen a stubborn chuck. C—In truly stubborn cases, reversing the chuck against a block of wood is often used.

A facing cut can be made in either direction. The tool may be started in the center and fed out, or the reverse may be done. The usual practice is to start from the center and feed outward. If the material is over 1 1/2" (38 mm) in diameter, automatic feed may be employed.

With the cutting tool on center a smooth face will result from the cut. A *rounded "nubbin"* (remaining piece of unmachined face material) will result if the tool is slightly above center, **Figure 13-93B**. A *square-shoulder "nubbin"* indicates that the cutter is below center, **Figure 13-93C**. Reposition the tool and repeat the operation if either condition is seen.

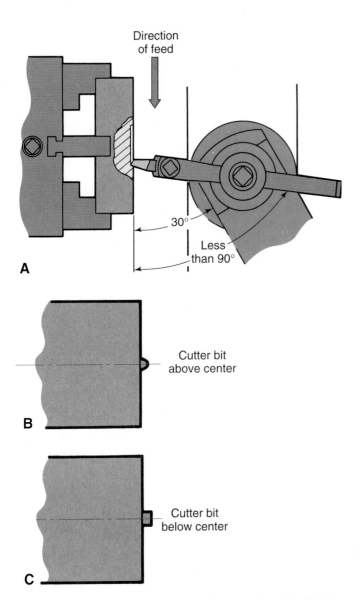

Figure 13-93. *Facing in a chuck. A—Correct tool and tool holder positions for facing. B—Rounded nubbin left by above-center cutter. C—Square-shoulder nubbin left by below-center cutter.*

13.12 PLAIN TURNING AND TURNING TO A SHOULDER

Work mounted in a chuck is machined in the same manner as if it were between centers. To prevent "springing" (flexing) while it is being machined, long work should be center drilled and supported with a tailstock center, **Figure 13-94**.

13.13 PARTING OPERATIONS

Parting is the operation of cutting off material after it has been machined, **Figure 13-95.** This is one of the more difficult operations performed on a lathe.

Cutting tools for parting or grooving are held in a straight or offset toolholder, **Figure 13-96.** They must be ground with the correct clearance (front, side, and end). A concave rake is ground on top of the cutter to reduce chip width, and prevent it from *seizing* (binding) in the groove.

Keep the tool sharp. This will permit easy penetration into the work. If the tool is *not* kept sharp, it may slip and as pressure builds up, dig in suddenly and break.

The cutoff blade is set at *exactly* 90° to the work surface, **Figure 13-97.** The cutting edge should be set *on center* when parting stock 1" (25.0 mm) in diameter. For larger pieces, the cutting edge should be positioned 1/16" (1.5 mm) *above center* for each 1" (25.0 mm) of diameter. The tool must be *lowered* as

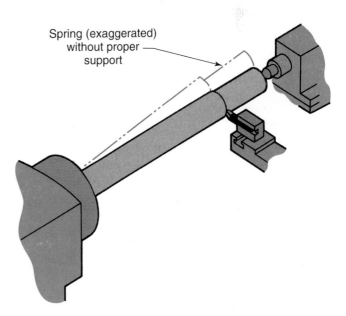

Figure 13-94. *For accurate turning, long work must be supported with tailstock center.*

Figure 13-95. Parting is one of the more difficult jobs performed on the lathe. This illustration shows parting of thick-wall tubing. The replaceable tool has a helical twisted geometry to prevent binding during parting operations. (Iscar Metals, Inc.)

Figure 13-96. Typical straight and offset tool holders for parting and grooving. They use replaceable carbide inserts, which are more wear resistant than conventional high speed tools. (Kaiser Tool Co., Inc.)

work diameter is reduced, unless the center of the piece has been drilled out.

Spindle speed is about one-third that employed for conventional turning. The compound rest and cross-slide must be tightened to prevent play. Do not forget to lock the carriage to the ways during a parting operation. Feed should be ample to

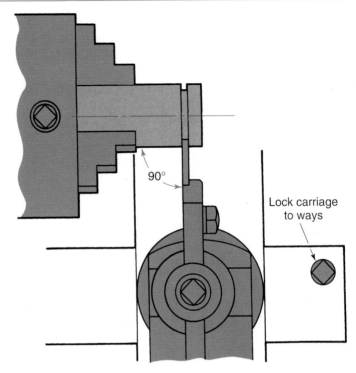

Figure 13-97. Work is held close in chuck for the parting operation. Parting tool blade is set at a 90° angle to cut, and carriage is locked to the ways.

provide a continuous chip. If feed is too slow, "*hogging*" (the cutter digging in and taking a very heavy cut) can result. The tool will not cut continuously, but will ride on surface of the metal for a revolution or two, then bite suddenly. If the machine is in good condition, automatic cross-feed may be employed.

When parting, apply ample quantities of cutting fluid. Whenever possible, hold the work "close" in the chuck and, if necessary, use an offset toolholder.

Never attempt to part work that is held between centers. Serious trouble will be encountered. See **Figure 13-98.**

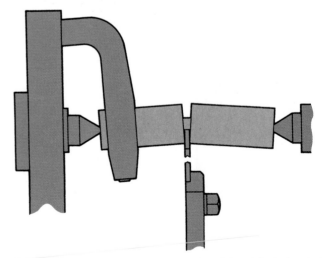

Figure 13-98. Work cannot be parted safely while being held between centers.

TEST YOUR KNOWLEDGE

Please do not write in the text. Write your answers on a separate sheet of paper.

1. The lathe operates on the principle of:
   a. The cutter revolving against the work.
   b. The cutting tool, being controllable, can be moved vertically across the work.
   c. The work rotating against the cutting tool, which is controllable.
   d. All of the above.
   e. None of the above.

2. The size of a lathe is determined by the _____ and the _____ of the _____.

3. The largest piece that can be turned between centers is equal to:
   a. The length of the bed minus the space taken up by the headstock.
   b. The length of the bed minus the space taken up by the tailstock.
   c. The length of the bed minus the space taken up by the headstock and the tailstock.
   d. All of the above.
   e. None of the above.

4. Into which of the following categories do the various parts of the lathe fall?
   a. Driving the lathe.
   b. Holding and rotating the work.
   c. Holding, moving, and guiding the cutting tool.
   d. All of the above.
   e. None of the above.

5. Explain the purpose of ways on the lathe bed.

6. Power is transmitted to the carriage through the feed mechanism to the quick-change gearbox which regulates the amount of _____ per _____.

7. The carriage supports and controls the cutting tool. Describe each of the following parts:
   a. Saddle.
   b. Cross-slide.
   c. Compound rest.
   d. Tool post.

8. Accumulated metal chips and dirt are cleaned from the lathe with a _____, *never* with _____.

9. Which of the following actions are considered dangerous when operating a lathe?
   a. Wearing eye protection.
   b. Wearing loose clothing and jewelry.
   c. Measuring with work rotating.
   d. Operating lathe with most guards in place.
   e. Using compressed air to clean machine.

10. In most lathe operations, you will be using a single-point cutting tool made of _____.

11. Cutting speeds can be increased 300% to 400% by using _____ tools.

12. What does cutting speed indicate?

13. _____ is used to indicate the distance that the cutter moves in one revolution of the work.

14. Calculate the cutting speeds for the following metals. The information furnished is sufficient to do so.
    a. Formula: $rpm = \dfrac{CS \times 4}{D}$
    b. CS = Cutting speed recommended for material being machined.
    c. D = Diameter of work in inches.
    Problem A: What is the spindle speed (rpm) required to finish-turn 2 1/2″ diameter aluminum alloy? A rate of 1000 fpm is the recommended speed for finish-turning the material.
    Problem B: What is the spindle speed (rpm) required to rough-turn 1″ diameter tool steel? The recommended rate for rough turning the material is 50 fpm.

15. Calculating the cutting speed for metric-size material requires a slightly different formula.
    a. Formula: $rpm = \dfrac{CS \times 1000}{D \times 3}$
    b. CS = Cutting speed recommended for particular material being machined (steel, aluminum, etc.) in meters per minute (mpm).
    c. D = Diameter of work in millimeters (mm).
    Problem: What spindle speed is required to finish-turn 200 mm diameter aluminum alloy? Recommended cutting speed for the material is 300 mpm.

16. Most work is machined while supported by one of four methods. List them.

17. Sketch a correctly drilled center hole.

18. A tapered piece will result, when the work is turned between centers, if the centers are not aligned. Approximate alignment can be determined by two methods. What are they?

19. Describe one method for checking center alignment if close tolerance work is to be done between centers.

20. It is often necessary to turn to a shoulder or to a point where the diameters of the work change. One of four types of shoulders will be specified. Make a sketch of each.
    a. Square shoulder.
    b. Angular shoulder.
    c. Filleted shoulder.
    d. Undercut shoulder.

21. What are the four types of lathe chucks most commonly used? Describe the characteristics of each.

22. When using the parting tool, the spindle speed of the machine is about _____ the speed used for conventional turning.

23. Why is a concave rake ground on top of the cutter when used for parting operations?

24. There are many safety precautions that must be observed when operating a lathe. List what *you* consider the five most important.

Chapter 14

Cutting Tapers and Screw Threads on the Lathe

LEARNING OBJECTIVES

After studying this chapter, you will be able to:
- Describe how a taper is turned on a lathe.
- Calculate tailstock setover for turning a taper.
- Safely set up and operate a lathe for taper turning.
- Describe the various forms of screw threads.
- Cut screw threads on a lathe.

IMPORTANT TERMS

external threads
internal threads
major diameter
minor diameter
offset tailstock method
pitch diameter
setover
taper attachment
thread cutting stop
three-wire method of
    measuring threads

14.1 TAPER TURNING

A section of material is considered to be *tapered* when it increases or decreases in diameter at a uniform rate, **Figure 14-1**. A cone is an example of a taper. The "wedging" action of a taper makes it ideal as a means for driving drills, milling arbors, end mills, and centers. In addition, it can be assembled and disassembled easily, and will automatically align itself in a similarly tapered hole each time. Taper can be stated in taper per inch, taper per foot, degrees, millimeters per 25 mm of length, or as a ratio, **Figure 14-2**.

There are five principal methods of machining tapers on a lathe. Each has its advantages and disadvantages. The five methods are listed in **Figure 14-3**.

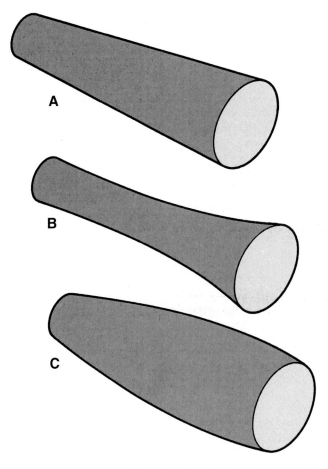

Figure 14-1. Taper. A—The diameter of a taper increases or decreases at a uniform rate. B and C—These pieces are "bell shaped," rather than tapered.

14.1.1 Taper turning with compound rest

The compound rest method of turning a taper is the easiest. Either internal or external tapers can be machined, as shown in **Figure 14-4**.

Taper length is limited, however, by the movement of the compound rest. Because the compound rest base is graduated in degrees, **Figure 14-5**, the taper must be converted to degrees. A conversion table may be used. See **Figure 14-6.**

A careful study of the print will show whether the angle given is from center, or is the included angle. **Figure 14-7** shows the difference in methods of measuring angles. If an *included angle* is given, it must be divided by two to obtain the *angle from the centerline.*

With the lathe's centerline representing 0°, pivot the compound rest to the desired angle and lock it in position. It is the usual practice to turn a taper from the smaller diameter to the larger diameter. Refer to **Figure 14-4A.**

As will be the case when turning all tapers, the cutting tool must be set on *exact center.* A toolholder that will provide ample clearance should be selected.

To machine a taper, bring the cutting tool into position with the work and lock the carriage to prevent it from shifting during the turning operation.

Since there is no power feed for the compound rest, the cutting tool must be fed evenly with both

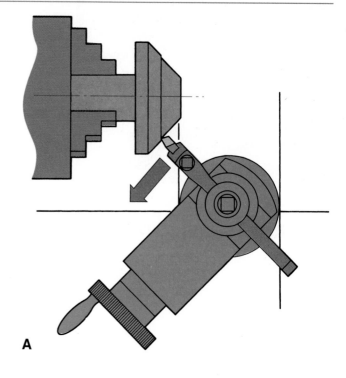

A

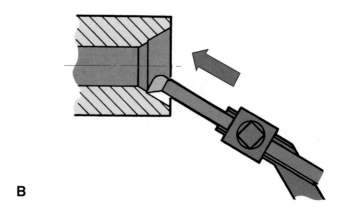

B

Figure 14-4. Cutting tapers using the compound rest. A—External taper. Note that the cut is being made from small diameter to large diameter. B—Internal taper being turned with the compound rest.

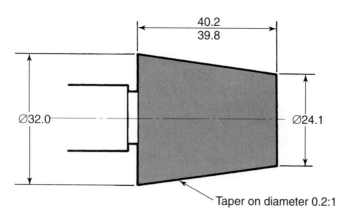

Taper on diameter 0.2:1

Figure 14-2. Taper may be stated as a ratio (0.2:1, in the example above), taper per inch, taper per foot, degrees, or in millimeters per 25 mm.

Ways of Machining Tapers		
Method	**Advantages and disadvantages**	**Information needed**
1. Compound	Length of taper limited. Will cut external and internal taper.	Must know the taper angle.
2. Offset tailstock	External taper only. Must work between centers.	Taper per inch or taper per foot.
3. Taper attachment	Best method to use.	Angle or taper per inch or foot.
4. Tool bit	Very short taper.	Taper angle.
5. Reamer	Internal only.	Taper number.

Figure 14-3. Methods by which tapers can be turned on a lathe.

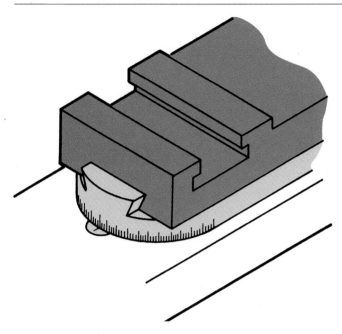

Figure 14-5. Base of the compound rest is marked in degrees to aid in precise positioning.

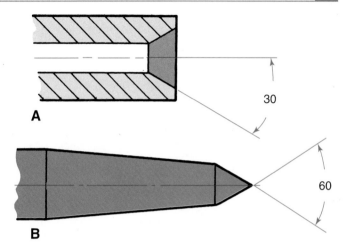

Figure 14-7. The two methods used to measure angles. A—Angle measured from the centerline of the workpiece. B—Measurement of the included angle.

Taper per Foot with Corresponding Angles		
Taper per foot	Included angle	Angle with centerline
1/16	0° 17′ 53″	0° 8′ 57″
1/8	0° 35′ 47″	0° 17′ 54″
3/16	0° 53′ 44″	0° 26′ 52″
1/4	1° 11′ 38″	0° 35′ 49″
5/16	1° 29′ 31″	0° 44′ 46″
3/8	1° 47′ 25″	0° 53′ 42″
7/16	2° 5′ 18″	1° 2′ 39″
1/2	2° 23′ 12″	1° 11′ 36″
9/16	2° 41′ 7″	1° 20′ 34″
5/8	2° 58′ 3″	1° 29′ 31″
11/16	3° 16′ 56″	1° 38′ 28″
3/4	3° 34′ 48″	1° 47′ 24″
13/16	3° 52′ 42″	1° 56′ 21″
7/8	4° 10′ 32″	2° 5′ 16″
15/16	4° 28′ 26″	2° 14′ 13″
1	4° 46′ 19″	2° 23′ 10″

Figure 14-6. You can use this table to convert taper per foot into corresponding angles for adjustment of the compound rest.

hands to achieve a smooth finish. The entire cut must be made without stopping the cutting tool. The compound rest is moved back to the starting point and positioned with the cross-slide for the next cut.

When tapers are cut with a compound rest, the work can be mounted between centers or held in a chuck. A suitable boring bar is needed when machining internal tapers. Some internal tapers are finished to size with a taper reamer.

14.1.2 Taper turning by offset tailstock method

The *offset tailstock method,* also known as the *tailstock setover method,* is also employed for taper turning, **Figure 14-8.** Jobs that can be turned between centers may be taper turned by this technique. Only *external* tapers can be machined in this way, however.

Most lathe tailstocks consist of two parts, which permits the upper portion to be shifted off center, **Figure 14-9.** This movement, referred to as *setover,* is accomplished by loosening the anchor bolt that locks the tailstock to the ways, then making the proper adjustments with screws on the tailstock. After the setover has been made, the screws are drawn up *snug,* but not tight.

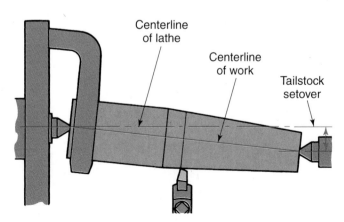

Figure 14-8. Machining a taper using the offset tailstock method.

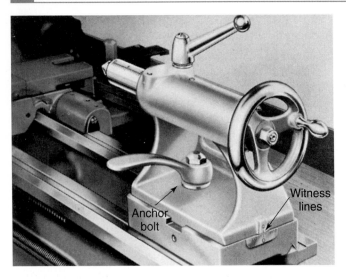

Figure 14-9. *The tailstock is usually constructed in two parts. This allows the section mounting the center to be shifted relative to the lathe's centerline. The distance off center, or setover, can be checked by observing the witness lines. (Rockwell International)*

14.2 CALCULATING TAILSTOCK SETOVER

Taper turning by this technique is *not* a precise method and requires some "trial and error" adjustments to produce an accurate tapered section. The approximate setover can be calculated when certain basic information is known.

Offset must be calculated for each job, because the length of the piece plays an important part in the calculations. When lengths of the pieces vary, different tapers will be produced with the same tailstock offset, **Figure 14-10**.

The following terms are used with calculating tailstock setover. See **Figure 14-11**.

$$D = \text{Diameter at large end}$$
$$d = \text{Diameter at small end}$$
$$\ell = \text{Length of taper}$$
$$L = \text{Total length of piece}$$
$$TPI = \text{Taper per inch}$$
$$TPF = \text{Taper per foot}$$

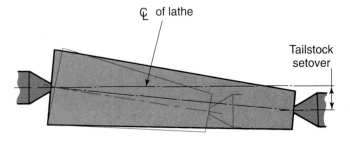

Figure 14-10. *Length of work causes taper to vary even though tailstock offset remains the same.*

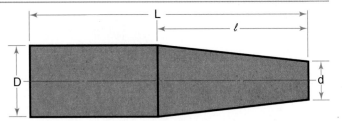

Figure 14-11. *Basic taper information. D = diameter at large end of taper; d = diameter at small end of taper; ℓ = length of taper; L = total length of piece.*

14.2.1 Calculating setover when taper per inch is known

Information needed:
$$TPI = \text{Taper per inch}$$
$$L = \text{Total length of piece}$$

Formula used: $\text{Offset} = \dfrac{L \times TPI}{2}$

Example: What will be the tailstock setover for the following job?

$$\text{Taper per inch} = 0.0125$$
$$\text{Total length of piece} = 8.000$$
$$\text{Offset} = \frac{L \times TPI}{2}$$
$$= \frac{8.000 \times 0.125}{2}$$
$$= 0.500''$$

Note: The same procedure would be followed when using metric units. However, all dimensions would be in millimeters.

14.2.2 Calculating setover when taper per foot is known

When taper per foot (TPF) is known, it must be converted to taper per inch (TPI). The following formula takes this into account:

$$\text{Offset} = \frac{TPF \times L}{24}$$

14.2.3 Calculating setover when dimensions of tapered sections are known but TPI or TPF is not given

Plans often do not specify TPI or TPF, but do give other pertinent information. Calculations will be easier if all fractions are converted to decimals. All dimensions must either be in inches or in millimeters.

Information needed:

D = Diameter at large end
d = Diameter at small end
ℓ = Length of taper
L = Total length of piece

Formula used: $\text{Offset} = \dfrac{L \times (D-d)}{2\ell}$

Example: Calculate the tailstock setover for the following job.

$$D = 1.250''$$
$$d = 0.875''$$
$$\ell = 3.000''$$
$$L = 9.000''$$

$$
\begin{aligned}
\text{Offset} &= \frac{L \times (D-d)}{2\ell} \\
&= \frac{9.000 \times (1.250 - 0.875)}{2 \times 3.000} \\
&= \frac{9.000 \times 0.375}{6} \\
&= 0.562''
\end{aligned}
$$

14.2.4 Calculating setover when taper is given in degrees

The space available in this text does not permit the introduction of basic trigonometry, which is necessary to make these calculations. However, any good machinist's handbook will provide this information. At least one such book should be part of every machinist's toolbox.

14.3 MEASURING TAILSTOCK SETOVER

When an ample tolerance is allowed, (±0.015″ or 0.05 mm), the setover can be measured with a steel rule. There are two ways to measure:

- Place a rule that has graduations on both edges between the center points, **Figure 14-12A**. Measure the distance between the center points.
- Measure the distance between the two witness lines on the tailstock base, **Figure 14-12B**.

Accurate work requires care in making the tailstock setover. An additional factor enters into the calculations—the distance that the center point enters the piece. Typically, 1/4″ (6.5 mm) is an ample allowance; it must be subtracted from the total length of the piece.

Use the appropriate method to calculate the offset. A precise setover may be made using

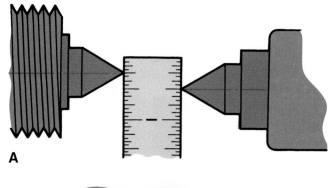

A

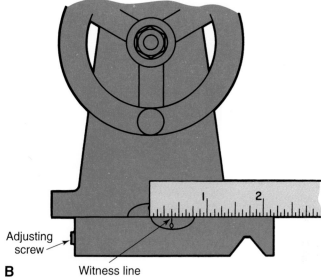

Adjusting screw

Witness line

B

Figure 14-12. Measuring setover. A—Approximate tailstock setover can be determined by measuring distance between center points. B—Approximate setover can also be determined by measuring distance between witness lines on the tailstock.

the *micrometer collar* on the lathe cross-slide. See **Figure 14-13**.

1. Clamp the toolholder in a reverse position in the tool post.

2. Turn the cross-slide screw back to remove all play.

3. Turn in the compound rest until the toolholder can be felt with a piece of paper between the toolholder and tailstock spindle.

4. Use the micrometer collar and turn out the cross-slide screw the distance the tailstock is to be set over.

5. Move the tailstock over until the spindle touches the paper in same manner described in Step 3.

6. Check the setting again after "snugging up" the adjusting screws.

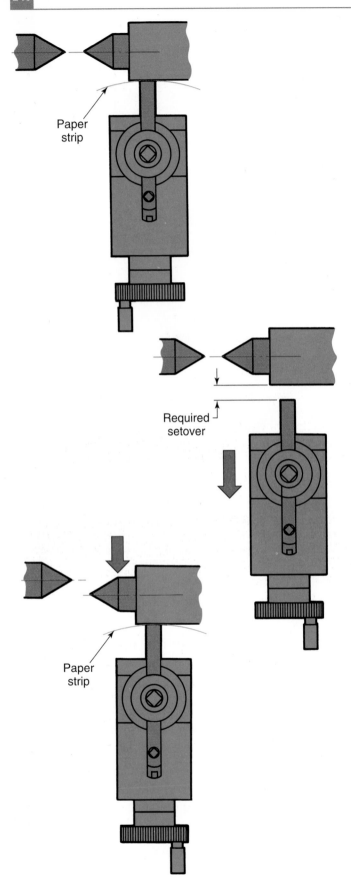

Figure 14-13. Using the micrometer collar of the compound rest to make the setover measurement.

In place of the toolholder and paper strip, a *dial indicator* can be employed to establish the offset. See **Figure 14-14**.

1. Mount the dial indicator in the tool post.

2. Position it with the cross-slide until the indicator reads zero when in contact with the tailstock spindle. There should be *no* "play" in the cross-slide.

3. Set the tailstock over the required distance using the dial indicator to make the measurement.

4. Recheck the reading after "snugging up" the adjusting screws. Make additional adjustments if any deviation in the indicator reading occurs.

14.4 CUTTING A TAPER

When cutting a taper, additional strain is imposed on the centers because they are out-of-line and do not bear true in the center holes. Because the pressures imposed are uneven, the work is more apt to heat up than when doing conventional turning between centers. It must be checked frequently for binding. A bell-type center drill offers some advantage in reducing strain. Some machinists prefer a center with a ball tip to produce an improved bearing surface. See **Figure 14-15**.

Make the cuts as in conventional turning. However, cutting should start at the small end of the taper.

14.4.1 Turning a taper with a taper attachment

A *taper attachment* is a guide that can be attached to most lathes. It is an accurate way to cut tapers and offers advantages over other methods of machining tapers.

Both internal and external tapers can be cut. This helps assure an accurate fit for mating parts. Once the attachment has been set, the taper can be machined on material of various lengths. Work can be held by any conventional means. One end of the taper attachment *swivel bar* is graduated in total taper in inches per foot. The other end is graduated to indicate the included angle of the taper in degrees.

The lathe does not have to be altered. The machine can be used for straight turning by locking out the taper attachment. No realignment of the lathe is necessary.

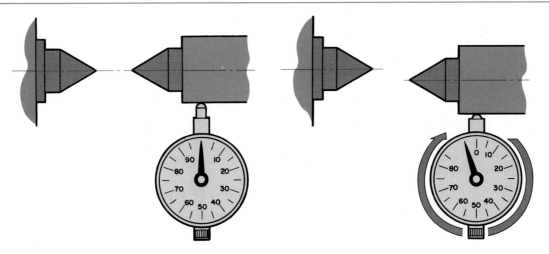

Figure 14-14. A dial indicator can also be used to measure amount of setover.

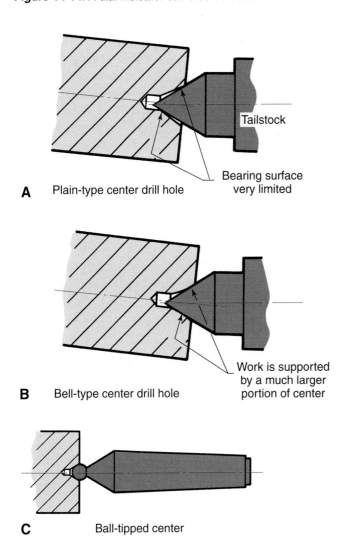

A   Plain-type center drill hole

Bearing surface very limited

Tailstock

B   Bell-type center drill hole

Work is supported by a much larger portion of center

C   Ball-tipped center

Figure 14-15. Taper turning done by the offset tailstock method is hard on the tailstock center. A—Center point does not bear evenly in conventional center hole. B—A center hole drilled with a bell-type center drill reduces the problem by providing more bearing surface. C—A ball-tipped center lessens pressure on tail center when turning tapers.

14.4.2 Types of taper attachments

There are two types of taper attachments, plain and telescopic. See **Figure 14-16**. The *plain taper attachment* requires disengaging the cross-slide screw from the cross-slide feed nut. The cutting tool is advanced by using the compound rest feed screw.

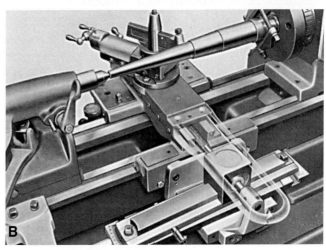

Figure 14-16. Taper attachments. A—Plain taper attachment. (South Bend Lathe Corp.) B—Telescopic taper attachment. (Clausing Industrial, Inc.)

The *telescopic taper attachment* is made in such a way that it is not necessary to disconnect the cross-slide feed nut. The tool can be advanced into the work with the cross-slide screw in the usual manner.

14.4.3 Setting a taper attachment

1. Study the plans and, if necessary, calculate the taper. Set the swivel bar as specified from the calculations.

2. Mount the work in the machine.

3. Slide the taper attachment unit to a position that will permit the cutting tool to travel the full length of the taper. Lock it to the ways.

4. Move the carriage to the right until the cutting tool is about 1" (25 mm) away from the end of the work. This will permit any play to be taken up before the tool starts to cut.

5. If the machine is fitted with a plain taper attachment, tighten the binding screw that engages the cross-slide feed to the attachment.

6. Oil the bearing surfaces of the taper attachment and make a trial cut. If necessary, readjust until the taper is being cut to specifications. Complete the cutting operation.

14.4.4 Turning a taper with a square-nose tool

Using a square-nose tool is a taper technique limited to the production of *short tapers*, **Figure 14-17**. The cutter bit is ground with a square nose and set to the correct angle with the protractor head and blade of a combination set.

The tool is positioned on center and fed into the revolving work. "Chatter" can be minimized by running the work at a slow spindle speed. The carriage must be locked to the ways.

Before using any of the taper-turning techniques on work mounted between centers, it is very important that centers be *"zeroed in"* (put in perfect alignment). Then the necessary adjustments (tailstock setover, taper attachment adjustment) can be made.

14.5 MEASURING TAPERS

There are two basic methods of testing the accuracy of machined tapers. One is a *comparison method;* the other involves *direct taper measurement.*

14.5.1 Measuring tapers by comparison

Taper plug gages and *taper ring gages* serve two purposes, **Figure 14-18**. They measure the basic diameter of the taper as well as the angle of slope. The angle is checked by applying *bluing* (usually a liquid known as "Prussian blue") to the machined surface or plug gage. The blued section is inserted into the mating part and slowly rotated. If the bluing rubs off *evenly*, it indicates that the taper is correct. If the bluing rubs off *unevenly*, **Figure 14-19**, the remaining material will show where the taper is incorrect and indicate what machine adjustments are needed.

Gages are also provided with notches to indicate the specified tolerance in taper diameter. The indentations show the *go* and *no-go* limits, **Figure 14-20**.

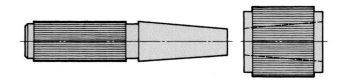

Figure 14-18. Left—Plug gage. Right—Ring gage.

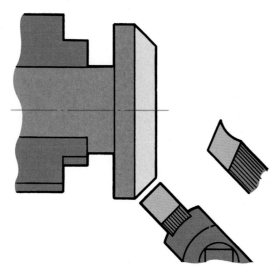

Figure 14-17. A short taper can be turned with a square-nose tool.

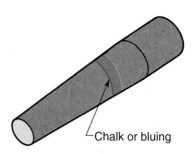

Chalk or bluing

Figure 14-19. When chalk or bluing does not rub off evenly, it indicates that taper does not fit properly and additional machine adjustments will have to be made.

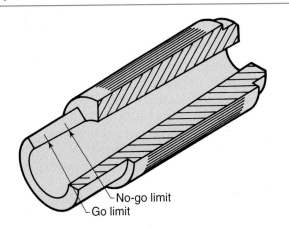

Figure 14-20. *Typical* go *and* no-go *ring gage for measuring tapers.*

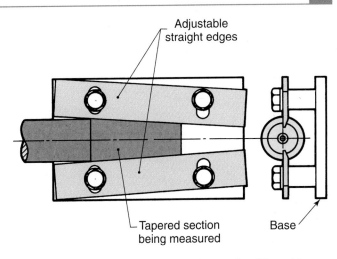

Figure 14-21. *A taper test gage can be set for different tapers.*

14.5.2 Direct measurement of tapers

A *taper test gage* is sometimes employed to check taper accuracy, **Figure 14-21**. It consists of a base with two adjustable straight edges. Slots in the straight edges permit adapting the gage to check different tapers. The taper test gage is set by using two discs of known size which are located the correct distance apart.

Another technique for checking and/or measuring tapers is to set the tapered section on a surface plate. Two gage blocks or ground parallels of the same height are placed on opposite sides of the taper. Two cylindrical rods (sections of drill rod are satisfactory) of the same diameter are placed on the blocks. See **Figure 14-22**. The distance across the rods is then measured with a micrometer.

Blocks 1", 3", or 6" (25, 75, or 150 mm) taller than those used for the first reading are substituted. The rods are the same diameter as those used to make the first reading. A second reading is made, **Figure 14-22B**. The taper per foot then can be determined. First, subtract to find the difference between the two measurements. Then multiply it by twelve (if the readings were made 1" apart), by four (if they were made 3" apart), or by two (if they were made 6" apart).

A *sine bar* is a very accurately machined bar with edges that are parallel, **Figure 14-23**. The bar is used in conjunction with gage blocks and sine tables to precisely measure angles.

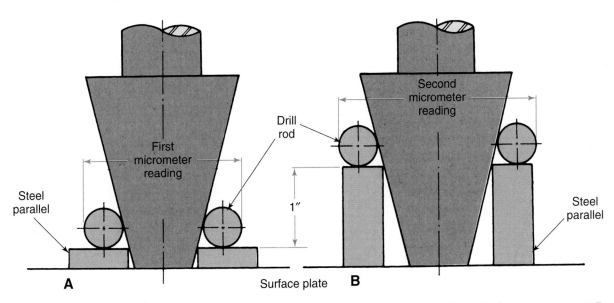

Figure 14-22. *Measuring a taper using parallels, drill rod, micrometer, and a surface plate. A—Setup for first measurement. B—Setup for second measurement.*

Figure 14-23. *A sine bar and precision gage blocks can also be used to measure a taper. (C.E. Johannson Co.)*

14.6 CUTTING SCREW THREADS ON THE LATHE

Screw threads are utilized for many applications. The more important are:
- Making adjustments (cross-feed on a lathe).
- Assembling parts (nuts, bolts, and screws).
- Transmitting motion (lead screw on a lathe).
- Applying pressure (clamps).
- Making measurements (micrometer).

14.6.1 Screw thread forms

The first screw threads cut by machine were square in cross-section. Since that time, many different thread forms have been developed, including American National, Unified, Sharp V, Acme, Worm threads, and others. Each thread form has a specific use and a formula for calculating its shape and size. See **Figure 14-24**. More than 75% of all threads cut in the United States are of the Unified (UN) 60° type.

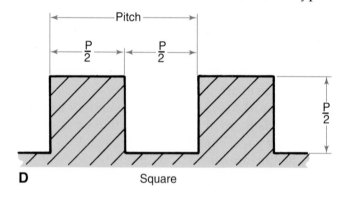

Unified Thread	Pitch = $\frac{1}{N}$	$d = \frac{0.866}{N}$
Sharp V Thread	Pitch = $\frac{1}{N}$	$d = \frac{0.866}{N}$
Acme Thread	Pitch = $\frac{1}{N}$	$d = \frac{P}{2 + 0.010}$
	Flat = 0.371P	
	Root = 0.71P–0.0052	
Square Thread	Pitch = $\frac{1}{N}$	$d = \frac{P}{2}$
	Flat or space = $\frac{P}{2}$	

Figure 14-24. *Common thread forms. A—Unified thread form, interchangeable with American National Thread. B—Sharp "V" thread form. C—Acme thread form. D—Square thread form. Note: In formulas above, N = Number of threads per inch; P = Pitch; d = depth of thread.*

The following terms relate to *screw threads,* as shown in **Figure 14-25**:
- *External threads* are cut on the outside surface of piece.
- *Internal threads* are cut on the inside surface of piece.
- *Major diameter* is the largest diameter of the thread.
- *Minor diameter* is the smallest diameter of the thread.
- *Pitch diameter* is the diameter of an imaginary cylinder that would pass through threads at such points to make width of thread and width of the spaces at these points equal.
- *Pitch* is the distance from one thread point to the next thread point, measured parallel to the thread axis. Pitch of inch-based threads is equal to 1 divided by the number of threads per inch.
- *Lead* is the distance that a nut will travel in one complete revolution of the screw. On a single thread, the lead and pitch are the same. Multiple thread screws have been developed to secure an increase in lead without weakening the thread. See **Figure 14-26**.

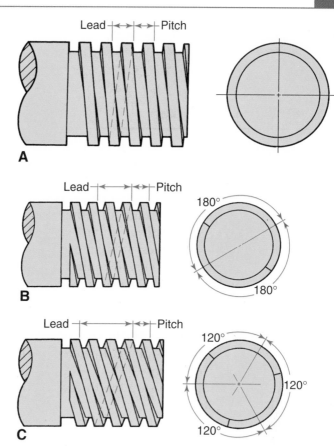

Figure 14-26. *The difference between lead and pitch. A— Single thread screw, the pitch and lead are equal. B—Double thread screw, the lead is twice the pitch. C—Triple thread screw, the lead is three times the pitch.*

14.6.2 Preparing to cut 60° threads on a lathe

Sharpen the cutting tool to the correct shape, including the proper clearance. The top is ground flat with no side or back rake, **Figure 14-27**. An oilstone is used to touch up the cutting edges and form the radius on the tip.

A *center gage* is used for grinding and setting the tool bit in position, **Figure 14-28**. The gage is often referred to as a *fishtail*.

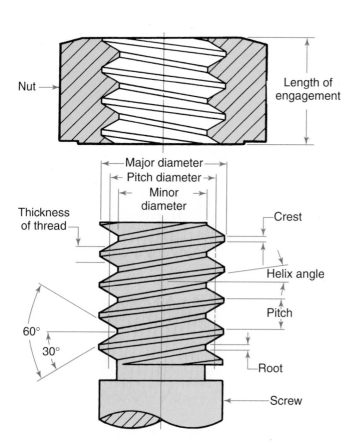

Figure 14-25. *Nomenclature of a thread.*

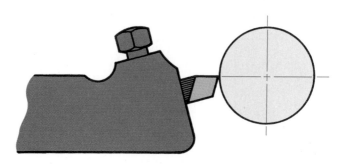

Figure 14-27. *Cutting tool positioned for cutting 60° threads. The tool is set on center as shown.*

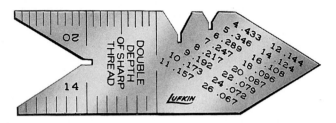

Figure 14-28. A center gage or "fishtail." (Lufkin Rule Co.)

The work is set up in the same manner as for straight turning. If mounted between centers, the centers must be precisely aligned; otherwise, a tapered thread will be produced. If this occurs, the thread will not be usable unless it is cut excessively deep at one end. The work must also run true with no "wobble." The tail of the lathe dog must have no play in the face plate slot.

A groove is frequently cut at the point where the thread is to terminate, **Figure 14-29**. The *thread end groove* is cut equal to the minor diameter of the thread and serves two purposes:

- It provides a place to stop the threading tool at the end of its cut.
- It permits a nut to be run up to the end of the thread.

Several methods may be employed to terminate a thread, as shown in **Figure 14-29**. Ordinarily, the beginner should use a groove until sufficient experience has been gained. However, the design of some parts does not permit a groove to be used. In such a case, the threads must be terminated by another method. They require perfect coordination and very rapid operation of the cross-slide to get the tool out of position at the end of the cut.

The gearbox is adjusted to cut the correct number of threads. Make apron adjustments to permit the half-nuts to be engaged. After the proper apron and gear adjustments have been made, pivot the compound rest to 29° to the right, **Figure 14-30**. Then set the threading tool in place.

It is essential that the tool be set on center with the tool axis at 90° to the centerline of the work.

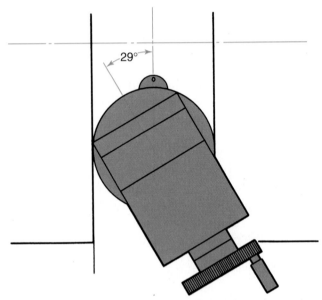

Figure 14-30. The compound rest is set up for machining right-hand external threads.

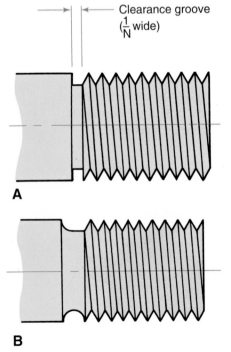

A

B

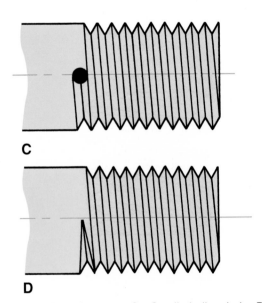

C

D

Figure 14-29. Techniques for terminating a screw thread. A—Square groove. B— Round groove. C—Small shallow hole. D—Tool withdrawn from thread at end of cut.

This is done with the aid of a center gage. Place the gage against the work while the tool is set into a V, **Figure 14-31**. Tool height can be set by using the centerline scribed on the tailstock spindle or with the center point.

The compound rest is set at 29° to permit the tool to shear the chip better than if it were fed straight into the work, **Figure 14-32**. Since the angle of the tool is 30° and it is fed in at an angle of 29°, the slight shaving action that results will produce a smooth finish on the right side of the thread. At the same time, not enough metal is removed to interfere with the main chip that is removed by the left edge of the tool.

Since the tool must be removed from the work after each cut and repositioned before the next cut can be started, a *thread cutting stop* may be used. After the point of the tool is set to just touch

the work, lock the stop to the saddle dovetail with the adjusting screw just bearing on the stop, **Figure 14-33**.

After a cutting pass has been made, move the tool back from the work with the cross-slide screw. Move the carriage back to start another cut. Feed the tool into the work until the adjusting screw again bears against the thread cutting stop. By turning the compound rest in a distance of 0.002″ to 0.005″ (0.05 mm to 0.12 mm), the tool will be positioned for the next cut.

A thread dial that meshes with the lead screw is fitted to the carriage of most lathes, **Figure 14-34**. The *thread dial* is used to indicate when to engage the half-nuts, which permit the tool to follow exactly in the original cut. The thread dial eliminates the need to reverse spindle rotation after each cut to bring the tool back to the starting point.

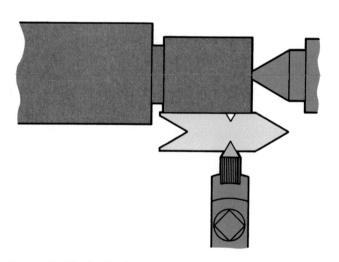

Figure 14-31. *Positioning a cutting tool for machining threads, using a center gage.*

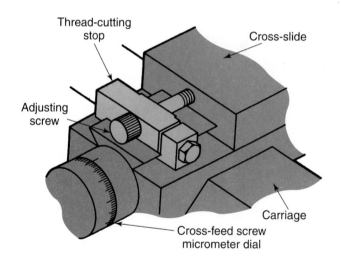

Figure 14-33. *After being properly adjusted, the thread cutting stop will let you start next cut in same location.*

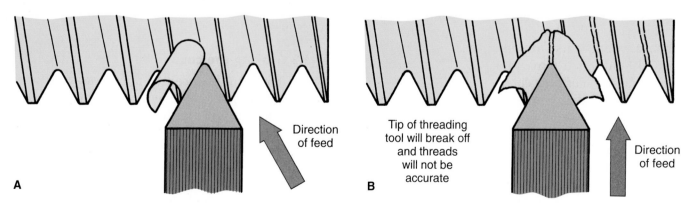

Figure 14-32. *Cutting action of tool. A—When the tool is fed in at 29° angle, note that only one edge is cutting, and that the cutting load is distributed evenly across the edge. B—When fed straight in, note that both edges are cutting and weakest part of tool, the point, is doing hardest work.*

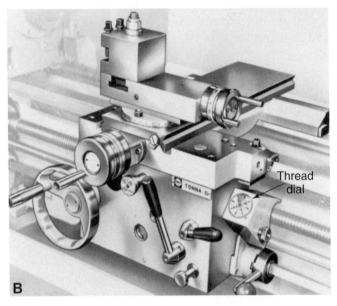

Figure 14-34. Thread dials. A—Thread dial for cutting inch-based threads. B—Dial used for cutting either inch-based or metric-based threads. The housing contains a series of gears, with gear selection depending upon threads being cut. (Clausing Industrial, Inc.)

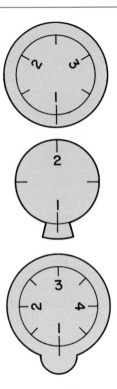

Figure 14-35. Typical thread dial faces.

The face of the thread dial, **Figure 14-35**, rotates when the half-nuts are *not* engaged. When the desired graduation moves into alignment with the index line, the half-nuts can be engaged.

The thread dial is used as follows for all inch-based threads:

- For all *even-numbered* threads, close the half-nuts at any line on the dial.
- For all *odd-numbered* threads, close the half-nuts at any numbered line on the dial.
- For all threads involving *one-half of a thread in each inch* (such as 11 1/2), close the half-nuts at any odd numbered line.

- For all threads involving *one-fourth of a thread in each inch* (such as 4 3/4), return to the original starting line before closing the half-nuts.

On lathes that have been *converted* to metric threading capability, the thread dial cannot be used. When thread cutting with such a lathe, the half-nuts (once closed) must not be opened until the thread is completely cut. The spindle rotation must be reversed after each cut to return the tool to its starting position.

The thread dial *can* be used, however, on lathes with full metric capabilities. The thread dial will vary with the lathe manufacturer and must be considered individually. To be sure of correct thread dial procedure, consult the manufacturer's handbook for the machine.

14.6.3 Making the cut

Set the spindle speed to about one-fourth the speed that is used for conventional turning. Feed in the tool until it just touches the work. Then, move the tool beyond the right end of the work and adjust it to take a 0.002″ (0.05 mm) cut.

Turn on the power and engage the half-nuts when indicated by the thread dial. This cut is made to check whether the lathe is producing the correct threads. Thread pitch can be checked with a rule or with a screw pitch gage, **Figure 14-36**. When

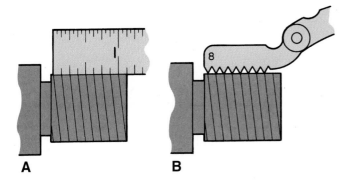

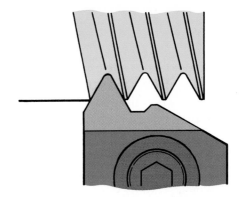

Figure 14-36. *Always check thread pitch after first light cut has been made. A—Checking with a rule. B—Checking with a screw pitch gage.*

Figure 14-37. *Cutting threads with a partial-profile insert. The major (outside) diameter of the thread must be cut to size before using this type insert.*

everything checks, make additional cuts, working in 0.005" (0.12 mm) increments, until the thread is almost to size. The last few cuts should be no more than 0.002" (0.05 mm) deep. Note that all advances of the cutting tool are made with the compound rest feed screw.

A liberal application of cutting oil, before each cut, will help to obtain a smooth finish.

14.6.4 Resetting tool in thread

It is sometimes necessary to replace a broken cutting tool, or to resharpen it for the finish cuts. After replacing the tool, you must realign it with the portion of the thread already cut. This can be done as follows:

1. Set the tool on center and position it with a center gage.

2. Engage the half-nuts at the proper thread dial graduation.

3. Move the tool back from the work and rotate the spindle until the tool reaches a position about halfway down the threaded section.

4. Using the compound rest screw and the cross-slide screw, align the tool in the existing thread. Reset the thread cutting stop after the tool has been aligned.

14.6.5 Cutting threads with insert-type cutting tools

There are two basic types of 60° threading inserts, the partial profile insert and the full-profile insert.

Partial-profile inserts, **Figure 14-37,** are most commonly used because they can cut a range of thread pitches. However, the major diameter (OD) of the thread must be cut to size prior to threading.

Deburring may be required when cutting threads on most metals.

Full-profile inserts, **Figure 14-38,** produce the best thread form and finish. The tool cuts the leading flank, the root, and the trailing flank simultaneously. The machinist needs only to check the pitch diameter to determine if the major and minor diameters of the thread are to size. No deburring is necessary since the insert trims the thread crest. The disadvantage of the full-profile insert is that a separate insert is required for each thread pitch.

14.6.6 Measuring threads

Measure threads at frequent intervals during the machining operation to assure accuracy. The easiest way to check thread size is to try fitting the threaded piece into a threaded hole or nut of the proper size. If the piece does not fit, it is too large and further machining is necessary. This technique is not very accurate, but is usually satisfactory when close tolerances are not specified.

A *thread micrometer* can be used to make quick, accurate thread measurements. It has a pointed

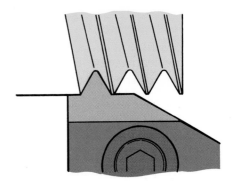

Figure 14-38. *Using a full-profile insert to cut a thread. A separate insert is required for each thread pitch.*

spindle and a double-V anvil to engage the thread. See **Figure 14-39**.

The micrometer reading given is the true pitch diameter. It equals the outside diameter of the screw minus the depth of one thread. Each micrometer is designed to read a limited number of thread pitches and is available in both inch and millimeter graduations.

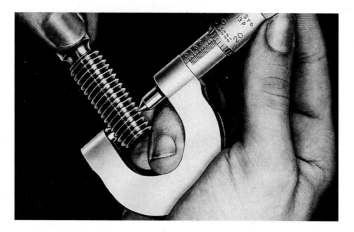

Figure 14-39. *A thread micrometer can be used to check cut threads precisely. (L.S. Starrett Co.)*

The *three-wire method of measuring threads* has proven to be quite satisfactory. As shown in **Figure 14-40A,** three wires of a specific diameter are fitted into the threads and a micrometer measurement is made over the wires. The formula in **Figure 14-40B** will provide the information necessary to calculate the correct measurement over the wires.

A three-wire thread measuring system has been developed to simplify and speed up the measuring process. It consists of a digital micrometer mounted in a special fixture that holds the threaded work-piece and the three wires. See **Figure 14-41.**

14.6.7 Cutting left-hand threads

Left-hand threads are cut in basically the same manner as right-hand threads. The major differences involve pivoting the compound to the *left* and changing the lead screw rotation so the carriage travels toward the tailstock (left to right), **Figure 14-42.**

14.6.8 Cutting square threads

Square threads are employed to transmit motion. They are more difficult to cut than 60° threads.

To cut a square thread, first calculate the width of the required tool bit (0.5 × thread pitch). If the

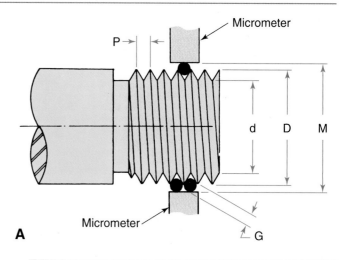

A

$$M = D + 3G - \frac{1.5155}{N}$$

Where: M = Measurement over the wires
D = Major diameter of thread
d = Minor diameter of thread
G = Diameter of wires

P = Pitch = $\frac{1}{N}$

N = Number of threads per inch

The smallest wire size that may be used for a given thread:

$$G = \frac{0.560}{N}$$

The largest wire size that can be used for a given thread:

$$G = \frac{0.900}{N}$$

The three-wire formula will work only if "G" is no larger or smaller than the sizes determined above. Any wire diameter between the two extremes may be used. All wires must be the same diameter.

B

Figure 14-40. *Three-wire method of measuring screw threads. A—Arrangement of the workpiece, wires, and micrometer. B—The three-wire thread measuring formula.*

Figure 14-41. *Thread measurements can be made in a fraction of the time normally needed with this new three-wire measuring system. Wires are mounted in individual holders that fit into the clamping fixture. (Mitutoyo/MTI Corp.)*

square thread is fairly coarse, a roughing tool is ground 0.010″ to 0.015″ (0.2 mm to 0.4 mm) smaller than the thread groove width. The cutting point of the finishing tool is ground 0.002″ to 0.003″ (0.05 mm to 0.08 mm) wider than the calculated groove width. Be sure adequate clearance is ground on the cutting tool, **Figure 14-43**.

14.6.9 Cutting Acme threads

On the *Acme thread,* the top and bottom are flat, but the sides have a 29° included angle. It was originally developed to replace the square thread. Its advantages are the strength and ease with which it can be cut, compared to the square thread. The thread form is employed in machine tools for precise control of component movement.

The *Acme screw thread gage* is the standard for grinding and setting Acme thread cutting tools. The tool angle is ground to fit a V in the thread gage. The width of the flat section varies with the pitch of the thread. This width is obtained by grinding back the tool point until it fits into the notch appropriate for the thread being cut. See **Figure 14-44.**

In cutting the threads, the groove is usually roughed out with a square nosed tool to

approximate depth, then finished with an Acme-shaped tool. The compound rest is set to 14° and the tool is positioned using the thread gage, **Figure 14-45**. Other than this, Acme threads are cut in the same manner as the Sharp V thread.

Figure 14-44. *The Acme screw thread gage and tool setup gage will allow you to check lathe settings.*

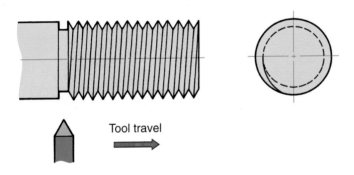

Tool travel

Figure 14-42. *Direction of tool travel for cutting left-hand threads.*

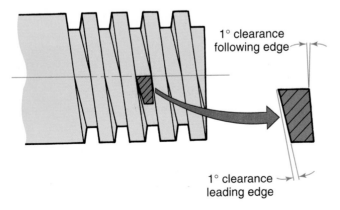

1° clearance following edge

1° clearance leading edge

Figure 14-43. *Allow adequate side clearance when sharpening a tool to cut square threads.*

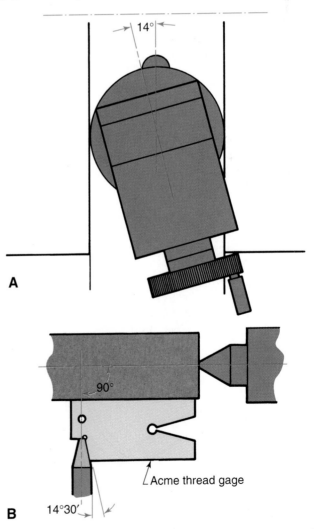

A

B

14°30′

90°

14°

Acme thread gage

Figure 14-45. *Cutting Acme threads. A—Compound setting for cutting Acme threads. B—Cutting tool is positioned with an Acme thread gage.*

14.6.10 Cutting internal threads

Internal threads, **Figure 14-46,** are made on the lathe with a conventional boring bar and a cutting tool sharpened to the proper shape.

Before internal threads can be machined, the work must be prepared. A hole is drilled and bored to correct size for the thread's minor diameter. A recess is then machined with a square-nosed tool at the point where the thread terminates, **Figure 14-47.** The diameter of the recess is equal to the major diameter of the thread.

To cut right-hand internal threads, pivot the compound rest 29° to the *left,* as shown in **Figure 14-48.** Mount the tool on center and align it, using a center gage, **Figure 14-49.**

Bring the tool up until it just touches the work surface. Adjust the micrometer collar on the

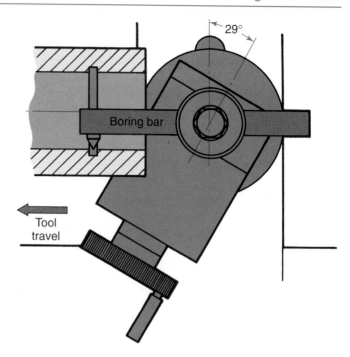

Figure 14-48. *Compound setting for cutting internal right-hand screw threads.*

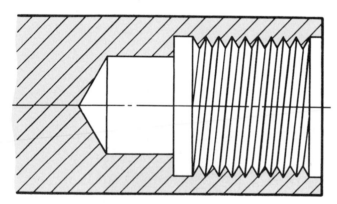

Figure 14-46. *Internal threads.*

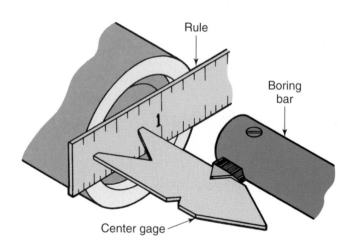

Figure 14-49. *How to position cutting tool for machining internal screw threads.*

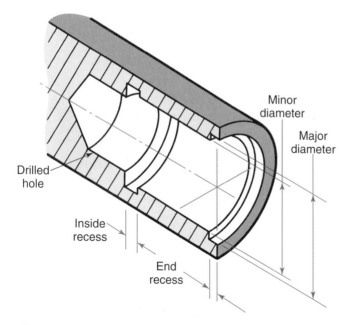

Figure 14-47. *Opening for internal screw threads has been drilled and grooves machined to the major diameter.*

cross-slide to zero with the tool in position. Using the compound rest screw, adjust the cutter to make a cut of 0.002″ (0.05 mm).

Remember that, when cutting *internal* threads, tool infeed and removal from the cut are the reverse of those used when cutting external threads.

A problem may arise in trying to determine when the tool has traveled far enough into the hole so the half-nuts can be disengaged. One method makes use of a line that has been lightly scribed in a blued area on the flat way of the lathe bed. The tool will have advanced far enough when the carriage reaches this point.

Another technique allows you to start at the back of the hole when cutting internal threads. Pivot the compound rest 29° to the *right*. Place the threading tool to the rear of the boring bar with the cutting edge up. See **Figure 14-50**.

The lathe spindle is run in reverse. To prevent the tool from being placed too far into the hole to start the cut, mount a micrometer carriage stop on the ways. See **Figure 14-51**. The carriage is returned until it touches the stop. For cutting the threads, follow the same general procedure previously described.

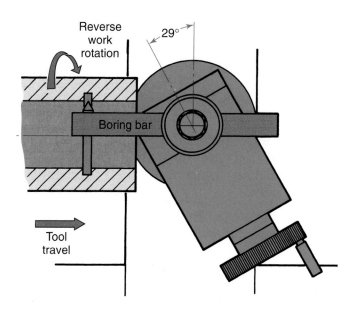

Figure 14-50. *An alternative setup for cutting internal right-hand threads. The work rotates in a direction opposite that of normal turning operations.*

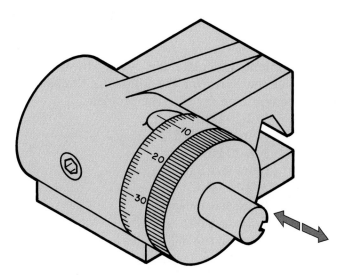

Figure 14-51. *When using alternate technique for cutting internal right-hand threads, mount a micrometer carriage stop on the ways. Adjust it to prevent the tool from being placed too far into hole when starting each cut.*

Continue making additional cuts until the threads are finished. Because the toolholder is not as rigid, lighter cuts must be taken when cutting internal threads than when machining external threads,. Keep the work flooded with cutting fluid.

14.6.11 Cutting threads on a taper surface

Tapered threads must be cut, at times, to obtain a fluid- or gas-tight joint. When this situation arises, the threading tool must be positioned in relation to the centerline of the taper rather than to the taper itself, **Figure 14-52**.

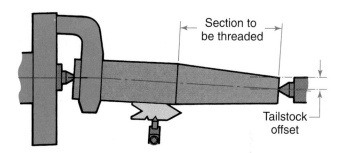

Figure 14-52. *Tool setup method for machining screw threads on a taper. Note that the tool is* not *positioned on the taper.*

TEST YOUR KNOWLEDGE

Please do not write in the text. Write your answers on a separate sheet of paper.

1. There are five ways of machining tapers on a lathe. List them, with their advantages and disadvantages.

2. When is a section of material considered tapered?

3. Machine adjustments must be calculated for each tapering job. The information given below will enable you to calculate the necessary tailstock setover for the problems given.

   Formula: When taper per inch is known,

   $$\text{Offset} = \frac{L \times TPI}{2}$$

   When taper per foot is known,

   $$\text{Offset} = \frac{L \times TPF}{24}$$

   When dimensions of tapered section are known but TPI or TPF is not given,

   $$\text{Offset} = \frac{L \times (D{-}d)}{2 \times \ell}$$

Where:

TPI = Taper per inch
TPF = Taper per foot
   D = Diameter at large end of taper
   d = Diameter at small end of taper
   ℓ = Length of taper
   L = Total length of piece

Note: These formulas, except for the TPF formula, can be used when dimensions are in mm.

Problem A: What will the tailstock setover be for the following job?

Taper per inch = 0.125″
Total length of piece = 4.000″

Problem B: What will the tailstock setover be for the following job?

D = 2.50″
d = 1.75″
ℓ = 6.00″
L = 9.00″

Problem C: What will the tailstock setover be for the following job?

D = 45.0 mm
d = 25.0 mm
ℓ = 175.0 mm
L = 275.0 mm

4. Screw threads are used for many reasons. List five or more important uses.

The following questions are of the matching type. Place the letter of the correct explanation on your paper.

5. _____ External thread.

6. _____ Internal thread.

7. _____ Major diameter.

8. _____ Minor diameter.

9. _____ Pitch diameter.

10. _____ Pitch.

11. _____ Lead.

   a. Smallest diameter of thread.
   b. Largest diameter of thread.
   c. Distance from one point on a thread to a corresponding point on next thread.
   d. Cut on outside surface of piece.
   e. Diameter of imaginary cylinder that would pass through threads at such points as to make width of thread and width of space at these points equal.
   f. Cut on inside surface of piece.
   g. Distance a nut will travel in one complete revolution of screw.

12. A groove is cut at the point where a thread is to terminate. It is cut to the depth of the thread and serves to:
   a. Provide a place to stop the threading tool after it makes a cut.
   b. Permits a nut to be run up to the end of the thread.
   c. Terminate the thread.
   d. All of the above.
   e. None of the above.

13. The tip of a cutting tool to cut a Sharp V thread is sharpened using a _____ to check that it is the correct shape. This tool is frequently called a _____.

14. The _____ is fitted to many lathe carriages. It meshes with the lead screw and is used to indicate when to engage the half-nuts to permit the thread cutting tool to follow exactly in the original cut.

15. The compound rest is set at _____ when cutting threads to permit the cutting tool to shear the material better than if it were fed straight into the work.

16. The three-wire thread measuring formula for inch-based threads is:

$$M = D + 3G - \frac{1.5155}{N}$$

Where: G = Wire diameter
   D = Major diameter of thread (Convert to decimal size)
   M = Measurement over the wires
   N = Number of threads per inch

Problem: Calculate the correct measurement over the wires for the following threads. Use the wire size given in the problem.
   a. 1/2-20 UNF (wire size 0.032″)
   b. 1/4-20 UNC (wire size 0.032″)
   c. 3/8-16 UNC (wire size 0.045″)
   d. 7/16-14 UNC (wire size 0.060″)

Chapter 15

Other Lathe Operations

LEARNING OBJECTIVES

After studying this chapter, you will be able to:
- ○ Safely set up and operate a lathe using various work-holding devices.
- ○ Properly set up steady and follower rests.
- ○ Perform drilling, boring, knurling, grinding, and milling operations on a lathe.
- ○ Demonstrate familiarity with industrial applications of the lathe.

IMPORTANT TERMS

automatic screw machine	*mandrel*
boring	*reaming*
boring mills	*steady rest*
follower rest	*turret*
knurling	*turret lathe*

15.1 BORING ON A LATHE

Boring is an internal machining operation in which a single-point cutting tool is used to enlarge a hole, **Figure 15-1**. Boring may be employed to enlarge a hole to a specified size where a drill or reamer will not do the job. When properly set up, it produces a hole that is concentric with the outside diameter of the work.

While the machining technique remains essentially the same as for external turning, several conditions will be encountered that could cause you difficulty. When boring on a lathe, you must make allowances for the following:

- Movement of the cross-slide screw is reversed.
- The machinist must work by "feel," since the cutting action cannot always be observed.
- Additional front clearance must be ground on the cutting tool to avoid rubbing, **Figure 15-2**. Otherwise, the shape of the cutting tool is identical to that used for external turning.
- Boring a deep or small-diameter hole requires a long, slender boring bar. The overhang makes the tool more likely to spring away from the surface being machined. It is also necessary to take several light cuts, instead of one heavy cut, to remove the same amount of material.
- Some people find that internal measuring tools are more difficult to use than those for making external measurements.

Figure 15-1. Boring or machining internal surfaces is sometimes done on lathe.

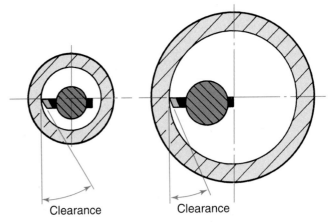

Clearance   Clearance

Figure 15-2. Tool used to bore small diameter holes requires greater front clearance to prevent rubbing.

15.1.1 Boring the hole

The hole size to be bored determines the type and size boring bar required, **Figure 15-3**. Always use the largest bar possible to give maximum tool support. The bar should extend from the holder only far enough to permit the tool to cut to the required hole depth, **Figure 15-4**.

The boring bar is set on center or slightly below center, with the bar parallel to tool travel, **Figure 15-5**. Check for adequate clearance when the tool is at maximum depth in the hole.

Begin by making a light cut in the same manner as you would for external machining. When the cut is completed, stop the machine. Set the cross-slide

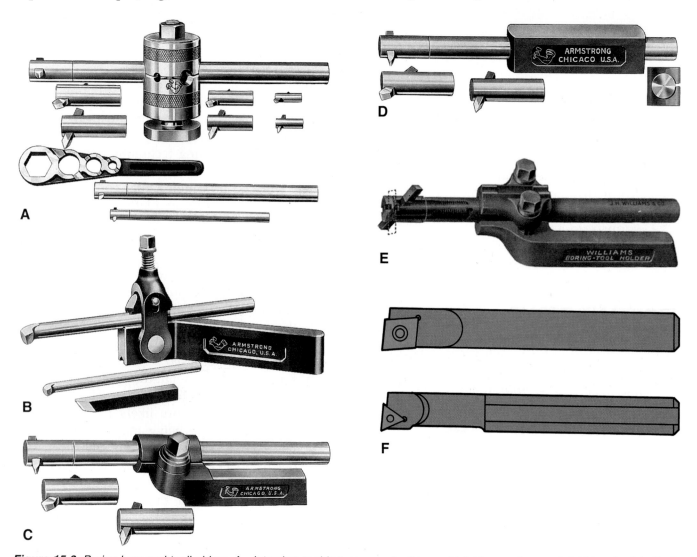

Figure 15-3. Boring bars and toolholders. A—Interchangeable type permits the machinist to use the most rigid bar for job. Body of this model replaces the tool post. B—Toolholder and boring bar for light internal machining operations. C—Boring bar with interchangeable tool ends. D—Heavy duty boring toolholder. The boring bar is held in the holder by pressure of the tool post screw. E—The clamp-type boring bar toolholder allows use of boring bars of different diameters. (Armstrong Bros. Tool Co.) F—Boring bars with indexable cutting tool inserts.

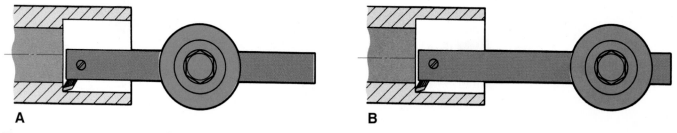

Figure 15-4. Keep the cutting tool as close to tool post as possible for maximum tool support. A—Properly positioned boring bar. B—Boring bar projecting too far from tool post. The resulting vibration and "chatter" could produce a rough machined surface.

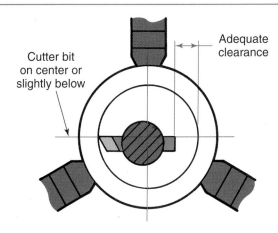

Figure 15-5. The tool is set on, or slightly below, center when boring. Be sure to check for adequate clearance between boring bar and hole.

micrometer dial to zero and back the tool away from the work. Remove the boring bar from the hole.

Check the hole diameter with an inside micrometer or with a telescoping gage and micrometer. After checking hole accuracy, bring the cross-slide back to zero, and advance the tool to make another cut. The amount of infeed will be determined by the boring bar in use and the material being bored. Make additional cuts, checking the hole size frequently, until the desired diameter is attained.

When making the final cut, it may be necessary to reverse tool travel after reaching the desired depth. *Reverse the carriage feed*, not *the spindle rotation*. Let the tool feed out of the hole without changing the tool setting. This will compensate for any tool spring.

When boring holes with long, slender boring bars, it may be necessary to run the tool into the hole without changing its setting after every second or third cut to compensate for tool spring.

With such long, slender boring bars, "chatter" is more likely to occur than when doing external work. This can usually be eliminated by:

- Using a slower spindle speed.
- Reducing tool overhang.
- Grinding a smaller radius on the cutting tool nose.
- Placing a weight on the back overhang of the boring bar.
- Placing the tool slightly below center.

15.2 DRILLING AND REAMING ON A LATHE

The lathe can perform many operations other than turning. It is sometimes used to drill or ream holes.

15.2.1 Drilling on a lathe

When a hole is to be cut in solid stock, the usual practice is to hold it in a suitable chuck and mount the drill in the tailstock. *Drilling* is accomplished on a lathe by feeding the stationary drill into the rotating workpiece. For holes that are 1/2" (12.5 mm) or less in diameter, a straight shank drill is placed in a Jacobs chuck, which is then fitted into the tailstock spindle, **Figure 15-6**. Holes larger than 1/2" (12.5 mm) in diameter are made with taper shank drills, **Figure 15-7**.

Drills that have taper shanks too large to be fitted in the tailstock can be used, if mounted as shown in **Figure 15-8**. A dog is fitted to the neck of the drill. The tool is set up to permit the tailstock center to press into the center hole in the drill tang.

The drill's cutting point bears against the rotating work. The drill is prevented from revolving by the dog bearing against the compound rest. The tailstock center keeps the drill aligned and enables it to be fed into the material by the tailstock handwheel.

Figure 15-6. Drilling is being done with straight shank drill held in Jacobs chuck. This arrangement is usually used for holes less than 1/2" (12.5 mm) in diameter.

Figure 15-7. Drills larger than 1/2" (12.5 mm) in diameter are usually fitted with a self-holding taper that fits into the tailstock spindle of the lathe.

Extreme care must be used to prevent the drill from slipping off the tailstock center after it breaks through the work.

The makeshift lathe dog setup can be avoided by use of a commercial drill holder, **Figure 15-9**.

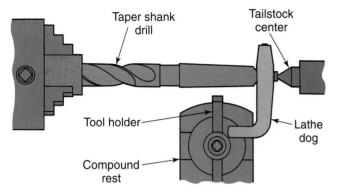

Figure 15-8. *When a drill shank is too large to be fitted into the tailstock, a lathe dog can be used to keep it from revolving. The tail of the lathe dog is supported by the compound rest. This type of drilling requires care to prevent the drill from slipping off the tailstock center when the full drill diameter breaks through the work.*

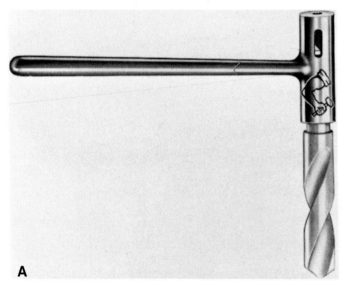

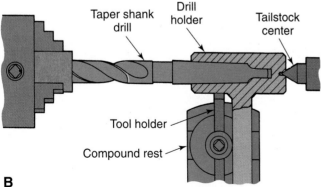

Figure 15-9. *Commercial drill holder. A—Large taper shank drills can be used on the lathe by fitting them in this holder. (Armstrong Bros. Tool Co.) B—How the commercial drill holder is used.*

Accuracy in drilling requires a centered starting point for the drill. A starting point made with a combination drill and countersink is adequate for most jobs. Holes over 1/2″ (12.5 mm) in diameter require a pilot hole. This hole should have a diameter equal to width of larger drill's dead center. See **Figure 15-10**.

Ample clearance must be provided in back of the work so that the drill will not strike the chuck or headstock spindle when it breaks through, **Figure 15-11**.

15.2.2 Reaming on a lathe

Reaming is an operation used to make a hole accurate in diameter and finish, **Figure 15-12**.

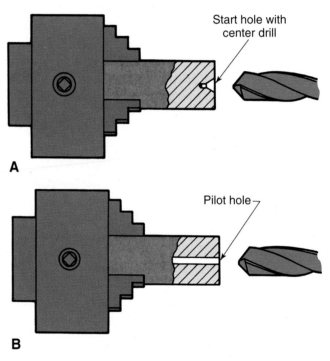

Figure 15-10. *Centering drill. A—The drill will cut exactly on center if the hole is started with a center drill. B—Holes larger than 1/2″ (12.5 mm) in diameter require drilling of a pilot hole.*

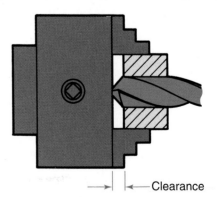

Figure 15-11. *There must be enough clearance between the back of the work and the chuck face to permit the drill to break through the work without damaging the chuck.*

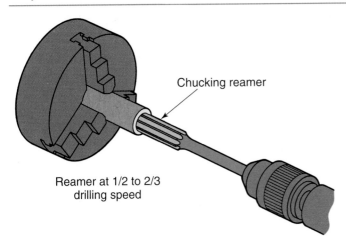

Figure 15-12. *A chucking reamer can be mounted in a Jacobs chuck to finish a hole.*

The hole is drilled slightly undersized to allow stock for reaming. The allowance for reaming depends upon hole size:

- With hole sizes ranging up to 1/4″ (6.5 mm) in diameter, allow 0.010″ (0.25 mm) of material for reaming.
- With a hole size from 1/4″ (6.5 mm) to 1/2″ (12.5 mm) in diameter, allow 0.015″ (0.4 mm).
- With a hole size from 1/2″ (12.5 mm) to 1.0″ (25.0 mm) in diameter, allow 0.020″ (0.5 mm).
- With a hole size from 1.0″ (25.0 mm) to 1.5″ (37.5 mm) in diameter, allow 0.025″ (0.6 mm).
- With a hole size above 1.5″ (37.5 mm) in diameter, allow 0.030″ (0.8 mm) for reaming.

When reaming, use a cutting speed about two-thirds the speed you would use for a similar size drill with the material being machined. Also, use a slow, steady feed with an adequate supply of cutting fluid. Remove the reamer from the hole before stopping the machine.

If a hand reamer is to be used, do *not* apply power to the workpiece mounted in the chuck. Fit the reamer into the hole, supporting the shank end with the tailstock center. Use an adjustable wrench to turn the reamer in a clockwise direction, **Figure 15-13.**

When removing a reamer from the hole, continue to rotate the tool *clockwise*. Avoid turning it *counterclockwise*, since that would ruin the tool's cutting edges.

15.3 KNURLING ON A LATHE

Knurling is the process of forming horizontal or diamond-shaped *serrations* (raised grooves or teeth) on the circumference of the work, **Figure 15-14.** Knurling is used to provide a gripping

surface, change the appearance of the work, or increase the work's diameter. It is done with a *knurling tool* mounted in the tool post, **Figure 15-15.** The knurled pattern is raised by rolling the knurls against the metal. This displaces the metal into the required pattern.

Angular knurls raise a diamond pattern, while a *straight knurl* produces a straight pattern along the length of the work. The patterns can be produced in coarse, medium, and fine pitch. See **Figure 15-16.**

Figure 15-13. *Using a hand reamer on the lathe. Never turn on the power when performing hand reaming operations.*

Figure 15-14. *Knurling rollers are being used to form serrations on a part.*

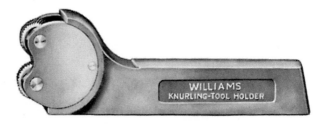

Figure 15-15. *One type of knurling tool. (Armstrong Bros. Tool Co.)*

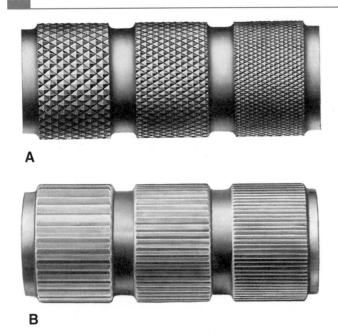

A

B

Figure 15-16. Knurling patterns. A—Diamond knurl in coarse, medium, and fine pitch. B—Straight knurl in coarse, medium, and fine pitch.

15.3.1 Knurling procedure

If a knurling tool setup is *not* made properly, the knurls will *not* track and will quickly dull. The following procedure is recommended:

1. Mark off section to be knurled.

2. Adjust the lathe to a slow back-geared speed and a fairly rapid feed.

3. Place the knurling tool in the tool post. Bring it up to the work. Both wheels must bear evenly on the work with their faces parallel with the centerline of the piece, **Figure 15-17**.

4. Start the lathe and slowly force the knurls into the work surface until a pattern begins to form. Tool travel should be *toward* the headstock whenever possible. Engage the automatic feed and let the tool travel across the work. Flood the work with cutting fluid.

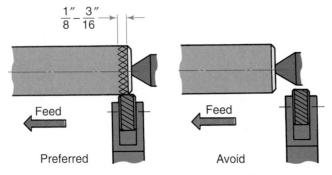

Figure 15-17. Always start the knurl on the work.

5. When the knurling tool reaches the proper position, reverse spindle rotation and allow the tool to move back across the work to the starting point. Apply additional pressure to force the knurls deeper into the work.

6. Repeat the operation until a satisfactory knurl is formed.

15.3.2 Knurling difficulties

If knurling is not performed properly, problems can arise and destroy the work. A common problem is the *double-cut knurl*, **Figure 15-18**. It occurs when one wheel of the knurling tool makes twice as many ridges as the other.

A double-cut knurl is usually caused by one wheel being dull. Raising or lowering the knurling tool to put more pressure on the dull wheel will frequently eliminate the trouble. Pivoting the tool slightly to allow the right side of the wheels to apply more pressure may also help.

Considerable side pressures are developed during knurling operations. Watch the tool carefully. Do not permit the work to slip into the chuck or loosen on the tailstock center. If a ball-bearing center is not used, keep the tailstock center well-lubricated.

Perform knurling before turning a shaft to a smaller diameter. If knurling is done after the smaller diameter has been machined, the work will spring away from the tool, giving the surface a superficial (light, nonpenetrating) knurl. It may also cause a permanent bend in the workpiece. See **Figure 15-19**.

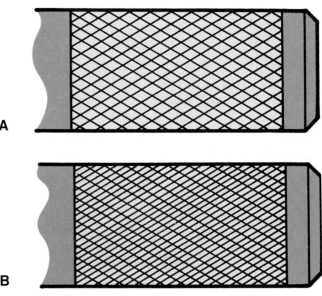

A

B

Figure 15-18. Double-cut knurl. A—This is a correctly made diamond knurl pattern. B—A double-cut diamond knurl. It results when one knurl wheel is slightly above or below center.

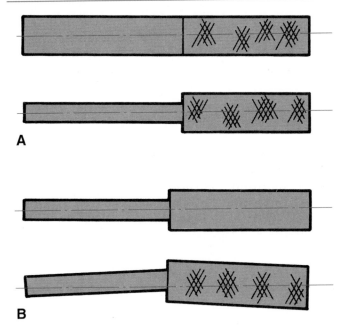

Figure 15-19. *Knurling problems. A—Do knurling before turning a shaft to a smaller diameter. B—If knurled after being turned to smaller diameter, shaft may take on a permanent bend and receive only a superficial (very light) pattern.*

Avoid applying too much pressure to the knurling tool. The work surface becomes hardened during the operation and the knurled section could "flake off." High pressure also tends to bend the shaft.

Never stop the lathe with the knurls engaged in the work. The piece will take on a permanent bend.

15.4 FILING AND POLISHING ON A LATHE

Lathe filing is done to remove burrs, to round off sharp edges, and blend-in form cut outlines. A file is *not* intended to replace a properly sharpened cutting tool and should not be used to improve the surface finish on a turned section.

15.4.1 Filing on a lathe

When filing on a lathe, avoid holding the tool stationary against the work. Keep it moving across the area being filed. If the file is held in one position, it will "load" with metal particles and score the surface of the work. An ordinary mill file will produce satisfactory results. However, a *long-angle lathe file* produces a superior cutting action. See **Figure 15-20.**

Operate the lathe at high spindle speed and apply long, even strokes. Release pressure on the return stroke. If uneven pressure is applied, out-of-round work will result. Clean the file often.

As simple as filing may appear, it can be quite dangerous if a few precautions are not observed.

- Move the carriage out of the way and remove the tool post.
- Use the left-hand method of filing, **Figure 15-21.** It involves holding the file handle in the left hand. The right hand is then clear of the revolving chuck or face plate.
- Avoid the right-hand method, **Figure 15-22.** This technique places your left arm over the chuck or faceplate.

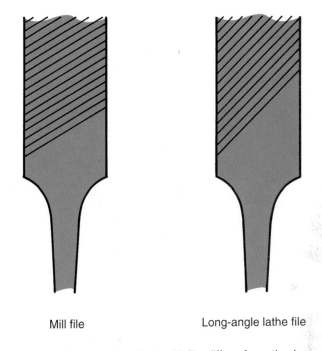

Mill file   Long-angle lathe file

Figure 15-20. *How a standard mill file differs from the long-angle lathe file.*

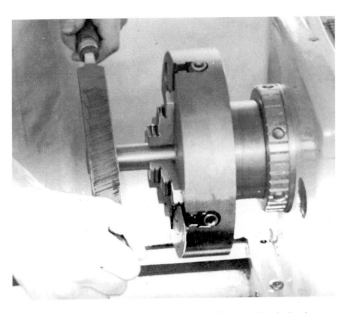

Figure 15-21. *The left-hand method of filing on the lathe is preferred. How does it differ from the right-hand method shown in Figure 15-22?*

15.4.2 Polishing on a lathe

Polishing is sometimes done on a lathe using a strip of abrasive cloth suitable for the material to be polished. The strip is grasped between your fingers and held across the work, **Figure 15-23**. If more pressure is required, mount the abrasive cloth on a strip of wood or on a file, **Figure 15-24**. A high spindle speed is used for polishing.

The finer the abrasive used, the finer the resulting finish. A few drops of machine oil on the abrasive will improve the finish. For the final polish, reverse the cloth so the cloth backing, rather than the abrasive, is in contact with the work surface. Like filing, polishing is *not* a substitute for a properly sharpened tool bit.

Carefully and thoroughly clean the lathe after polishing operations. If not removed, abrasive particles from the cloth will cause rapid wear of the machine's moving parts.

15.5 STEADY AND FOLLOWER RESTS

The *steady rest* and the *follower rest* are needed to provide additional support when the workpiece is long and thin. This additional support keeps the work from springing or bending away from the cutting tool. The support is also needed to reduce "chattering" when long shafts are machined. See **Figure 15-25**.

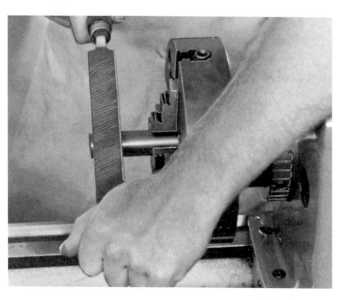

Figure 15-22. *In the right-hand filing method, your left hand and arm must be over the revolving chuck. Avoid this method, which has the potential for injury.*

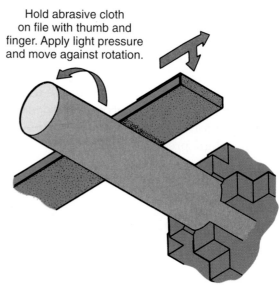

Hold abrasive cloth on file with thumb and finger. Apply light pressure and move against rotation.

Figure 15-24. *More pressure can be applied if abrasive cloth is supported by a file or block of wood.*

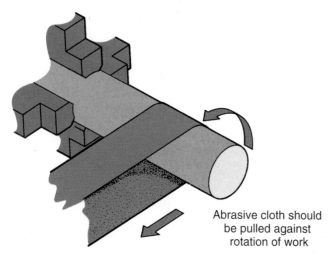

Abrasive cloth should be pulled against rotation of work

Figure 15-23. *Polishing with abrasive cloth held in hands. Keep hands away from revolving chuck or dog.*

Figure 15-25. *When the nature or shape of work prevents it from being mounted between centers, special devices called "rests" are used to provide support. This lathe is equipped with both a steady rest and a follower rest. (South Bend Lathe Corp.)*

The steady rest, sometimes called a *center rest*, is bolted directly to the ways. It is provided with three adjustable jaws, each with individual locking screws. The upper portion of the attachment is fitted with a single jaw. It can be opened to permit the work to be placed in position, **Figure 15-26**.

15.5.1 Steady rest setup

To set up a steady rest:

1. Bolt the attachment to the ways at the desired position.

2. Back off all jaws and open the upper section.

3. Mount the work between centers or in a chuck. Support the free end with the tailstock center.

4. Lower and lock the upper segment in place.

5. Adjust the jaws up to the work and lock them into position. The jaws act as bearing surfaces where they contact the work. They must be well lubricated.

6. If the shaft being machined is unsuitable as a bearing surface (rough surface, out-of-round, square, etc.), a cat head is employed. Care must be taken to center the shaft within the cat head. **See Figure 15-27**.

7. When machining at the end of a long shaft which cannot be supported with the tailstock, center the work in the chuck. Adjust the center rest to the work as close to the chuck as possible, then move it to a point where the support is needed. The same technique is employed when performing drilling, reaming, tapping, or other operations on the end of a long shaft.

15.5.2 Follower rest setup

The follower rest operates on the same principle as the steady rest and is used in a similar manner. The follower rest differs slightly in that it provides support directly in back of the cutting tool and follows along during the cut. See **Figure 15-28**.

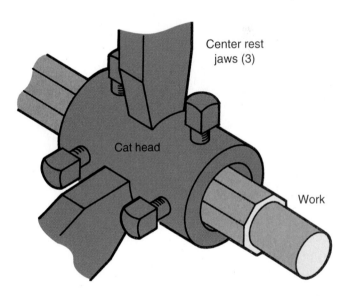

Figure 15-27. A cat head will provide a bearing surface when needed.

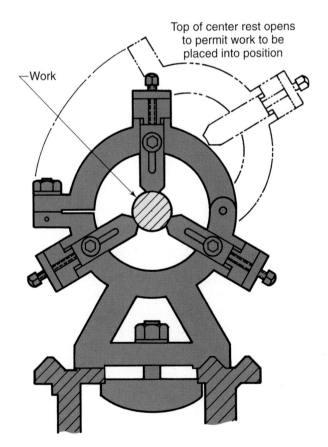

Figure 15-26. To permit easy installation of work, the top of the steady or center rest swings open. Care must be taken to have work accurately centered.

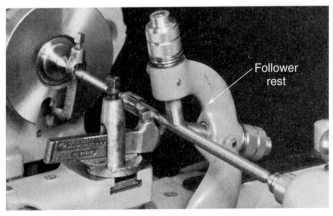

Figure 15-28. A follower rest is being used to support a long slender shaft while threads are being machined on it. (South Bend Lathe Corp.)

The follower rest bolts directly to the carriage and the jaws adjust in the same way as on the steady rest. Note that the jaws must be readjusted after each cut.

15.6 MANDRELS

At times, it is necessary to machine the outside diameter of a piece concentric with a hole that has been previously bored or reamed. This can be a simple operation if the material can be held in the lathe by conventional means. There are, however, times when the material cannot be gripped solidly to permit accurate machining. In such cases, the work is mounted on a *mandrel* and turned between centers, **Figure 15-29**.

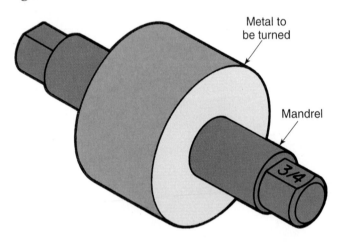

Figure 15-29. This lathe mandrel has work mounted on it.

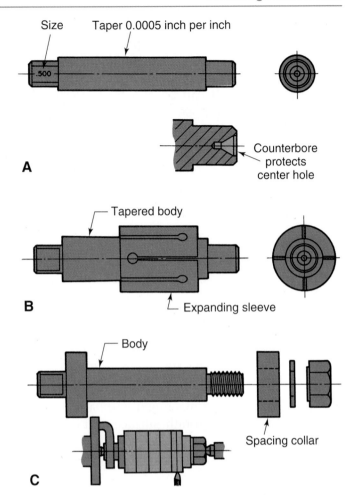

Figure 15-30. Different kinds of mandrels. A—Solid type. B—Expansion type. C—Gang mandrel that allows mounting of several identical pieces.

A *solid mandrel* is made from a section of hardened steel that has been machined with a slight taper (0.0005" per inch), **Figure 15-30A**. These mandrels are made in standard sizes starting at 1/8" in diameter. The size is stamped on the large end. The other end is slightly smaller than specified size to permit easy installation in the work.

An *expansion mandrel* permits work with openings that vary from standard sizes to be turned. See **Figure 15-30B**. The shaft and sleeve have corresponding tapers and are machined from hardened steel. The sleeve is slotted so it can expand when forced onto the tapered shaft.

A *gang mandrel*, **Figure 15-30C,** is helpful when many pieces of the same configuration must be turned. Several pieces are mounted on the mandrel and separated with spacing collars. They are locked in place by tightening a nut.

15.6.1 Installing a mandrel

Work is pressed on a mandrel with a mechanical arbor press. The work must first be checked for burrs and cleaned. Lubricate the work with a light oil to prevent it from "freezing" on the mandrel.

The mandrel is mounted between centers and driven by a lathe dog. Use care so the tool does not come into contact with the mandrel during the machining operation. In an emergency, a mandrel can be machined from a section of mild steel.

15.7 GRINDING ON THE LATHE

The tool post grinder permits the lathe to be used for internal and external grinding, **Figure 15-31**. With a few simple attachments, it is possible to sharpen reamers and milling cutters on the lathe. You can also grind shafts and true lathe centers.

Since steel parts sometimes warp during heat treatment, it is common to machine the piece to within 0.010" to 0.015" (0.2 mm to 0.3 mm) of finished size. After heat treatment, the metal is mounted on the lathe for grinding to finished size. A light grinding cut is made on each pass. When grinding is done properly, a very smooth finish results.

A *diamond wheel dresser,* consisting of an industrial diamond tip mounted on a steel shank, is used for the truing operation, **Figure 15-32.** It is mounted solidly to the lathe, on center or slightly below the center of the grinding wheel. The rotating wheel is moved back and forth across the diamond, removing about 0.001″ (0.02 mm) on each pass. Remove only enough material to true the wheel.

15.7.3 External grinding

External grinding, **Figure 15-33,** is done to finish the exterior surface of the piece. The following steps are recommended to complete the job with the least amount of difficulty:

1. Mount the work solidly in the lathe. Provide adequate clearance.

2. Adjust lathe spindle speed for 80–100 rpm, and set a feed of 0.005″– 0.007″ (0.12 mm–0.17 mm).

3. Turn on power for lathe and grinder. The work turns into the grinding wheel, **Figure 15-34.**

4. Feed the grinding wheel into the work until it just begins to "spark."

A

B

Figure 15-31. Tool post grinders. A—This is a typical light tool post grinder. (The Dumore Co.) B—A reciprocating type tool post grinder. (NSK America)

15.7.1 Preparing a lathe for grinding

Particles of the grinding wheel wear away during the grinding operation. Abrasive particles can cause excessive wear should they get into moving parts, so it is important to protect the lathe from them. When preparing to grind, cover the lathe bed, cross-slide, and other parts with canvas or heavy kraft paper to protect them from abrasive dust and grit. It is also good practice to place a small tray of water or oil just below the grinding wheel to collect as much grit and dust as possible.

When placing protective covering on the lathe, be sure the covering material cannot become entangled in the lead screw or other moving parts.

15.7.2 Preparing the grinder

Select the grinding wheel best suited for the job. It must be balanced and run true if a smooth, accurately sized job is to be obtained.

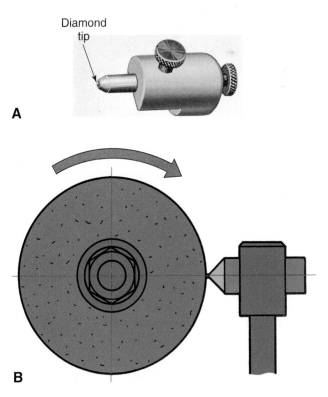

Diamond tip

A

B

Figure 15-32. Diamond wheel dresser. A—A diamond tip on the wheel dresser is used to true tool post grinder wheels. (Black and Decker) B—Using a diamond dressing tool to true a wheel on a tool post grinder before grinding on lathe.

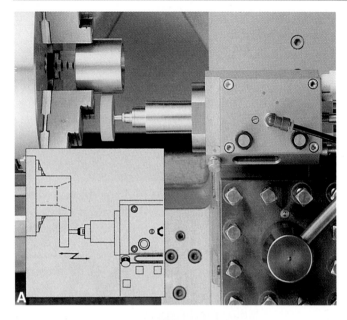

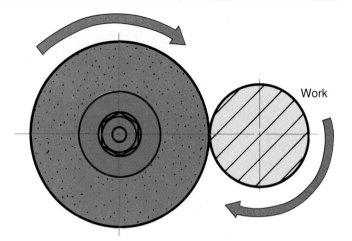

Figure 15-34. With external grinding, the work turns into grinding wheel.

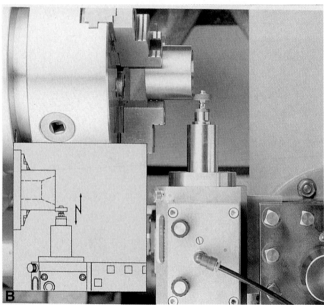

Figure 15-33. External grinding on a lathe. A—Grinding the circumference of a workpiece. B—Grinding the end of a workpiece. (NSK America)

5. Engage the automatic longitudinal feed.

6. Check work diameter frequently with a micrometer. Use light cuts; otherwise, the piece might overheat and warp.

7. Dress the grinding wheel again before making the final pass over the work. Allow the work to "spark out" (reach the point where the grinding wheel no longer cuts).

15.7.4 Internal grinding

Internal grinding is done in much the same manner as external grinding but on the inside of the

work, **Figure 15-35**. The work and grinding wheel must rotate in opposite directions, **Figure 15-36**. Because the quill (shaft for mounting the grinding wheel for internal work) is quite slender, use very light cuts and slow feeds to prevent the hole from "bell mouthing," as shown in **Figure 15-37**. For the same reason, it is suggested that the grinding wheel be allowed to "spark out" on the last cut.

15.8 MILLING ON A LATHE

Some lathes can be fitted with a vertical milling attachment, **Figure 15-38**. Such machines are primarily designed for home workshops, but are often used in model and experimental shops. A vise is mounted to the cross-slide which also provides traverse (in and out) movement while longitudinal (back and forth) feed is furnished by the carriage.

A special horizontal milling attachment is available for some lathes. It permits limited milling operations to be performed, **Figure 15-39**. The cutter is mounted on an arbor or fitted into the headstock. Cutter depth is controlled by the adjusting screw on the device. Cutter movement is controlled by carriage and cross-slide movements.

15.9 SPECIAL LATHE ATTACHMENTS

A *tracing* or *duplicating unit* should be used when several identical pieces must be produced. The duplicating unit improves the quality of the part because each is an exact duplicate of the master template. However, *computer-controlled lathes* are now doing the work formerly done by these units.

Several types of duplicating or tracing units are available. One type makes use of flat templates.

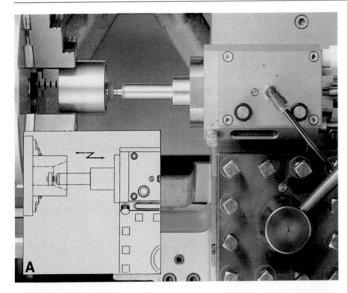

A

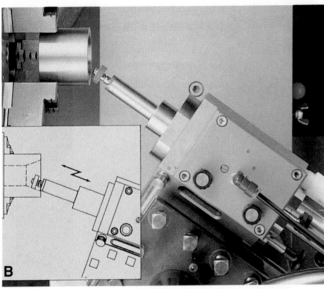

B

Figure 15-35. *Internal grinding operations on the lathe. A—Internal grinding. B—Taper grinding. (NSK America)*

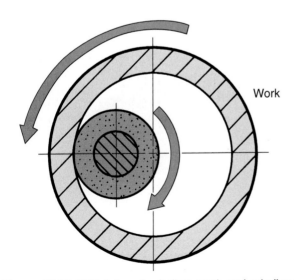

Figure 15-36. *With internal grinding, work and grinding wheel turn in opposite directions.*

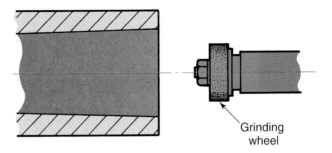

Grinding wheel

Figure 15-37. *Bell mouthing (grinding a hole larger at its mouth) is caused by taking too deep a cut with the grinder, or grinding with a feed that is too rapid.*

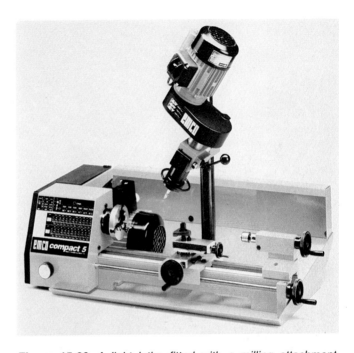

Figure 15-38. *A light lathe fitted with a milling attachment. Work-holding devices for milling operations are fitted on the cross-slide after removing the toolholder. (Emco-Maier Corp.)*

Figure 15-39. *Horizontal milling can be done on a lathe by using a special attachment.*

Another type employs a three-dimensional template or pattern. Most units are hydraulically operated. See **Figure 15-40.**

15.10 INDUSTRIAL APPLICATIONS OF THE LATHE

Industry makes wide use of variations of the basic lathe, **Figure 15-41.** The super-precision toolroom lathe is required to meet the close tolerances and fine surface finish specifications of toolrooms, model shops, and research and development laboratories. See **Figure 15-42.**

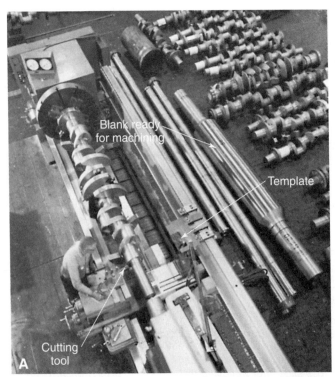

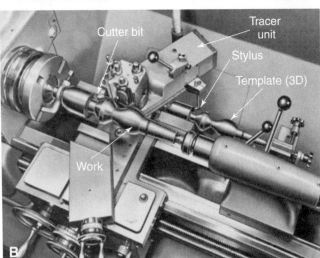

Figure 15-40. Duplicating units. A—A flat template is used to guide cutting tool as it machines the bearing area of a crankshaft for a large diesel engine. B—This unit makes use of three-dimensional templates. (Clausing Industrial, Inc.)

Limited production runs (usually less than 250 pieces) are sometimes produced on a manually operated *turret lathe*, **Figure 15-43.** This is a conventional lathe equipped with a six-sided tool

A

B

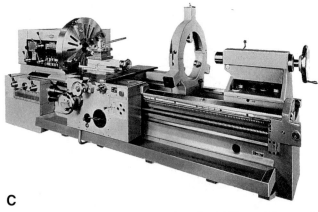

C

Figure 15-41. Engine lathes. A—A typical 10″ precision engine lathe. (South Bend Lathe Corp.) B—This 15″ geared head lathe offers a wide variety of precisely controlled spindle speeds for different applications. (Clausing Industrial, Inc.) C—Heavy-duty engine lathes like this one are used for turning very large workpieces. (Clausing Industrial, Inc.)

holder called a *turret*. **Figure 15-44** illustrates how a number of different cutting tools are fitted to the turret. Stops control the length of tool travel and rotate the turret to bring the next cutting tool into position automatically.

A *cross-slide* unit is fitted for turning, facing, forming, and cutoff operations, **Figure 15-45**. Turret lathes range in size from the small *precision instrument turret lathe* to the more versatile *automatic turret lathe*. See **Figures 15-46** and **15-47**.

The *automatic screw machine*, **Figure 15-48**, is a variation of the lathe that was developed for high speed production of large numbers of small parts. The machine performs a maximum number of operations, either simultaneously, or in a very rapid sequence.

Increasingly, industry is coming to rely on automatic turning centers to produce tiny precision parts in quantity. These centers, referred to as "Swiss-type" machines because they were originally used in the Swiss watchmaking industry, use computer control to perform a number of operations in sequence, producing a finished part. See **Figure 15-49**.

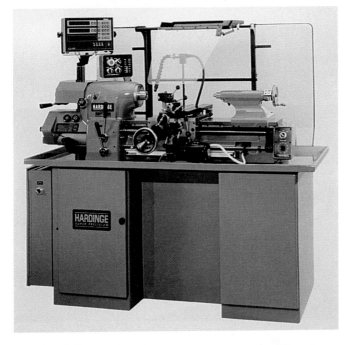

Figure 15-42. The Hardinge Super Precision HLV-DR toolroom lathe. (Hardinge Super Precision HLV-DR is a registered trademark of Hardinge, Inc.)

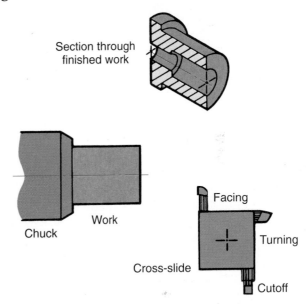

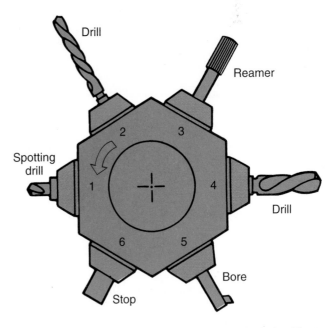

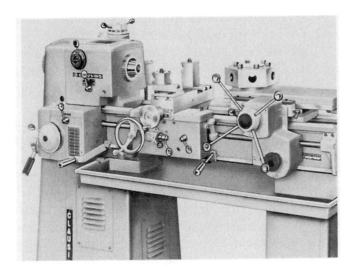

Figure 15-43. A manually operated turret lathe. (Clausing Industrial, Inc.)

Figure 15-44. Turret in relation to other parts of a lathe. The turret rotates to bring tool (drill, reamer, etc.) into position. Stops control depth of tool cuts.

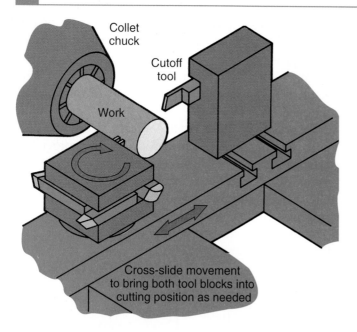

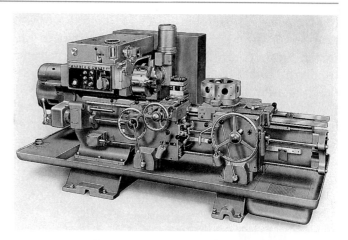

Figure 15-47. A large, versatile turret lathe. (Warner & Swasey Co.)

Figure 15-45. The cross-slide on a turret lathe is similar to the cross-slide on a conventional lathe. However, it is fitted with several cutting tools that can be brought into position as needed.

Figure 15-48. The multiple spindle automatic screw machine is used for precision high speed production. (Warner & Swasey Co.)

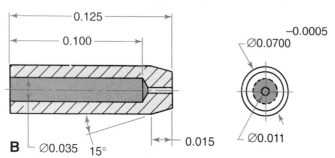

Figure 15-46. Turret lathe operation. A—This machinist is using a magnifying lens to check a drilling operation on a precision instrument turret lathe. (Louis Levin and Son, Inc.) B—This part was machined from stainless steel and produced in quantity on the turret lathe shown above. The part is only about 1/8″ (3.0 mm) long.

Work that is too large or too heavy to be turned in a horizontal position, is machined on a *vertical boring machine*, **Figure 15-50.** These huge machines, known as *boring mills,* are capable of turning and boring work with diameters up to 40′ (12 m).

Conventional metalworking lathes are manufactured in a large range of sizes from the tiny jeweler's lathe to large machines that turn forming rolls for the steel industry, **Figure 15-51.**

Portable turning equipment is available for work in the field, such as chamfering the ends of large pipe prior to welding. See **Figure 15-52.**

Computer numerically controlled (CNC) lathes and turning machines are widely used for industrial production. With proper programming, these machine tools are capable of producing complex work with great accuracy and repeatability. *Note: A detailed description of CNC machine tools and automated manufacturing operations can be found in Chapters 21 and 22.*

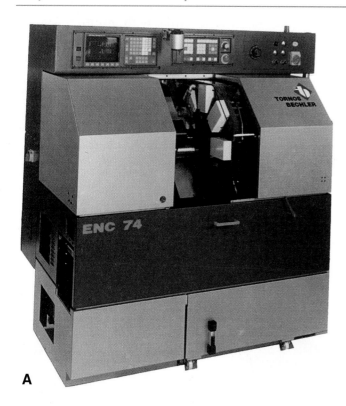

A

B

Figure 15-49. *Swiss-type automatic turning center. A—This high-precision turning center machines small parts from bar stock ranging from 1 mm to 10 mm in diameter at high production rates. B—These tiny components, produced by a Swiss-type automatic turning center, are used in precision instruments. (Tornos-Bechler S.A.)*

A

B

Figure 15-50. *Boring mills. A—A twin-spindle boring mill. B—This 17′ (5.1 m) vertical boring mill features high cutting speeds and the capability of infinite speed variations while the machine is operating. Workpieces are mounted on the large turntable and rotated into position for machining. (Simmons Machine Tool Corp.)*

Figure 15-51. *Some idea of the size of this huge lathe can be gained by comparing the workpiece with the machinist in the photo.*

Many types of CNC lathes and turning machines are in use today. Some examples include:
- Small machines used primarily for training purposes, **Figure 15-53**.
- Conventional lathes with two-axis CNC movement, **Figure 15-54**.
- CNC slant bed lathe with multiple-station tool turrets, **Figure 15-55**.
- CNC lathes designed for specific applications, **Figure 15-56**.
- Completely automated turning centers, **Figure 15-57**. These centers often feature robotic loading/unloading, automatic gaging of the workpiece, and tool monitoring.

Figure 15-52. A portable lathe that can be taken into the field. This worker is shown turning the end of a high-pressure gas pipeline to prepare it for being welded. (Tri-Tool, Inc.)

Figure 15-54. A conventional lathe with two-axis CNC control. (Bridgeport Machines, Inc.)

Figure 15-53. This small CNC turning center is used for training. The controller (not shown) is a separate unit that connects to the machine tool. (Light Machines Corp.)

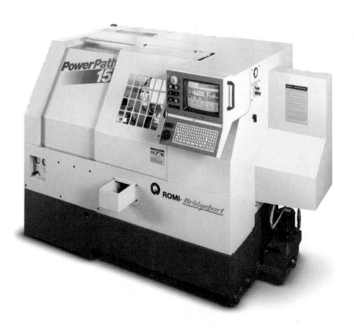

Figure 15-55. This two-axis CNC slant bed lathe has a turret that can mount 12 different tools for various turning operations. The menu permits the operator to simply fill in the finished dimensions of the part. The software calculates the cutting path to produce that part. An on-screen graphic preview of tool path allows the operator to verify the path prior to cutting. (Bridgeport Machines, Inc.)

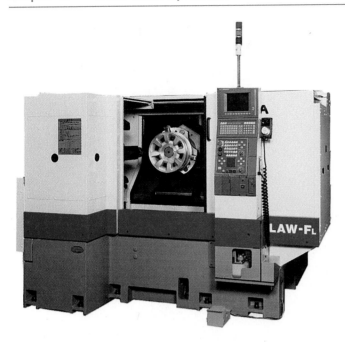

Figure 15-56. A CNC lathe that is designed specifically for aluminum wheel production. (Okuma America Corporation)

Figure 15-57. This twin-spindle CNC turning cell has an automatic loader for material to be turned and two tool-holding turrets. It has two facing spindles that, in combination with the upper and lower turrets, permit simultaneous 4-axis or parallel machining operations. (Okuma America Corporation)

TEST YOUR KNOWLEDGE

Please do not write in the text. Write your answers on a separate sheet of paper.

1. Drills that are used on the lathe are fitted with _____ shanks or _____ shanks.

2. Boring is:
   a. A drilling operation.
   b. An internal machining operation in which a single-point cutting tool is employed to enlarge a hole.
   c. An external machining operation in which a single-point cutting tool is employed to reduce the diameter of a hole.
   d. All of the above.
   e. None of the above.

3. When is reaming done?

4. The process of forming horizontal or diamond-shaped serrations on the circumference of the work is called _____. It is commonly done to provide a _____.

5. When is filing on the lathe usually done?

6. Polishing is an operation used to produce a _____ on the work.

7. When is a steady rest used?

8. What is the difference between a steady rest and a follower rest?

9. There are times when a shaft is unsuitable as a bearing surface and cannot be used with a steady rest. When this occurs, a _____ can be employed so the shaft can be supported with the steady rest.

10. What is a mandrel? When is it used?

11. A mandrel is usually pressed into the work with an _____.

12. Internal and external grinding can be done on a lathe with a_____.

13. What should be done to protect the lathe from the abrasive particles that wear away from the grinding wheel?
   a. Use a nonabrasive grinding wheel.
   b. Cover the bed and moving parts with a heavy cloth.
   c. Use a soft abrasive grinding wheel.
   d. All of the above.
   e. None of the above.

Broaching can be used for flat, round, and contoured surfaces such as the ones shown on these parts. Both internal and external surfaces can be broached. (Metal Powder Industries Federation)

Broaching Operations

IMPORTANT TERMS

broach	*pot broaching*
broaching	*pull broach*
burnishing	*roughing teeth*
finishing teeth	*semifinishing teeth*
keyway	*slab broach*

Broaching is a manufacturing process for machining flat, round, and contoured surfaces. Both internal and external surfaces can be shaped by this process. Internal broaching requires a starting hole so that the cutting tool can be inserted. There are three basic types of broaching operations. Internal broaching makes use of a *pull broach*. External broaching uses the *slab broach*, a flat toothed strip that is usually held (singly or in groups) in a slotted fixture. In *pot broaching*, the tool is stationary and the work is pulled through it.

A variety of broaching equipment is available. The machines range in size from large multifunctional broaching machines, **Figure 16-1**, to small manually operated units using an arbor press to cut keyways in gears, pulleys, and similar components. The manual units will be explained more fully later in this chapter.

With broaching, a multitoothed cutting tool is pushed or pulled across the work, **Figure 16-2**. Each tooth on the *broach* (cutting tool) removes only a small portion of the material being machined, **Figure 16-3**.

A broach has three kinds of teeth: *roughing teeth*, *semifinishing teeth*, and *finishing teeth*. See **Figure 16-4**. In many industrial applications, the broach is assembled from units with different tooth

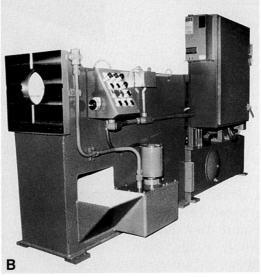

Figure 16-1. Three types of broaching machines. A—A front 42″ stroke vertical internal pull-down broaching machine. B—A 36″ internal horizontal broaching machine. C—This high-speed vertical surface single-ram broaching machine has a tilting fixture table. (Broaching Machine Specialties)

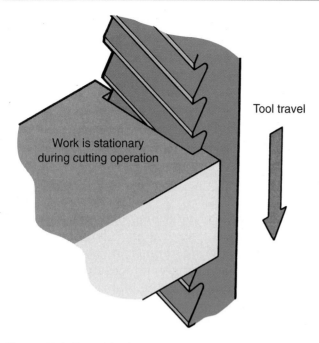

Figure 16-2. *Broaching involves the use of a multitooth cutting tool (the broach) that moves against the stationary work. The operation may be on a vertical or horizontal plane, and may involve making internal or external cuts.*

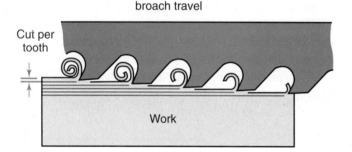

Figure 16-3. *Each tooth on a broaching tool removes only a small portion of the material being machined.*

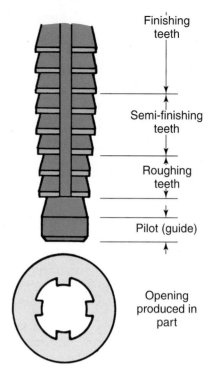

Figure 16-4. *This drawing shows a greatly shortened section of an internal broaching tool and a cross-section of the splines it cuts in a part. The pilot guides the cutter into a cut or hole previously made in the work. Each tooth of the broach increases slightly in size until the specified size is attained.*

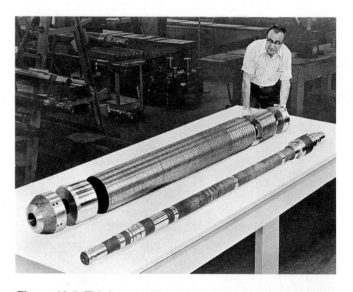

Figure 16-5. *This large pull broach is assembled from separate units in a variety of tooth sizes. The ring-like units are stacked on the mandrel in the foreground to be pulled through the workpiece. (National Broach & Machine Co.)*

sizes. As shown in **Figure 16-5,** the tooth units are stacked on a mandrel to be pulled through the work.

The machining operation usually can be completed in a single pass of the cutting tool, **Figure 16-6**. When properly employed, broaching can remove metal faster than almost any other machining technique. Small parts can be stacked and shaped in a single pass, **Figure 16-7**. Larger units, such as compressor cylinder heads, may require several passes to machine all surfaces, **Figure 16-8**.

Almost any material that can be machined by other techniques can be broached. Consistently close tolerances can be maintained by the broaching process. While the surface finishes produced by broaching are smooth compared to many other machining processes, they can be further improved by adding *burnishing* (noncutting) elements to the finishing end of the broach.

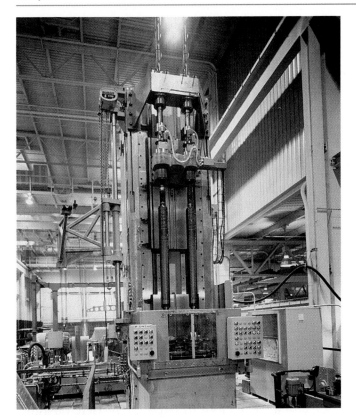

Figure 16-6. *Two pull broaches, similar to the one shown in Figure 16-5, are mounted on this broaching machine. Parts are typically completed in a single pass of the broach. (National Broach & Machine Co.)*

16.1 ADVANTAGES OF BROACHING

The broaching process offers several manufacturing advantages:

- High productivity.
- Capability of maintaining close tolerances.
- Production of good surface finishes.
- Economy (even though initial tooling costs can be high unless standard tooling is used).
- Long tool life, since only a small amount of material is removed by each tooth.
- Capability of using semiskilled workers, since equipment is automated.

16.2 KEYWAY BROACHING

Cutting a *keyway*, in a gear, pulley, or similar component, is a simple broaching operation that can be done in the average machine shop. See **Figure 16-8.** A typical keyway broach set, **Figure 16-9**, contains an assortment of precision broaches, slotted bushings, necessary shims, instructions, and a lubrication guide.

Figure 16-7. *Typical small parts machined by broaching. (LaPoint Machine Tool Co.)*

Broach

Figure 16-8. *Vertical broach machining the flat surface of a cylinder head for a compressor. (The Association for Manufacturing Technology)*

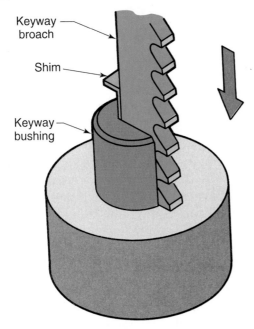

Figure 16-9. Cutting a keyway using an arbor press to push a broach through the work. Several passes may be required. Different size shims move cutter into the work.

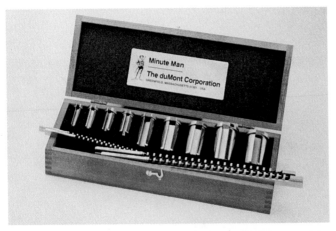

Figure 16-10. A typical keyway broach set. It contains precision broaches, slotted bushings, and necessary shims. Ample lubrication is necessary. (duMont Corp.)

First, measure the bore into which the keyway is to be cut. Then complete the following steps:

1. Select the bushing that fits the hole and the required broach. Handle the broach with care. Its sharp teeth can cause a serious injury to the hand.

2. Place the bushing into the hole and insert the broach.

3. Set the assembly into position on the *arbor press*, making sure there is ample clearance for the broach to pass through the work. Also, be sure that the broach is centered on the arbor press ram. Otherwise, the broach may be damaged by being pushed to one side. A loose or worn arbor press ram also can damage a broach by pushing it to one side.

4. Lubricate the broach as instructed by the broach manufacturer.

5. Push the broach through.

6. Clean the broach and insert the second pass shim.

7. Lubricate the broach again, and push the broach through.

8. Repeat the sequence until the keyway is the correct depth.

9. Use a clean cloth to wipe the broach, bushing, and shims clean. Apply a thin coating of oil to prevent rusting and return them to storage.

10. Remove any burrs from the keyway.

TEST YOUR KNOWLEDGE

Please do not write in the text. Write your answers on a separate sheet of paper.

1. Broaching is a manufacturing process for machining _____ surfaces.

2. What does internal broaching require that external broaching does not?

3. What is unique about the cutting tool used on a broaching machine?

4. List three advantages offered by broaching.

5. With broaching, the machined surface can be further improved by adding _____ to the finishing end of the broach.

Chapter 17

The Milling Machine

LEARNING OBJECTIVES

After studying this chapter, you will be able to:
- ◯ Describe how milling machines operate.
- ◯ Identify the various types of milling machines.
- ◯ Select the proper cutter for the job to be done.
- ◯ Calculate cutting speeds and feeds.

IMPORTANT TERMS

arbor
climb milling
column and knee milling machine
face milling
horizontal spindle milling machine

peripheral milling
rate of feed
side milling cutters
traverse
vertical spindle milling machine

A *milling machine* rotates a multitoothed cutter into the workpiece to remove material, **Figure 17-1**.

Each tooth of the cutter removes a small individual chip of material. A wide variety of cutting operations can be performed on a milling machine. The milling machine is capable of machining flat or contoured surfaces, slots, grooves, recesses, threads, gears, spirals, and other configurations.

Milling machines are available in more variations than any other family of machine tools. These machines are well suited to computer-controlled operation, **Figure 17-2**. Work may be clamped

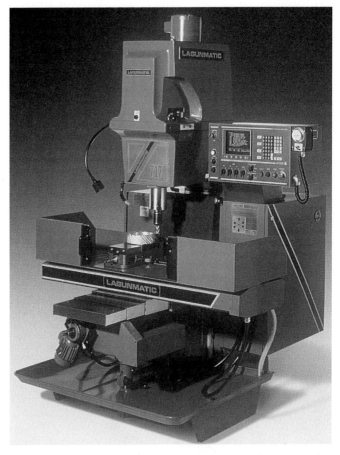

Figure 17-2. A CNC knee-type vertical machining center that incorporates the latest advances in technology. It features quick toolchanging capabilities. It is easily programmed for machining a single part or for production work. (Republic-Lagun CNC Corp.).

Figure 17-1. A milling machine rotates a multitoothed cutter into the workpiece. (Giddings & Lewis, Inc.)

directly to the machine table, held in a fixture, or mounted in or on one of the numerous work-holding devices available for milling machines.

17.1 TYPES OF MILLING MACHINES

It is difficult to classify the various categories of milling machines, because their designs tend to merge with one another. For practical purposes, however, milling machines may be grouped into two large families:

- Fixed-bed type.
- Column and knee type.

Both groups are made with horizontal or vertical spindles. On a *horizontal spindle milling machine,* the cutter is fitted onto an arbor mounted in the machine on an axis parallel with the worktable. See **Figure 17-3.** Multiple cutters may be mounted on the spindle for some operations.

The cutter on a *vertical spindle milling machine* is normally *perpendicular* (at a right angle) to the worktable, **Figure 17-4.** However, on many vertical spindle machines, the spindle can be tilted to perform angular cutting operations.

17.1.1 Fixed-bed milling machines

Fixed-bed milling machines are characterized by very rigid worktable construction and support, **Figure 17-5.** The worktable moves only in a *longitudinal* (back and forth/X-axis) direction, and can vary in length from 3' to 30' (0.9 to 9.0 m). *Vertical* (up and down/Z-axis) and *cross* (in and out/Y-axis) movements are obtained by moving the cutter head.

Figure 17-3. *This 50 hp horizontal milling machine can make a cut 1/8" deep by 12" wide. Note alignment of arbor on which the cutter is mounted. (National Machine Tool Builders Assoc.)*

Figure 17-4. *The cutter on a vertical milling machine is perpendicular to the surface being machined. (Heidenhain Corp.)*

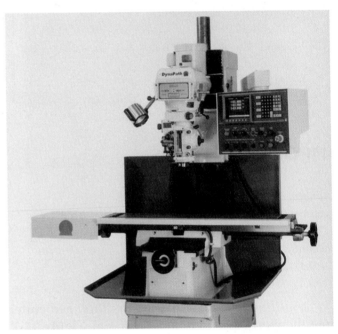

Figure 17-5. *Fixed bed or bed mills have a very rigid worktable that moves only in a longitudinal direction. The machine shown can be manually or CNC operated. (Autocon Technologies, Inc.)*

Bed-type milling machines can be further classified as horizontal, vertical, or planer type machines. The type of bed permits heavy cutting on large workpieces, **Figure 17-6.**

17.1.2 Column and knee milling machines

The *column and knee milling machine* is so named because of the parts that provide movement to the workpiece. They consist of a *column* that supports and guides the knee in *vertical* (up and down/Z-axis) movement, and a *knee* that supports the mechanism for obtaining table movements. These movements are *traverse* (in and out/Y-axis) and *longitudinal* (back and forth/X-axis). See **Figure 17-7.**

Figure 17-6. This large, multispindle 3-axis fixed-bed milling machine is shaping upper wing skins for the Boeing 727 aircraft. (Northrop-Grumman Aerospace Corp.)

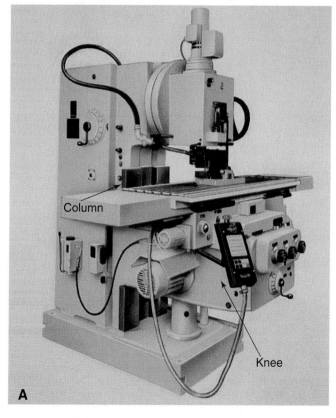

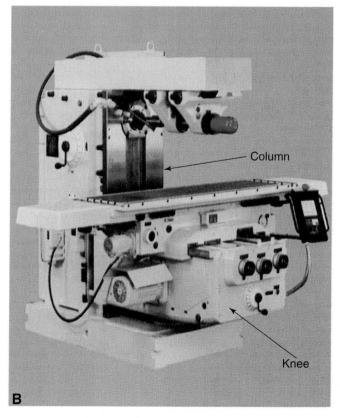

Figure 17-7. Column and knee-type milling machines. A—Vertical. B—Horizontal. (WMW Machinery Company, Inc.)

These machines are commonly referred to as "knee-type milling machines." There are three basic types:

- Plain (horizontal spindle) milling machine.
- Universal milling machine.
- Vertical spindle milling machine.

17.1.3 Plain milling machine

On the *plain milling machine*, the cutter spindle projects horizontally from the column, **Figure 17-8**. The worktable has three movements: *vertical, cross,* and *longitudinal* (X-, Y-, and Z-axes), **Figure 17-9**.

17.1.4 Universal milling machine

A *universal milling machine*, **Figure 17-10**, is similar to the plain milling machine, but the table has a fourth axis of movement. On this type of machine, the table can be swiveled on the saddle through an angle of 45° or more, **Figure 17-11**. This makes it possible to produce spiral gears, spiral splines, and similar workpieces, **Figure 17-12**.

17.1.5 Vertical spindle milling machine

A *vertical spindle milling machine* differs from the plain and universal machines by having the cutter spindle in a vertical position, at a right angle to

Figure 17-8. Plain-type milling horizontal machine. (Sharp Industries, Inc.)

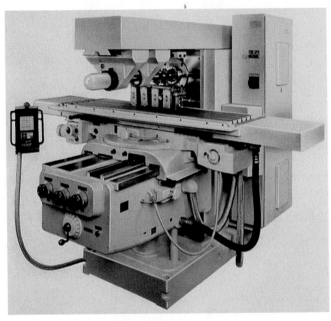

Figure 17-10. On a universal type horizontal milling machine, the table can be swiveled 45° or more. (WMW Machinery Company, Inc.)

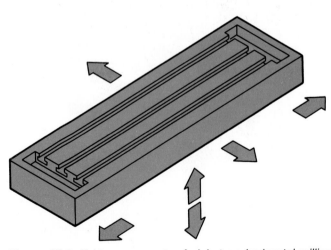

Figure 17-9. Table movements of plain-type horizontal milling machine.

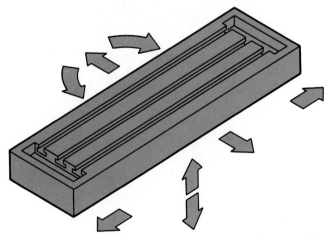

Figure 17-11. Table movements of the universal type milling machine.

the top of the worktable. See **Figure 17-13**. The cutter head can be raised and lowered by hand or by power feed.

This type of milling machine is best suited for use with an end mill or face mill cutter. Vertical mills include swivel head, sliding head, and rotary head types.

Figure 17-12. Note how table is swiveled at an angle to cutting tool. This feature makes it possible to machine helical-flute workpieces like one being cut. (WMW Machinery Company, Inc.)

A *swivel head milling machine*, **Figure 17-14**, is the type often found in vocational-technical school training programs. The spindle can be swiveled for angular cuts.

On the *sliding head milling machine*, the spindle head is fixed in a vertical position. The head can be moved in a vertical direction by hand or under power, **Figure 17-15**.

The spindle on the *rotary head milling machine* can be moved vertically and in circular arcs of adjustable radii about a vertical center line. It can be adjusted manually or under power feed, **Figure 17-16**.

17.1.6 Methods of milling machine control

The method employed to control table movement is another way of classifying milling machines, and all machine tools in general. Basically, there are four methods of control:

- *Manual*—All movements are made by hand lever control.
- *Semi-automatic*—Movements are controlled by hand and/or power feeds.

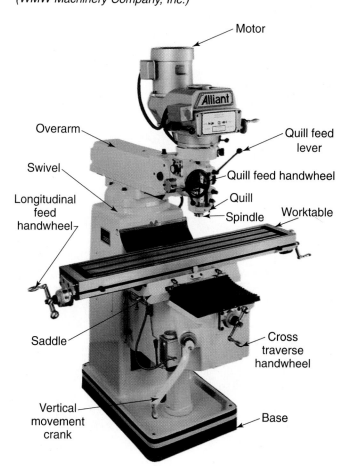

Figure 17-13. The vertical spindle milling machine. (Rem Sales, Inc.)

Figure 17-14. A typical swivel-head milling machine (Republic-Lagun Machine Tool Co.)

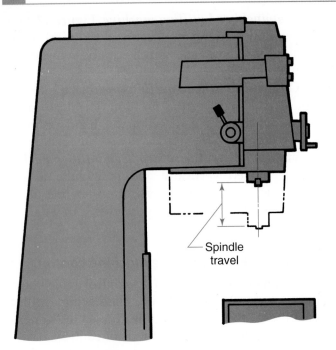

Figure 17-15. Spindle head is fixed in a vertical position on sliding head milling machine. Entire head is moved to make cutting adjustments.

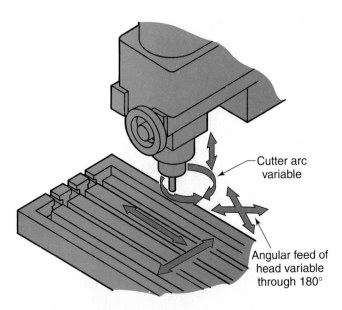

Figure 17-16. Movements possible on a rotary head vertical milling machine. Milling machines with CNC capabilities are replacing this type machine.

- **Fully automatic**—A complex hydraulic feed arrangement that follows two or three dimensional templates to automatically guide one or more cutters. Specifications can also be programmed on magnetic or perforated tape to guide the cutters and table through the required machining operations.

- **Computerized** (CNC)—Machining coordinates are entered into a master computer or computer on the machine, using a special programming language. Instructions from the computer operate actuators (electric, hydraulic, or pneumatic devices) that move the table and cutter or cutters through the required machining sequence.

Small machines, such as the **bench type milling machine** shown in **Figure 17-17**, have power feed available only for longitudinal table movement. On larger machines, automatic feed or power feed is used for all table movements, **Figures 17-18** and **17-19**.

Figure 17-17. Light plain type milling machine is sometimes referred to as a bench mill.

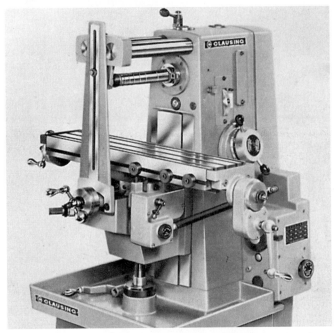

Figure 17-18. The vertical table feed handle. Power feed to raise the table is also available on this machine. It is activated by the knobbed handle to the right.

Table movement (feed) can be engaged at cutting speed; however, there is a *rapid traverse feed* that allows fast power movement in any direction of feed engagement. This permits work to be positioned at several times the fastest rate indicated on the feed chart. It operates by positioning the automatic power feed control lever to give the desired directional movement and activating the rapid traverse lever. See **Figure 17-20**. Never activate rapid traverse while the cutter is positioned in a cut.

17.1.7 Adjusting cutting speed and feed for milling

Depending upon the make of the milling machine, *cutting speed* (cutter rpm) and *rate of feed* (table movement speed) may be changed by:

1. Shifting V-belts, **Figure 17-21**.

2. Adjusting variable speed pulleys, **Figure 17-22**.

3. Utilizing a quick change gear box and shifting or dialing to the required speed and feed setting, **Figure 17-23**. On some machines, speed and feed changes are made hydraulically or electronically.

Always make speed changes as specified in the instruction manual for the specific machine being used! Make sure the machine has come to a complete stop before attempting to adjust V-belts.

Figure 17-21. Spindle speeds on a V-belt drive are changed by using various pulley ratios. (South Bend, Inc.)

Figure 17-19. The crank provides longitudinal table movement. The small handle to the left activates the power feed. (Clausing Industrial, Inc.)

Figure 17-20. A machinist is engaging the rapid traverse on a #2 plain milling machine. The #2 denotes machine size.

Figure 17-22. Spindle speed on this machine is controlled by adjusting a split pulley. Pulley can be spread open or narrowed in width to control where belt rides in pulley. This is done by dialing speed selector to proper position. (Cincinnati Lathe & Tool Co.)

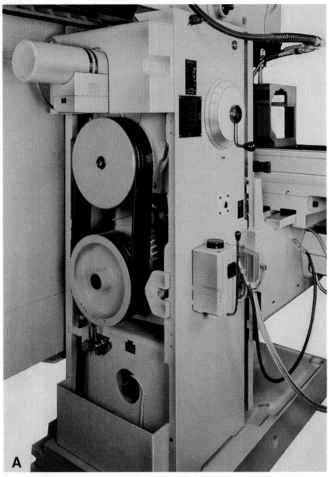
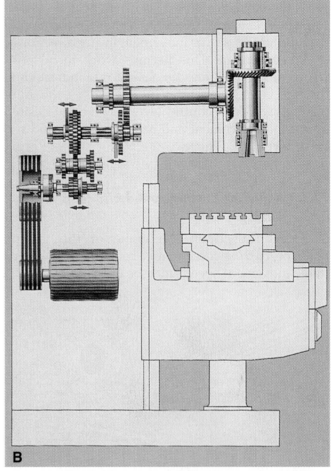

Figure 17-23. *Spindle speed control. A—Main belt drive and spindle speed control dial on a vertical milling machine. B—Cutaway showing sliding gear mechanism that permits dialing any of 18 speed choices. (WMW Machinery Company, Inc.)*

17.2 MILLING SAFETY PRACTICES

Milling machines, like all machine tools, should be cleaned after each work session. A medium width paintbrush may be used to remove accumulated chips, **Figure 17-24**.

Chips are razor-sharp; never use your *hand* to remove them. Never remove chips with compressed air. The flying chips may injure you or a nearby person. Also, if cutting oil was used in the machining operation, the compressed air will create a highly flammable oily mist. If ignited by an open flame, the mist can explode. **Finish by wiping down the machine with a soft cloth.**

The following procedures are suggested for the safe operation of a milling machine.

- Become thoroughly familiar with the milling machine before attempting to operate it. When in doubt, obtain additional instructions.
- Never attempt to operate a milling machine while your senses are impaired by medication or other substances.

- Wear appropriate clothing and approved safety glasses.
- Stop the machine before:
  - attempting to make adjustments or measurements.
  - trying to remove accumulated chips.
  - before opening or removing guards and covers. Be sure all power to the machine is turned off.
- Use a piece of heavy cloth or gloves for protection when handling milling cutters. Avoid using your bare hands.
- Get help to move any heavy machine attachment, such as a vise, dividing head, rotary table, or large work.
- Be sure the work-holding device is mounted solidly to the table, and the work is held firmly. Spring or vibration in the work can cause thin cutters to jam and shatter!
- Never reach over or near a rotating cutter!
- Avoid talking with anyone while operating a machine tool, do not allow anyone to turn on your machine for you.

- Never "fool around" when operating a milling machine. Keep your mind on the job and be ready for any emergency.
- Keep the floor around your machine clear of chips and wipe up spilled cutting fluid immediately! Place sawdust or special oil-absorbing compound on slippery floors. Place all oily rags in an approved metal container that can be closed tightly.
- Be thoroughly familiar with the placement of the machine's "stop" switch or lever, **Figure 17-25**.
- Treat any small cuts and skin punctures as potential infections. Clean them thoroughly.

Apply antiseptic and cover injury with a bandage. Report any injury, no matter how minor, to your instructor or supervisor.
- Launder work clothes frequently. Greasy clothing is a fire hazard.

17.3 MILLING OPERATIONS

There are two main categories of milling operations:
- *Face milling* is done when the surface being machined is parallel with the cutter face, **Figure 17-26**. Large, flat surfaces are machined with this technique.
- *Peripheral milling* is done when the surface being machined is parallel with the periphery of the cutter, **Figure 17-27**.

17.4 MILLING CUTTERS

The typical *milling cutter* is circular in shape with a number of cutting edges (teeth) located around its circumference. Milling cutters are manufactured in a large number of stock shapes, sizes,

Figure 17-24. *Use a brush to remove metal chips;* never *use your hand, since sharp chips can cause serious cuts.*

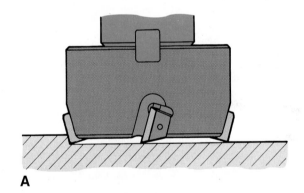

A

B

Figure 17-26. *Face milling. A—With face milling, the surface being machined is parallel with the cutter face. B—An example of face milling. (Sandvik Coromant Co.)*

Figure 17-25. *Make it a "first point of business" to know the location of the* stop switch *or* stop lever. *Wear safety glasses and stop the machine before attempting to make measurements, adjustments, or to clear away chips. (Cincinnati Lathe & Tool Co.)*

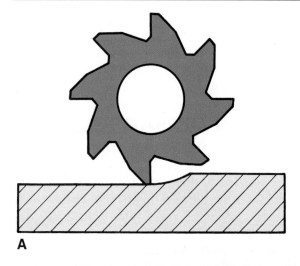

A

B

Figure 17-27. Peripheral milling. A—In this milling method, the surface being machined is parallel with the periphery of the cutter. B—An example of peripheral milling.

and kinds, because they can't be economically ground for a particular job as can a lathe cutter bit. See **Figure 17-28**.

17.4.1 Types of milling cutters

There are two general types of milling cutters:
- A *solid cutter* has the shank and body made in one piece, **Figure 17-29**.
- The *inserted-tooth cutter* has teeth made of special cutting material, which are brazed or clamped in place, **Figure 17-30**. Worn and broken teeth can be replaced easily instead of discarding the entire cutter.

17.4.2 How milling cutters are classified

Milling cutters are frequently classified by the method used to mount them on the machine:
- *Arbor cutters* have a suitable hole for mounting to an arbor, **Figure 17-31**.
- *Shank cutters* are fitted with either a straight or taper shank that is an integral part of the cutter, **Figure 17-32**. They are held in the machine by collets or special sleeves.
- *Facing cutters* can be mounted directly to a machine's spindle nose or on a stub arbor. See **Figure 17-33**.

17.4.3 Milling cutter material

Considering the wide range of materials that must be machined, the ideal milling cutter should have:
- *High abrasion resistance*, so the cutting edges do not wear away rapidly due to the abrasive nature of some materials.
- *Red hardness*, so the cutting edges are not affected by the terrific heat generated by many machining operations.
- *Edge toughness*, so the cutting edges do not readily break down due to the loads imposed upon them by the cutting operation.

Since no single material can meet these requirements in all situations, cutters are made from materials that are, by necessity, a compromise.

High-speed steels (*HSS*) are the most versatile of the cutter materials. Cutters made from high-speed steels are excellent for general purpose work or where vibration and chatter are problems. They are preferred for use on machines of low power.

HSS milling cutters can be improved by the application of surface lubricating treatments, surface hardening treatments, or by the application of coatings (such as chromium, tungsten, or tungsten carbide) to the cutting surfaces. While the treated tools cost 2 to 6 times as much as conventional HSS tools, they may last 5% to 10% longer or provide 50% to 100% percent higher metal removal rates with the same tool life.

Cemented tungsten carbides include a broad family of hard metals. They are produced by powder metallurgy techniques and have qualities that make them suitable for metal cutting tools. Cemented carbides can, in general, be operated at speeds 3 to 10 times faster than conventional HSS cutting tools.

Normally, only the cutting tips are made of cemented carbides rather than the entire cutter. They are brazed or clamped to the cutter body. See **Figures 17-34** and **17-35**.

Figure 17-28. A selection of HSS (high speed steel) milling cutters. (Brown & Sharpe Mfg. Co.)

Figure 17-29. With the exception of the cutter in the back row, these are solid milling cutters. (Brown & Sharpe Mfg. Co.)

Figure 17-30. A variety of cutters with indexable carbide inserts. When the cutting edges dull, new cutting edges are rotated into position. The inserts are available in different grades depending upon the material being machined. (Dapra Corporation)

Figure 17-31. An arbor-type milling cutter.

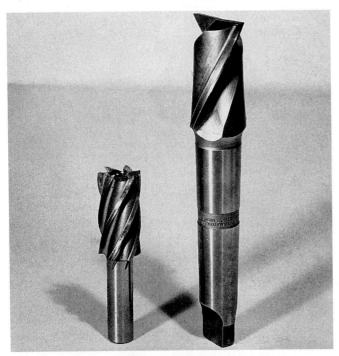

Figure 17-32. Shank-type milling cutters.

Figure 17-33. This face-type milling cutter has an unusual design, making use of bearing-mounted inserts that rotate as they cut. The manufacturer claims better heat dissipation provided by the rotating inserts will increase cutter life and permit higher cutting speeds. (Valenite, Inc.)

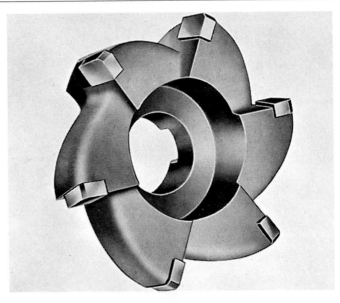

Figure 17-34. Teeth on this inserted-tooth milling cutter are brazed to cutter body. (Brown & Sharpe Mfg. Co.)

Figure 17-35. These inserted-tooth cutters use teeth that are clamped to the cutter body. The cutter teeth (colored gold in this photo) can be indexed four times to present a fresh cutting edge as they wear. (Mitsubishi Materials USA Corporation)

Most newer inserted-tooth cutters make use of indexable inserts. Each insert has several cutting edges at various corners. When they become dull, the inserts are indexed (turned) so a new cutting edge contacts the metal.

Cemented carbide cutters are excellent for long production runs and for milling materials with a scale-like surface (cast iron, cast steel, bronze, etc.)

17.5 TYPES AND USES OF MILLING CUTTERS

The following are the more commonly used milling cutters, with a summary of the work to which they are best suited.

17.5.1 End mills

End milling cutters are designed for machining slots, keyways, pockets, and similar work, **Figure 17-36**. The cutting edges are on the circumference and end. End mills may have straight or helical flutes, **Figure 17-37,** and have straight or taper shanks, **Figure 17-38**. Straight shank end mills are available in single and double end styles, **Figure 17-39.**

The term *hand* is used to describe the direction of cutter rotation and the helix of the flutes, **Figure 17-40**. When viewed from the cutting end, a *right-hand cutter* rotates counterclockwise, and a *left-hand cutter* rotates clockwise.

Ball nose end mills, **Figure 17-41,** are utilized for tracer milling, computer-controlled contour milling, die-sinking, fillet milling, and other radius work. A cut with a depth equal to one-half the end mill diameter can generally be taken in solid stock. See **Figure 17-42.**

Figure 17-36. An indexable-insert end mill machining clearance slots on a face milling cutter. (Kennametal Inc.)

Figure 17-38. Straight shank and taper shank end mills.

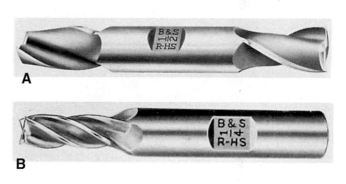

Figure 17-39. Single-end-type and double-end- type end mills.

Figure 17-37. End mill with multiple indexable inserts are made in both helical fluted and straight-fluted types. Two face-type cutters are also shown. (Mitsubishi Materials USA Corporation)

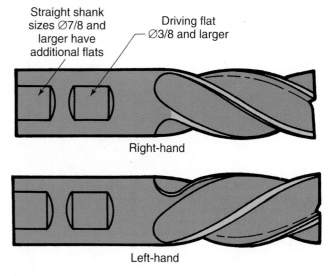

Straight shank sizes ⌀7/8 and larger have additional flats

Driving flat ⌀3/8 and larger

Right-hand

Left-hand

Figure 17-40. Cutter is right-hand if it rotates counterclockwise when viewed from cutting end. It is left-hand if rotation is clockwise.

Several end mill styles are available:

- A *two-flute end mill* can be fed into the work like a drill. There are two cutting edges on the circumference, with the end teeth cut to the center, **Figure 17-43**.
- The *multiflute end mill* can be run at the same speed and feed as a comparable two-flute end mill, but it has a longer cutting life and will produce a better finish. It is recommended for conventional milling where plunge cutting (feeding into the work like a twist drill) is not necessary. See **Figure 17-44**.
- A *shell end mill*, **Figure 17-45**, has teeth similar to the multiflute end mill but is mounted on a stub arbor. The cutter is designed for

both face and end milling. Shell end mills are made with right-hand cut, right-hand helix, or with left-hand cut, left-hand helix.

17.5.2 Face milling cutters

Face milling cutters are intended for machining large flat surfaces parallel to the face of the cutter, **Figure 17-46**. The teeth are designed to make the roughing and finishing cuts in one operation. Because of their size and cost, most face milling cutters have inserted cutting edges.

Figure 17-43. The two-flute end mill can be fed into work like a drill.

Figure 17-44. Multifluted end mills. The peripheral grooves in the cutter at right reduce chip size, lowering cutting forces. Most modern cutters have a nitride coating that improves resistance to abrasive wear and corrosion.

Figure 17-41. A ball nosed end mill has a rounded tip. These mills have two replaceable cutting inserts. (Mitsubishi Materials USA Corporation)

Figure 17-42. Three-dimensional milling is being done with a ball tip end mill. (Delcam International)

Figure 17-45. A shell end mill being used on a vertical milling machine.

A *fly cutter* is a single-point cutting tool used as a face mill. An example is shown in **Figure 17-47**.

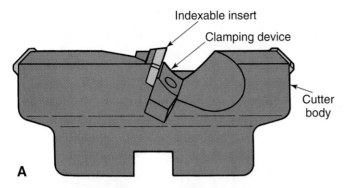

A

B

Figure 17-46. *Face mill with indexable inserts. A—The replaceable inserts are mechanically clamped in place on the cutter body. As cutting edges wear, the inserts can be turned (indexed) to present a fresh cutting edge. B—An example of a face mill with indexable inserts. Insert selection is based on material to be machined. (Mitsubishi Materials USA Corporation)*

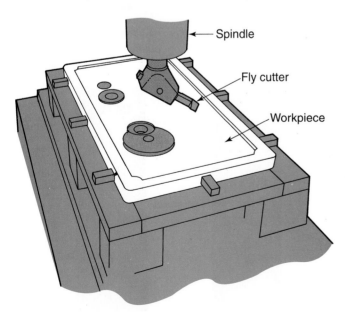

Figure 17-47. *The fly cutter is a single-point cutting tool used for face milling.*

17.5.3 Arbor milling cutters

The more common arbor milling cutters and the work for which they are best adapted include:
- Plain milling cutter
- Side milling cutter
- Angle cutters
- Metal slitting saws
- Formed milling cutters.

Plain milling cutter

Plain milling cutters are cylindrical, with teeth located around the circumference. See **Figure 17-48**. Plain milling cutters less than 3/4″ (20 mm) are made with straight teeth. Wider plain cutters, called

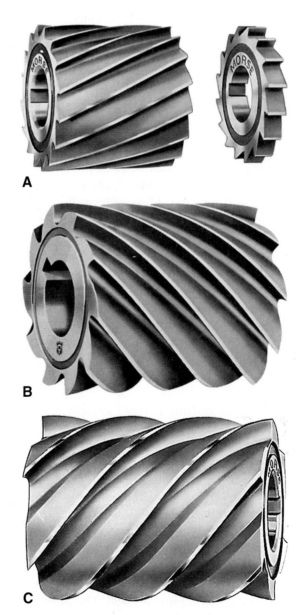

A

B

C

Figure 17-48. *Plain milling cutters. A— Light-duty cutters. B—Heavy-duty plain milling cutter. (Standard Tool Co.) C—Helical plain milling cutter. (Morse Cutting Tools)*

slab cutters, are made with helical teeth designed to cut with a shearing action. This reduces the tendency for the cutter to chatter.

- The *light-duty plain milling cutter* is used chiefly for light slabbing cuts and shallow slots.
- The *heavy-duty plain milling cutter* is recommended for heavy cuts where considerable material must be removed. It has fewer teeth than a comparable light-duty cutter. The cutting edges are better supported and chip spaces are ample to handle the larger volume of chips.
- A *helical plain milling cutter* has fewer teeth than either of the two previously mentioned cutters. The helical cutter can be run at high speeds and produces exceptionally smooth finishes.

Side milling cutter

Cutting edges are located on the circumference and on one or both sides of *side milling cutters.* They are made in solid form or with inserted teeth. See **Figure 17-49.**

- *Plain side milling cutter* teeth are on the circumference and on both sides of the cutter. It is recommended for side cutting, straddle milling, and slotting. Plain side milling cutters are available in diameters ranging from 2″ (50 mm) to 8″ (200 mm), and in widths from 3/16″ (5 mm) to 1″ (25 mm).
- A *staggered-tooth side milling cutter* has alternating right-hand and left-hand helical teeth. They aid in reducing chatter while providing adequate chip clearance for higher operating speeds and feeds than are possible with the plain side milling cutter. This type cutter is especially good for machining deep slots.
- The *half side milling cutter* has helical teeth on the circumference but side teeth on only one side. It is made as a right-hand or left-hand cutter, and is recommended for heavy straddle milling and milling to a shoulder.
- The *interlocking side milling cutter* is ideally suited for milling slots or bosses, and for making other types of cuts that must be held to extremely close tolerances. The unit is made as two cutters with interlocking teeth that can be adjusted to the required width by using spacers or collars, **Figure 17-50.** The alternating right and left shearing action eliminates side pressures, producing a good surface finish.

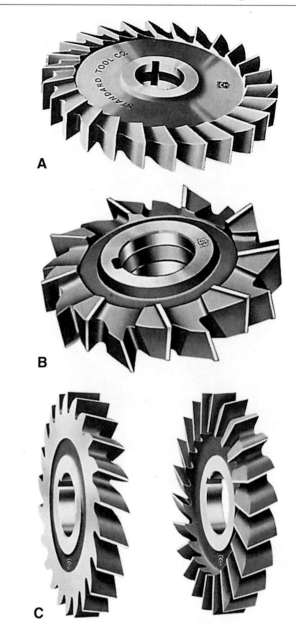

A

B

C

Figure 17-49. Side milling cutters. A—Plain side milling cutter. B—Staggered-tooth side milling cutter. C—Half side milling cutters. (Standard Tool Co.)

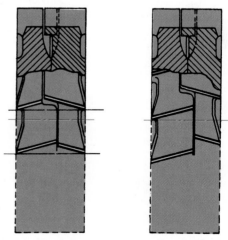

Figure 17-50. Tooth pattern on interlocking side milling cutters. (Morse Tool Co.)

Angle cutters

Angle cutters differ from other cutters in that the cutting edges are neither parallel, nor at right angles to the cutter axis. See **Figure 17-51**.

- On a *single-angle milling cutter*, the teeth are on the angular face and on the side adjacent to the large diameter. Single-angle cutters are made in both right-hand and left-hand cut, with included angles of 45° and 60°.
- The *double-angle milling cutter* is used for milling threads, notches, serrations, and similar work. Double-angle cutters are manufactured with included angles of 45°, 60°, and 90°. Other angles can be special ordered.

Metal slitting saws

Metal slitting saws are thin milling cutters that resemble circular saw blades, **Figure 17-52**. They are employed for narrow slotting and cutoff operations. Slitting saws are available in diameters as small as 2 1/2″ (60 mm) and as large as 8″ (200 mm).

- The plain metal slitting saw is essentially a thin plain milling cutter. It is used for ordinary slotting and cutoff operations. Both sides are ground concave for clearance. The hub is the same thickness as the cutting edge. It is stocked in thicknesses ranging from 1/32″ (0.8 mm) to 3/16″ (5 mm).
- A side chip clearance slitting saw is similar to the plain side milling cutter. It is especially suitable for deep slotting and sawing applications because of its ample chip clearance.

Formed milling cutters

Formed milling cutters are employed to accurately duplicate a required contour. A wide range of

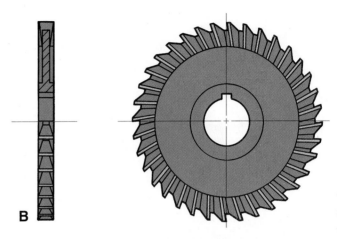

A

B

Figure 17-52. Slitting saws. A—Plain metal slitting saw is used for slotting and cutoff operations. B—The construction of the side chip clearance slitting saw allows it to be used for deep slotting applications.

shapes can be machined with standard cutters available. See **Figure 17-53**. Included in this cutter classification are the *concave cutter, convex cutter, corner rounding cutter,* and *gear cutter.*

17.5.4 Miscellaneous milling cutters

Included in this category are cutters that do not fit into any of the previously mentioned groups.

- The *T-slot milling cutter* has cutting edges for milling the bottoms of T-slots after cutting with an end mill or side cutter, **Figure 17-54**.
- A *Woodruff key seat cutter* is used to mill the semicircular keyseat for a Woodruff key, **Figure 17-55**.
- A *dovetail cutter* mills dovetail-type ways and it is used in much the same manner as the T-slot cutter. See **Figure 17-56**.

17.5.5 Care of milling cutters

Milling cutters are expensive and easily damaged if care is not taken in their use and storage. The following recommendations will help extend cutter life:

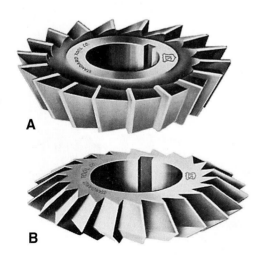

A

B

Figure 17-51. Angle cutters. A—Single-angle milling cutter. B—Double-angle milling cutter.

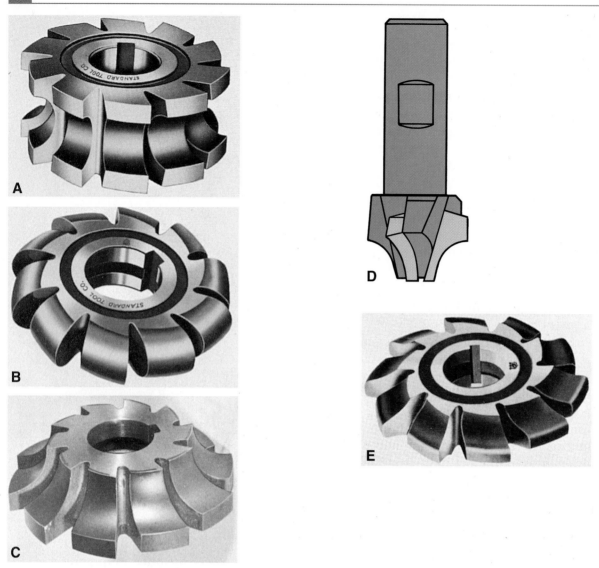

Figure 17-53. Formed milling cutters. A—Concave milling cutter will produce an accurate shape along cut. B—Convex milling cutter will produce a curved slot. C—Corner rounding milling cutter is available as left and right-hand cut. D—Corner-rounding end mill with replaceable tungsten carbide cutting edges. E—Gear cutter. (Standard Tool Co.)

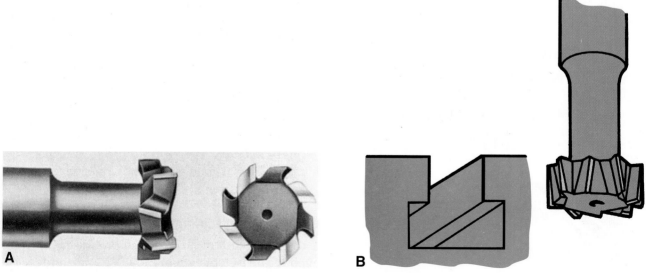

Figure 17-54. A T-slot milling cutter and the cut it produces. (Morse Cutting Tools)

- Use sharp cutting tools! Machining with dull tools results in low-quality work, and eventually damages the cutting edges beyond the point where they can be salvaged by grinding.
- Properly support tools and make sure the work is held rigidly.
- Use the correct cutting speed and feed for the material being machined.
- Assure an ample supply of cutting fluid, **Figure 17-57**.
- Employ the correct cutter for the job.
- Store cutters in individual compartments or on wooden pegs. They should never come in contact with other cutters or tools, **Figure 17-58**.
- Clean cutters before storing them. If they are to be stored for any length of time, it is best to give them a light protective coating of oil.

- Never hammer a cutter onto an arbor! Examine the arbor for nicks or burrs if the cutter does not slip onto it easily. Do not forget to key the cutter to the arbor.
- Place a wooden board under an end mill when removing it from a vertical milling machine. This will prevent cutter damage if it is dropped accidentally. Protect your hand with a heavy cloth or gloves. See **Figure 17-59**.

Figure 17-57. *Rigidly supported cutters will permit heavier cuts and prolong cutter life. An ample supply of cutting fluid is essential. (Sharnoa Corp.)*

Figure 17-55. *A Woodruff keyseat cutter.*

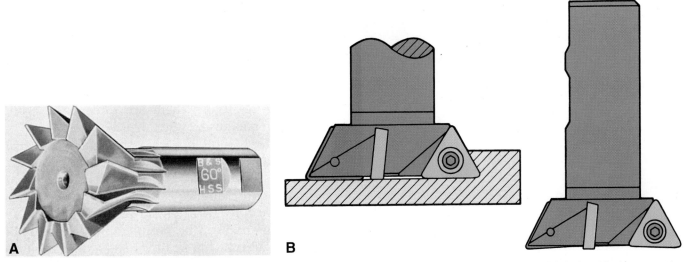

A   **B**

Figure 17-56. *Dovetail cutter. A—This cutter will make a slot with angled sides. B—A dovetail cutter with indexable insert cutting edges.*

Figure 17-58. Store cutters so they cannot come into contact with other cutters.

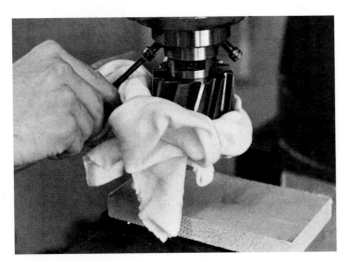

Figure 17-59. When removing or replacing a cutter on a milling machine, protect your hand with a heavy cloth. On vertical milling machines, also place a piece of wood under the cutter.

17.6 METHODS OF MILLING

Milling operations can be classified into one of two distinct methods:

- With *conventional* or *up-milling*, the work is fed into the rotation of the cutter, **Figure 17-60A**. The chip is at minimum thickness at the start of the cut. The cut is so light that the cutter has a tendency to slide over the work until sufficient pressure is built up to cause the teeth to bite into the material. This initial sliding motion, followed by the sudden breakthrough as the tooth completes the cut, leaves the "milling marks" so familiar on many milled surfaces. The marks and ridges can be kept to a minimum by keeping the table gibs properly adjusted.

- With *climb* or *down-milling*, the work moves in the same direction as cutter rotation, **Figure 17-60B**. Full engagement of the cutter tooth is instantaneous. The sliding action of conventional milling is eliminated, resulting in a better finish and longer tool life. The main advantage of climb milling is the tendency of the cutter to press the work down on the worktable or holding device.

Climb milling is *not recommended* on light machines, nor on large older machines that are not in top condition or are not fitted with an antibacklash device to take up play. There is danger of a serious accident if there is play in the table, or if the work or work-holding device is not mounted securely.

17.7 HOLDING AND DRIVING CUTTERS

The *arbor* is the most common method employed to hold and drive cutters. It is made in a number of sizes and styles. Arbors with *self-holding tapers* are used on some small hand milling machines and older models of larger millers, **Figure 17-61**. Today, there are three basic arbor styles in general use, **Figure 17-62.**

- *Style A* is fitted with a small pilot end that runs in a bronze bearing in the arbor support. This style is best used when maximum arbor support clearance is required.

- *Style B* is characterized by the large bearing collar that can be positioned on any part of the arbor. This feature makes it possible to mount the bearing support as close to the cutter as possible for maximum cutter support. This permits heavy cuts.

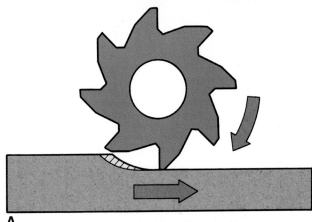

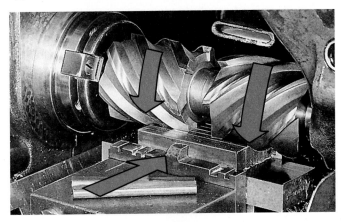

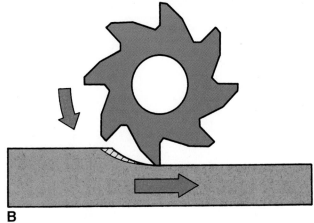

Figure 17-60. *Milling methods. A—Cutter and work movement with conventional (up) milling. B—Cutter and work movement with climb (down) milling.*

Figure 17-61. *The self-holding taper arbor is seldom used today.*

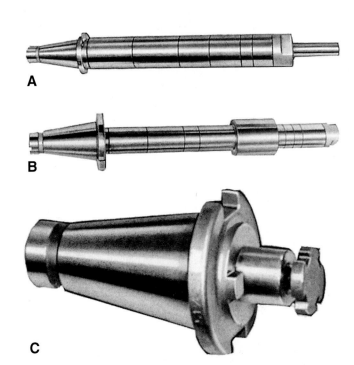

Figure 17-62. *Basic arbor styles. A—Style A arbor. B—Style B arbor. C—Style C arbor.*

- *Style C* is used to hold smaller sizes of shell end and face milling cutters that cannot be mounted directly to the spindle nose.

In general, use the *shortest arbor* possible that will permit adequate clearance between the arbor support and the work.

Both style A and style B arbors have a keyway milled their entire length, allowing a key to be employed to prevent the cutter from revolving on the arbor. See **Figure 17-63**.

Spacing collars allow the cutter to be positioned on the arbor, **Figure 17-64**. Accurately made in a number of widths, collars permit two or more cutters to be precisely spaced for *gang* and *saddle milling*. A left-hand threaded nut tightens the cutter and collars on the arbor. The nut should not be tightened directly against the bearing collar on the style B arbor, because the bearing may be damaged.

A *draw-in bar* is used on most vertical and horizontal milling machines, **Figure 17-65**. It fits through the spindle and screws into the arbor or collet to hold it firmly on the spindle. *Drive keys* on the nose of the spindle fit into corresponding slots on

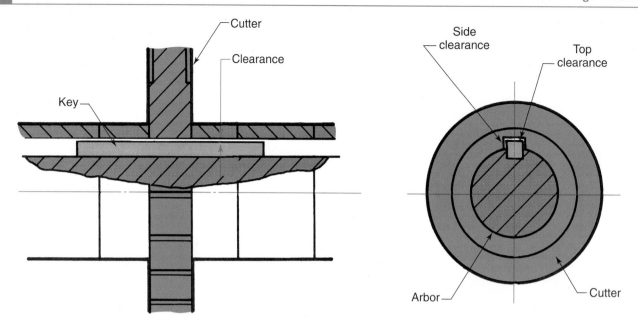

Figure 17-63. *Keying the cutter to the arbor prevents it from slipping during the cutting operation.*

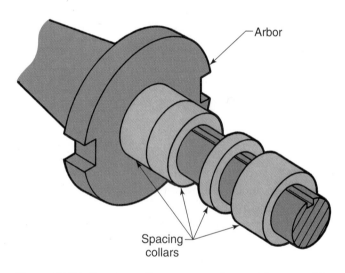

Figure 17-64. *Spacing collars are manufactured in many different widths. They are used to position one or more cutters on the arbor.*

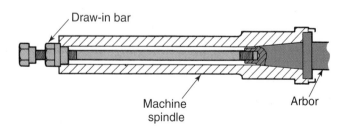

Figure 17-65. *The draw-in bar holds the arbor onto the spindle. Avoid operating a milling machine if the arbor is not held in place with a draw-in bar.*

the arbor, collet, or collet holder to provide positive (nonslip) drive, **Figure 17-66**.

End mills may be mounted in *spring collets, adapters, shell end mill holders,* or *stub arbors,* depending upon the type of work to be done. See **Figure 17-67**.

Spring collets accommodate straight shank end mills and drills. Some collets must be fitted in a *collet chuck. Adapters* are used for taper shank end mills and drills. *Shell end mill holders* enable shell end mills, used for face and side milling, to be fitted to a vertical milling machine. *Stub arbors* are short arbors that permit various side cutters, slitting saws, formed cutters, and angle cutters to be used on a vertical mill.

Figure 17-66. *Clearly shown are spindle drive keys, style A arbor support, and style B arbor support. (Cincinnati Lathe & Tool Co.)*

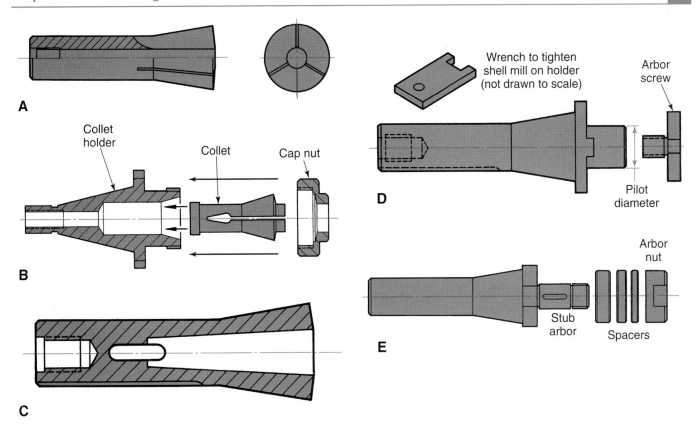

Figure 17-67. *Mounting devices. A—Spring collet (R-8 taper type). B—Collet chuck and collet. C—Adapter used with taper shank cutting tools. D—Shell end mill holder (R-8 taper). E—Stub arbor (R-8 taper).*

17.7.1 Care of cutter holding and driving devices

To maintain precision and accuracy during a milling operation, care must be taken to prevent damage to the cutter holding and driving devices.

- Keep the taper end of the arbor clean and free of nicks. The same applies to the spindle taper.
- Clean and lubricate the bearing sleeve before placing the arbor support on it. Also make sure the bearing sleeve fits snugly.
- Clean the spacing collars before slipping them onto an arbor, **Figure 17-68**. Otherwise, cutter runout will occur, making it difficult to make an accurate cut.
- Store arbors separately and in a vertical position.
- Never loosen or tighten an arbor nut unless the arbor support is locked in place, because this could spring the arbor so that it will not run true.
- Use a wrench of the correct size on the arbor nut, **Figure 17-69**. Make sure at least *four*

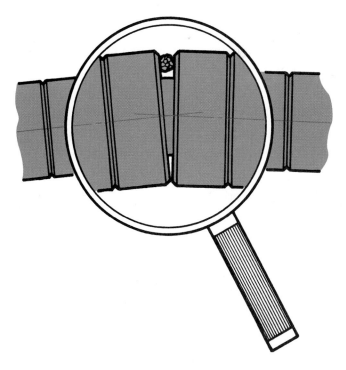

Figure 17-68. *A chip between spacing collars will cause an arbor to be sprung out of true. Bend in the arbor is greatly exaggerated in this drawing.*

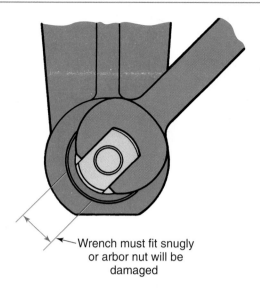

Figure 17-69. *Use a wrench of correct type and size to loosen an arbor nut.*

Wrench must fit snugly or arbor nut will be damaged

threads are engaged before tightening the arbor nut.

- Avoid tightening an arbor nut by striking the wrench with a hammer or mallet. This can crack the nut and/or distort the threads.
- Do not *force* a cutter onto an arbor. Check to see what is making it difficult to slide on. Correct any problem.
- Key all cutters to the arbor.

To remove an arbor or adapter from the machine:

1. Loosen the draw-in bar nut a few turns. Do not remove it from the arbor completely.

2. Tap the draw-in bar with a lead hammer to loosen the arbor in the spindle.

3. Hold the loosened arbor with one hand and unscrew the draw-in bar with the other.

4. Remove the arbor from the spindle. Clean and store it properly.

17.8 MILLING CUTTING SPEEDS AND FEEDS

The time required to complete a milling operation and the quality of the finish is almost completely governed by the cutting speed and feed rate of the cutter.

Milling cutting speed refers to the distance, measured in feet or meters, that a point (tooth) on the cutter's circumference will move in one minute. It is expressed in feet per minute (fpm) or meters per minute (mpm). Milling cutting speed is directly dependent on the revolutions per minute (rpm) of the cutter.

Milling feed is the rate at which work moves into the cutter. It is given in *feed per tooth per revolution (ftr)*. Proper feed rate is probably the most difficult setting for a machinist to determine. In view of the many variables (width of cut, depth of cut, machine condition, cutter sharpness, etc.), feed should be as coarse as possible, consistent with the desired finish.

17.8.1 Calculating cutting speeds and feeds

Considering the previously mentioned variables, the speeds listed in **Figure 17-70**, and the feeds listed in **Figure 17-71**, are suggested. The usual procedure is to start with the midrange figure and increase or decrease speeds until the most satisfactory combination is obtained, consistent with cutter life and surface quality.

In general, speed is *reduced* for hard or abrasive materials, deep cuts, or high alloy content metals. Speed is *increased* for soft materials, better finishes, and light cuts. Refer to **Figure 17-72** to calculate the cutting speed and feed for a specific material.

Material	High-speed steel cutter		Carbide cutter	
	Feet per minute	Meters per minute*	Feet per minute	Meters per minute*
Aluminum	550–1000	170–300	2200–4000	670–1200
Brass	250–650	75–200	1000–2600	300–800
Low carbon steel	100–325	30–100	400–1300	120–400
Free cutting steel	150–250	45–75	600–1000	180–300
Alloy steel	70–175	20–50	280–700	85–210
Cast iron	45–60	15–20	180–240	55–75

Reduce speeds for hard materials, abrasive materials, deep cuts, and high alloy materials. Increase speeds for soft materials, better finishes, light cuts, frail work, and setups. Start at midpoint on the range and increase or decrease speed until best results are obtained.
*Figures rounded off.

Figure 17-70. *Recommended cutting speeds for milling. Speed is given in surface feet per minute (fpm) and in surface meters per minute (mpm).*

Type of cutter	Material				
	Aluminum	Brass	Cast iron	Free cutting steel	Alloy steel
End mill	0.009 (0.22) 0.022 (0.55)	0.007 (0.18) 0.015 (0.38)	0.004 (0.10) 0.009 (0.22)	0.005 (0.13) 0.010 (0.25)	0.003 (0.08) 0.007 (0.18)
Face mill	0.016 (0.40) 0.040 (1.02)	0.012 (0.30) 0.030 (0.75)	0.007 (0.18) 0.018 (0.45)	0.008 (0.20) 0.020 (0.50)	0.005 (0.13) 0.012 (0.30)
Shell end mill	0.012 (0.30) 0.030 (0.75)	0.010 (0.25) 0.022 (0.55)	0.005 (0.13) 0.013 (0.33)	0.007 (0.18) 0.015 (0.38)	0.004 (0.10) 0.009 (0.22)
Slab mill	0.008 (0.20) 0.017 (0.43)	0.006 (0.15) 0.012 (0.30)	0.003 (0.08) 0.007 (0.18)	0.004 (0.10) 0.008 (0.20)	0.001 (0.03) 0.004 (0.10)
Side cutter	0.010 (0.25) 0.020 (0.50)	0.008 (0.20) 0.016 (0.40)	0.004 (0.10) 0.010 (0.25)	0.005 (0.13) 0.011 (0.28)	0.003 (0.08) 0.007 (0.18)
Saw	0.006 (0.15) 0.010 (0.25)	0.004 (0.10) 0.007 (0.18)	0.001 (0.03) 0.003 (0.08)	0.003 (0.08) 0.005 (0.13)	0.001 (0.03) 0.003 (0.08)

US Customary value expressed in inches per tooth. Metric value (shown in parentheses) expressed in millimeters per tooth.
Increase or decrease feed until the desired surface finish is obtained.
Feeds may be increased 100 percent or more depending upon the rigidity of the machine and the power available, if carbide tipped cutters are used.

Figure 17-71. Recommended feed rates in inches per tooth and millimeters per tooth for high speed steel (HSS) milling cutters.

Rules for determining speed and feed			
To find	Having	Rule	Formula
Speed of cutter in feet per minute (fpm)	Diameter of cutter and revolutions per minute	Diameter of cutter (in inches) multiplied by 3.1416 (π) multiplied by revolutions per minute, divided by 12	$fpm = \dfrac{\pi D \times rpm}{12}$
Speed of cutter in meters per minute	Diameter of cutter and revolutions per minute	Diameter of cutter multiplied by by 3.1416 (π) multiplied by revolutions per minute, divided by 1000	$mpm = \dfrac{D(mm) \times \pi \times rpm}{1000}$
Revolutions per minute (rpm)	Feet per minute and diameter of cutter	Feet per minute, multiplied by 12, divided by circumference of cutter (πD)	$rpm = \dfrac{fpm \times 12}{\pi D}$
Revolutions per minute (rpm)	Meters per minute and diameter of cutter in millimeters (mm)	Meters per minute multiplied by 1000, divided by the circumference of cutter (D)	$rpm = \dfrac{mpm \times 1000}{\pi D}$
Feed per revolution (FR)	Feed per minute and revolutions per minute	Feed per minute, divided by revolutions per minute	$FR = \dfrac{F}{rpm}$
Feed per tooth per revolution (ftr)	Feed per minute and number of teeth in cutter	Feed per minute (in inches or millimeters) divided by number of teeth in cutter $\times$ revolutions per minute	$ftr = \dfrac{F}{T \times rpm}$
Feed per minute (F)	Feed per tooth per revolution, number of teeth in cutter, and rpm	Feed per tooth per revolution multiplied by number of teeth in cutter, multiplied by revolutions per minute	$F = ftr \times T \times rpm$
Feed per minute (F)	Feed per revolution and revolutions per minute	Feed per revolution multiplied by revolutions per minute	$F = FR \times rpm$
Number of teeth per minute (TM)	Number of teeth in cutter and revolutions per minute	Number of teeth in cutter multiplied by revolutions per minute	$TM = T \times rpm$

rpm = Revolutions per minute
T = Teeth in cutter
D = Diameter of cutter
π = 3.1416 (pi)
frm = Speed of cutter in feet per minute

TM = Teeth per minute
F = Feed per minute
FR = Feed per revolution
ftr = Feed per tooth per revolution
mpm = Speed of cutter in meters per minute

Figure 17-72. Rules for determining cutting speed and feed.

Example Problem: Determine the approximate cutting speed and feed for a 6″ (152 mm) diameter side cutter (HSS) with 16 teeth, when milling free cutting steel.

Information Available:

Recommended cutting speed for free
cutting steel (midpoint in range) . . . . . 200 fpm
Recommended feed per tooth
(midpoint in range). . . . . .0.008″
Cutter diameter . . . . .6″
Number of teeth on cutter 16

To determine *speed setting* (cutter rpm), the following is given in **Figure 17-72.**

Rule: Divide the feet per (fpm) by the circumference of the cutter, expressed in feet.

$$\text{Formula:} \quad \text{rpm} = \frac{\text{fpm} \times 12}{\pi D}$$

$$= \frac{200 \times 12}{3.14 \times 6}$$

$$= \frac{2400}{18.84}$$

$$= 127.39 \text{ rpm*}$$

To determine *feed setting* (feed in inches per minute or F)

Rule: Multiply feed per tooth per revolution by number of teeth on cutter and by speed (rpm).

$$\text{Formula:} \quad F = \text{ftr} \times T \times \text{rpm}$$

$$= 0.008 \times 16 \times 127$$

$$= 16.25*$$

* The speed and feed are only approximate. Set machine to closest setting, either higher or lower. Increase or reduce speed until satisfactory cutting conditions are achieved.

17.9 CUTTING FLUIDS

Cutting fluids serve several purposes:
- They carry away the heat generated during the machining operation.
- They act as a lubricant.
- They prevent the chips from sticking to or fusing with the cutter teeth.
- They flush away chips.
- They influence the finish quality of the machined surface.

As this list shows, it is important that the correct cutting fluid be used for the material being machined. See **Figure 17-73.**

17.10 MILLING WORK-HOLDING ATTACHMENTS

One of the more important features of the milling machine is its adaptability to a large number of work-holding attachments. Each of these attachments increases the usefulness of the milling machine.

17.10.1 Vises

The *vise* is probably the most widely employed method of holding work for milling. The jaws are hardened to resist wear, and are ground for accuracy. A milling vise, like other work-holding attachments, is keyed to the table slot with *lugs*, **Figure 17-74.**

Aluminum and its alloys	Kerosene, kerosene and lard oil, soluble oil
Plastics	Dry
Brass, soft	Dry, soluble oil, kerosene and lard oil
Bronze, high tensile	Soluble oil, lard oil, mineral oil, dry
Cast Iron	Dry, air jet, soluble oil
Copper	Soluble oil, dry, mineral lard oil, kerosene
Magnesium	Low viscosity neutral oils
Malleable iron	Dry, carbonated water
Monel metal	Lard oil, soluble oil
Slate	Dry
Steel, forging	Soluble oil, sulfurized oil, mineral lard oil
Steel, manganese	Soluble oil, sulfurized oil, mineral lard oil
Steel, soft	Soluble oil, mineral lard oil, sulfurized oil, lard oil
Steel, stainless	Sulfurized mineral oil, soluble oil
Steel, tool	Soluble oil, mineral lard oil, sulfurized oil
Wrought iron	Soluble oil, mineral lard oil, sulfurized oil

Figure 17-73. *Recommended cutting fluids for various materials.*

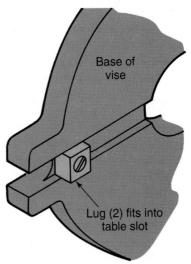

Figure 17-74. *Lugs on the base of a work-holding attachment position the device on the worktable.*

A *flanged vise* has slotted flanges for fastening the vise to the table, **Figure 17-75**. The slots permit the vise to be mounted on a horizontal milling machine either parallel to, or at right angles to, the spindle.

The body of a *swivel vise* is similar to a flange vise but is fitted with a circular base, graduated in degrees. This permits it to be locked at any angle to the spindle. See **Figure 17-76**.

The *toolmaker's universal vise* permits compound or double angles to be machined without complex or multiple setups. See **Figure 17-77**.

A *magnetic chuck*, shown in **Figure 17-78**, is ideally suited for many milling operations. The magnet eliminates the need for time-consuming hold-down clamps to mount the work to the table. The magnetic chuck can be used only with ferrous metals.

A *rotary table*, **Figure 17-79**, can perform a variety of operations, such as: cutting segments of circles, circular slots, cutting irregular-shaped slots,

and similar operations. A dividing attachment can be fitted to many rotary tables in place of the hand wheel, **Figure 17-80**.

The table is graduated in degrees around its circumference and adjustments can be made accurately with the hand wheel to 1/30 of a degree (2 minutes).

An *index table* permits the rapid positioning of work, **Figure 17-81**. Indexing is usually in 15° increments. However, a clamping device allows the table to be locked at any setting.

A *dividing head* will divide the circumference of circular work into equally spaced units. It is one of the more important milling attachments, **Figure 17-82**. This feature makes a dividing head

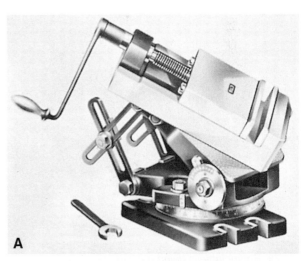

A

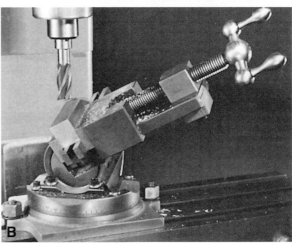

B

Figure 17-77. Toolmaker's universal vise.   A—The universal vise can be pivoted on several planes. B—Universal vise is being used on light vertical milling machine to cut a compound angle (double angle). Note how vise is tilted.

Figure 17-75. This is a typical flanged vise.

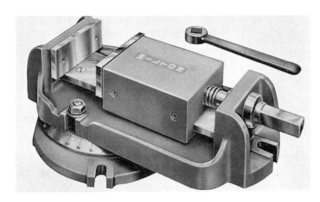

Figure 17-76. The swivel vise is handy and will rotate to align work easily. Note that the base of the vise is graduated in degrees. It can be positioned and locked at an angle to the machine spindle. (Wilton Tool Mfg. Co.)

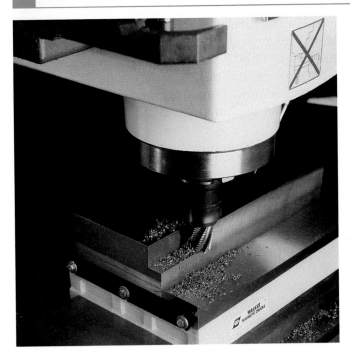

Figure 17-78. Milling a workpiece held with a magnetic chuck. (O.S. Walker Co.)

Figure 17-81. An index table permits work to be rapidly positioned. The 3-jaw chuck is used for round workpieces. Note that the unit can be mounted on the milling machine vertically or horizontally. (Yukiwa Seiko USA, Inc.)

Figure 17-79. Rotary table is being used to machine a part with round and curved shapes.

Figure 17-80. Rotary table with handwheel replaced by a dividing attachment. (WMW Machinery Company, Inc.)

indispensable when milling gear teeth, cutting splines, and spacing holes on a circle. It also makes possible the milling of squares, hexagons, etc., when required.

The dividing head consists of two parts, the dividing unit and the foot stock. Work may be mounted between centers, in a chuck, or in a collet. See **Figure 17-83**.

An *index plate*, identified by circles of holes on its face, and the *index crank*, which revolves on the index plate, are fundamental in the dividing operation. See **Figure 17-84**.

Rotating the index crank causes the dividing head spindle (to which work is mounted) to rotate. The standard ratios for the dividing head are five turns of the index crank for one complete revolution of the spindle (5:1); or 40 turns of the index crank for one revolution of the spindle (40:1). See **Figure 17-85**.

The ratio between index crank turns and spindle revolution, plus the index plate with its series of equally spaced hole circles, makes it possible to divide the circumference of the work into the required number of equal spaces.

For example, if 10 teeth were to be cut on a gear, it would require 1/10 of 5 turns (assuming dividing head has a 5:1 ratio), or one-half turn of the index crank for each tooth.

For 25 teeth, the number of crank turns would be 1/25 of 5, or dividing 5 by 25, or 5/25. By further

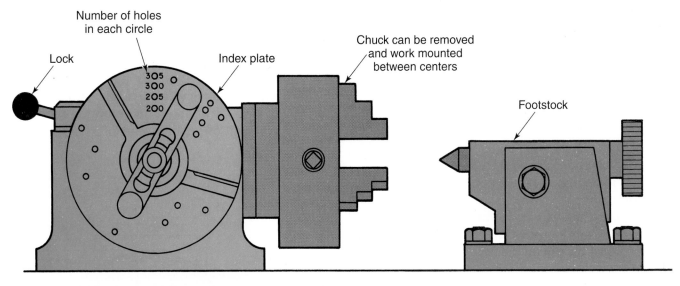

Figure 17-82. Dividing head and foot stock.

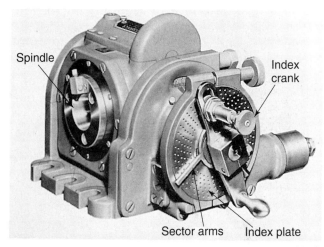

Figure 17-84. Location of index plate, sector arms, and index crank. (Kearney & Trecker Corp.)

Figure 17-83. Dividing head applications. A— Work has been mounted between centers for a milling operation. B—A cam is being milled while held in a collet. (Cincinnati Lathe & Tool Co.)

Figure 17-85. Internal mechanism of a dividing head uses precision gears. On this dividing head, 40 turns of the index crank rotates the spindle (work) one complete revolution. (Kearney & Trecker Corp.)

reduction, this becomes 1/5 of a turn of the crank for each tooth. This is where the holes in the index plate come into use; they allow fractional turns to be made accurately.

Select an index plate with a series of holes divisible by 5. On one such plate, the circles have 47, 49, 51, 53, 54, 57, and 60 holes. In this situation, 60 is divisible by 5. Thus, indexing would be through 12 holes on the 60-hole circle for each tooth.

When indexing, it is not necessary to *count* 12 holes each time the work is repositioned after a tooth has been cut. Two arms, called *sector arms* or *index fingers,* are loosened and positioned. One is located against the pin on the index crank. The other is moved clockwise until the arms are 12 holes apart, *not including* the hole that the pin is in. See **Figure 17-86**.

To index, first move the workpiece clear of the cutter. Disengage the crank by withdrawing the pin from the index plate and rotating it clockwise through the section marked by the sector arms. Drop the pin into the hole at the position of the second sector arm and lock the dividing head mechanism. Next, move the sector arms in the same direction as crank rotation to catch up with the pin in the index crank. For each cut, repeat the

operation. The dividing head spindle typically can be moved through an arc of 100°. It swivels 5° below horizontal, and 5° beyond perpendicular. See **Figure 17-87**.

A precision device, a dividing head has an indexing accuracy of about one minute of arc. This is the equivalent of 1/21,600 part of a circle.

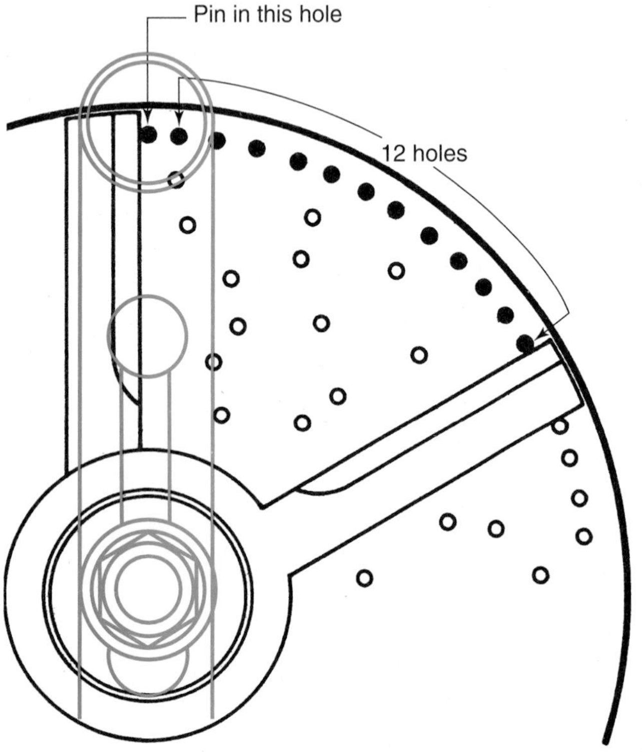

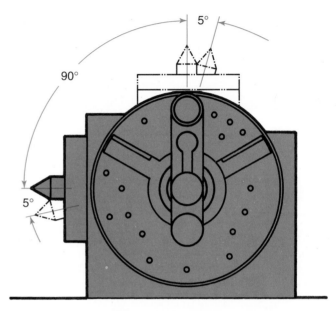

Figure 17-87. *Most dividing heads can pivot the spindle through an arc of 100°.*

Figure 17-86. *Positioning sector arms for proper movement of the spindle. When positioning sector arms, do not count the hole that the pin on index crank is in.*

TEST YOUR KNOWLEDGE

Please do not write in the text. Write your answers on a separate sheet of paper.

1. Milling machines fall into two broad classifications: _____ and _____ types.

2. There are three basic types of milling machines.
   a. A _____ type has a horizontal spindle and the worktable has three movements.
   b. A _____ type is similar to the above machine but a fourth movement has been added to the worktable to permit cutting helical shapes.
   c. A _____ type has the spindle perpendicular or at right angles to the worktable.

3. List the four methods of machine control. Briefly describe each of them.

4. Stop the machine before making _____ and _____.

5. Metal chips must never be removed with your _____. Use a _____.

6. Treat all small cuts and skin punctures as potential sources of infection. The following should be done:
   a. Clean them thoroughly.
   b. Apply antiseptic and cover with a bandage.
   c. Promptly report the injury to your instructor.
   d. All of the above.
   e. None of the above.

7. Milling cutters are sharp. Protect your hands with a _____ or _____ when handling them.

8. Milling operations fall into two main categories:
   a. _____ milling, in which the surface being machined is parallel with the cutter face.
   b. _____ milling, in which the surface being machined is parallel with the periphery of the cutter.

9. What are two general types of milling cutters?

10. What is the term "hand" used to describe, in reference to an end mill?

Match each term with the correct sentence below.

11. _____ Two-flute end mill.

12. _____ Multiflute end mill.

13. _____ Fly cutter.

14. _____ Shell end mill.

15. _____ Face milling cutter.

16. _____ Plain milling cutter.

17. _____ Slab cutter.

18. _____ Side milling cutter.

19. _____ Staggered-tooth side cutter.

20. _____ Metal slitting saw.
   a. Has cutting teeth on the circumference and on one or both sides.
   b. Cutter with helical teeth designed to cut with a shearing action.
   c. A facing mill with a single-point cutting tool.
   d. Can be fed into work like a drill.
   e. Cutter with teeth located around the circumference.
   f. Recommended for conventional milling where plunge cutting (going into work like a drill) is *not* required.

   g. Intended for machining large flat surfaces parallel to the cutter face.
   h. Thin milling cutter designed for machining narrow slots and for cutoff operations.
   i. Mounts on a stub arbor.
   j. Has alternate right-hand and left-hand helical teeth.

21. Flat surfaces are machined with _____ or _____ tooth milling cutters.

22. Make a sketch that illustrates climb milling.

23. Make a sketch illustrating conventional milling.

24. In climb milling:
   a. The work is fed into the rotation of the cutter.
   b. The work moves in the same direction as the rotation of the cutter.
   c. Neither of the above.

25. In conventional milling:
   a. The work is fed into the rotation of the cutter.
   b. The work moves in the same direction as the rotation of the cutter.
   c. Neither of the above.

26. What is a draw-in bar, and how is it used?

27. _____ refers to the distance, measured in _____ or _____, that a point (tooth) on the circumference of a cutter moves in _____.

28. _____ is the rate at which the work moves into the cutter.

29. Calculate machine speed (rpm) and feed (F) for a 1.5" diameter tungsten carbide 5 tooth (T) end mill when machining cast iron. Recommended cutting speed is 190 fpm. Feed per tooth (ftr) is 0.004". Use the following formulas:

$$rpm = \frac{fpm \times 12}{\pi D}$$

and

$$F = ftr \times T \times rpm$$

30. Determine machine speed (rpm) and feed (F) for a 2.5" diameter HSS shell end mill with 8 teeth (T), machining aluminum. Recommended cutting speed is 550 fpm. Feed per tooth (ftr) is 0.010". Use the formulas given in Problem 29.

31. Calculate machine speed (rpm) for machining aluminum with a 6″ diameter HSS side milling cutter. Recommended cutting speed is 550 fpm. Use the appropriate formula given in Problem 29.

32. Determine machine speed (rpm) and feed (F) for a 4″ diameter HSS side milling cutter with 16 teeth (T) milling free cutting steel. Recommended cutting speed is 200 fpm. Feed per tooth (ftr) is 0.005″. Use the formulas given in Problem 29.

33. Calculate machine speed (rpm) and feed (F) for a 2.5″ diameter HSS slab milling cutter with 8 teeth (T) machining brass. Recommended cutting speed is 250 fpm. Feed per tooth (ftr) is 0.006″. Use the formulas given in Problem 29.

34. Cutting fluids serve several purposes. List at least three of them.

Match each term with the correct sentence below.

35. _____ Flanged vise.

36. _____ Swivel vise.

37. _____ Universal vise.

38. _____ Rotary table.

39. _____ Dividing head.

40. _____ Indexing table.

41. _____ Magnetic chuck.

42. _____ Vise lug.

    a. Keys vise to a slot in worktable.

    b. Can only be mounted parallel to or at right angles on worktable.

    c. Needed when cutting segments of circles, circular slots, and irregular-shaped slots.

    d. Can only be used with ferrous metals.

    e. Has a circular base graduated in degrees.

    f. Permits rapid positioning of circular work in 15° increments and can be locked at any angular setting.

    g. Used to divide circumference of round work into equally spaced divisions.

    h. Permits compound angles (angles on two planes) to be machined without complex or multiple setups.

Chapter 18

Milling Machine Operations

LEARNING OBJECTIVES

After studying this chapter, you will be able to:
○ Describe how milling machines operate.
○ Set up and safely operate horizontal and vertical milling machines.
○ Perform various cutting, drilling, and boring operations on a milling machine.
○ Make the needed calculations and cut spur gears.
○ Make the needed calculations and cut a bevel gear.
○ Point out safety precautions that must be observed when operating a milling machine.

IMPORTANT TERMS

addendum	*gang milling*
bevel gear	*slitting*
circular pitch	*slotting*
dedendum	*spur gear*
diametral pitch	*straddle milling*

18.1 MILLING OPERATIONS

The versatility of the milling machine family permits many different machining operations to be performed. Because of the number of varied operations that can be performed, it is not possible to cover all of them in a book of this type. Only basic operations will be described.

18.2 VERTICAL MILLING MACHINE

The vertical milling machine is capable of performing milling, drilling, boring, and reaming operations, **Figure 18-1**. It differs from the horizontal mill in that the spindle is mounted in a vertical position.

The *spindle head* swivels 90° left or right for machining at any angle, **Figure 18-2**. The *ram,* on which it is mounted, can be adjusted in and out. On many vertical mills, it also revolves 180° on a horizontal plane. Both swivels are graduated in degrees with a vernier scale to assure accurate angular settings. See **Figure 18-3**.

18.2.1 Cutters for vertical milling machine

Although adapters are available that permit the use of side and angle cutters, face mills and end mills are the cutters normally used in vertical milling machines, **Figure 18-4**.

Taper shank end mills and drills are fitted in an adapter, **Figure 18-5A**. Some machine spindles have a Brown & Sharpe taper. When taper shanks are large enough, they are mounted directly, **Figure 18-5B**. When a taper is too small to fit directly into the spindle, a sleeve must be employed. Straight shank end mills are held in a spring collet, **Figure 18-5C**, or in an end mill adapter, **Figure 18-5D**. Small drills, reamers, and similar tools are held in a standard Jacobs chuck fitted to the spindle by one of the above methods.

18.3 VERTICAL MILLING MACHINE OPERATIONS

In addition to the usual precautions that must be observed when getting a machine tool ready for a job, the spindle head alignment must be checked. Make sure that the spindle head is at an *exact right angle* (perpendicular) to the worktable. If the spindle is not perpendicular, it is not possible to machine a flat surface, **Figure 18-6**.

Milling head perpendicularity can be checked with the use of a dial indicator, as shown in **Figure 18-7**. The device holding the indicator may be shop made or purchased.

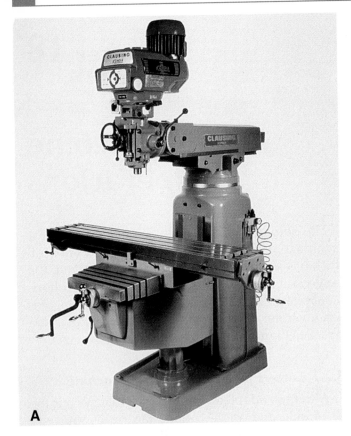

A

B

Figure 18-1. Vertical milling machine has the mounted spindle in a vertical position. A—Manually operated vertical milling machine. B—CNC 2-axis vertical milling machine. *(Clausing Industrial, Inc.)*

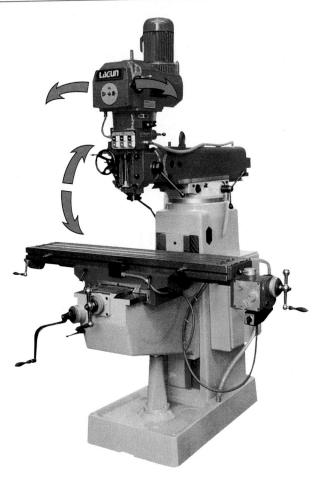

Figure 18-2. Angular head adjustments are possible on many vertical milling machines. *(Republic-Lagun Machine Tool Co.)*

If a vise is used to mount the work, wipe the vise base and worktable clean. Inspect for burrs and nicks. They prevent the vise from seating properly on the table. Bolt it firmly to the machine.

If extreme accuracy is required, the next step is to align the vise with a dial indicator, **Figure 18-8.** However, for many jobs the vise can be aligned with a square, as shown in **Figure 18-9.** Angular settings can be made with a protractor, **Figure 18-10,** or by using the degree divisions on the base of a swivel vise.

If you have not already done so, wipe the vise jaws and bottom clean of chips and dirt. Place clean parallels in the vise and place the work on them.

Tighten the jaws and tap the work onto the parallels with a mallet or soft-faced hammer. Thin paper strips can be employed to check whether the work is firmly on the parallels, **Figure 18-11.**

Never strike the vise handle with a hammer or mallet to put additional holding pressure on the jaws. If the workpiece is rough, protect the vise jaws and parallels with soft metal strips.

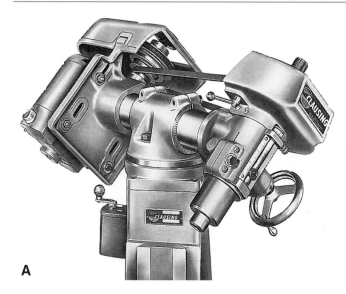

A

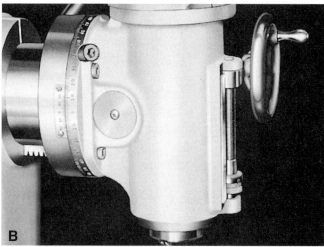

B

Figure 18-3. Head adjustments. A—Closeup shows how the head of this light vertical milling machine can be swiveled left or right. (Clausing Industrial, Inc.) B—A Vernier scale permits the spindle head to be set with extreme accuracy. Note the partial extension of the ram. (South Bend Lathe, Inc.)

Figure 18-4. Adapters permit arbor and other type cutters to be mounted on a vertical milling machine. (Parlec, Inc.)

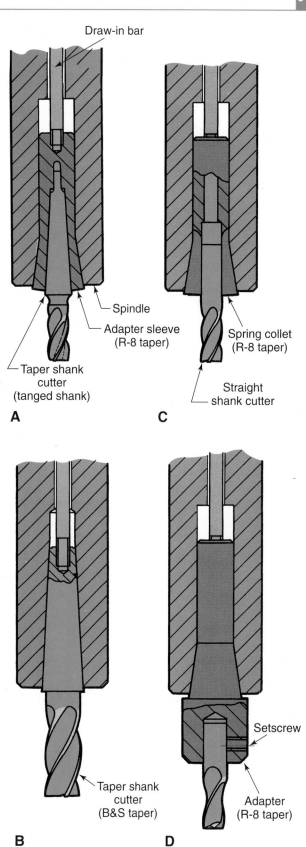

Figure 18-5. Four of the most common methods employed to mount end mills in a vertical milling machine. A—Adapter sleeve with taper shank cutter. B—B&S taper mounted directly in the spindle. C—Spring collet with straight shank cutter. D—Adapter with setscrew on straight shank cutter.

Figure 18-6. Irregular machined surface that results when the spindle head is not perpendicular to the table.

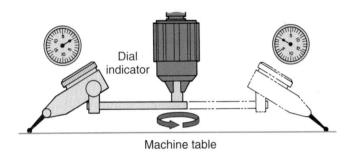

Figure 18-7. A dial indicator can be employed, as shown, to check whether the milling head is perpendicular to the table.

Figure 18-8. Use of a dial indicator permits extreme accuracy in aligning a solid vise jaw.

18.3.1 Squaring stock

A specific sequence must be followed to machine several surfaces of a piece square with one another, as shown in **Figure 18-12.**

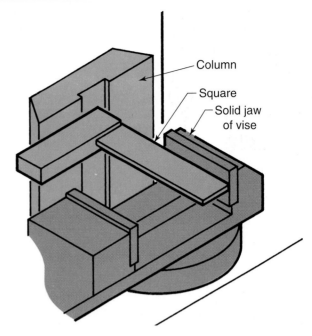

Figure 18-9. Procedure for squaring a solid vise jaw using a machinist's steel square.

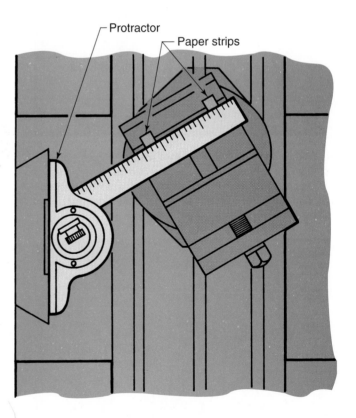

Figure 18-10. Angular settings can be made with protractor head and steel rule of a combination set. Paper strips aid in determining whether setting is accurate.

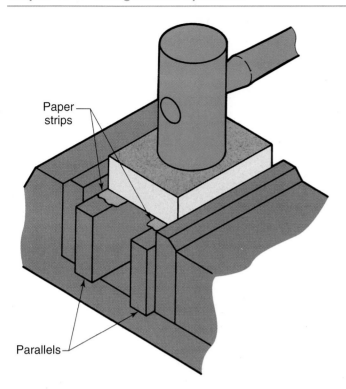

Figure 18-11. *Setting work on parallels with a soft-faced hammer. Thin paper strips are used to check whether the work is firmly on the parallels.*

1. Machine the first surface. Remove the burrs and place the first machined surface against the *fixed* vise jaw. Insert a piece of soft metal rod between the work and *movable* jaw if that portion of the work is rough or not square.

2. Machine the second surface.

3. Remove the burrs and reposition the work in the vise to machine the third side. This side must be machined to dimension. Take a light cut and "mike" for size. The difference between this measurement and the required thickness is the amount of material that must be removed. Always stop the machine before attempting to make measurements.

4. Repeat the above operation to machine the fourth side.

5. If the piece is short enough, the ends may be machined by placing it in a vertical position with the aid of a square, **Figure 18-13A**. Otherwise, it may be machined as shown in **Figure 18-13B**.

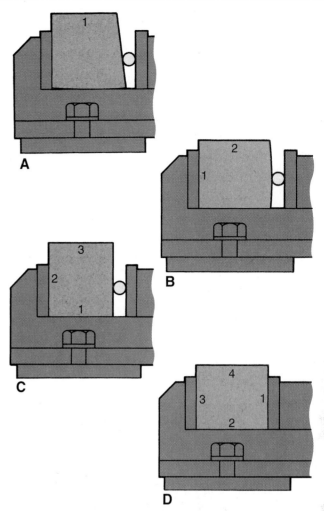

Figure 18-12. *Sequence for squaring work on a milling machine. A—Square top. B—Square side. C—Square bottom. D—Square other side.*

18.3.2 Machining angular surfaces

Angular surfaces (bevels, chamfers, and tapers) may be milled by tilting the spindle head assembly to the required angle. They may also be made by setting the work at the specified angle in a vise. See **Figure 18-14**.

A *fixture*, **Figure 18-15**, is often used when many similar pieces must be milled. *Compound angles* (angles on two planes) are made in a *universal vise*, **Figure 18-16**.

When the pivoted spindle head is used for angular cutting, it is essential that the vise be aligned with a dial indicator. Make a layout of the desired angle on the work and clamp it in the vise. Position the cutter and machine to the line.

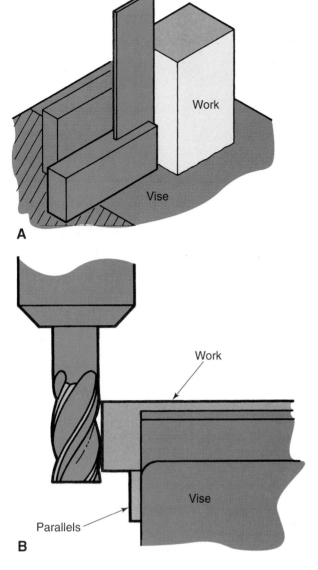

A

B

Figure 18-13. Squaring ends. A—Using a square to position short pieces for machining ends. Movable jaw is not shown for clarity. B—Another technique for squaring ends of work. Jaw must be checked with a dial indicator to be sure it is at a right angle to the column.

Work mounted at an angle in the vise for machining must be set up carefully. Alignment may be made with a protractor head fitted with a spirit level, **Figure 18-17**, or with a surface gage, **Figure 18-18**.

18.3.3 Milling a keyseat or slot

An end mill may be used to cut a keyseat or slot. After aligning the vise with a dial indicator, the workpiece is clamped in the vise or to the machine table. If mounted directly to the table, a piece of paper between the table and the work will seat the work more solidly and prevent slippage.

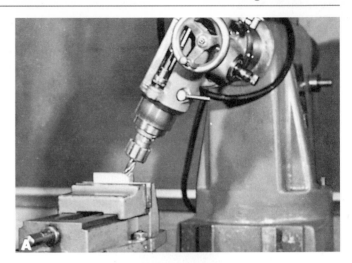

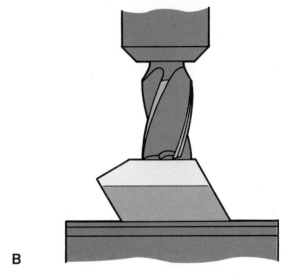

B

Figure 18-14. Cutting angular surfaces. A—Surface is being cut with spindle head set to required angle. B—Making an angular cut by positioning work at desired angle.

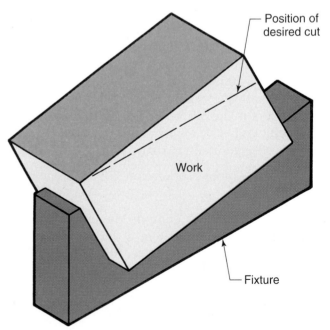

Figure 18-15. If many pieces are to have angular surfaces machined, considerable time can be saved by using a fixture.

Figure 18-16. An angular cut is being made on work held in a universal vise. With this setup, compound angles are easily cut.

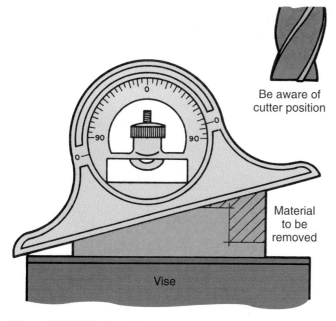

Figure 18-17. Work can be quickly set at the desired angle with aid of a protractor head.

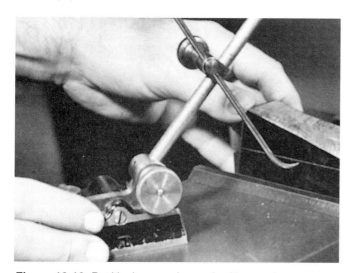

Figure 18-18. Positioning angular work with a surface gage.

A sharp cutter, equal in diameter to the keyseat or slot, must be used. A two-flute end mill is employed when a blind keyseat or slot is to be machined, **Figure 18-19.** Otherwise, a multiflute end mill is used.

18.3.4 Locating end mill to cut keyseat or slot on round work

After the milling machine has been set up, the work is secured in a vise, between centers, in V-blocks, or in a fixture. Before machining begins, precise centering of the end mill must be done as follows:

1. Lock the knee to the column.

2. Lay out the slot length, as shown in **Figure 18-20.**

3. Hold the end of a long, narrow strip of paper between the cutter and the work. Carefully move the cutter toward the work until the paper strip is pulled lightly from your fingers by the rotating cutter. Pay close attention when using this alignment technique. Use a paper strip long enough to keep your fingers well clear of the cutter. Release the paper as soon as you feel the cutter "grabbing" it.

4. Unlock the knee from the column and lower the table until the cutter is slightly above the work. Move the cutter inward half the work

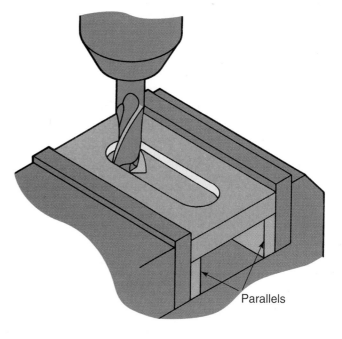

Figure 18-19. Blind slot being machined with a two-flute end mill.

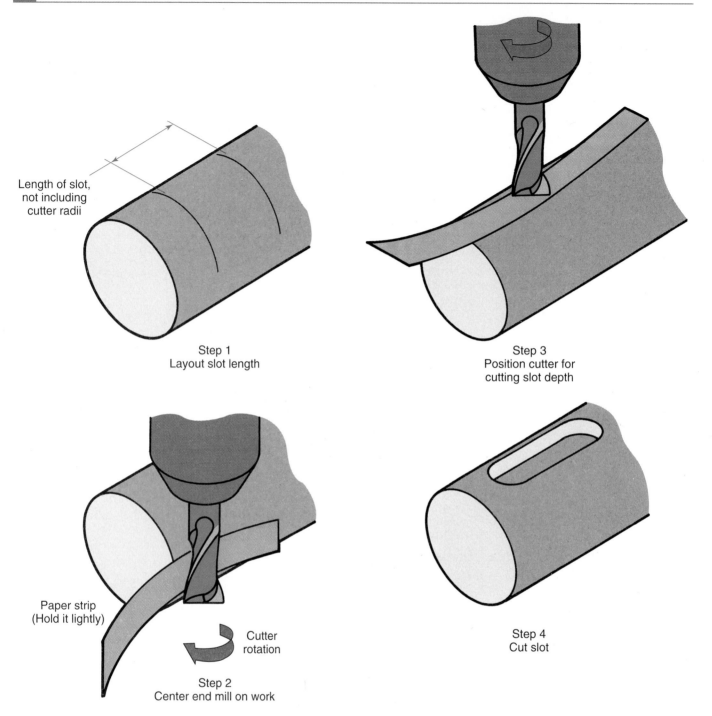

Length of slot,
not including
cutter radii

Step 1
Layout slot length

Step 3
Position cutter for
cutting slot depth

Paper strip
(Hold it lightly)

Cutter
rotation

Step 2
Center end mill on work

Step 4
Cut slot

Figure 18-20. *Procedure for using a paper strip to position cutter on the exact center of round stock for cutting a keyseat. Keep your fingers clear of the rotating cutter!*

diameter, plus half the cutter diameter, plus the paper thickness.

5. Using another long, narrow strip of paper, employ the same technique to get the required depth.

Correct keyseat depth may be obtained from tables in a machinist's handbook.

18.3.5 Machining internal openings

Internal openings are easily machined with a vertical milling machine, **Figure 18-21.** A two-flute end mill must be used if the cutter is to make the initial opening. It can be fed directly into the material in much the same manner as a drill.

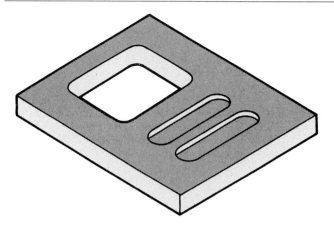

Figure 18-21. *Internal openings have been milled in this aircraft part.*

When the slot is wider than the cutter diameter, it is important that the direction of feed, in relation to cutter rotation, be observed. Feed direction is normally *against cutter rotation*, **Figure 18-22.** This applies only when the cutter is removing metal from one side of the opening.

18.3.6 Machining multilevel surfaces

Milling a multilevel surface is probably the easiest of the milling operations, **Figure 18-23.** A layout of the various levels is made on the work's surface.

Cuts are made until the lines are reached. For accuracy, the cut must be checked with a depth micrometer (remove any burrs before making the measurement), **Figure 18-24.** Make necessary table adjustments accordingly.

18.3.7 Milling and boring

Holes may be located to very close tolerances for drilling, reaming, or boring on a vertical milling machine. The first hole can be located with a *wiggler*, **Figure 18-25,** a *centering scope,*

Figure 18-23. *A stepped (multilevel) surface being milled.*

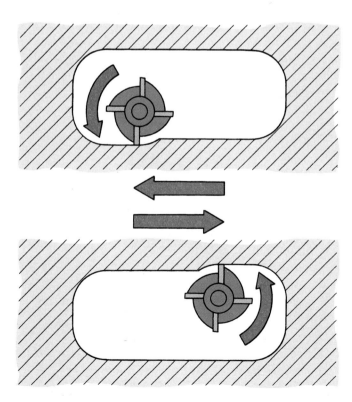

Figure 18-22. *When making this type cut, remember that feed direction is always against cutter rotation.*

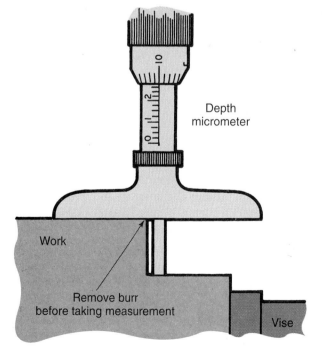

Figure 18-24. *Check depth of cut with a depth micrometer before making final cutter adjustments. Remove any burrs before making the measurement. Always stop the machine before attempting to remove burrs and making a measurement.*

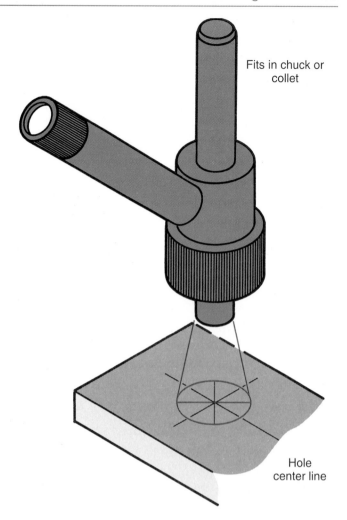

Figure 18-26. *A centering scope is another tool that aids in centering work for drilling or boring on vertical milling machines. It is fitted in the chuck (the chuck must run true) or collet. The machinist sights through the optical system and positions work using the crosshairs in the scope. It is capable of locating true position within ±0.0001" (0.0025 mm).*

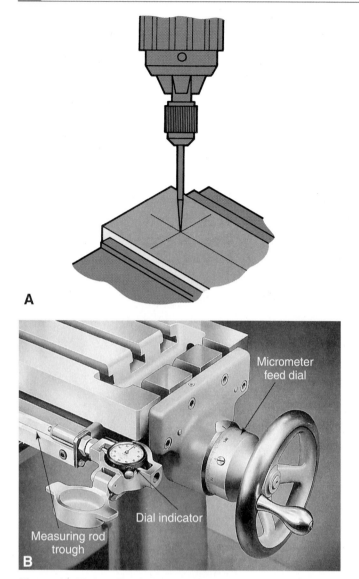

Figure 18-25. *Locating holes. A—To ensure accuracy, align first hole with "wiggler." B—Remaining holes can be located with micrometer dials on cross and longitudinal feeds of machine. Micrometer feed dials make it possible to maintain very close positional tolerances when drilling a series of holes. (South Bend Lathe, Inc.)*

Figure 18-26, or an *edge finder,* **Figure 18-27.** Then, it is possible to locate any remaining holes within 0.001" (0.025 mm) using the micrometer feed dials. Tolerances of 0.0001" (0.0025 mm) are possible with a *measuring rod and dial indicator* attachment, **Figure 18-28,** or a *digital readout gaging system,* **Figure 18-29,** fitted to the machine.

Boring permits holes to be machined accurately with fine surface finishes, **Figure 18-30.** A single-point tool is fitted to the boring head, which in turn is mounted in the spindle, **Figure 18-31.** Hole diameter is obtained by offsetting the tool point from center. The adjustment is graduated for a direct reading, **Figure 18-32.** A hole must first be drilled in the work before boring can be started.

Other highly specialized attachments are available for special jobs and long production runs. See **Figures 18-33** through **18-35.**

18.4 MILLING MACHINE CARE

- Check and lubricate the machine with the recommended lubricants.
- Clean the machine thoroughly after each job. Use a brush to remove chips, **Figure 18-36.** Never attempt to clean the machine while it is running.
- Keep the machine clear of tools.
- Check each setup for adequate clearance between the work and the various parts of the machine.
- *Never* force a cutter into a collet or holder. Check to see why it does not fit properly.

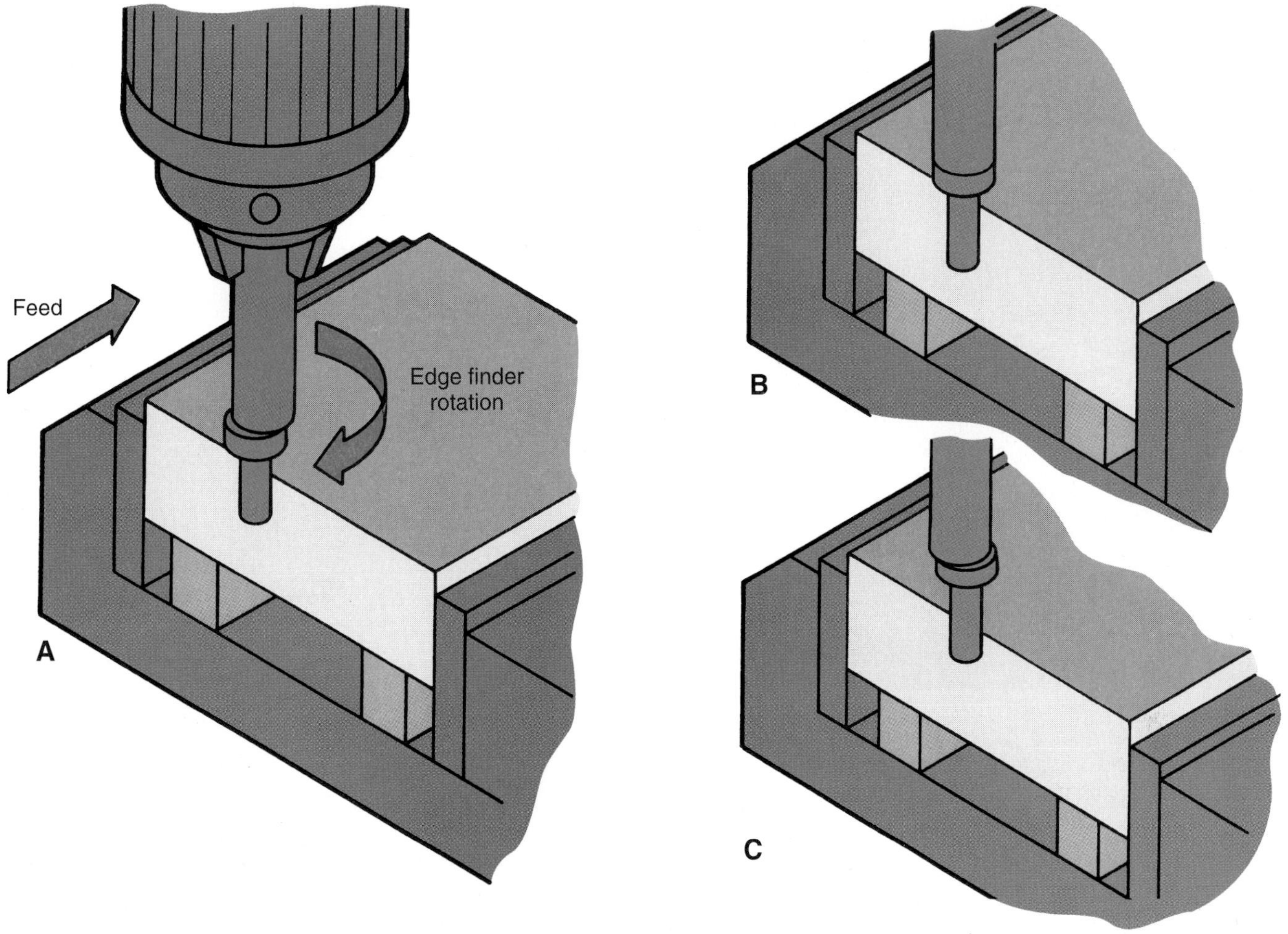

Feed

Edge finder
rotation

A

B

C

Figure 18-27. *The edge finder is a precision positioning tool that will locate the edge of the work in relation to center of the spindle with 0.0002″ (0.005 mm) accuracy. A—With spindle rotating at moderate speed, and with edge finder tip as shown, slowly feed tip of tool against work. B—Edge finder tip will gradually become centered with its shank. C—When the tip becomes exactly centered, it will abruptly jump sideways about 1/32″ (0.8 mm). When this occurs, stop table movement immediately. Center of the spindle will be exactly one-half tip diameter away from edge of work. Set the micrometer dial to "0" and, with edge finder clear of work, move table longitudinally the required distance plus one-half the tip diameter. Follow the same procedure to get traverse measurement.*

- Use a sharp cutter. Protect your hands when mounting it.
- Check the machine to determine whether it is level. Checking should be done at regular intervals.
- Have all guards in place before attempting to operate a milling machine.
- Check coolant level and condition if the reservoir is built into the milling machine. Change it if it becomes contaminated.
- Start the machining operation only after you are sure that everything is in satisfactory working condition. It may be necessary to make special fixtures to hold odd-shaped or difficult-to-mount work.
- Use attachments designed for the machine.

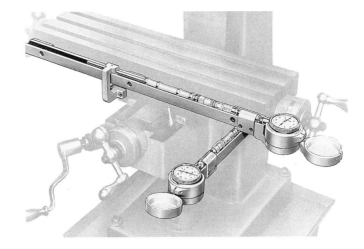

Figure 18-28. *Measuring rods and dial indicator attachments will allow precise location of work. (Clausing Industries, Inc.)*

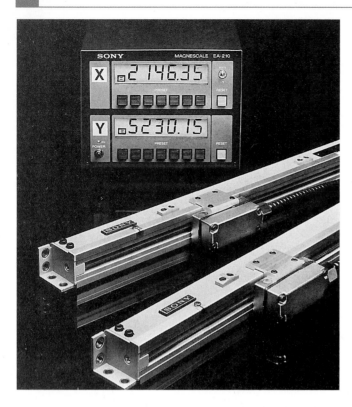

Figure 18-29. *Digital readout gaging systems are linear scales that employ magnetic or photoelectric scanning devices to display absolute distance from a predetermined datum point. Capabilities include inch/metric conversion. (National Machine Systems)*

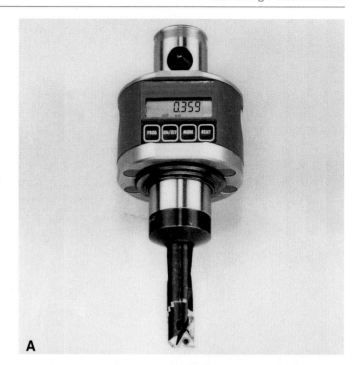

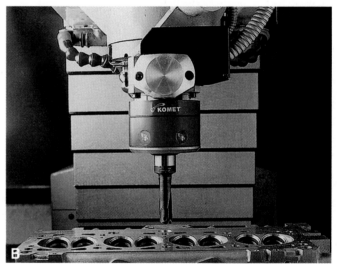

Figure 18-31. *Boring heads. A—Micro-adjustable boring head with digital display. This unit has an accuracy of 0.00004″ (0.01 mm), and has a decimal or metric readout display. B—Boring head with a programmable electronic controller interfaced to the machine control. It relates measurement data to preset tolerance limits, and signals tool to make size adjustments to compensate for tool wear. The boring head incorporates a small servomotor that adjusts the boring slide in extremely fine limits. Communication to, and powering of, the tool are entirely wireless. (Marposs Corp./Komet of America, Inc.)*

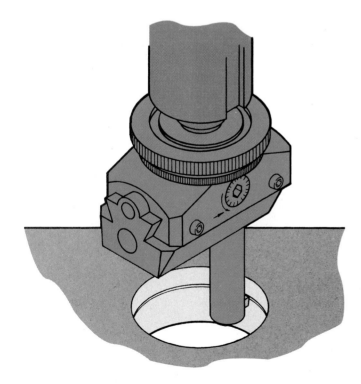

Figure 18-30. *Boring on vertical milling machine produces a very accurately sized hole.*

18.5 HORIZONTAL MILLING MACHINE OPERATIONS

Like the vertical milling machine, **Figure 18-37,** the horizontal mill is a very versatile machine. Many different machining operations can be performed on it.

Figure 18-32. Dial on the adjusting screw of this boring head permits accurate cutting tool settings. Some boring head dials indicate actual tool movement, while others indicate actual material removal.

Figure 18-33. A slotting attachment mounted on a vertical milling machine. Rotary motion in the spindle head is changed to reciprocating (up-and-down) movement of the cutting tool. (Bridgeport Machines, Inc.)

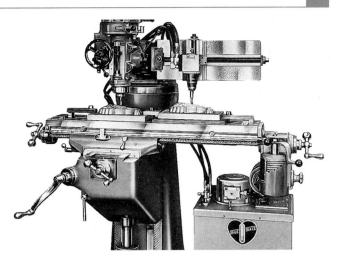

Figure 18-34. This vertical mill fitted with a hydraulic tracer unit is machining an irregular shape. (Bridgeport Machines, Inc.)

Figure 18-35. A right angle milling attachment permits the machine to be employed for internal and confined-area milling.

Figure 18-36. To remove metal chips, use a brush — never your hand!

Machining Fundamentals

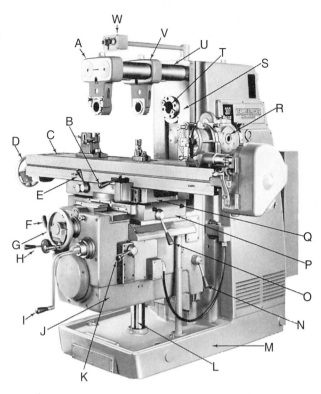

Figure 18-37. Parts of horizontal milling machine: A—Arbor support. B—Table feed lever. C—Table. D—Table (longitudinal) hand wheel. E—Table clamp lever (lock). F—Saddle power feed lever. G—Saddle (in-out) hand wheel. H—Knee power feed lever. I—Knee (up-down) crank. J—Knee. K—Rapid traverse lever. L—Telescopic coolant return and elevating screw. M—Base. N—Knee clamp lever. O—Saddle clamp lever (lock). P—Saddle. Q—Saddle plate. R—Universal dividing head. S—Column. T—Spindle. U—Overarm. V—Inner arbor support. W—Spindle stop-start and master switch (arm also engages clutch). (Kearney & Trecker Corp.)

18.5.1 Milling flat surfaces

A careful study of the part drawing will let you determine what operation is to be performed, what cutter is best suited for the job, and the most advantageous way to hold the workpiece. Flat surfaces may be milled with a plain cutter or slab cutter mounted on an arbor (peripheral milling), or with an inserted tooth face or shell milling cutter (face milling). The method employed will be determined by the size and shape of the work.

After the milling method and cutter have been selected, the following sequence of operations is recommended.

1. Check and lubricate the machine. Wipe the worktable clean and examine it for nicks and burrs. Nicks or burrs will prevent the workpiece or holding attachments from seating properly on the table.

2. Mount the work directly on the table, if possible. **Figure 18-38** illustrates one method for mounting long work. If the work cannot be mounted to the table, use a vise. Clean its base and bolt it firmly in place. Locate the vise as close to the machine column as workpiece shape and the arbor support will permit. When possible, pivot the vise so that the solid jaw supports the work against cutting rotation, **Figure 18-39**.

3. If extreme accuracy is required, align the vise using a dial indicator, **Figure 18-40**. Otherwise, a square or machine arbor will do, **Figure 18-41**. Angular vise settings can be made using the vise base graduations or a protractor.

4. Wipe the vise jaws and bottom clean.

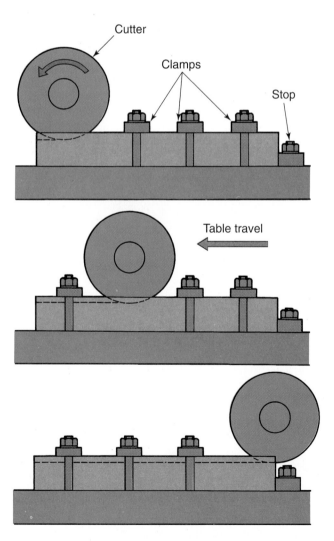

Figure 18-38. A method used to mount long work. Reposition clamps as the cut progresses across the workpiece.

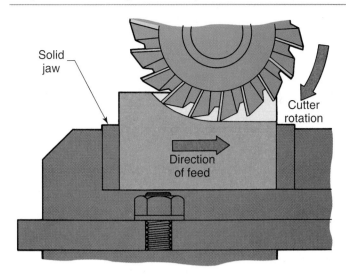

Figure 18-39. *Whenever the setup permits, the solid jaw of the vise should be in this position.*

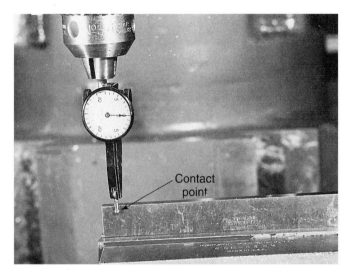

Figure 18-40. *Aligning the vise jaw by using a dial indicator will ensure work accuracy.*

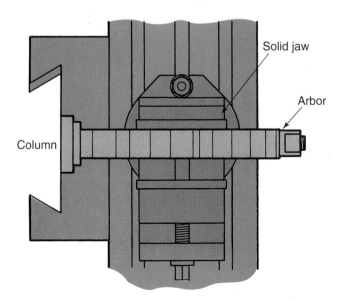

Figure 18-41. *An arbor can also be used to align a vise.*

5. Place clean parallels in the vise with the work seated on them. Tighten the jaws and tap the work onto the parallels with a mallet or soft-faced hammer. Thin paper strips can be used to check that the work is seated tightly on top of the parallels.

6. Select an arbor that is as short as the job will permit. Wipe the taper section of the arbor and the spindle opening with a dry cloth. Insert the arbor and draw it tightly into place with the draw-in bar.

7. When possible, the cutter should be wide enough to machine the area in one pass. It should have the smallest diameter possible, while being large enough to provide adequate clearance. See **Figure 18-42**.

8. Key the cutter to the arbor. Position it as close to the column as the work will permit.

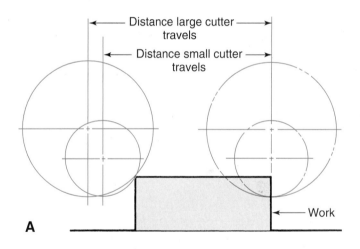

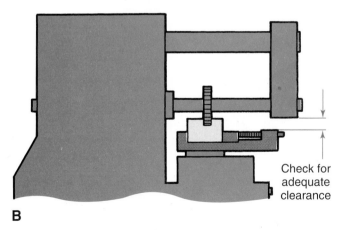

Figure 18-42. *Cutter diameter. A—A small-diameter cutter is more efficient than a large-diameter cutter because it travels less distance while doing the same amount of work. B—Use the smallest cutter diameter possible, but be certain it is large enough for adequate clearance.*

To protect your hands when mounting a cutter on the arbor, use a piece of cloth or gloves, **Figure 18-43.** If a helical slab mill is used, mount it so the cutting pressure forces it *toward* the column.

9. Position and lock the arbor support into place, **Figure 18-44.** Then, tighten the arbor nut.

10. Adjust the machine to the proper cutting speed and feed.

11. Turn on the machine and check cutter rotation and direction of power feed. If satisfactory, loosen all worktable and knee locks. Position the workpiece under the rotating cutter until the cutter just touches its surface. Set the

micrometer dial to 0. Back the work away from the rotating cutter. Make a light cut with ample cutting fluid flooding the surface. Make a measurement, then raise the table the required distance.

Should a previously machined surface require additional machining, it will be best to position the cutter in the following manner. Hold a long, narrow strip of paper with its loose end between the work and the cutter, **Figure 18-45.** Raise the table until the paper is pulled lightly from your fingers.

Once the cutter has been positioned, it is only necessary to move the cutter clear of the work and raise the table the required distance, plus the thickness of the paper. Tighten all locks (except longitudinal) and feed the work into the cutter. As soon as cutting starts, turn on the coolant and power feed.

Unless there is an emergency, do not stop the work during the machining operation. This will cause a slight depression to be cut in the machined surface, **Figure 18-46.**

Figure 18-43. *Use a cloth to protect your hands when mounting or removing a cutter from an arbor.*

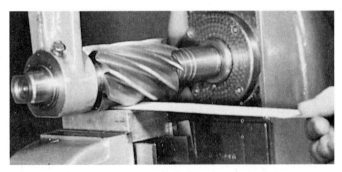

Figure 18-45. *Positioning a cutter on the workpiece, employing the paper strip technique. Pay close attention when positioning a cutter by this method. Use a long strip of paper, hold it lightly, and keep your fingers well clear of the rotating cutter.*

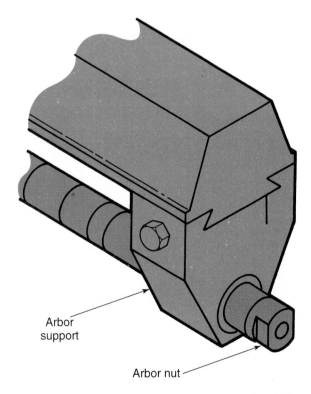

Arbor support

Arbor nut

Figure 18-44. *The arbor nut must not be tightened until after the arbor support has been positioned and locked in place.*

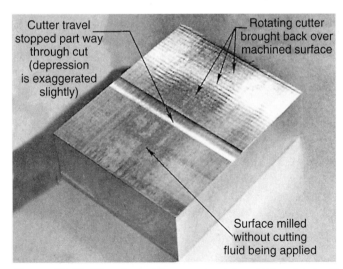

Cutter travel stopped part way through cut (depression is exaggerated slightly)

Rotating cutter brought back over machined surface

Surface milled without cutting fluid being applied

Figure 18-46. *Stopping work movement in middle of cut will cause a slight depression in machined surface. Also, note ridges that were made when revolving cutter was brought back over machined surface.*

Complete the cut and stop the cutter. Return the work to the starting position. Avoid feeding the work back to the starting position while the cutter is rotating. This will cause a series of depressions to be made on the newly machined surface.

Do not attempt to feel the machined surface while the cut is in progress or while the cutter is rotating.

Repeat the above operations if additional metal must be removed to bring the work to size.

18.5.2 Squaring stock

The sequence for squaring stock on a horizontal milling machine is the same as that used on a vertical milling machine. Whenever possible, the cutter should be wide enough to make a full-width cut on the material in one pass. If the material is short enough, the ends can be machined by placing it in a vertical position with the aid of a square, as shown in **Figure 18-13A**. If too long for this technique, the ends can be squared as shown in **Figure 18-47**.

18.5.3 Face milling

Face milling makes use of a cutter that machines a surface at right angles to the spindle axis and parallel to the face of the tool, **Figure 18-48**. Face milling cutters over 6″ (150 mm) in diameter are usually of the inserted tooth type and mount directly to the spindle nose. They are used to mill large, flat surfaces.

Face mills smaller than 6″ (150 mm) are called *shell end mills* and are held on a Style C arbor.

1. Select a cutter that is 3/4″ (20 mm) to 1″ (25 mm) larger in diameter than the width of the surface to be machined, **Figure 18-49**.

2. The work should project about 1″ (25 mm) beyond the edge of the table to provide adequate clearance, **Figure 18-50**. In face milling, it is frequently necessary to mount the work on an angle plate, **Figure 18-51**.

Figure 18-48. In face milling, both the cutter and the workpiece face being machined are at a right angle to the axis of the spindle. (Mazak Corp.)

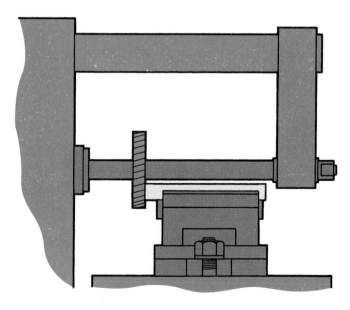

Figure 18-47. Another technique for squaring work ends. The solid vise jaw must be checked with a dial indicator to assure that it is properly aligned. Be sure there is adequate clearance between the work and the arbor.

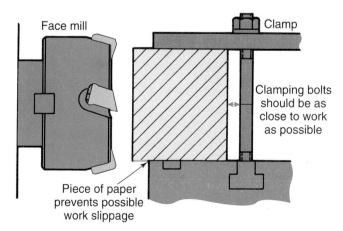

Figure 18-49. Select cutter that is 3/4″ (20 mm) to 1″ (25 mm) larger in diameter than width of surface to be machined.

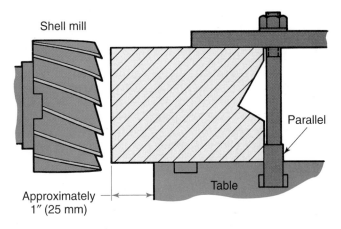

Figure 18-50. Work should project approximately 1" (25 mm) beyond the table edge to provide adequate clearance.

Shell mill

Parallel

Table

Approximately 1" (25 mm)

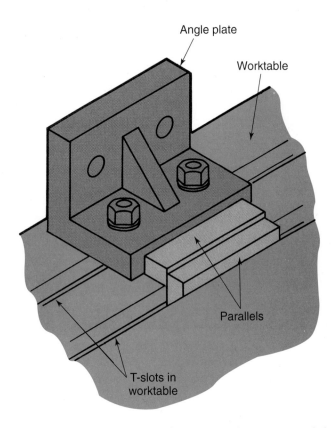

Angle plate

Worktable

Parallels

T-slots in worktable

Figure 18-51. An angle plate is often used to mount work for face milling. Check that mounting clamps clear the cutter. Note the use of parallels to align the angle plate.

3. Adjust the machine for correct speed and feed.

4. Slowly feed the work into the cutter until it starts to remove material. Roughing cuts up to 1/4" (6 mm) may be taken. Use adequate cutting fluid.

5. When the cut is complete, stop the cutter. Return the work to the starting position for additional machining, if needed.

6. Make the finishing cut and tear down the setup. Use a brush to remove chips. Clean and store the cutter.

18.5.4 Side milling

Side milling refers to any milling operation that involves the use of half side and side milling cutters. When cutters are employed in pairs to machine opposite sides of a piece at the same time, the setup is called *straddle milling,* **Figure 18-52.**

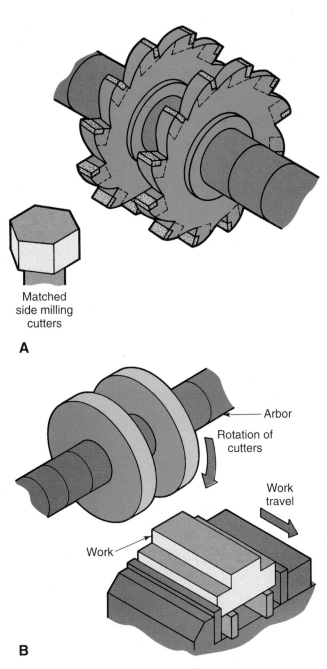

Matched side milling cutters

A

Arbor

Rotation of cutters

Work travel

Work

B

Figure 18-52. Straddle milling. A—This machining method uses a spacer between cutters. These two cutters are machining a hexagon on a machine part. B—An example of straddle milling on flatwork.

Cutters used for this operation should be kept in matched pairs. That is, they should be sharpened at the same time to maintain equal diameters. Shoulder width of the machined surface is determined by the thicknesses of the spacers between the cutters, **Figure 18-53.**

Gang milling involves mounting several cutters on an arbor to machine several surfaces in a single pass, **Figure 18-54.** It is a variation of straddle milling. Gang milling is used when many identical pieces must be made.

A *side milling cutter* can also be utilized to machine grooves, keyseats, and when used with a dividing head or rotary table, squares, hexagons, etc., on round stock.

18.5.5 Locating side cutter for milling a slot in square or rectangular work

The machine is set up in much the same manner as it was for milling flat surfaces. Use a dial indicator to check the vise to ensure accuracy. Exercise the same care in placing the side cutter on the arbor as was followed with the slab cutter. A plain side milling cutter may be used if the slot is not too deep. Otherwise, a *staggered-tooth side milling cutter* should be employed.

Make a layout on the end of the work, as shown in **Figure 18-55,** then position the cutter by using one of the following methods:

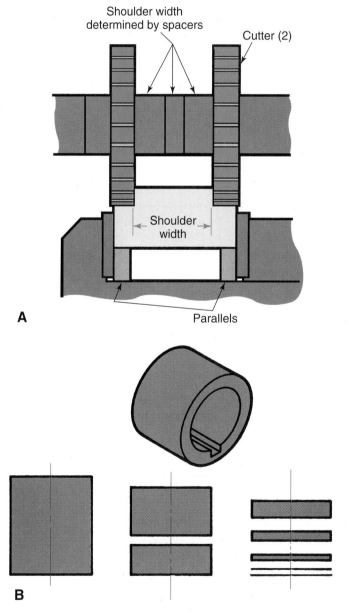

A

B

Figure 18-53. Arbor spacers. A—Spacers are used to set distance between cutters. B—Spacers are available in a large selection of sizes. Special sizes can be made by surface grinding standard sizes to needed dimensions. Shim stock spacers can be used to build up standard size spacers to desired dimension.

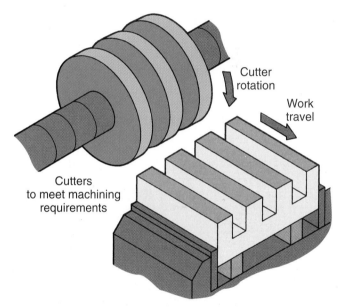

Figure 18-54. Gang milling involves the use of two or more milling cutters mounted on a single arbor.

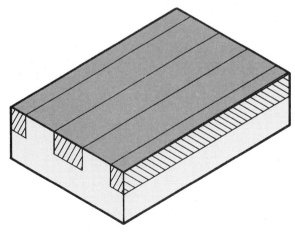

Figure 18-55. An example of work that has been laid out for milling.

- Use a steel rule, **Figure 18-56**. Make a light cut part way up the piece and remove the burrs. Measure the cut depth with a depth micrometer. The difference between this measurement and the required depth equals the amount of material that must be removed.
- Use the paper strip technique to bring the side of the cutter against the side of the work, **Figure 18-57**. Move the cutter inward the

required distance, *plus* the thickness of the paper. The paper strip or depth micrometer positioning technique may be used to set the cutter to the desired depth. When using the paper strip technique to position a cutter, remember to use a long paper strip and keep your fingers clear of the revolving cutter

18.5.6 Locating side cutter for milling a slot or keyseat in round stock

There are many situations that require keyseats for the standard square key to be cut in round stock. The keyseat must be kept precisely on center if it is to be in alignment with the keyway in the mating piece.

After the milling machine has been set up and work positioned in a vise, between centers, in V-blocks, or in a fixture, you must center the cutter. Precise centering of the cutter may be accomplished by one of the following methods:

- Center the cutter visually on the work. With the aid of a steel square and rule, adjust the table until both sides measure the same, **Figure 18-58**. Due to the difficulty of obtaining precise measurements with a rule, most machinists prefer to use a depth micrometer in place of the rule.
- Short pieces cannot always be centered by the above method. For situations of this type, the work is positioned under the rotating cutter and brought lightly into contact with it.

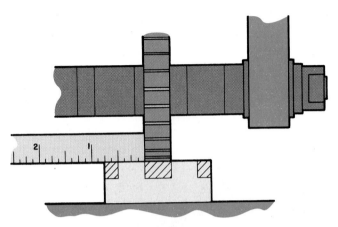

Figure 18-56. A cutter being positioned with aid of a steel rule.

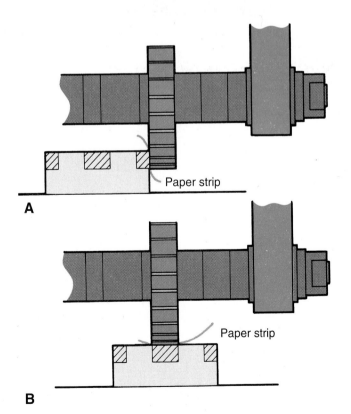

A

B

Figure 18-57. *Paper strip method of positioning the cutter. A—Using a paper strip to secure internal dimension. Read micrometer to move cutter the correct distance over the work. B—Using a paper strip to position the cutter for depth. Read micrometer dial as cutter is lowered.*

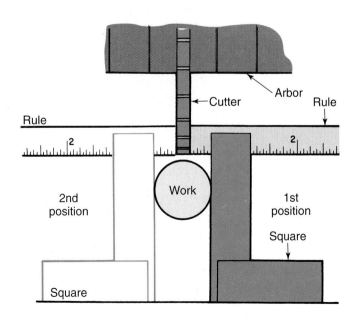

Figure 18-58. *How to center cutter on round stock with a steel rule and machinist's square.*

Traverse (in/out) feed is used to pass the work under the cutter. Because the work is circular in shape, an oval-shaped cut will result, and the oval will be perfectly centered. To center the cutter, position it on the oval, **Figure 18-59**.

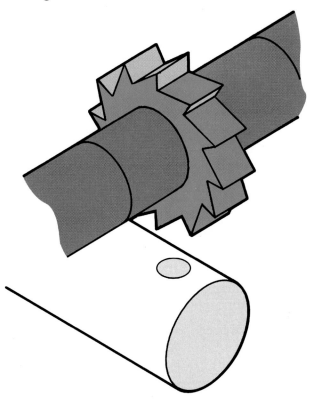

Figure 18-59. *Cutter is being positioned on center using an oval made in the work with cutter as a guide.*

- The previously mentioned narrow paper strip technique may be used to center the cutter. Hold the strip between the work and the cutter. Carefully move the work toward the cutter until it causes the paper to be lightly pulled from between your fingers. Lower the table until the cutter is slightly above the work. Move the cutter inward half the diameter of the work, plus half the cutter thickness, plus the paper thickness, **Figure 18-60**. The same technique may be employed to center a Woodruff keyseat cutter.

Lock the saddle to prevent traverse table movement after the cutter has been centered.

Correct keyseat depth can be obtained from tables in one of the many machinist's handbooks. The paper strip technique is used to set the cutter to the required depth. Tighten the knee locks after the depth setting has been made. Cutting fluid should be applied liberally during the cutting operation.

When the Woodruff keyseat cutter is used, it must also be positioned longitudinally on the work. Slowly feed into the piece until the required depth is attained. This can be checked by placing a key in the cut and "miking" the section.

18.6 SLITTING

Slitting thin stock into various widths for the production of flat gages, templates, etc., is a fairly

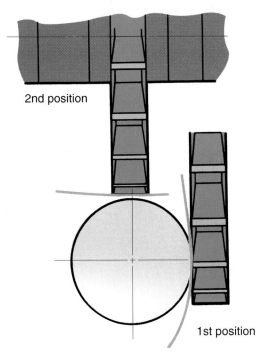

2nd position

1st position

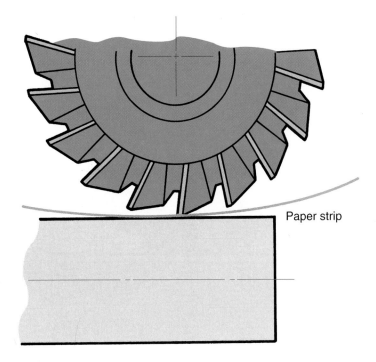

Paper strip

Figure 18-60. *Using the paper strip technique to position round stock. Use a long strip of paper. Hold lightly between your fingers and keep them well clear of the revolving cutter.*

common milling operation, **Figure 18-61**. It is performed with a *slitting saw* and is likely to give considerable trouble if extreme care is not exercised.

A slitting saw of the smallest diameter permitting adequate clearance is used. It must be keyed to the arbor (the key should also fit into spacers on either side of cutter). Best results can be obtained if the cutter is mounted for *climb milling.* That is, the work and cutter move in the same direction at the point of contact, **Figure 18-62**. Cutting pressure is

downward and will tend to press the work onto the table or holding device.

Adjust the table gibs until there is heavy drag felt when the table is moved by hand. This will remove table "play" and prevent the cutter from jumping in the cut.

If the section is narrow enough, the piece may be clamped in a vise, **Figure 18-63**. It should be well supported on parallels. Do not permit the parallels to project out into the cutter path.

Long strips must be clamped to the worktable. The shop-made angle iron clamp shown in **Figure 18-64** is recommended. The work is aligned with the column face and must be positioned to permit the saw to make the cut over the center of a table T-slot, **Figure 18-65.**

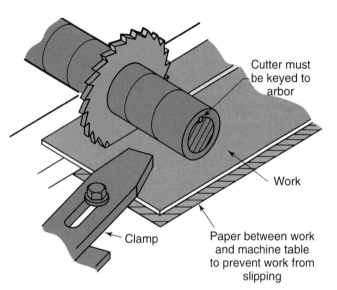

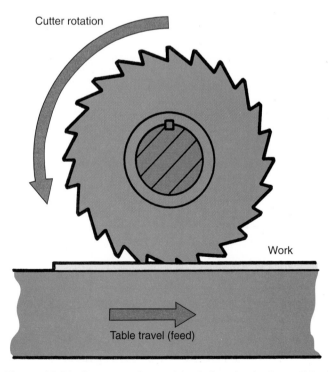

Figure 18-61. *Typical slitting or sawing operation setup. Work must be positioned over a table slot and clamped securely. Be sure the clamping bolts clear the arbor.*

Figure 18-62. *Cutter rotation and feed direction for best slitting results. Use a slow feed.*

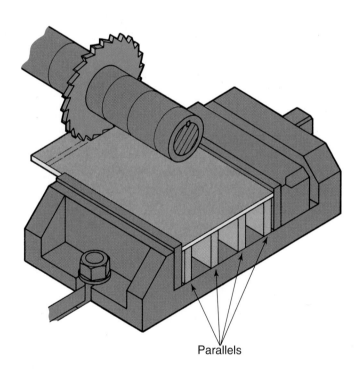

Figure 18-63. *Slitting work is held in vise. Be sure parallels do not project into cutter path!*

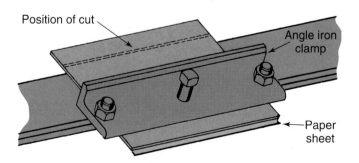

Figure 18-64. *A worktable clamp made from angle iron. Paper sheet prevents work movement.*

A sheet of paper between the work and table will prevent the metal from slipping during the slitting operation. The cutter is set to a depth equal to the work thickness plus 1/16″ (1.5 mm). Always use a sharp cutter!

18.7 SLOTTING

Slotting is similar to slitting, except that the cut is made only part way through the work, **Figure 18-66**. The slot in a screw head is an example of slotting.

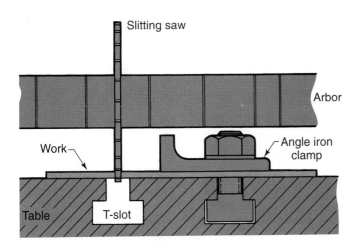

Figure 18-65. *Position work so that the cut is made over a T-slot.*

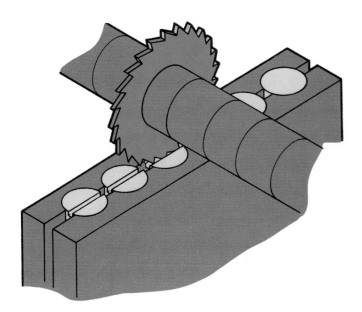

Figure 18-66. *Screw heads being slotted with a slitting saw. Special slotting saw is available which has many more teeth than a slitting saw of a similar diameter.*

1. Mount the cutter as for conventional milling. Use a sharp cutter of a width suitable for the job. Note the difference between a slotting cutter and a slitting cutter, **Figure 18-67**.

2. Set the machine for the correct cutting speed. Use the slowest feed possible, increasing feed rate if conditions warrant.

3. Align the vise and mount the work.

4. Position the cutter and make a light cut. Check the trial cut and make adjustments if necessary. Stop the machine before making measurements or adjustments.

5. Adjust the work for proper cut depth.

6. Apply cutting fluid and make the cut. Avoid standing directly in line with the cutter. Despite all precautions, saws shatter occasionally and can cause serious injury.

Figure 18-67. *How slitting and slotting cutters differ. A slitting saw is at left; a slotting saw at right.*

18.8 DRILLING AND BORING ON A HORIZONTAL MILLING MACHINE

The machinist often finds it necessary to produce accurately spaced and drilled holes in work. The milling machine offers a convenient way to make these holes in a specified and precise alignment.

Small drills are held in a standard Jacobs-type chuck mounted in the machine spindle, **Figure 18-68**, or in collet chucks, **Figure 18-69**. Larger, taper shank drills are fitted into an adapter sleeve, **Figure 18-70**.

Boring is done with a single point cutting tool fitted in a ***boring head***, **Figure 18-71**. The boring head may be equipped with a taper shank and mounted directly in the spindle, **Figure 18-72**, or a straight shank and held in a collet or adapter. See **Figure 18-73**.

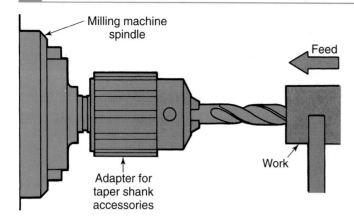

Figure 18-68. Drilling can be done on a horizontal mill as shown.

Figure 18-69. Drills can also be mounted in collets. Collet holders are available in a range of sizes. (Parlec, Inc.)

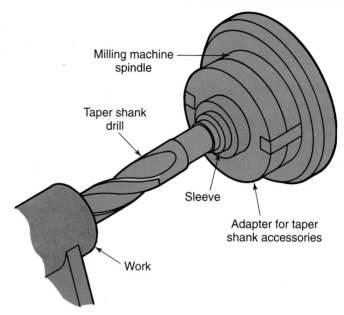

Figure 18-70. Taper shank drills can be used by fitting an adapter to the spindle.

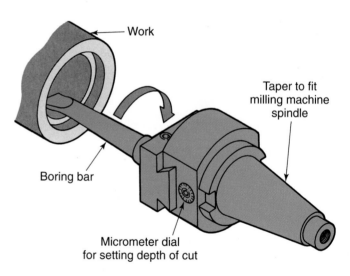

Figure 18-71. Boring permits large holes to be machined to close tolerances.

reamed. A dial indicator must be used to realign previously made holes for boring to final size, **Figure 18-75.**

18.9 CUTTING A SPUR GEAR

A *gear,* **Figure 18-76,** is a toothed wheel, usually fitted to a shaft. It typically engages a similar toothed wheel to smoothly transmit power or

A wiggler, **Figure 18-74,** will aid in aligning the machine for drilling the hole prior to boring, or when holes are small enough to be drilled and

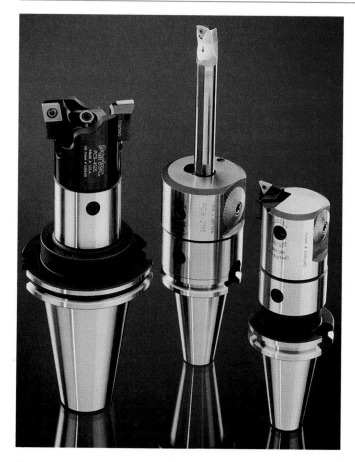

Figure 18-72. Boring tool holder and boring heads that mount directly into the spindle. (Parlec, Inc.)

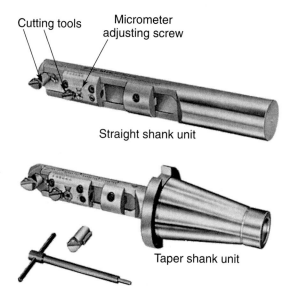

Figure 18-73. Multipurpose boring bars permit multistep simultaneous boring. Adjustments can be made without removing boring bar from machine. (Aloris Tool Co., Inc.)

Cutting tools

Micrometer adjusting screw

Straight shank unit

Taper shank unit

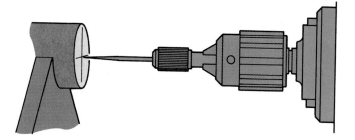

Figure 18-74. Holes can be located with a wiggler.

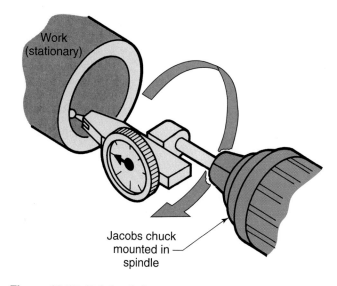

Work (stationary)

Jacobs chuck mounted in spindle

Figure 18-75. Existing holes can be realigned on a horizontal milling machine with the aid of a dial indicator.

motion at a definite ratio between the shafts. The teeth are shaped so that contact between the mating gears is continually maintained while they are in operation.

The **spur gear** has teeth that run straight across the face and are perpendicular to the sides. It is the simplest gear and is widely used, **Figure 18-77**.

Gear cutting requires a knowledge of *gear nomenclature* (terminology) to aid in determining the proper gear cutter to use, the depth of the teeth, and the dividing head setup.

18.9.1 Gear nomenclature

The following information is necessary to calculate data needed to machine a simple inch-based spur gear. Inch-based gears and metric-based gears are *not* interchangeable. The various gear parts are shown in **Figure 18-78**.

Figure 18-76. *A few of the many types of gears available. A—Worm gear. B—Crossed helical gears. C—Spiral miter gears. D—Bevel gears (small gear is called a pinion). E—Gear and pinion. F—Rolled pinion gears. G—Spur gear. H—Internal gear and pinion. I—Double gear. J—Rack and pinion. K—Miter gears. (Boston Gear Co.)*

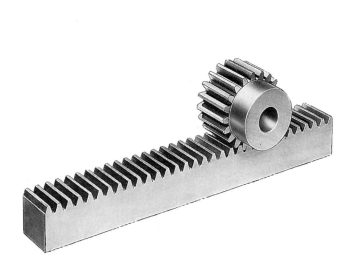

Figure 18-77. *Spur gear meshes in rack gear. The spur gear is the simplest of gears. Teeth are cut straight across gear face. Rack is a flat section of metal with teeth cut into it. Combination of spur gear and rack converts rotary motion to linear motion. (Boston Gear Co.)*

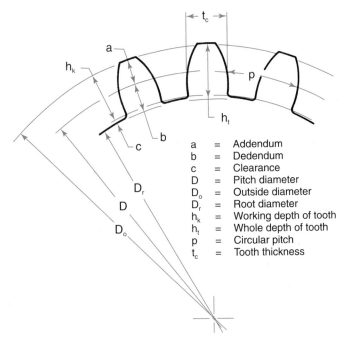

a	=	Addendum
b	=	Dedendum
c	=	Clearance
D	=	Pitch diameter
D_o	=	Outside diameter
D_r	=	Root diameter
h_k	=	Working depth of tooth
h_t	=	Whole depth of tooth
p	=	Circular pitch
t_c	=	Tooth thickness

Figure 18-78. *Gear nomenclature.*

- **Pitch diameter** (D): The diameter of the pitch circle.

$$D = \frac{N}{P} \text{ or } D = 0.3183pN \text{ or } D = \frac{D_o N}{N+2}$$

- **Diametral pitch** (P): The number of teeth per inch of pitch diameter.

$$P = \frac{N}{P} \text{ or } P = \frac{N+2}{D_o} \text{ or } P = \frac{\pi}{P}$$

- **Circular pitch** (p): The distance, measured on the pitch circle, between similar points on adjacent teeth.

$$p = \frac{\pi}{P} \text{ or } p = \frac{\pi D}{N} \text{ or } p = \frac{\pi D_o}{N+2}$$

- **Number of teeth** (N): The number of teeth on a gear.

$$N = DP \text{ or } N = D_o P - 2 \text{ or } N = \frac{\pi D}{P}$$

- **Outside diameter** (D_o): Diameter or size of the gear blank.

$$D_o = D + 2a \text{ or } D_o = \frac{N}{P} + 2\left(\frac{1}{P}\right) \text{ or } D_o = \frac{N+2}{P}$$

- **Whole depth of tooth** (h_t): Total depth of a tooth space, equal to the addendum (a) plus dedendum (b), or the depth to which each tooth is cut.

$$h_t = a + b \text{ or } h_t = \frac{2.250}{P} \text{ or } h_t = \frac{2.157}{P}$$

- **Working depth** (h_k): The sum of the addendum's of the two mating gears.

$$h_k = a_1 + a_2$$

- **Clearance** (c): The difference between the working depth and the whole depth of a gear tooth. The amount by which the dedendum on a given gear exceeds the addendum of the mating gear.

$$c = \frac{0.157}{P}$$

- **Addendum** (a): The distance the tooth extends above the pitch circle.

$$a = \frac{1}{P} \text{ or } a = \frac{D}{N} \text{ or } a = \frac{D_o}{N+2}$$

- **Dedendum** (b): The distance the tooth extends below the pitch circle.

$$b = \frac{1.157}{P}$$

- **Tooth thickness** (t_c): Thickness of the tooth at the pitch circle. The dimension used in measuring tooth thickness with vernier gear tooth caliper.

$$t_c = \frac{1.5708}{P}$$

- **Pitch circle**: An imaginary circle located approximately half the distance from the roots and tops of the gear teeth. It is tangent to the pitch circle of the mating gear.
- **Pressure angle** (θ): The angle of pressure between contacting teeth of mating gears. It represents the angle at which the forces from the teeth of one gear are transmitted to the mating teeth of another gear. Pressure angles of 14 1/2°, 20°, and 25° are standard. However, the 20° is replacing the older 14 1/2°.
- **Distance between centers of two mating gears** (C): This distance may be calculated by adding the number of teeth of both gears and dividing one-half that sum by the diametral pitch.

$$C = \frac{N_1 + N_2}{2 \div P}$$

N_1 = Number of teeth on first gear.
N_2 = Number of teeth on second gear.

18.9.2 Gear cutters

No one gear cutter, **Figure 18-79**, can be employed to cut all gears. Gear cutters are made with eight different forms for each diametral pitch (P), depending upon the number of teeth for which the cutter is to be used. **Figure 18-80** illustrates the comparative sizes for gear teeth. The cutter range is as follows:

No. of Cutter	Range of Teeth
1	135 to a rack
2	55 to 134
3	35 to 54
4	26 to 34
5	21 to 25
6	17 to 20
7	14 to 16
8	12 to 13

With the information furnished, it is possible to calculate the data needed to cut a simple inch-based spur gear.

Figure 18-79. A typical gear cutter. (Standard Tool Co.)

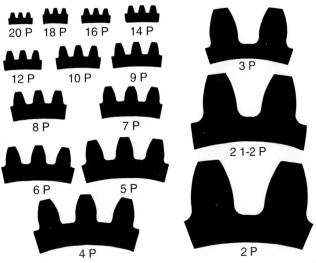

Figure 18-80. Comparative sizes of gear teeth. Diametral pitch is shown.

Example Problem: Calculate the data needed to cut a 40 tooth, 10 diametral pitch gear.

1. Diameter (D_o) of gear blank needed.

   Having: Diametral pitch (P) = 10

   Number of teeth (N) = 40

   Formula: $D_o = \dfrac{N+2}{P} = \dfrac{40+2}{10} = \dfrac{42}{10}$

   $= 4.200$

   $= 4.200''$ diameter

2. Whole depth of tooth (h_t) needed. This will be the depth of the cut.

   Having: Diametral pitch (P) = 10

   Formula: $h_t = \dfrac{2.157}{P} = \dfrac{2.157}{10} = 0.216$

   $= 0.216''$

3. The dimension of the addendum (a) is needed to measure the gear tooth for determining whether it is being machined to specifications.

Having: Diametral pitch (P) = 10

Formula:

$$a = \frac{1}{P} = \frac{1}{10} = 0.100$$
$$= 0.100''$$

4. Tooth thickness (t_c) is needed to determine whether the gear is being machined to specifications.

   Having: Diametral pitch (P) = 10

   Formula:

   $$t_c = \frac{1.5708}{P} = \frac{1.5708}{10} = 0.157$$
   $$= 0.157''$$

5. Reference to the gear cutter chart indicates that a No. 3 cutter, with a range of 35 to 54 teeth, must be used to cut 40 teeth.

6. Using a 40:1 ratio dividing head for this job means that the index crank must be turned through one complete revolution to position the gear blank for each cut. A 5:1 ratio dividing head would require the use of an index plate that would permit a setting of one-eighth turn for each cut.

18.9.3 Cutting the gear

A few simple precautions, carefully followed, will greatly reduce the possibility of an inaccurately machined gear.

1. Set up the milling machine as previously described. Check center alignment of the dividing head and foot stock.

2. Press the gear blank onto a mandrel and mount the unit to the dividing head. Cutting is done *toward* the dividing head, **Figure 18-81**.

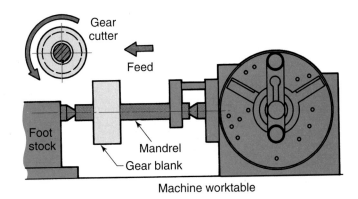

Figure 18-81. Cutting is done toward dividing head.

3. Use a dial indicator on the gear blank longitudinally and turn the blank through one complete revolution. Make any adjustments that are necessary.

4. Center the cutter on the gear blank. Use a depth micrometer and a steel square, **Figure 18-82.** Position the cutter for depth and use the paper strip technique. Set the micrometer dial to 0; then raise the table to within 0.040″ of finished depth. For the example problem given earlier, it would be raised to make a cut of 0.216″ – 0.040″, or 0.176″.

5. Move the work until the cutter just begins removing metal. Back it away from the cutter. Using the dividing head, bring the next cut into position. Repeat this sequence around the gear blank until you are back to the original cutting position. If there is exact alignment with the first cut, you are ready to cut the gear.

6. Make the roughing cuts. Use liberal quantities of cutting fluid.

7. The finish cut requires more care. Set the vertical scale on a gear tooth vernier caliper to the distance calculated for the addendum (a), **Figure 18-83.** Raise the work to within a few thousandths of the calculated whole depth of the tooth (h_t) and make cuts at two positions.

Make your measurement and adjust until the reading equals the distance calculated for tooth thickness (t_c). Make the finish cuts. Press the completed gear from the mandrel. Remove all burrs and cut the keyway, if required.

Spur gears can also be measured using *Van Kuren wires*, **Figure 18-84.** A table furnished with

Figure 18-83. *Machinist is measuring gear tooth with gear tooth Vernier caliper. Tool is read in same manner as a Vernier caliper and Vernier height gage. (L.S. Starrett Co.)*

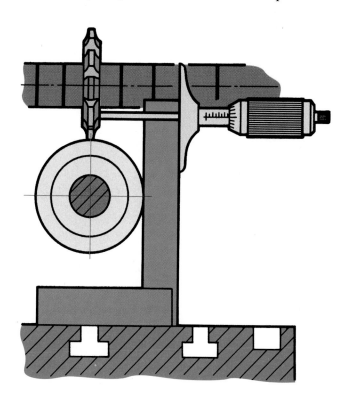

Figure 18-82. *A depth micrometer can be used to center the gear cutter on a gear blank.*

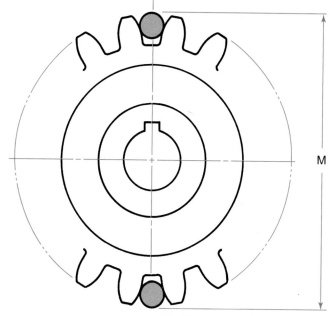

M = Measurement over wires

Figure 18-84. *Gears can also be measured using Van Kuren wires. Tables furnished with the wire set provide information needed when measuring gears with even number and odd number of teeth.*

the wire set specifies the wire diameter to use according to the diametral pitch of the gear. The table also includes dimensions for checking external spur gear measurement (M) over the wires. Measurement is made with outside micrometers.

Spiral gears and helical gears (refer to **Figure 18-76**) are cut on a universal type milling machine utilizing a *universal dividing head* geared to the table lead screw, **Figure 18-85**.

Figure 18-85. Setup for cutting a spiral gear. Universal dividing head is coupled to automatic feed mechanism of the worktable. Operation is similar to cutting threads on a lathe. Note how table is angled with respect to cutter. (Kearney & Trecker Corp.)

18.10 CUTTING A BEVEL GEAR

Bevel gears, **Figure 18-86**, are employed to change the angular direction of power between shafts. The teeth are either straight or curved. The procedure for cutting a straight tooth bevel gear is illustrated and explained in this section.

Since tooth space at the pitch diameter is narrower at the small end than at the large end, special form relieved cutters have been designed to cut bevel gears. See **Figure 18-87**. To achieve the required gear tooth dimensions, additional material must be removed from the flanks of the teeth after the preliminary roughing operation.

Measurements of the finished gear are made of the blank size and shape, tooth thickness, and depth. However, there is no simple method for checking tooth surfaces. Final inspection is made by running the mating gears and checking for quietness and shape of the tooth contact.

18.10.1 How to mill a bevel gear

The following information was obtained using the dimensions shown in **Figure 18-88**, and the formulas given in this chapter.

- Pitch cone distance = 3.535″
- Pitch diameter at large end = 5.000″

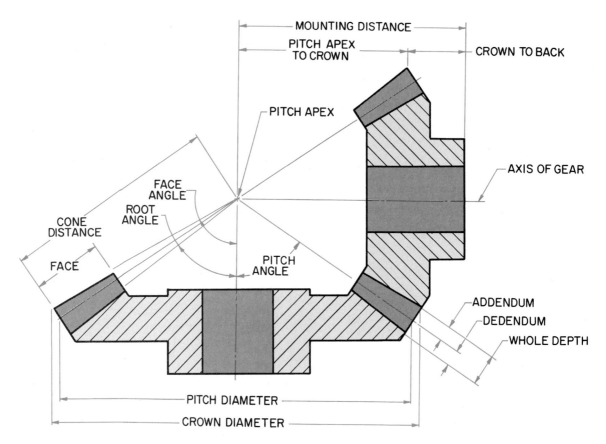

Figure 18-86. Nomenclature of bevel gears. The smaller gear is called the pinion.

- Pitch diameter at small end = 3.585″
- Circular pitch at large end = 0.5236″
- Circular pitch at small end = 0.3756″
- Tooth thickness and tooth space at large end = 0.2618″
- Tooth thickness and tooth space at small end = 0.1878″
- Whole depth of tooth at large end = 0.3595″
- Whole depth of tooth at small end = 0.2588″

- Addendum at large end of gear = 0.166″
- Addendum at small end of gear = 0.1195″
- Dedendum at large end of gear = 0.193″
- Dedendum at small end of gear = 0.139″

The tooth parts at the small end of the gear are in exact proportion to those at the large end. Dimensions at the small end can be found by multiplying the dimensions at the large end by the ratio

$$\frac{C_s}{C_r}$$

of the respective cone distances, or $C_r = 3.535″$ and $C_s = 3.535″ - 1″ = 2.535″$.

The ratio: $\frac{C_s}{C_r} = 0.72$

18.10.2 Preparing to cut a bevel gear

1. Mount the dividing head and tilt it to 45° 53′. See **Figure 18-89**.

2. Calculate the correct index plate to cut 30 teeth.

3. Secure the gear blank on an arbor.

4. Mount the gear blank and arbor in the dividing head with the large end of the gear toward the dividing head.

5. Select the proper bevel cutter and mount it on the arbor. Cutting is *toward* the dividing head. Position the arbor bearing as close to the cutter as possible, allowing for adequate clearance.

Figure 18-87. *A variety of form-relieved milling cutters developed for special applications (Pfauter-Maag Cutting Tools)*

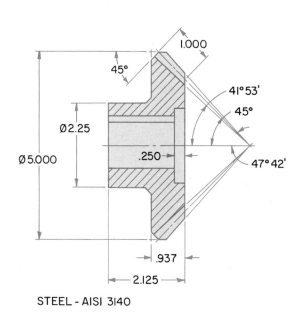

STEEL - AISI 3140

Figure 18-88. *Dimensions of bevel gear to be cut.*

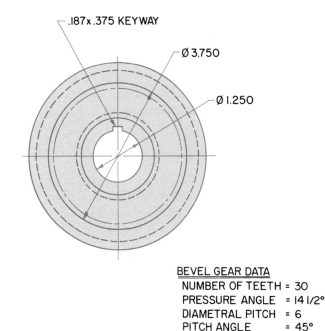

.187 x .375 KEYWAY

Ø 3.750

Ø 1.250

BEVEL GEAR DATA
NUMBER OF TEETH = 30
PRESSURE ANGLE = 14 1/2°
DIAMETRAL PITCH = 6
PITCH ANGLE = 45°

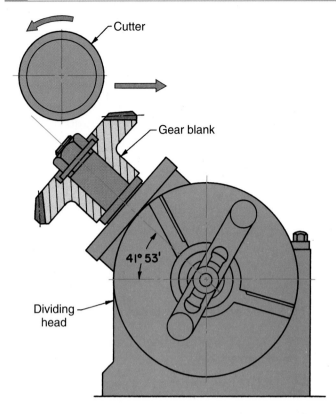

Figure 18-89. *Tilt dividing head to the required angle (45° 53')
and mount the gear blank.*

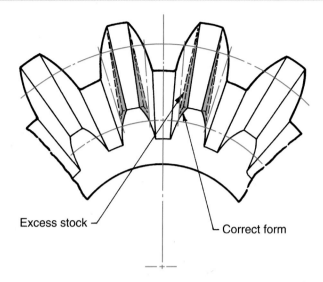

Figure 18-90. *Excess material that must be removed when
using the described technique to cut a bevel gear.*

12. To remove this additional material, the rough
finished blank must be rotated slightly (2° for
this gear) and the table set over (0.044"). The
following formulas were used to make the
calculations.

Determining angle of roll:

$$C = \frac{57.3}{P_d}\left[\frac{P_c}{2} - \frac{C_r}{W}(T_L - T_s)\right]$$

Where:

C = Angle of roll in degrees.
P_d = Pitch diameter at large end of gear.
P_c = Circular pitch at large end of gear.
C_r = Pitch cone distance at large end of gear.
T_s, T_L = Chordal thickness of gear cutter
corresponding to pitch line at small
and large ends of gear, respectively.
57.3 = Degrees per radian.
W = Width of gear tooth face.

Determining table setover:

$$n = \frac{T_L}{2} - T_L - \frac{T_s}{2} \times \frac{C_r}{W}$$

Where: n = Table setover.

Note: The direction of roll and setover *must
always be made in opposite directions.* See **Figure
18-91**. Remove all backlash before making work
movements.

6. Center the cutter on the gear blank. Lock the
cross slide. Set the graduated collar to zero.

7. Position the cutter for depth, using the paper
strip technique. Set the knee graduated collar
to zero.

8. Clear the cutter and raise the table to the whole
tooth depth at the large end (0.3595").

9. Set the machine for the proper cutting speed
and feed. *Note:* Many machinists prefer to
make the cut in two runs before making the
final cut.

10. Cut all teeth by plain indexing. Use adequate
coolant.

11. The correct dimensions for the finished gear
teeth are 0.2618" and 0.1878". To obtain these
dimensions, an additional amount of material
must be removed on each side of the teeth, as
shown in **Figure 18-90**.

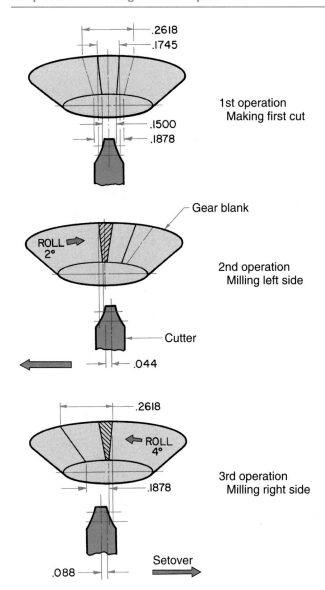

.2618
.1745

1st operation
Making first cut

.1500
.1878

Gear blank

ROLL
2°

2nd operation
Milling left side

Cutter

.044

.2618

ROLL
4°

3rd operation
Milling right side

.1878

.088

Setover

Figure 18-91. Sequence to be followed when milling the teeth of a bevel gear.

13. Finish machining the gear. Remove all burrs. Measure tooth thickness at small and large ends. It may be necessary to remove a slight amount of material at the small end to get proper meshing. Refer to Figure 18-90.

This method of making bevel gears does not produce a tooth form that is accurate throughout the length of the tooth face. This is especially true at the small end of the gear even though the tooth form is correct at the large end. Remove this small amount of excess material by rotating the blank through a small angle with the dividing head and taking light cuts until proper meshing is attained.

18.11 PRECAUTIONS WHEN OPERATING A MILLING MACHINE

- Avoid performing any machining operation on the milling machine until you are thoroughly familiar with how it should be done.
- Some materials that are machined can produce chips, dust, and fumes that are dangerous to your health. Never machine materials that contain asbestos, fiberglass, beryllium, or beryllium copper unless you are fully aware of the precautions that must be taken.
- Make sure there is adequate ventilation when performing jobs where dust and fumes are a hazard.
- If the area where you work is extremely noisy, wear hearing protectors. Take no chances. Protect your hearing and sight at all times in the shop.
- Maintain cutting fluids properly. Discard them when they become rancid or contaminated. Never pour used coolants or solvents down the drain.
- Carefully read instructions when using the new synthetic oils, solvents, and adhesives. Many of them are dangerous if not handled correctly. Return all oils and solvents to proper storage. Always wipe up spills.
- Never start a cut until you are sure there is adequate clearance on all moving parts!
- Be sure the cutter rotates in the proper direction. Expensive cutters can be quickly ruined.
- Carefully store milling cutters, arbors, collets, adapters, etc., after use. They can be damaged if not stored properly.
- Exercise care when handling long pieces of metal. Accidentally contacting a light fixture or busbar can cause severe electrical burns and even electrocution!

18.12 INDUSTRIAL APPLICATIONS

The milling machines found in industry operate on the same basic principles as those found in training programs. In many cases, the same equipment is utilized, **Figure 18-92.**

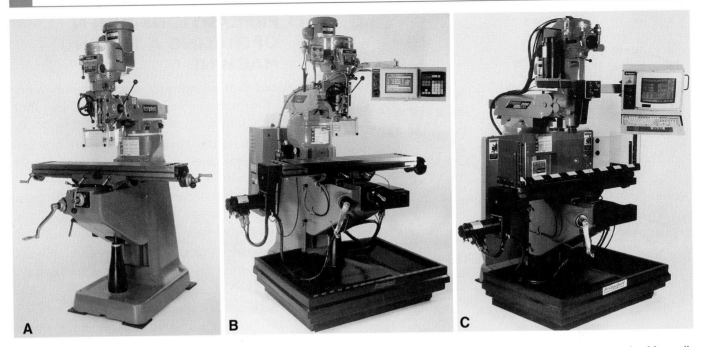

Figure 18-92. *Three types of vertical milling machines found in both industry and training programs. A—Manually operated vertical milling machine. B—CNC 2-axis vertical milling machine. C—CNC 3-axis vertical milling machine. (Bridgeport Machines, Inc.)*

There are many types of milling machines in use. They range in size from small tabletop models to machines capable of handling work that weighs many tons. Today, an estimated 80% to 90% of the milling machines in use have CNC capability.

A number of the machines used in industry are illustrated in **Figures 18-93** through **18-98**.

Figure 18-93. *A tabletop CNC vertical machining center. It has industrial applications as well as being used by training programs. (Light Machines Corp.)*

Figure 18-94. *This double-housing milling machine with three cutting heads is milling a large aircraft wing section. (Lockheed-Martin)*

Figure 18-95. *Double-column CNC machining center designed to machine body dies in one setup for the automotive industry. (Okuma America Corp.)*

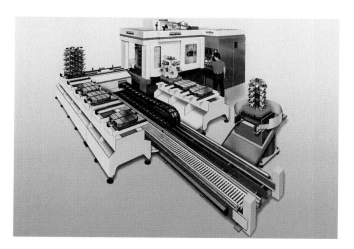

Figure 18-96. *CNC horizontal machining center (HMC) that uses an automated rail-guided changer to move pallets into and out of the milling area. (Cincinnati Milacron)*

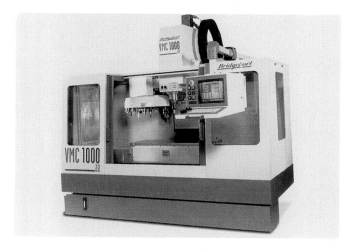

Figure 18-97. *A vertical machining center (VMC) equipped with an automatic toolchanger holding a large number of cutters that might be needed for various operations. (Bridgeport Machines, Inc.)*

Figure 18-98. *The CNC milling machine of the future has six degrees of freedom provided by three pairs of variable-length legs. The legs move the multiaxis cutting head through three-dimensional space for rapid and precise positioning. Note that the machine has no conventional controls. (Giddings & Lewis, Inc.)*

TEST YOUR KNOWLEDGE

Please do not write in the text. Write your answers on a separate sheet of paper.

1. _____ mills and _____ mills are the cutters normally used on a vertical milling machine.

2. The _____ end mill is used when the cutter must be fed into the work like a drill.

3. Blind holes or closed keyseats are made with a _____ end mill.

4. Face milling cutters over 6″ (150 mm) in diameter are usually of the _____ type.

5. A _____ scale on the spindle head of a vertical milling machine assures accurate angular settings.

6. List three methods for machining chamfers, bevels, and tapered sections on a vertical milling machine.

7. An _____ or a _____ can be used to locate the first hole of a series to be drilled on a vertical milling machine.

8. The most accurate way to align a vise on milling machine is with a _____.

9. Explain how to center an end mill on round stock for the purpose of machining a keyseat. Use the paper strip technique.

10. Gang milling means:
    a. Several cutters being used at the same time to machine a job.
    b. Two or more cutters straddling the job.
    c. Several side cutters being used at the same time to machine a job.
    d. All of the above.
    e. None of the above.

11. Why should a milling cutter be keyed to the arbor?

12. When sawing (slitting) thin stock, the _____ diameter cutter that provides adequate clearance should be used.

13. In general, use the _____ arbor possible that permits adequate clearance between the arbor support and the work.

14. Describe how to safely remove or mount a milling cutter on an arbor.

15. How would a dividing head be set up to cut a 100-tooth gear? The dividing head has a 40:1 ratio and the index plate has the following series of holes: 33, 37, 41, 45, 49, 53, 57.
    Number of full turns. _____
    Hole series used. _____
    Number of holes in sector arm spacing. _____

16. What is a spur gear?

17. How does a rack differ from a spur gear?

18. List five precautions to be observed when operating a milling machine.

Chapter 19

Precision Grinding

LEARNING OBJECTIVES

After studying this chapter, you will be able to:
- ○ Explain how precision grinders operate.
- ○ Identify the various types of precision grinding machines.
- ○ Select, dress, and true grinding wheels.
- ○ Safely operate a surface grinder using various work-holding devices.
- ○ Solve common surface grinding problems.
- ○ List safety rules related to precision grinding.

IMPORTANT TERMS

centerless grinding
creep grinding
diamond dressing tool
form grinding
internal grinding
magnetic chuck

planer-type surface
   grinders
plunge grinding
tooth rest
universal tool and cutter
   grinder

Grinding, like milling, drilling, sawing, planing, and turning, is a *cutting operation.* However, instead of using one, two, or several cutting edges, grinding makes use of an abrasive tool composed of thousands of cutting edges. See **Figure 19-1.** Since each of the abrasive particles is actually a separate cutting edge, the grinding wheel might be compared to a many-toothed milling cutter, **Figure 19-2.**

In precision grinding, each abrasive grain removes a relatively small amount of material, permitting a smooth, accurate surface to be generated. It is also one of the few machining operations that can produce a smooth, accurate surface on material regardless of its hardness. Grinding is frequently used as a finishing operation.

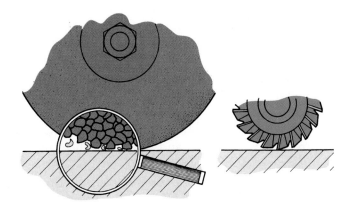

Figure 19-1. *A grinding wheel removes material in the same manner as a milling cutter, but the chips of metal removed are much smaller.*

Figure 19-2. *The abrasive grains that make up a typical grinding wheel are magnified about 50x in this closeup view. (Cincinnati Milacron)*

19.1 TYPES OF SURFACE GRINDERS

While all grinding operations might be called surface grinding because all grinding is done on the surface of the material, industry classifies *surface grinding* as the grinding of *flat surfaces*. There are two basic types of surface grinding machines:

- *Planer-type surface grinders* make use of a reciprocating motion to move the worktable back-and-forth under the grinding wheel, **Figure 19-3**. Three variations of planer-type surface grinding are illustrated in **Figure 19-4**.
- *Rotary-type surface grinders* have circular worktables that revolve under the rotating grinding wheel, **Figure 19-5**. Two variations of the technique are shown in **Figure 19-6**.

The planer-type surface grinder is frequently found in training situations. It slides the work back-and-forth under the edge of the grinding wheel. Table movement can be controlled manually or by means of a mechanical or hydraulic drive mechanism.

A manually operated machine is shown in **Figure 19-7A**. All work and grinding wheel movements are made by hand.

The large *traverse handwheel* controls the left-and-right movement of the table. Cross-feed (in-and-out motion) is controlled by the smaller *cross-feed handwheel*. The *down-feed handwheel* controls the up-and-down adjustment of the grinding wheel. This handwheel is located on the top of the vertical column.

A variation of this type of surface grinder can be run manually or automatically, **Figure 19-7B**. To prepare the machine for automatic operation, the operator simply fills in the blanks when requested by the menu prompts. No computer or CNC experience is needed. As the operator creates the part manually, pressing a button after each move, the machine "memorizes" how the part is made. The machine can run subsequent parts automatically.

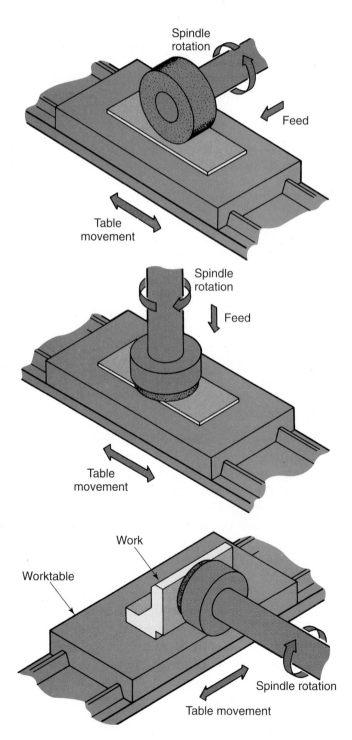

Figure 19-4. *Three variations of the planer-type surface grinder.*

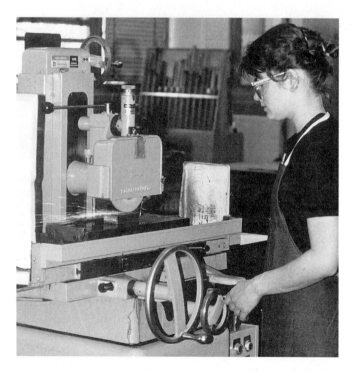

Figure 19-3. *This student is using a planer-type surface grinder to machine a flat surface. (William L. Schotta, Millersville University)*

Figure 19-5. Rotary surface grinder. A—This rotary fine grinding system can handle work from 5/32" to 3 5/8" (0.4 mm to 90 mm) thick and 13/32" to 13 5/8" (10 mm to 340 mm) in length. It is a rapid, safe, clean, and economical way to finish material to close tolerances. B—Ceramic pieces in place for grinding to required thickness. Note that not only does the grinding head rotate, but the work-holders also rotate to provide a superior finish. (Peter Wolters of America, Inc.)

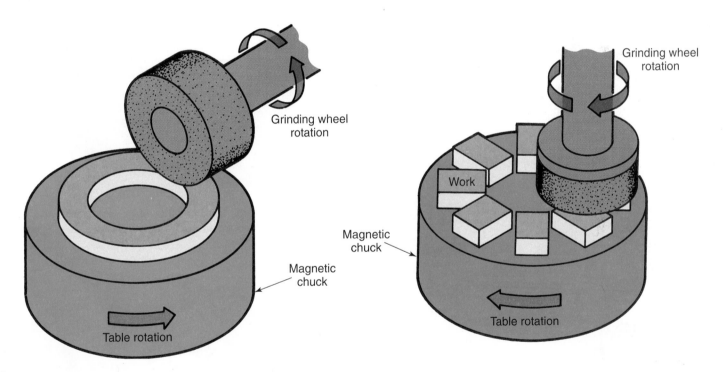

Figure 19-6. Two variations of the rotary-type surface grinder.

The machine shown in **Figure 19-8** makes use of hydraulic traverse feed and cross-feed. The grinder is also fitted with a coolant attachment.

Both manual and automatic machines operate in much the same manner. However, the person using a manually operated machine must develop a rhythm to get a smooth, even cutting stroke. Spring stops act as cushions at the end of maximum table travel, **Figure 19-9.**

Adjustable table stops on the hydraulically activated traverse feed permit the operator to establish exact table positioning, **Figure 19-10.** At the end of the stroke, table direction is reversed automatically. Automatic cross-feed moves the work in or out a predetermined distance at the completion of each cutting cycle.

A control console, **Figure 19-11,** is located on the front of the machine. Table travel is started and stopped from this station. Table speed is also controllable from this location. Some grinding machines have a control for *dwell*—a hydraulic cushion at the end of each stroke.

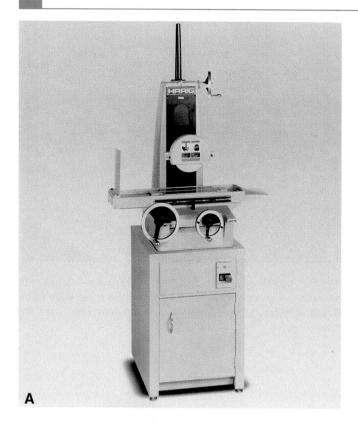

A

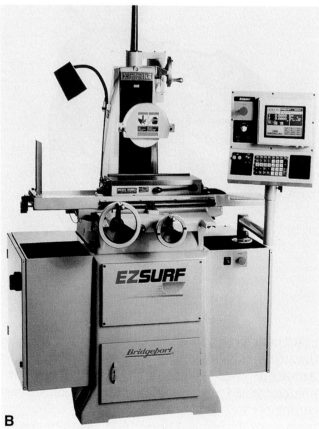

B

Figure 19-7. Planer-type surface grinders. A—A manually-operated surface grinder. B—A surface grinder that operates either manually or automatically. It permits an operator in training to step up to automatic grinding when ready to do so. (Harig Div. of Bridgeport Machines Inc.)

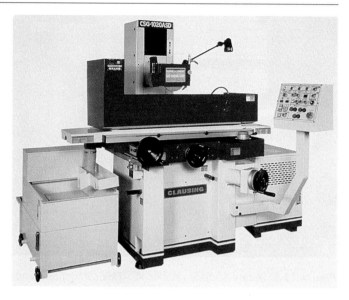

Figure 19-8. A modern CNC surface grinder. The cross-feed and longitudinal table movements are hydraulically actuated. (Clausing Industrial, Inc.)

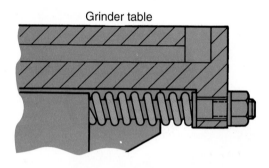

Grinder table

Figure 19-9. Springs on worktable guide are often used to cushion the end of stroke on manually operated surface grinders.

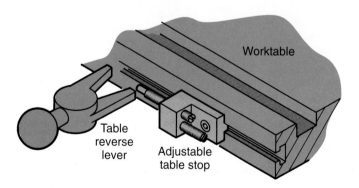

Worktable

Table reverse lever

Adjustable table stop

Figure 19-10. An adjustable table stop is employed to regulate length of the worktable stroke.

19.2 WORK-HOLDING DEVICES

Much of the work done on a surface grinder is held in position by a *magnetic chuck*, **Figure 19-12.** This holds the work by exerting a magnetic force.

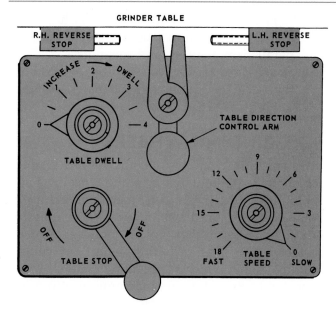

Figure 19-11. *Typical surface grinder control console. Table dwell sets up a cushioning action at end of each table stroke.*

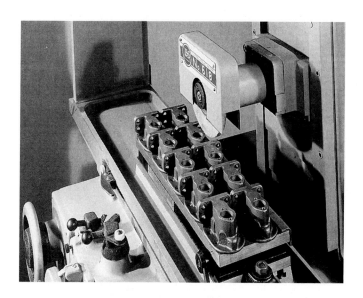

Figure 19-12. *A magnetic chuck being used to hold multiple pieces for surface grinding. (Brown & Sharpe Mfg. Co.)*

Nonmagnetic materials (aluminum, brass, etc.) can be ground by bracing with steel blocks or parallels to prevent movement.

An *electromagnetic chuck* utilizes an electric current to create a strong magnetic field. See **Figure 19-13.**

Another type of magnetic chuck makes use of a permanent magnet, **Figure 19-14.** This eliminates cords needed for electromagnets and the danger of the work flying off the chuck if the electrical connection is accidentally broken.

Frequently, work mounted on a magnetic chuck becomes magnetized and must be demagnetized

before it can be used. A *demagnetizer* like the one shown in **Figure 19-15** may be employed to neutralize the magnetic field.

Other ways to mount work on a surface grinder are:

- A *universal vise*, **Figure 19-16A**.
- An *indexing head* with centers, **Figure 19-16B**.
- *Clamps* to hold the work directly on worktable, **Figure 19-16C**.

Figure 19-13. *An electromagnetic chuck makes use of an electric current to create a magnetic field. (O.S. Walker Co., Inc.)*

Figure 19-14. *Magnetic chuck with a permanent-type magnet.*

Figure 19-15. *Demagnetizer can be used to neutralize magnetized part after it has been clamped in a magnetic chuck. (O.S. Walker Co., Inc.)*

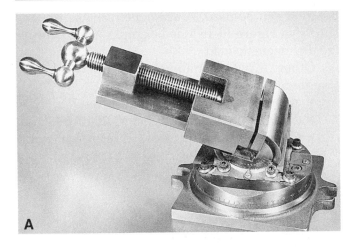

Figure 19-16. Work-holding devices. A—A universal vise can be used for grinding operations. B—Centers and an indexing head are used for grinding tasks when shape of work permits. An indexing head is used in much the same manner as the dividing head in milling. C—Work can also be clamped directly to table for grinding. Coolant flow has been stopped for clarity in this photo. (Brown & Sharpe Mfg. Co.)

- A *precision vise*, **Figure 19-17**, is hardened and ground within 0.0002″ (0.0050 mm) of parallelism, flatness, and squareness.
- *Double-faced masking tape* can be used to hold thin sections of nonmagnetic materials. Refer to **Figure 19-18**.

19.3 GRINDING WHEELS

As mentioned at the opening of this chapter, each abrasive particle in a grinding wheel is a cutting tooth. As the wheel cuts, metal chips dull the abrasive grains and wear away the *bonding material* (medium that holds abrasive particles together). The ideal grinding wheel, of course, would be one in which the bonding material wears away slowly enough to get maximum use from the individual abrasive grains. However, it would also wear rapidly enough to permit dulled abrasive particles to drop off and expose new particles.

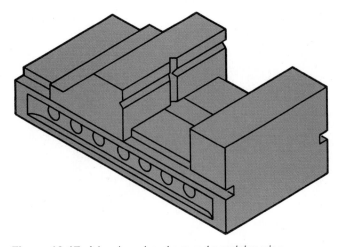

Figure 19-17. A hardened and ground precision vise.

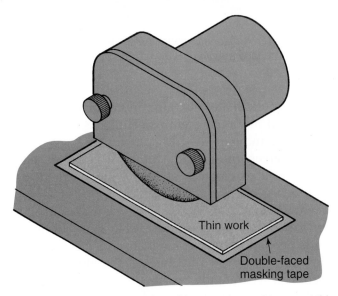

Figure 19-18. Double-faced masking tape is used to mount thin nonferrous material when it is ground.

Because so many factors affect grinding wheel efficiency, the wheel eventually dulls and must be dressed with a *diamond dressing tool.* See **Figure 19-19.** Failure to dress the wheel of a precision grinding machine will, in time, result in the wheel face becoming loaded or glazed so it cannot cut freely, **Figure 19-20.**

Only manufactured abrasives are suitable for modern high-speed grinding wheels. The properties and spacing of abrasive particles and composition of the bonding medium can be controlled to get the desired grinding performance, **Figure 19-21.**

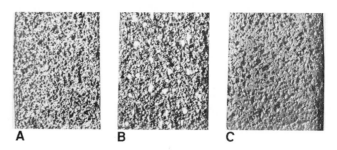

Figure 19-20. *Grinding wheels in various conditions: A—Properly dressed. B—Loaded. C—Glazed. (Norton Co.)*

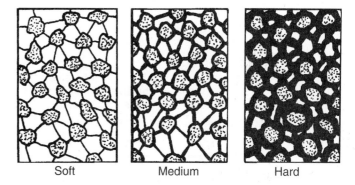

Figure 19-21. *Wheel hardness is determined by the type and percentage of bond and grain spacing.*

19.3.1 Grinding wheel marking system

To aid in achieving consistent grinding performance, a standard system of marking grinding wheels has been defined by American National Standards Institute (ANSI). Called ANSI Standard B74.13-1977, it is used by all grinding wheel manufacturers. Five factors were considered:

- *Abrasive-type* classifies the abrasive material in the grinding wheel. Manufactured abrasives fall into two main groups identified by letter symbols:
  - A = aluminum oxide
  - C = silicon carbide

  An optional prefix number may be employed to designate a particular type of aluminum oxide or silicon carbide abrasive.
- *Grain size* is indicated by a number, usually from 8 (coarse) to 600 (very fine).
- *Grade* is the strength of the bond holding the wheel together, ranging from A (soft) to Z (hard).

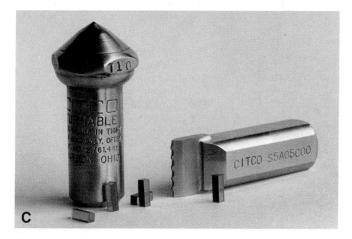

Figure 19-19. *Diamond wheel dressing. A—Dressing tool will true and clean a grinding wheel face. B—Closeup of a natural diamond chip on a grinding wheel dresser. The diamond should be rotated a partial turn each time it is used. This will put a new edge of the diamond into position. C—Dressing tools manufactured from manmade diamonds eliminate the irregularities of natural diamonds, resulting in consistent diamond exposure and longer wear. (CITCO Div., Western Atlas, Inc.)*

- *Structure* refers to grain spacing or the manner in which the abrasive grains are distributed throughout the wheel. It is numbered 1 to 16; the higher the number, the more "open" the structure (wider the grain spacing). The use of this number is optional.
- *Bond* indicates the type of material that holds the abrasive grains (wheel) together. Eight types are used:
  - B = Resinoid
  - BF = Resinoid reinforced
  - E = Shellac
  - 0 = Oxychloride
  - R = Rubber
  - RF = Rubber reinforced
  - S = Silicate
  - V = Vitrified

An additional number or one or more letters may be used as the manufacturer's private marking to identify the grinding wheel. Its application is optional.

The adoption of a standardized grinding wheel marking system has guaranteed, to a reasonable degree, duplication of grinding performance. The wheel marking system is shown in **Figure 19-22**.

19.3.2 Grinding wheel shapes

Grinding wheels are made in many standard shapes, **Figure 19-23**. While twelve basic face shapes are generally available, the face may be changed to suit specific job requirements, **Figure 19-24**. Wheels used for internal grinding are manufactured in a large selection of shapes and sizes, **Figure 19-25**.

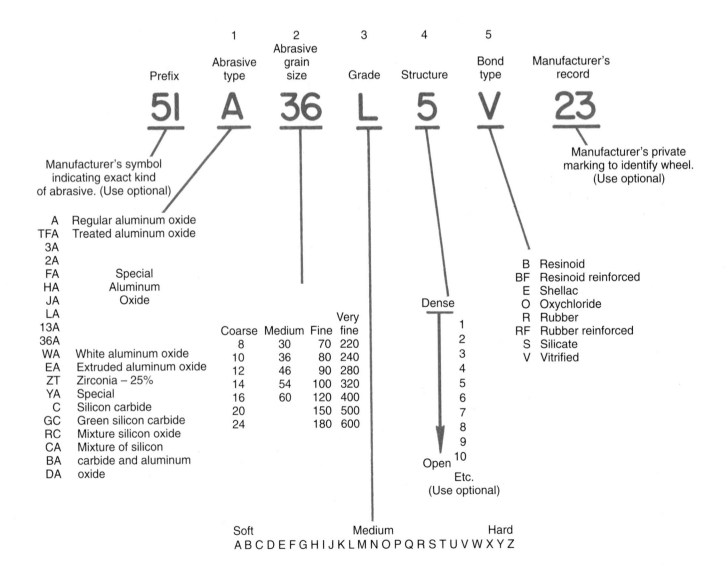

Figure 19-22. Standard system for marking grinding wheels.

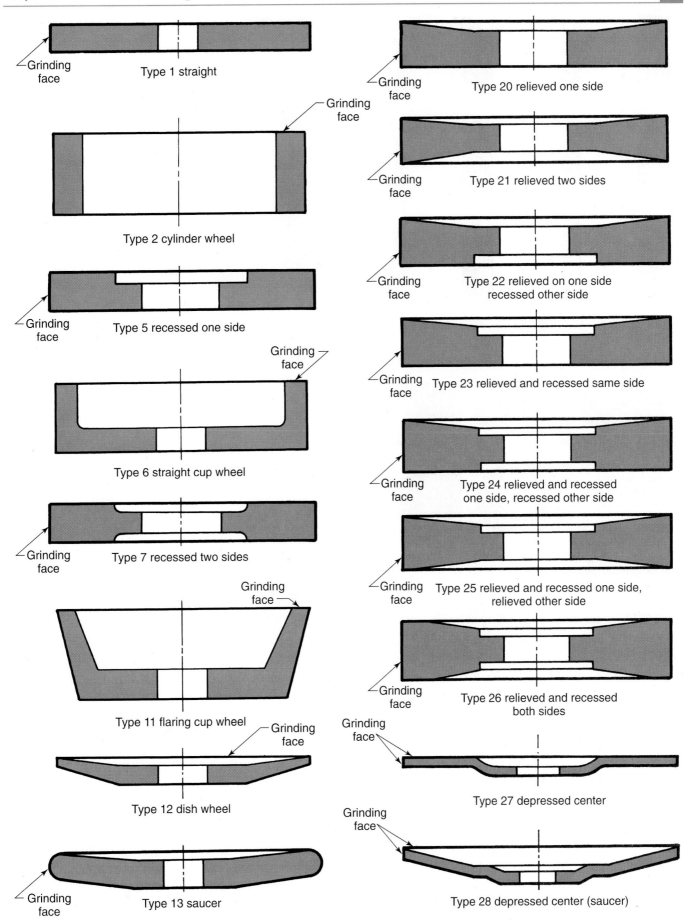

Figure 19-23. *Standard grinding wheel shapes.*

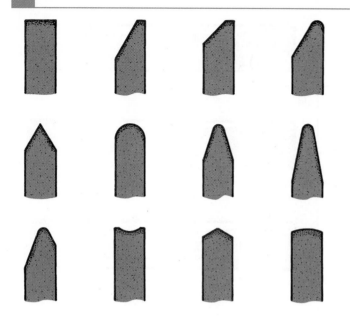

Figure 19-24. *The twelve basic face shapes that are generally available.*

Figure 19-26. *Check the soundness of a grinding wheel before mounting it on machine. A sound grinding wheel will give a clear "metallic ring" when tapped as shown.*

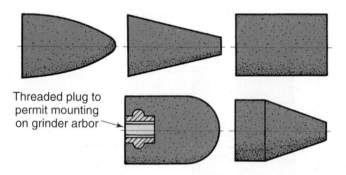

Threaded plug to permit mounting on grinder arbor →

Figure 19-25. *A few of the many grinding wheel shapes available for internal grinding.*

19.3.3 How to mount grinding wheels

Select a grinding wheel recommended for the job. Check its soundness by lightly tapping the wheel as shown in **Figure 19-26**. A sound wheel will give a clear "metallic ring." If the wheel is cracked, the tone will be "flat," rather than a clear ringing sound.

Always discard unsound grinding wheels. If possible, break them into several pieces to ensure they are not used.

Unbalanced wheels will cause irregularities on the finished ground surface. They should be statically balanced as shown in **Figure 19-27A**. On CNC grinders, automatic wheel balancing systems can lengthen the life of the wheel and provide improved surface finishes. Automatic systems like the one shown in **Figure 19-27B** use vibration sensors and ultrasound wheel contact sensors to monitor operation of the wheel. A microprocessor-based controller

signals a flange or spindle-mounted balancing head to make necessary adjustments.

Mount the wheel on the spindle. It should fit snugly. Never *force* a grinding wheel on a shaft. The blotter rings or compressible washers should be large enough to extend beyond the wheel flanges, **Figure 19-28**. It is essential that the wheel be mounted properly. If it is not, excessive strains will develop during the grinding operation and the wheel could shatter. Avoid standing in line with the grinding wheel, especially during the first few passes across the work.

19.4 CUTTING FLUIDS (COOLANTS)

Cutting fluids are an important factor in lessening wear on the grinding wheel. They help to maintain accurate dimensions, and are important to the quality of the surface finish produced. As a coolant, the cutting fluid must remove the heat generated during the grinding operation. Heat must be removed as fast as it is generated.

Several types of cutting fluids are utilized in grinding operations:

- *Water-soluble chemical fluids* are solutions that take advantage of the excellent cooling ability of water. They are usually transparent and include a rust inhibitor, water softeners, detergents to improve the cleaning ability of water, and bacteriostats (substances that regulate and control the growth of bacteria).
- *Polymers* are added to water to improve lubricating qualities.

A

B

Figure 19-27. *Grinding wheel balancing. A—Static balance method. The wheel nut shown features a series of threaded holes on a bolt circle. By adding/subtracting setscrews of different lengths opposite the wheel's heavy/light sections, the wheel may be statically balanced. (Revolution Tool Company) B—Automatic grinding wheel balancing system uses a flange- or spindle-type balancing head, vibration and ultrasound sensors, and a microprocessor-based controller to keep wheel in balance. (Marposs Corp.)*

- *Water-soluble oil fluids* are coolants that are usually "milky white," since they consist of a mixture of oil and water. They are also less expensive than most chemical-type fluids. Bacteriostats are added to control bacteria growth.

Coolant can be applied by flooding the grinding area, **Figure 19-29.** The fluid recirculates by means of a pump and holding tank built into the machine. A *mist system* forces the coolant over the wheel or applies it to the work surface under pressure (air). It cools by evaporation. A coolant can also be applied manually by pumping the fluid from a pressure-type oil pump can. If coolant is applied manually, keep the tip of the oil pump can a safe distance from the wheel.

For safety, long equipment life, and quality control, a coolant system should be cleaned at regular intervals. Cleaning means removing all dirt and sludge from the holding tank, **Figure 19-30.** Discard the fluid when it becomes contaminated.

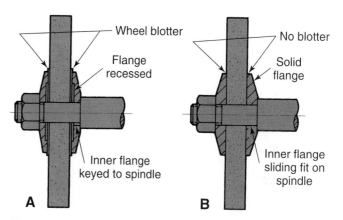

A   B

Figure 19-28. *Do not operate a grinder unless the wheel is properly mounted. A—Correctly mounted wheel. B—Wheel incorrectly and dangerously mounted.*

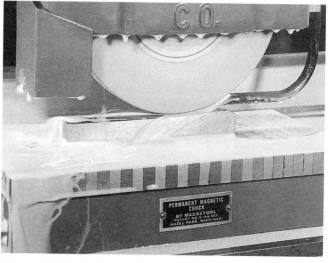

Figure 19-29. *Coolant must flood area being ground.*

Figure 19-30. *The coolant tank must be cleaned frequently. Chips and grinding wheel residue in the coolant can mar the ground surface of the workpiece.*

19.5 GRINDING APPLICATIONS

The following procedure is recommended to produce a surface that is flat or free of waviness:

1. Select and mount a suitable grinding wheel.

2. True and dress the wheel with a diamond dressing tool, **Figure 19-31**.

3. Mount the work-holding device. If a magnetic chuck is used, it should be "ground-in" to assure a surface that is true and parallel to table travel, **Figure 19-32**. Grind off as little material

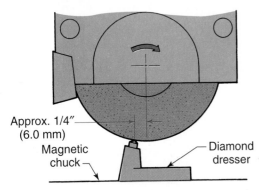

Figure 19-31. *For best results when cleaning or truing a grinding wheel, position the diamond wheel dresser as shown.*

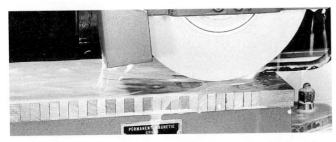

Figure 19-32. *A machine operator is "grinding in" the surface of a magnetic chuck to true it before use. Note how the surface is flooded with coolant.*

as possible to true the surface. For high-precision work, this should be done each time the chuck is remounted on the machine.

4. Check the coolant system to be sure it is operating satisfactorily.

5. Locate the work, and energize the chuck. If the work is already ground on one surface, protect it and the chuck surface by fitting a piece of oiled paper between them before energizing the chuck, **Figure 19-33**.

6. Adjust the table stops, **Figure 19-34**.

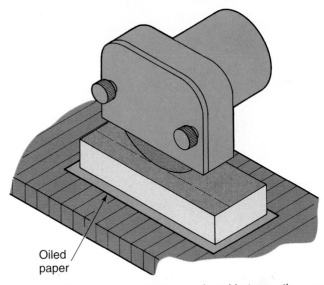

Figure 19-33. *A piece of oiled paper placed between the magnetic chuck and a newly ground surface will protect the finish of the work when it is removed from the chuck.*

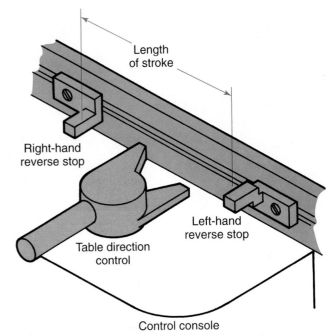

Figure 19-34. *Adjustable stops regulate length of the table stroke. Care must be taken to be sure stop adjustment permits the entire work surface to be ground.*

7. Check the holding power of the magnetic chuck by trying to move the work.

8. Down-feed the grinding wheel until it just touches the highest point on the work surface. The grinding wheel can be set to the approximate position by down-feeding until it just touches a sheet of paper placed between the wheel and the work surface.

9. Turn on the coolant, spindle, and hydraulic pump motors.

10. Set the cross-feed to move the table in or out about 0.020" (0.5 mm) at the end of each cycle.

11. With the wheel clear of the work, down-feed about 0.001" to 0.003" (0.025 mm to 0.075 mm) for average roughing cuts per pass.

12. Use light cuts of 0.0001" (0.0025 mm) for finishing the surface. It is wise to redress the wheel for finishing cuts.

When the work surface has been ground to the required dimension and finish, use the following procedure to turn off the machine:

1. Move the grinding wheel clear of the work.

2. Turn off table travel.

3. Turn off coolant.

4. Let the grinding wheel run for a few moments after the coolant has been turned off. This will permit the wheel to free itself of all traces of fluid, **Figure 19-35**; otherwise, the wheel can absorb some of the coolant and become out of balance.

5. Use a squeegee to remove excess coolant from the work. De-energize the chuck and remove the work. Be careful of the sharp edges on newly ground work when removing it from the machine.

6. Clean the machine. Apply a light coating of oil on the chuck's work surface to prevent possible rusting.

7. Place all tools in proper storage.

19.5.1 Grinding edges square and parallel with face sides

Most rectangular work requires the edges to be parallel to each other and square with the finished face sides. In one commonly used technique, **Figure 19-36,** edge 1 is ground while being held in a precision vise. After burrs are removed, the adjacent edge 2 is ground. Its squareness is checked vertically with a dial indicator.

Figure 19-35. Sequence for turning off a surface grinder.

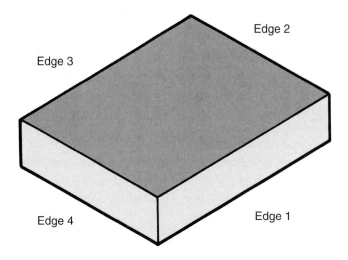

Figure 19-36 The sequence for squaring edges of a rectangular workpiece after the faces have been ground flat and parallel.

After edges 1 and 2 are ground square with each other, they will serve as reference planes to grind edges 3 and 4 to required dimensions. Use oiled paper in the vise to protect the ground faces and edges from stray metal and abrasive particles. The vise must be carefully wiped clean and burrs removed after each edge is ground.

An *angle plate* can also be used when grinding edges square and parallel with the finished faces. See **Figure 19-37**. A parallel may be used to set the work in approximate position.

The same positioning and grinding sequences are followed as previously described.

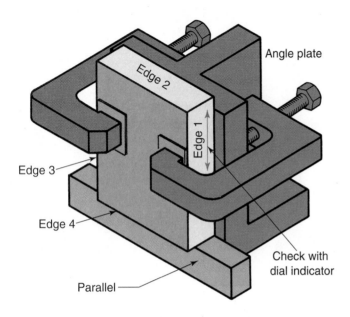

Figure 19-37. An angle plate can also be used to hold work for grinding the edges square and parallel.

19.5.2 Creep grinding

Creep grinding is a relatively new production-type machining technique that makes a deep cut into the work. It is also sometimes known as *deep grinding.* Special grinding machines are required for this type of work.

Creep grinding is a surface grinding operation that is often performed in a single pass with an unusually large depth of cut, **Figure 19-38**. In comparison to conventional surface grinding, the depth of cut is increased 1000-10,000 times and the work speed is reduced in the same proportion. Machining time can be reduced by 50% to 80%. The tools (grinding wheels, work-holding devices, etc.) must be designed for this heavy-duty work.

19.6 GRINDING PROBLEMS

There are many problems peculiar to precision surface grinding. A few of the more common difficulties, with suggestions for their solution, are as follows:

Irregular table movement or *no table movement* (on hydraulic-type machines) may be caused by clogged hydraulic lines, insufficient hydraulic fluid,

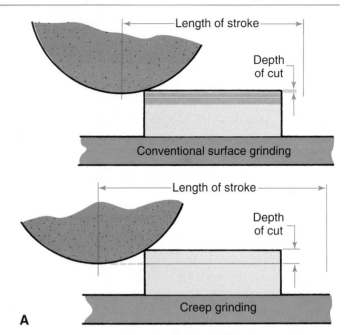

Figure 19-38. Creep grinding. A—The difference between conventional and creep grinding. Creep grinding equipment must be specially designed for this heavy-duty work. B—A heavy-duty CNC creep grinding machine. (Jones & Shipman, Inc.)

a hydraulic pump that is not functioning properly, or inadequate table lubrication. A cold hydraulic system may also cause these symptoms. Let the machine warm up for at least 15 minutes before use. Air in hydraulic lines can cause erratic table movement. Make corrections as recommended by the manufacturer of the machine.

Irregular scratches, of no identifiable pattern, are frequently caused by a dirty coolant system, or by particles becoming loosened in the wheel guard. This could also mean that the grinding wheel is too soft, and the abrasive particles are carried to the wheel by the coolant system.

Work surface waviness can be caused by a wheel being out of round. It can be corrected by truing the wheel.

Chatter or *vibration marks* may be caused by a glazed or loaded grinding wheel. There is a slipping action between the wheel and the work. The wheel cuts until the glazed section comes into position and slides over the work, rather than cutting. Correct the problem by redressing the grinding wheel. The same effect can also be caused by a grinding machine that is not mounted solidly, or by a wheel that is loose on the spindle. Check for these conditions and make corrections if necessary.

Burning or *work surface checking* may be the result of too little coolant reaching the work surface, a wheel that is too hard, or a wheel with grain that is too fine. Make needed corrections as indicated by inspection.

Wheel glazing or *loading* often indicates that the wrong coolant is being used. A dull diamond on the wheel dresser can also cause this problem.

Deep, irregular marks on the work surface may be caused by a loose grinding wheel.

Work that is not flat may be caused by insufficient coolant, a nicked or dirty chuck surface, or a wheel that is too hard. Check and make any necessary corrections.

Work that is not parallel is frequently caused by a chuck that has not been "ground in" since the last time it was mounted on the machine. A nicked or dirty chuck can also cause the same problem. Insufficient coolant may allow the work to heat up and expand in the center of the cut, permitting more material to be removed in that area than at each end. As the piece cools, the center will become depressed, **Figure 19-39.** Correct the problem by directing more fluid to the cutting area.

19.7 GRINDING SAFETY

- Never attempt to operate a grinder until you have been instructed in its proper and safe operation. When in doubt, consult your instructor.
- Do not use a grinder unless all guards and safety devices are in place and securely attached.
- Never try to operate grinding machines while your senses are impaired by medication or other substances.
- Always wear approved-type eye protection when performing any grinding operation.
- Never place a wheel on a grinder before checking it for soundness. Destroy faulty wheels!

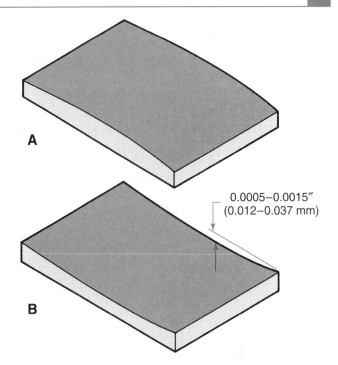

0.0005–0.0015"
(0.012–0.037 mm)

Figure 19-39. *Effect of insufficient coolant. A—If not enough coolant is used, frictional heat may cause the work to expand in center of cut, causing more material to be removed in center than at each end. B—As piece cools, center becomes depressed.*

- Check the wheel often to prevent it from becoming glazed or loaded. Dress the wheel when required.
- Make sure the grinding wheel is clear of the work before starting the machine.
- Never operate a grinding wheel at a speed higher than specified by the manufacturer.
- Change coolant fluid before it becomes contaminated. It is good practice to set up a schedule for replacing the fluid at regular intervals or adding chemicals to control bacterial growth.
- Have any cuts and grinding "burns" treated promptly — major infections can result from untended minor injuries.
- Immediately wipe up any spilled coolants from the floor around the machine.
- Stop the machine before making measurements or performing major machine and work adjustments.
- If a magnetic chuck is used, make sure it is holding the work solidly before starting to grind.
- Remove your watch, rings, etc., before using a magnetic chuck to prevent them from becoming magnetized or pulled toward the machine.

- If automatic feed is used, run the work through one cycle by hand to be sure there is adequate clearance and that the dogs are adjusted properly.
- Keep all tools clear of the worktable.

19.8 UNIVERSAL TOOL AND CUTTER GRINDER

The *universal tool and cutter grinder* is a grinding machine designed to support cutters (primarily milling cutters) while they are sharpened to specified tolerances. See **Figure 19-40.** Special attachments permit straight, spiral, and helical cutters to be sharpened accurately. Other attachments enable the machine to be adapted to all types of internal and external cylindrical grinding. See **Figure 19-41.**

19.9 TOOL AND CUTTER GRINDING WHEELS

The wheel shapes most frequently employed for tool and cutter grinding are shown in **Figure 19-42.** Charts prepared by the various grinding wheel manufacturers are used to determine the correct wheel composition (abrasive, grain size, and bond) for the job at hand.

Figure 19-41. *Tool and cutter grinder applications. A—Limited cylindrical grinding can be done on a tool and cutter grinder. B—Tool and cutter grinder can also be employed for internal grinding. C—Center is being trued on tool and cutter grinder. Work is rotated by a powered work head. (K.O. Lee Co.)*

Figure 19-40. *Universal tool and cutter grinder. (K.O. Lee Co.)*

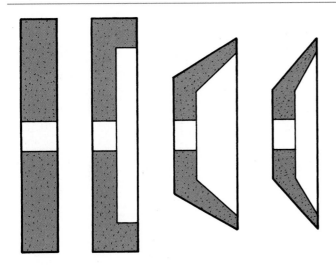

Figure 19-42. *Typical wheel shapes used for grinding cutters.*

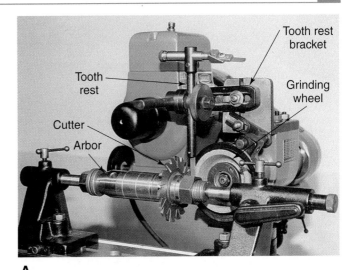

A

Keep the grinding wheel clean and sharp by frequent dressing with a diamond tool. Use light cuts to avoid drawing the temper out of the tooth cutting edge.

Crowding the wheel into the cutter is a common mistake when grinding cutters. The cutters are made from materials (HSS and cemented carbides) that do not give off a brilliant shower of sparks when in contact with a grinding wheel. This creates the illusion that the cut being made is too light.

19.9.1 Sharpening cutters

A *tooth rest* locates each tooth quickly and accurately into position. Several types are employed to permit different cutter types to be sharpened. The *supporting bracket* can be mounted to the worktable or on the grinding wheel housing. **See Figure 19-43.**

Grinding plain milling cutters

1. Select the correct wheel for the job. True it with a diamond tool.

2. Mount the cutter on a suitable arbor and place the unit between centers, **Figure 19-44.**

3. Mount the tooth rest to the wheel head. Position the edge about 1/4″ (6.0 mm) above the center line of the grinding wheel, **Figure 19-45.** This will produce a 5° to 6° clearance angle on the tooth cutting edge of a 6″ (150 mm) diameter cutter. Adjust to suit the cutter being ground.

4. The setup should permit the wheel to grind away from the tooth cutting edge. While requiring more machining care than grinding

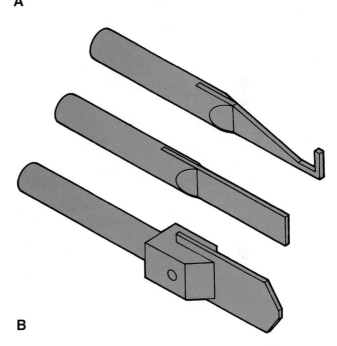

B

Figure 19-43. *Tooth rest. A—The tooth rest being used to position teeth of cutter. Note the support bracket and the cup-shaped wheel that does the grinding. (K.O. Lee Co.) B— Several-types of tooth rests.*

into the cutting edge of the tooth, there is less chance of drawing the temper. Also, no burr is formed that must be oilstoned off to secure a sharp edge. **See Figure 19-46.**

5. Flare cup wheels are also used for cutter and tool grinding. How they are set up is shown in **Figure 19-47.** Since there is a greater area of contact when using a flare cup wheel, *lighter cuts* should be taken than with straight grinding wheels.

6. Start the machine and feed the cutter into the wheel. Take a light cut. A bit of thinned layout bluing should be applied to the back of the

Figure 19-44. *Cutters must be mounted on an arbor for sharpening. Approved-type eye protection is a must. If machine is not equipped with a vacuuming system, a respirator mask should also be worn. (Norton Co.)*

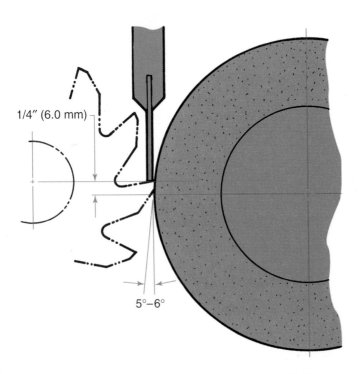

1/4″ (6.0 mm)

5°–6°

Figure 19-45. *Setup is for grinding a 6″ (150 mm) diameter cutter.*

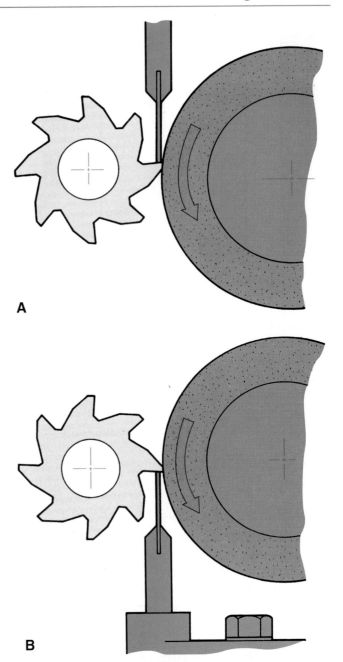

A

B

Figure 19-46. *Tooth grinding setup. A—Wheel grinds away from cutting edge of tooth. With this technique, there is less chance of drawing temper out of tooth and no burr is formed. B— When wheel grinds into cutting edge of tooth, there is some danger that a burr will be formed or temper drawn.*

tooth. This will allow a visual check of how the grinding operation is progressing and whether the setup is producing the proper clearance angle.

7. When satisfied with the setup, bring the next tooth into position on the tooth rest and grind that tooth.

8. Repeat the operation until all of the teeth are sharpened. Make necessary adjustments to

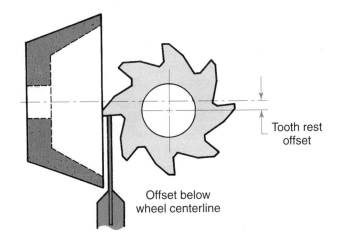

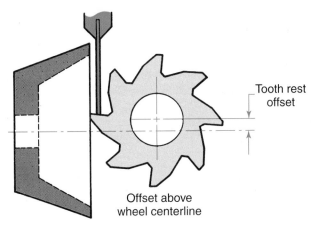

Figure 19-47. Milling cutter is being ground with a flare cup wheel.

assure tooth *concentricity* (the cutting surfaces of all teeth must be the same distance from arbor hole center line).

After a cutter has been sharpened several times, the clearance angle *flat* (land) will become too wide. Then, it becomes necessary to grind in a secondary clearance angle, **Figure 19-48.**

If it becomes apparent that more material is being removed from some teeth than others, a quick check must be made to determine the cause:

- The grinding wheel may be too soft and wearing down too rapidly. As the wheel wears, less material is removed from the cutter tooth.
- The tooth rest may not be mounted solidly, allowing it to move during the grinding operation.
- The arbor may not be running true on the centers. Test the arbor runout with an indicator as it is rotated.

When the cause of the problem has been identified, make the necessary corrections and continue the operation.

An *indexing disc* may also be used to position each tooth for sharpening, **Figure 19-49.** It is mounted on the arbor. The divisions are normal to each other, plus or minus 4 minutes (1/15°). They are available in a range of graduations.

Teeth on a side-milling cutter must also be sharpened. This is done by mounting the cutter on a stub arbor and fitting the unit into a *workhead*, rather than positioning it between centers. Facing mills are sharpened in the same manner. See **Figure 19-50.**

Grinding cutters with helical teeth

Slabbing cutters and other cutters that have helical teeth are sharpened in much the same manner as plain milling cutters, **Figure 19-51.** However, these cutters must be held against the tooth rest as the

Figure 19-48. This cutter has been sharpened so many times that a secondary clearance angle is needed to permit the tool to cut properly.

Figure 19-49. An indexing disc is often used to position each tooth for sharpening. (K.O. Lee Co.)

Figure 19-50. *A grinder being used to sharpen the side of teeth of a face milling cutter.*

Figure 19-51. *Setup for sharpening a cutter with helical teeth. (K.O. Lee Co.)*

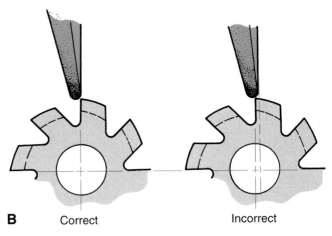

B   Correct   Incorrect

Figure 19-52. *Form tooth cutters. A—This convex cutter is typical of form tooth cutters. (Standard Tool Co.) B—Form tooth cutters must be ground radially; otherwise, the form or shape that the cutter was designed to machine will be altered.*

table is traversed. This will impart a twisting motion to keep the tooth correctly located against the grinding wheel.

Grinding end mills

End mills are sharpened in much the same way as helical teeth cutters, with the end mill mounted in a workhead rather than between centers. The end teeth are sharpened with the same technique used to sharpen the side teeth on a side milling cutter.

Grinding form cutters

Form tooth cutters must be ground radially to preserve the tooth shape, **Figure 19-52.** An index disc may be employed, or a special form cutter grinder may be utilized.

Grinding taps

A universal tool and cutter grinder may also be used to resharpen taps. Normally, a tap becomes dull when the leading edges of the starting chamfer become worn. The chamfer can be reground by mounting the tap in a workhead, **Figure 19-53.** *Flutes* are reground using a straight wheel with an edge that has been shaped to fit the flutes.

Grinding reamers

The cutting action of a machine reamer takes place at the front end of the teeth, **Figure 19-54.** Sharpen the reamer in the same manner employed to sharpen a face milling cutter. The worktable is pivoted at a 45° angle. Using a cup wheel, adjust the tooth rest and/or grinding head to give the correct clearance.

Figure 19-53. Grinding taps. A—Setup for regrinding chamfer on a tap. B—Flutes are being reground on tap to renew cutting edges of teeth. (K.O. Lee Co.)

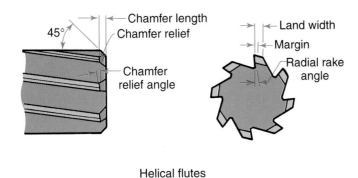
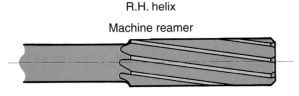

Figure 19-54. Cutting edges of a machine reamer.

19.10 CYLINDRICAL GRINDING

With a cylindrical grinder, it is economically feasible to machine hardened steel to tolerances of 0.00001" (0.0002 mm) with extremely fine surface finishes. See **Figures 19-55** and **19-56**.

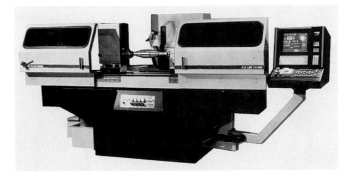

Figure 19-55. A CNC cylindrical grinder capable of producing extremely fine surface finishes while meeting tolerances of 0.00001" (0.0002 mm). (K.O. Lee Co.)

Figure 19-56. This heavy-duty 60" x 228" (1500 mm x 5700 mm) cylindrical grinder is being used to recondition forming rolls used by steel mills. (Simmons Machine Tool Corp.)

On this machine, work is mounted between centers and rotates while in contact with the grinding wheel, **Figure 19-57.** Straight, taper, and form grinding operations are possible with this technique. Two variations of cylindrical grinding are traverse grinding and plunge grinding.

- In *traverse grinding*, a fixed amount of material is removed from the rotating workpiece as it moves past the revolving grinding wheel. Work wider than the face of the grinding wheel can be ground. See **Figure 19-58**.
- In *plunge grinding*, the work still rotates; however, it is not necessary to move the grinding wheel across the work surface. The area being ground is no wider than the wheel face. Grinding wheel infeed is continuous rather than incremental (minute changes at end of each cut), **Figure 19-59**. Grinding to a shoulder by both techniques is shown in **Figure 19-60**.

Figure 19-57. *Closeup of a cylindrical grinding operation. (Landis Div. of Western Atlas)*

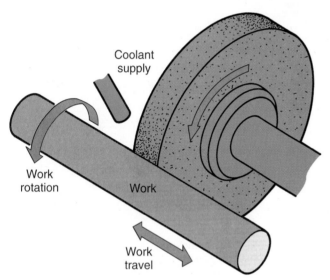

Figure 19-58. *The principle of traverse grinding. The rotating work moves past the rotating grinding wheel.*

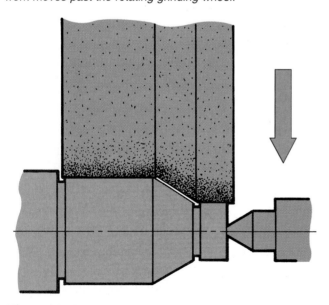

Figure 19-59. *With plunge grinding, grinding wheel is fed into rotating work. Since work is no wider than grinding wheel, reciprocating motion is not needed.*

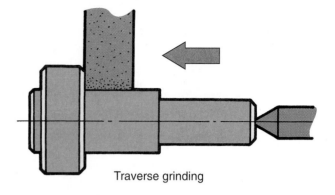

Traverse grinding

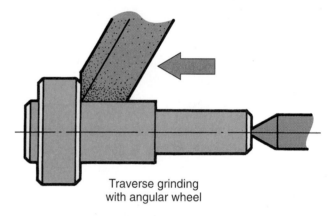

Traverse grinding
with angular wheel

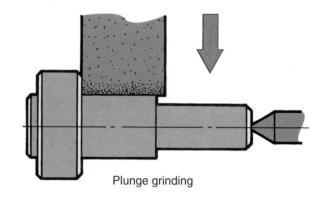

Plunge grinding

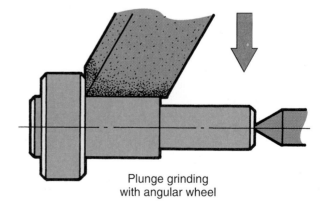

Plunge grinding
with angular wheel

Figure 19-60. *Grinding to a shoulder by traverse and plunge grinding techniques.*

19.10.1 Holding and driving the work

As the work rotates on centers, it is extremely important that the centers be free of dirt and nicks. They must also run absolutely true. If possible, the head center should be ground in place. The center holes must also be clean, of the correct shape and depth, and well-lubricated.

Long work is best supported by *work rests*, **Figure 19-61.** These support the workpiece from the back and bottom and are adjustable to compensate for material removed in the grinding operation.

Work rotation is accomplished through the use of a drive plate that revolves around the headstock center and an adjustable drive pin and dog, **Figure 19-62.** Work may also be mounted in a chuck.

19.10.2 Machine operation

To assure a good finish and size accuracy, it is vital that work rotation and traverse table movement (back-and-forth in front of the grinding wheel) be smooth and steady.

Table movement should be adjusted so that the wheel will *overrun* the work end by about one-third the width of the wheel face, **Figure 19-63.** This permits the grinding wheel to do a more accurate grinding job. *Insufficient runoff* will result in work that is oversize. *Complete runoff* of the grinding wheel will cause the piece to be undersize.

The grinding wheel must be trued and balanced. Otherwise, vibration will cause chatter marks on the work and may cause it to be out-of-round.

Cutting speeds and feeds can be determined from information available on charts furnished by the grinding machine and grinding wheel manufacturers. They will also specify what coolant will give the best results.

19.11 INTERNAL GRINDING

Internal grinding is done to secure a fine surface finish and accuracy on inside diameters, **Figure 19-64.** Work is mounted in a chuck and

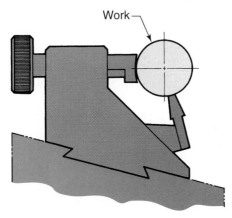

Figure 19-61. *Work rests should be placed every four or five diameters along the work for support. They must be adjusted after each grinding pass.*

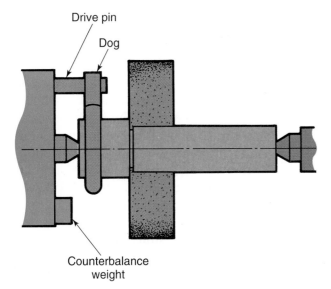

Figure 19-62. *One method used to rotate work on a cylindrical grinder.*

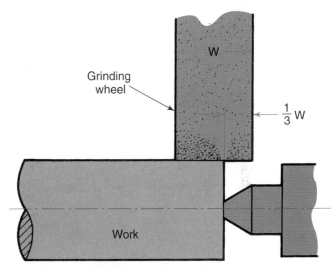

Figure 19-63. *Table movement should permit about one-third of wheel's width to run beyond the end of the work. This permits the wheel to do a more accurate job.*

Figure 19-64. *Internal grinding operation is being performed on a universal grinding machine. Note how extended work piece is supported. (Norton Co.)*

rotates. During the grinding operation, the revolving grinding wheel moves in and out of the hole.

A special grinding machine that finishes holes in pieces too large to be rotated by the conventional machine is shown in **Figure 19-65.** Hole diameter is controlled by regulating the diameter of the circle in which the grinding head moves.

19.12 CENTERLESS GRINDING

In *centerless grinding*, the work does not have to be supported between centers because it is rotated against the grinding wheel, **Figure 19-66.** Instead, the piece is positioned on a work support blade, and fed automatically between a regulating or feed wheel and a grinding wheel. See **Figure 19-67.** Basically, the regulating wheel causes the piece to rotate, and the grinding wheel does the cutting. Feed through the wheels is obtained by setting the regulating wheel at a slight angle.

There are four variations of centerless grinding, **Figure 19-68:**

- *Through feed grinding* can only be employed to produce simple cylindrical shapes. Work is fed continuously by hand, or from a feed hopper, into the gap between the grinding wheel and the regulating wheel. The finished pieces *drop off* the work support blade.

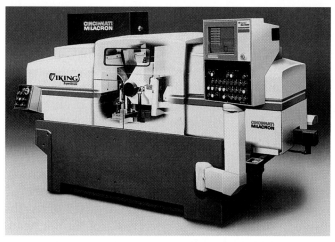
Figure 19-66. A CNC centerless grinding machine. (Cincinnati Milacron)

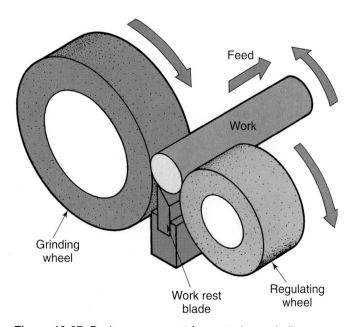
Figure 19-67. Basic arrangement for centerless grinding.

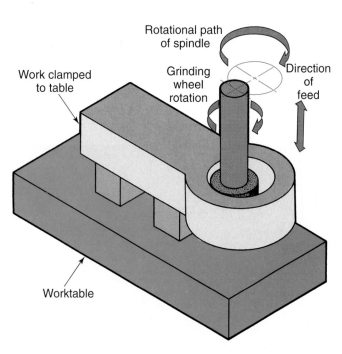
Figure 19-65. Internal grinding technique employed when work is too large or odd-shaped to be rotated.

- *Infeed grinding* is a centerless grinding technique that feeds the work into the wheel gap until it reaches a *stop*. The piece is ejected at the completion of the grinding operation. Work diameter is controlled by adjusting the width of the gap between the regulating wheel and the grinding wheel. Work with a shoulder can be ground using this technique.
- *End feed grinding* is a form of centerless grinding ideally suited for grinding short tapers and spherical shapes. Both wheels are dressed to the required shape and work is fed in from the side of the wheel to an end stop. The finished piece is ejected automatically.

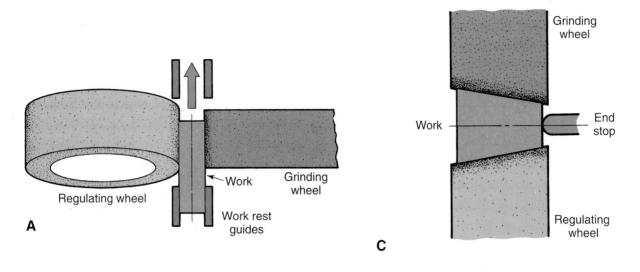

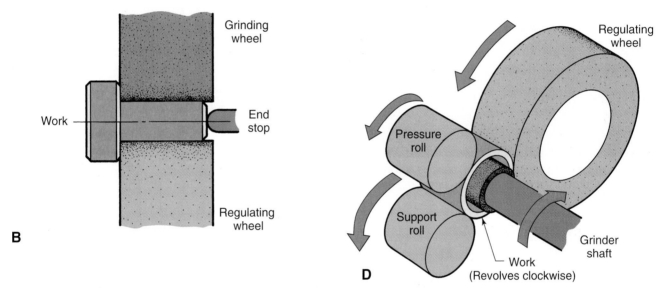

Figure 19-68. Centerless grinding variations. A—In through feed centerless grinding, angle of regulating wheel pulls work over the grinding wheel. B—Infeed centerless grinding. The work is fed into the wheel gap until it reaches a stop. The piece is ejected at completion of the grinding operation. C—End feed centerless grinding. It is best suited for grinding short tapers and spherical shapes. D—Setup for internal centerless grinding.

- *Internal centerless grinding* minimizes distortion in finishing thin-wall work and eliminates reproduction of hole-size errors and waviness in the finish.

Centerless grinding is utilized when large quantities of the same part are required. Production is high and costs are relatively low, because there is no need to drill center holes nor to mount work in a holding device. Almost any material can be ground by this technique.

19.13 FORM GRINDING

In *form grinding,* the grinding wheel is shaped to produce the required contour on the work. **Figure 19-69** shows this principle.

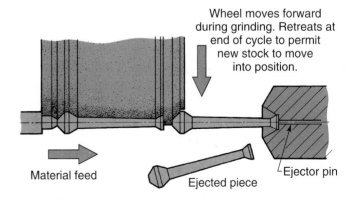

Figure 19-69. Form grinding of this engine part is done at rate of 200 pieces an hour. Material was heat-treated before grinding.

Thread grinding is an example of form grinding, **Figure 19-70**. A form or template guides a diamond particle wheel that dresses the wheel used to grind the required thread shape, **Figure 19-71**. There is automatic compensation on the grinding machine for the material removed from the grinding wheel when it is dressed.

19.14 OTHER GRINDING TECHNIQUES

In addition to the grinding techniques already described, industry makes considerable use of other abrasive-type processes.

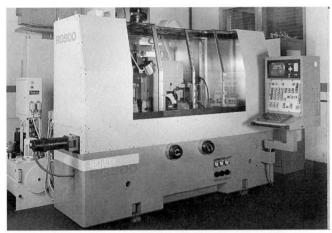

Figure 19-70. This CNC thread grinding machine has electronic controls that provide thread accuracy and cost-efficiency. (Reishauer Elgin)

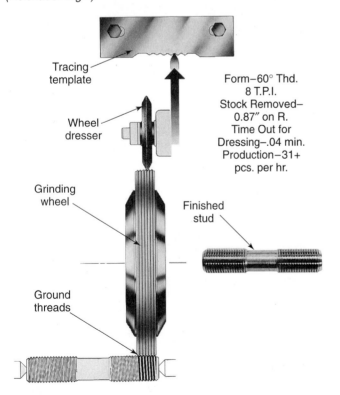

Tracing template

Wheel dresser

Form–60° Thd.
8 T.P.I.
Stock Removed–
0.87" on R.
Time Out for
Dressing–.04 min.
Production–31+
pcs. per hr.

Grinding wheel

Finished stud

Ground threads

Figure 19-71. Precision threads are being form-ground on a special stud. (Jones & Lamson Machine Co.)

19.14.1 Abrasive belt machining

Abrasive belt grinding was first employed for light stock removal and polishing operations. However, the capability of the technique has advanced to the stage where high-rate metal removal to close tolerances is possible. This is primarily due to tougher and sharper abrasive grains, improved adhesives, and stronger backings. Several abrasive grinding machine applications are shown in **Figure 19-72**.

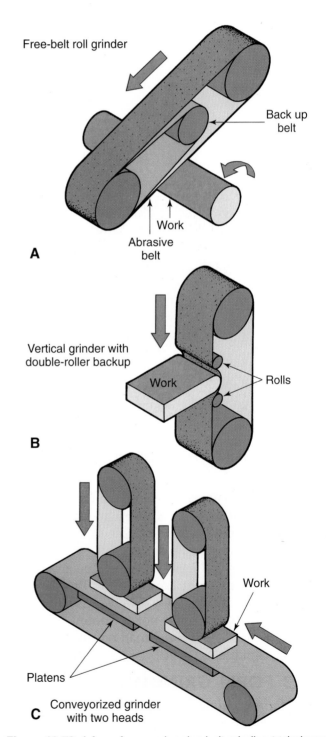

Free-belt roll grinder

Back up belt

Work

Abrasive belt

A

Vertical grinder with double-roller backup

Work

Rolls

B

Work

Platens

C Conveyorized grinder with two heads

Figure 19-72. A few of many abrasive belt grinding techniques.

Abrasive belts, because of their length, run cool and require light contact pressure, thus reducing the possibility of metal distortion caused by heat. Soft contact wheels and flexible belts conform to irregular shapes. Belts may be used dry or with a coolant. The most satisfactory belt speed for grinding ferrous and nonferrous metals is between 5000-9000 *sfm* (surface feet per minute). Slower speeds of 1500-3000 sfm are required for tougher materials like titanium.

Abrasive belt grinding usually requires support behind the belt. This may be in the form of contact wheels or platens.

- *Contact wheels* are usually made of cloth or rubber. Hardness and/or density of the contact wheel affects stock removal and finish. Serrated or slotted wheels improve cutting action and prolong abrasive belt life.
- *Platens* are made of metal (some have cemented carbide inserts) and are usually not as effective as contact wheels. They are flat, but can be shaped to conform to the contour required on the work. Jets of air or water may be applied between the belt and platen to reduce friction.

A major advantage of abrasive belt grinding is its versatility. A machine can be converted quickly from heavy stock removal to finishing operations, or for grinding a different material, by simply changing the abrasive belt.

19.14.2 Electrolytic grinding

Electrolytic grinding is actually a form of electrochemical machining, **Figure 19-73.** Applications of the technique include rapid removal of stock from alloy steel parts, sharpening carbide tools, and machining heat-sensitive work.

An electric current is passed between a metal-bonded grinding wheel (cathode) and the work (anode) through a conductive electrolyte, **Figure 19-74.** The surface of the work is attacked electrochemically and dissolved (a process similar to electroplating, but in reverse).

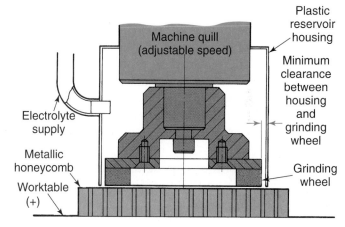

Figure 19-73. *Electrolytic grinding process. It is actually an electrochemical machining process.*

Figure 19-74. *Nomenclature (parts) of an electrochemical grinding machine designed to sharpen carbide lathe tools. (Hammond Machinery Builders)*

The dissolved material is removed by the wheel. No burr is developed, making it possible to machine fragile work like stainless steel and exotic metal honeycomb sections. No heat is generated, and there is no metallurgical change in the metal.

An electrochemical grinding machine for sharpening carbide tools is shown in **Figure 19-75.**

19.14.3 Computer controlled (CNC) grinders

Many types of computer numerical control (CNC) grinders are available, **Figure 19-76.** They are designed to operate automatically. Such functions as positioning, spindle start and stop, vertical feed motion, and linear feed rates are programmed into the machine's computer. The operator is then relieved from the responsibility of controlling and monitoring the numerous coordinate settings and related machine functions.

The grinding wheel path representing the contours of the part (at a selected distance from the part edge) is programmed directly to its dimensional specifications. Grinding wheel wear (wheel diameter decreases each time it is dressed) is compensated for automatically.

The part's contour is generated in a continuous motion by the X- and Y-axes slides of the machine. See **Figure 19-77.**

A

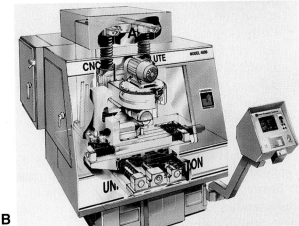

B

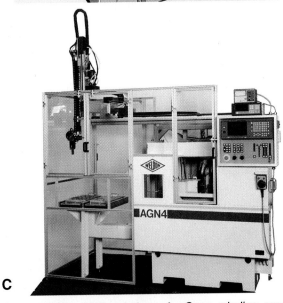

C

Figure 19-76. *CNC grinders. A—Gear grinding machine is capable of grinding spur or helical gears to close tolerances and precise tooth geometry at high stock removal rates. (Reishauer Corp.) B—Cutaway of CNC grinder for remanufacturing and sharpening fluted cutting tools such as drills, reamers, and end mills. (Unison Corp.) C—CNC cylindrical grinder with automatic loading system. The machine is capable of plunge, contour, and traverse operations. (Weldon Machine Tool, Inc.)*

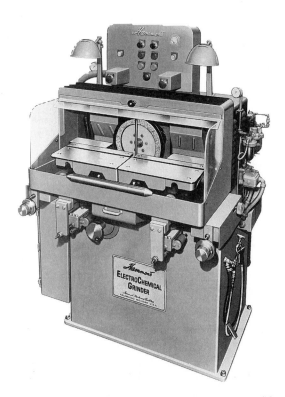

Figure 19-75. *An electrochemical grinding machine. (Hammond Machinery Builders)*

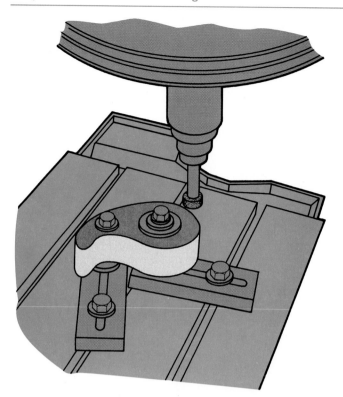

Figure 19-77. Contour grinding using CNC to shape a cam. In this illustration, it appears that the grinding wheel is moving around the cam edge. In reality, the part moves along its programmed path around the vertical machine spindle holding the rotating wheel.

TEST YOUR KNOWLEDGE

Please do not write in the text. Write your answers on a separate sheet of paper.

1. Industry classifies surface grinding as the grinding of _____ surfaces.

2. Surface grinding operations fall into two categories. List them.

3. Various work-holding devices are used to hold work for surface grinding. Name three of them.

4. List five (5) factors that are distinguishing characteristics of a grinding wheel.

5. The ideal grinding wheel will:
   a. Wear away as the abrasive particles become dull.
   b. Wear away at a predetermined rate.
   c. Wear away slowly to save money.
   d. All of the above.
   e. None of the above.

6. A solid grinding wheel will give off a _____ when struck lightly with a metal rod.

7. List the two conditions that commonly prevent a grinding wheel from cutting efficiently.

8. Why are cutting fluids or coolants necessary for grinding operations?

9. List the basic types of cutting fluids.

10. A _____ wheel dressing tool is usually used to true and dress wheels for precision grinding.

11. Chatter and vibration marks are caused on the work when the grinding wheel is _____ or _____.

12. The problems in Question 11 can be corrected by _____ the _____.

13. Irregular scratches on the work are usually caused by a _____ system. How can this problem be corrected?

14. What is the difference between conventional grinding and creep grinding?

15. A _____ and _____ is a grinding machine designed to support cutters (usually milling cutters) while they are being sharpened.

16. List the two variations of cylindrical grinding.

17. With _____ grinding, it is not necessary to support work between centers or mount work in a chuck while it is being rotated against the grinding wheel.

18. Make sketches of nine standard grinding wheel shapes.

19. The grinding technique that employs a belt on which abrasive particles are bonded for stock removal, finishing, and polishing operations is known as _____.

20. _____ or _____ grinding is actually an electrochemical machining process. Describe how is it done.

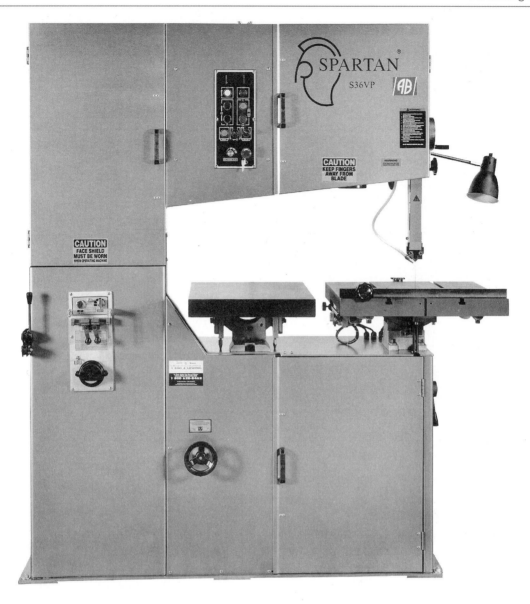

Vertical contour band saws are used for various band machining operations. This saw is available with throat depths of up to 36″ (91 cm), and a blade speed of 5000 fpm. The table is able to tilt 45° right and 10° left, and can be power-operated for high-capacity production. (Armstrong-Blum Manufacturing Company)

Chapter 20

Band Machining

LEARNING OBJECTIVES

After studying this chapter, you will be able to:
- ○ Describe how a band machine operates.
- ○ Explain the advantages of band machining.
- ○ Select the proper blade for the job to be done.
- ○ Weld a blade and mount it on a band machine.
- ○ Safely operate a band machine.

IMPORTANT TERMS

blade guide inserts	mist coolant
diamond-edge band	raker set
file band	straight set
internal cuts	tooth form
knife-edge blade	wavy set

Band machining is a widely employed machining technique that makes use of a continuous saw blade, **Figures 20-1**. Each tooth is a precision cutting tool, **Figure 20-2,** so accuracy can be held to close tolerances. This eliminates or minimizes many secondary machining operations.

20.1 BAND MACHINING ADVANTAGES

Band machining offers several major advantages over other machining techniques, **Figure 20-3**.
- Band machining *maintains sharpness*. Wear is distributed over many teeth. Chip load is uniform and constant on each tooth, minimizing tool wear.
- Band machining provides *unrestricted cutting geometry*. Cutting can be done at any angle, in any direction, and the length of the cut is unlimited.
- Band machining is *efficient*. Excess chip production wastes power. Band machining

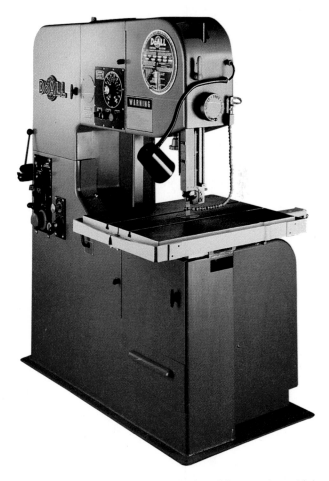

Figure 20-1. *Vertical band saw designed for metal machining. (DoALL Co.)*

produces the desired shape with a minimum of chips. There is *little waste*, since band machining cuts directly to shape, and unwanted material is removed in solid sections.
- Band machining provides a *built-in workholder*. Cutting action is downward, so cutting forces hold the workpiece to the table. In most situations, work need not be clamped.

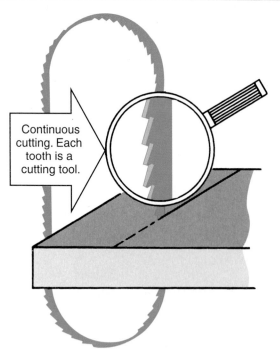

Figure 20-2. Band machining makes use of a continuous saw blade. Each tooth is a cutting tool.

20.2 BAND BLADE SELECTION

Some blade manufacturers list more than 500 different band saw blades. Points that must be considered by the machinist when selecting the correct blade for a specific job are the *blade type* and *blade characteristics*.

There are six basic band saw blade types. **Figure 20-4** lists them, along with their applications:

- Tungsten carbide.
- Bimetal (high-speed steel cutting edge with a flexible carbon steel back).
- High-speed steel.
- Shock resistant high-speed steel.
- Hard edge with spring-tempered back.
- Carbon steel flexible back.

Blade characteristics include *width*, *pitch*, *set*, *gage*, and *tooth form*.

- *Width* of the blade is important—the wider the blade, the greater its strength and the more accurately it will cut, **Figure 20-5**.

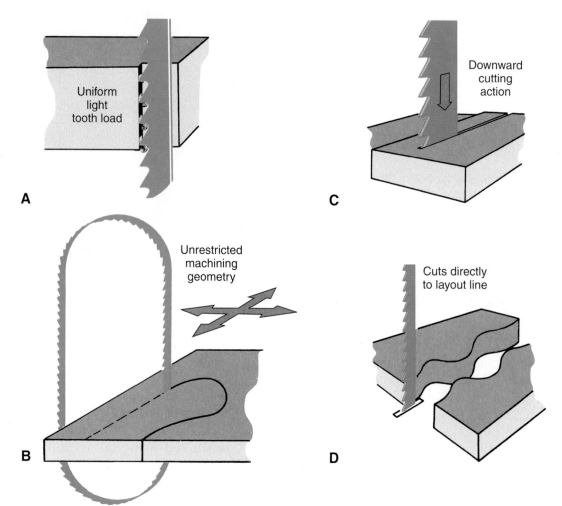

Figure 20-3. Band machining advantages. A—Wear is distributed over many cutting edges (teeth) with band machining. B—Band machining permits machining at any angle or direction. Cut length is almost unlimited. C—Band machining is very efficient and produces little waste. Unwanted material is removed in solid sections. D—Cutting action helps hold work on table.

Match the Tools to the Job		
Type of blade	**Applications**	**Band machine**
T/C Inserted tungsten carbide teeth on fatigue-resistant blade.	Heavy production and slabbing operations in tough materials.	Horizontal cutoff machines over 5 hp with positive feed. Vertical contour machines over 5 hp with positive feed.
Imperial bimetal HSS cutting edge with flex-resistant carbon-alloy back.	Mild to tough production and cutoff applications.	Horizontal cutoff machines over 1 1/2 hp with controlled feed, generally with variable-speed drives and with coolant system. Vertical contour machines over 1 1/2 hp with coolant system.
Demon M–2 HSS blade.	Heavy-duty toolroom and maintenance shop work. Full-time production applications.	Horizontal cutoff machines over 1 1/2 hp with controlled feed, generally with variable-speed drives and with coolant system. Vertical contour machines over 1 1/2 hp with coolant system.
Demon shock-resistant M–2 HSS blade specially processed for greater shock resistance.	Structurals, tubing, materials of varying cross section.	Horizontal cutoff machines over 1 1/2 hp with controlled feed, generally with variable-speed drives and with coolant system. Vertical contour machines with 1 1/2 hp with coolant system.
Dart Carbon-alloy, hard-edge, spring-tempered back blade.	Superior accuracy for light toolroom and maintenance shop applications as well as light manufacturing.	Horizontal cutoff machines under 1 1/2 hp with coolant system. Vertical contour machines under 1 1/2 hp with coolant system.
Standard carbon All-purpose, hard-edge, flexible back blade.	Light toolroom and maintenance shop applications.	Horizontal cutoff machines under 1 1/2 hp with weight feed and without coolant. Generally step speeds. Vertical contour machines under 1 1/2 hp without coolant.

Figure 20-4. *Recommendations for using basic blade types. (DoALL Co.)*

Figure 20-5. *Blade width is from tooth tip to other edge or back.*

Width of blade	Smallest radius	The width of the blade is determined by the smallest radius to be cut
1/16	1/16	
3/32	1/8	
1/8	7/32	
3/16	3/8	
1/4	5/8	
5/16	7/8	
3/8	1 1/4	
1/2	3	

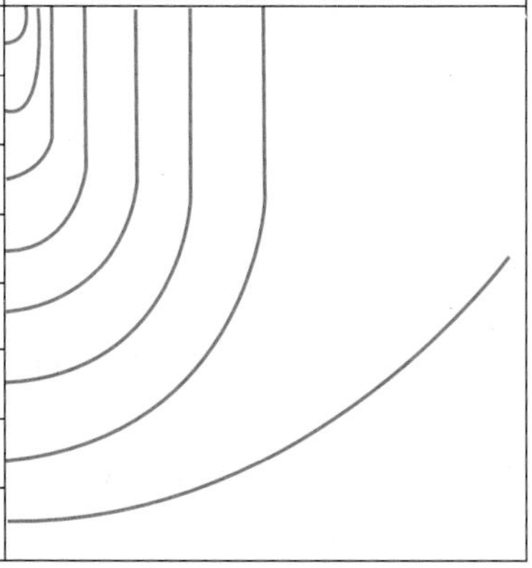

Figure 20-6. *How blade width affects smallest radius that can be cut.*

Use the *widest* blade the machine will accommodate when making *straight* cuts. *Contour* cutting should utilize the widest blade that will cut the required radius, **Figure 20-6**. Widths from 1/16″ to 2″ (1.5 mm to 50 mm) are available.

- *Pitch* refers to the number of teeth per inch or the distance between teeth measured in millimeters, **Figure 20-7**. The *thickness* of the material to be cut determines the proper pitch blade to use. At least *three* teeth should be in contact with the work for best performance. Blades in pitches from 2 to 32 teeth per inch are manufactured.
- *Set* provides clearance for the blade back, **Figure 20-8**. *Raker set* is recommended for cutting *large* solids or *thick* plate and bar stock. *Wavy set* should be used for work with *varying thicknesses*, such as pipe, tubing, and structural materials. *Straight set* is specified for *free cutting* materials, such as aluminum and magnesium.
- *Gage* refers to blade thickness, **Figure 20-9**. Extra strength can be obtained when using narrow blades by securing a blade of a heavier gage.
- *Tooth form* is the shape of the tooth, **Figure 20-10**.

Material thickness	Band pitch
Less than 1″ (25 mm)	10 or 14
1 to 3″ (25 to 75 mm)	6 or 8
3 to 6″ (75 to 150 mm)	4 to 6
6 to 12″ (150 to 300 mm)	2 or 3

Figure 20-7. Recommended band pitches to saw various thicknesses of material.

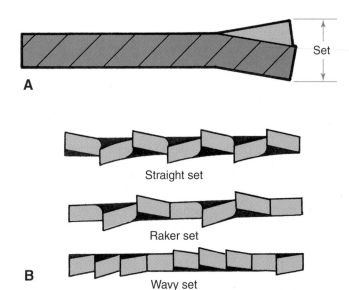

Figure 20-8. Blade set. A—The term "blade set" refers to side angle of the teeth. B—The different types of blade set.

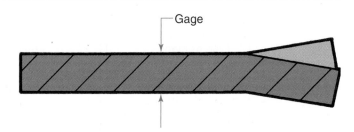

Figure 20-9. Blade thickness is referred to as gage.

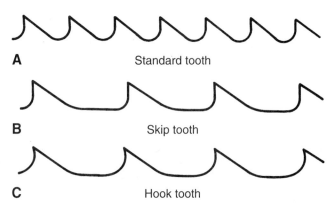

Figure 20-10. Tooth forms. A—Standard tooth blades, with their well-rounded gullets, are best for most ferrous metals, hard bronzes, and brasses. B—Skip tooth blades provide more gullet and better chip clearance without weakening blade body. This type blade is best-suited for aluminum, copper, magnesium, and soft brasses. C—The hook tooth blade offers two advantages over skip tooth blade: blade design makes it feed easier and its chip breaker design prevents it from gumming up.

There are three basic forms. Each has its specific application.

- *Standard tooth blades*, with well rounded gullets, are usually best for most ferrous metals, hard bronze, and brass.
- *Skip tooth blades* provide more gullet and better chip clearance without weakening the blade body. They are recommended for most aluminum, magnesium, and brass alloys.
- *Hook tooth blades* offer two advantages over the skip tooth blade. Blade design makes it feed easier and its chip breaker design prevents gumming up.

The parts of a blade are shown in **Figure 20-11**. Many band machines have built-in blade selection devices, **Figure 20-12**. With them, it is a simple matter to dial in the various bits of information necessary to determine the best blade for the job. Following this recommendation will result in the job being done faster and with a better finish.

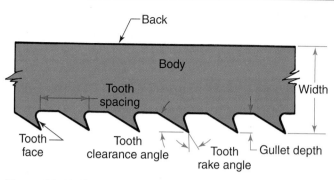

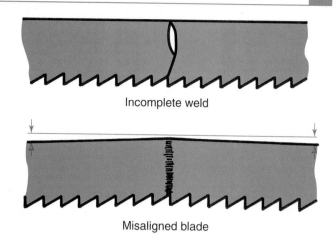

Figure 20-11. Saw blade terminology.

Incomplete weld

Misaligned blade

Figure 20-13. Common problems encountered when welding band saw blades. They are usually the result of poorly squared blade ends or dirty blade material.

Figure 20-12. Selector dials used to determine the best blade and cutting speed for a specific job. (DoALL Co.)

20.3.2 Making the blade weld

For convenience, most band machines have a welder built-in. This is a resistance-type butt welder, **Figure 20-14.** Blades up to 1/2″ (12.5 mm) wide can be welded in a light-duty welder. A heavy-duty "flash" butt welder is required for heavier blades. Refer to **Figure 20-15.** Clean the welder jaws before and after making a weld.

The blade ends are butted together and clamped in the jaws of the welder. The saw teeth should be placed against the aligning plates. Pressure (determined by width and thickness of blade being joined) is applied to the band ends. Check the blade to be sure the ends are touching across their entire width and are in the center of the gap between the welder jaws.

After checking, stand to one side to be clear of any "flash" that might result. Press the welder switch as far in as it will go; then release it immediately. The weld will be made automatically. The resulting weld should look like the one shown in **Figure 20-16.**

20.3 WELDING BLADES

Band saw blade stock can be purchased as ready-to-use *welded bands*. However, it is more economical to buy it in 100′, 300′, or 500′ (30 m, 90 m, or 150 m) *strip-out containers*. The desired length of blade material is withdrawn from the container, the ends squared, and the blade welded. Extreme care must be taken to make a good weld (one that is as strong as the blade).

When handling band saw blade stock or welded blades, always wear leather gloves and approved eye protection.

20.3.1 Preparing blade for welding

Use snips or blade cutoff shears to trim the blade to length. Cuts in blade stock must be square or the problems shown in **Figure 20-13** will result.

After squaring the blade ends, it will be necessary to remove several teeth (depending upon blade pitch) by grinding. Since about 1/4″ (6.5 mm) of the blade is consumed in the welding process, teeth must be ground off to assure uniform tooth spacing after the weld has been made. Remove only the teeth. Do *not* grind into the back of the blade.

Figure 20-14. Resistance butt welder is used for blades up to 1/2″ (12.5 mm) in width. (DoALL Co.)

Figure 20-15. *A heavy-duty "flash" welder in operation. It is evident how the welder got its name. Eye protection is a must!*

Figure 20-16. *Flash buildup at point of the weld should be uniform across blade. Do not flex blade at the weld until the welded section has been annealed. It is very brittle and will break if flexed. (DoALL Co.)*

Wear approved tinted glass-type eye protection when welding band saw blades. The bright flash can cause eye injury.

The weld, at this point, is brittle and must be *annealed* before it can be used. Follow the recommendations for annealing furnished by the manufacturer of the machine being used. Avoid overheating the blade, or it will remain brittle. Let it cool slowly after heating.

Remove the "flash" formed during welding on the grinder built into the welder. A finished weld should look like **Figure 20-17**. Use care when grinding to prevent dulling the teeth. The blade will also be weakened if it becomes "dished" during the grinding, **Figure 20-18**.

Figure 20-17. *A properly made and cleaned weld. If done properly, the weld will be as strong as blade itself.*

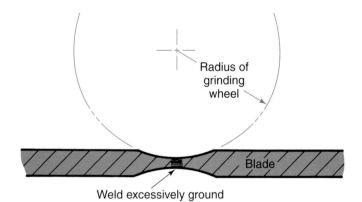

Figure 20-18. *A blade is seriously weakened if it becomes "dished" during grinding operation that removes weld flash.*

20.4 BAND MACHINE PREPARATION

As with all other machine tools, a band machine must be made ready with care if the tool is to operate at maximum efficiency.

20.4.1 Band machine lubrication

Use the grades of lubricants specified in the manufacturer's manual for the machine. Develop a definite lubrication sequence. It will reduce the possibility of missing a vital point.

20.4.2 Band blade guides

Select and install blade guides suitable for the job at hand, **Figure 20-19**. Use *blade guide inserts* for light sawing. *Roller guides* are recommended for continuous high-speed sawing.

Guides must be the proper width, **Figure 20-20**. If they are *too wide*, the saw teeth will be damaged. If they are *too narrow,* the blade will tend to twist in the work, making it difficult to follow the desired path or cut.

Before sawing, inspect and clean the upper and lower guide units. Make sure the backup bearings are not clogged with chips. Typically, there should be 0.001" to 0.002" (0.025 mm to 0.050 mm) clearance between the guide and blade, **Figure 20-21**. For best results, follow the manufacturer's recommendations.

20.4.3 Blade tracking

Adjust the band carrier wheels so the blade will track correctly. Again, it is important that the manufacturer's recommendations be carefully followed. If the manufacturer's information is not available, observing the following points will usually permit satisfactory operation:

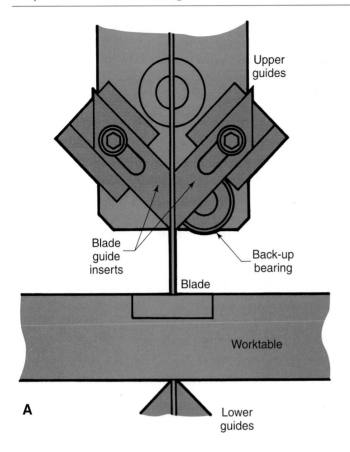

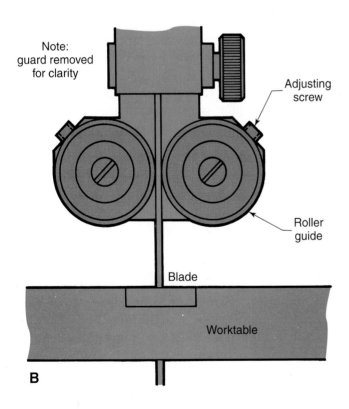

Figure 20-19. *Blade guides. A—Blade guide inserts are used for light sawing. B—Roller guides are recommended for continuous high-speed cutting.*

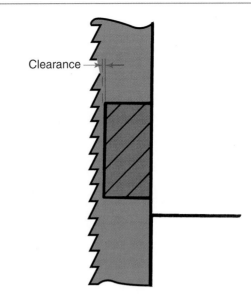

Figure 20-20. *Blade guides must be wide enough to prevent the blade from twisting, but narrow enough so they will not damage the teeth.*

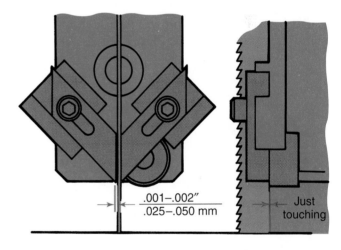

Figure 20-21. *How blade guides are adjusted to produce long blade life.*

- The center of the band should ride directly over the center of the wheel crown on the rubber tire, **Figure 20-22**. Replace the tire if it becomes frayed or damaged.
- There should be no noticeable gap between the back of the band and the back-up bearings of the saw guides.
- Remember that the blade must be installed with the teeth *downward*.

Observe extreme caution when handling saw blades. They are sharp and can cause serious injury. Be sure that power to the band machine has been turned off at the master switch before attempting to install a blade.

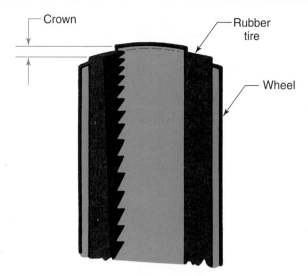

Figure 20-22. Blade should ride directly on center line of wheel crown.

20.4.4 Band blade tension

Blade tension refers to the pressure put on the saw band to keep it taut and tracking properly. On smaller machines, tension is usually applied by means of a hand crank, **Figure 20-23**. On large

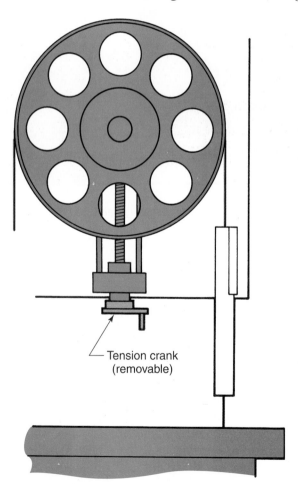

Figure 20-23. Blade tension is determined by the width and pitch of the blade. Use the band tension chart furnished with the machine.

heavy-duty machines, it is applied hydraulically. The amount of blade tension is determined by the *width* and *pitch* of the blade. Use the band tension chart furnished with the machine.

Many band machines have built-in tension meters that make it easy to adjust and maintain proper blade tension. This is especially important when a new blade is installed. A new blade has a tendency to stretch slightly when first used. This can create a safety problem if tension is not readjusted before it falls off too far.

20.4.5 Band cutting speed

As with other machining operations, best results will be obtained if recommended cutting speeds are maintained. See **Figure 20-24** for band speeds of a few selected materials.

20.4.6 Band cutting fluids

Cutting fluid can be applied on a band machine by flooding, in the form of a mist, or as a solid-type lubricant.

- *Flooding* is recommended for heavy-duty band machining.
- *Mist coolant* is used for high-speed sawing of free machining nonferrous metals. It is also employed when tough, hard-to-machine materials are cut.
- *Solid lubricants* are applied when the machine does not have a built-in coolant system.

Upon request, coolant manufacturers will furnish a coolant chart with recommendations for band machining operations.

20.5 BAND MACHINING OPERATIONS

The vertical band machine is designed to do the following sawing operations.

20.5.1 Straight sawing

Straight, two-dimensional sawing is band machining in its simplest form, **Figure 20-25**. The operator just follows a straight layout line. Slitting, shown in **Figure 20-26**, is another job that can be done rapidly on a band machine.

Exercise extreme care when the blade breaks through the work at the completion of a cut. If possible, a piece of backup metal should be between your hand and the point where the band will break through the work. The sharp moving blade can cause serious injury! See **Figure 20-27**.

Material	Thickness Inches (millimeters)	Band speed Surface feet per minute (meters per minute)
Low-to-medium carbon steels	Under 1″ (25 mm)	345–360 sfm (105–110) mpm
	1″–6″ (25 mm–150 mm)	295–345 sfm (90–105) mpm
Medium-to-high carbon steels	Under 1″ (25 mm)	225–250 sfm (70–75) mpm
	1″–6″ (25 mm–150 mm)	200–225 sfm (60–70) mpm
Free machining steels	Under 1″ (25 mm)	260–395 sfm (80–120) mpm
	1″–6″ (25 mm–150 mm)	260–345 sfm (80–105) mpm
Titanium, pure and alloys	Under 1″ (25 mm)	100–115 sfm (30–35) mpm
	1″–6″ (25 mm–150 mm)	90–110 sfm (30–35) mpm

Figure 20-24. Recommended cutting speeds for selected metals and alloys.

Figure 20-25. An example of two-dimensional band machining. Two pieces 16″ long, 6″ wide, and tapering from 3/4″ at one end to 1/4″ at the other, are cut quickly and accurately from a 16″ × 6″ × 1″ section of steel. (DoALL Co.)

Figure 20-26. Slitting operation. Note how this bearing is held in place for machining. (DoALL Co.)

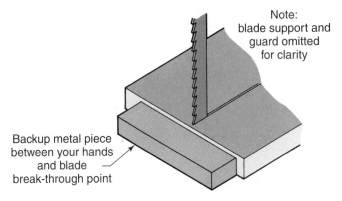

Note: blade support and guard omitted for clarity

Backup metal piece between your hands and blade break-through point

Figure 20-27. To protect your hands, be sure that a piece of backup metal is in place a the point where the saw blade breaks through.

20.5.2 Contour sawing

Contour sawing is possible on a vertical band machine, **Figure 20-28**. Machine size is the limiting factor on the work dimensions that can be cut.

20.5.3 Angular sawing

The table on the vertical band machine is usually mounted on trunnions, permitting it to be tilted. This makes it possible to machine compound angular cuts, **Figure 20-29**.

20.5.4 Internal cuts

Precision *internal cuts* can be made on a vertical band machine, **Figure 20-30**. The band is threaded through a hole drilled in the piece, welded, and the work maneuvered along the prescribed line. Additional holes must be drilled if sharp corners are specified, **Figure 20-31**. It may be necessary to use the blade as a file to get the work into cutting position.

After completing the internal cut, cut the band close to the weld so that the entire weld can be cut away prior to rewelding. It is recommended that there be no more than *one* weld in the band.

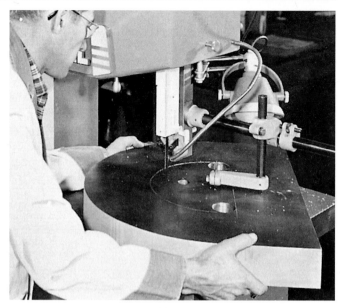

Figure 20-30. *Precision internal cuts can be made on band machine. The blade is threaded through holes drilled in the workpiece, then welded. Cutting is done by guiding the work along prescribed lines. (DoALL Co.)*

Figure 20-28. *Contour sawing. Power feed and handles on the work-holding cradle provide the machinist with excellent control of the work during the cutting operation. (DoALL Co.)*

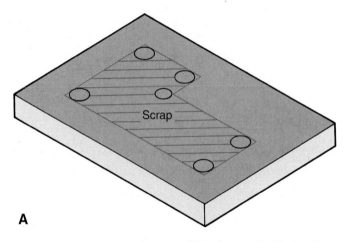

A

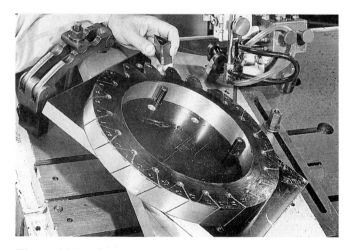

Figure 20-29. *Compound angular cuts, such as those around the edge of this workpiece, can be made by tilting the worktable. (DoALL Co.)*

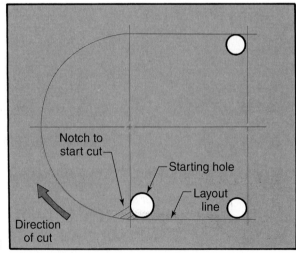

B

Figure 20-31. *Making internal cuts. A—Typical layout of drilled holes when sharp corners are specified on an internal cut. B— Start internal cuts by using blade as a file, and notch work until blade can be positioned to start cut.*

20.6 BAND MACHINE POWER FEED

Power feed or mechanical pressure attachments are available for band machines. The simplest attachment makes use of weights to pull work into the blade, **Figure 20-32**. Both hands of the operator are free to guide the work.

Several types of hydraulic power feed attachments have been devised. On some vertical band machines, the worktable is hydraulically actuated and feeds the work into the blade at a constant rate. Accidental overfeeding is eliminated, greatly extending band life.

A hand wheel, connected to the work by a sprocket and chain, guides work along the layout line, **Figure 20-33**. A servomechanism on the lower blade guide senses changes in feed force on the work and automatically counteracts them. To maintain a constant feeding force on the work, the device advances, slows, stops, or reverses the worktable movement. To leave the operator's second hand free, a foot switch permits moving the table by remote control.

Another type feed mechanism is a self-contained unit attached to the machine. A hydraulic cylinder applies and maintains constant pressure on the work through a sprocket and chain system. **Figure 20-34** shows the parts of this type of system.

20.7 OTHER BAND MACHINING APPLICATIONS

The great versatility of the band machine is further utilized by the addition of accessories and/or minor tool modifications.

20.7.1 Band filing

A smooth, uniformly finished surface may be obtained rapidly and with considerable accuracy on a band machine fitted for filing, **Figure 20-35**.

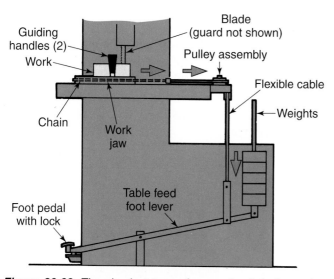

Figure 20-32. The simplest type of power feed device makes use of weights to pull the work into the blade.

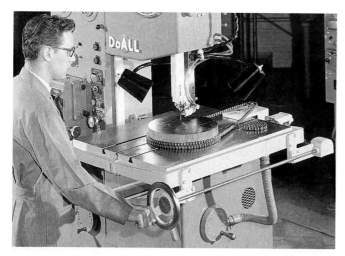

Figure 20-33. Band machine is fitted with a servo-contour feed attachment. Only one hand is needed to guide cutting operation. Table movement is controlled by a foot switch. (DoALL Co.)

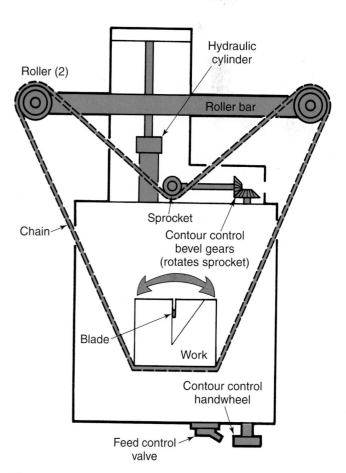

Figure 20-34. A hydraulically actuated power feed unit.

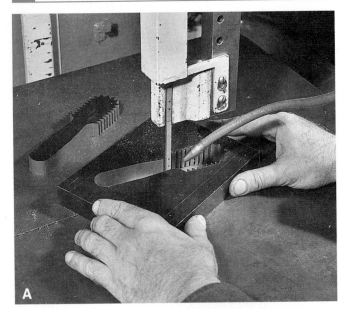

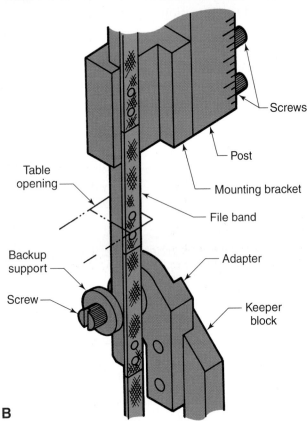

B

Figure 20-35. Band filing. A—Filing is being done on this band machine. (DoALL Co.) B—Guides are set up for band filing. Worktable has been removed for clarity.

A series of small file segments make up the *file band.* The individual units interlock and form a continuous file. The segments are fitted to a flexible back. File guides replace the regular saw guides when a band file is used. A variety of file shapes and cuts are available.

20.7.2 Band polishing

Parts can be polished on a band machine with a polishing attachment, **Figure 20-36**. A continuous *band abrasive cloth* replaces the saw blade. Best results can be obtained if the back of the abrasive band is lubricated with graphite powder. Abrasive band life will also be greatly extended.

20.7.3 Friction sawing

Friction sawing makes use of extremely high cutting speeds of between 6000 and 15,000 feet per minute (fpm) or 1800 to 4500 meters per minute (mpm) and heavy pressure to cut ferrous metals.

In friction sawing, the band operates at high speed, generating sufficient heat to soften the metal just ahead of the blade. The band removes the softened metal as it moves through the work. Only a small area on either side of the blade is affected by the heat.

The blade teeth do not actually cut — they are used to scoop out the softened metal. As a matter of fact, dull teeth are superior to sharp teeth, since they generate heat better. Friction sawing is a spectacular operation, as can be testified to by the shower of sparks produced.

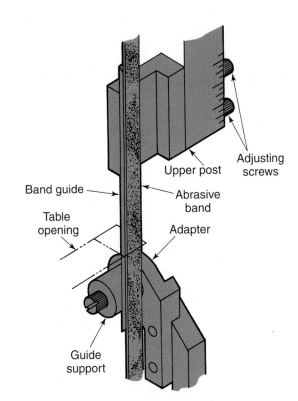

Figure 20-36. Polishing can be done on a band machine by replacing the saw blade with an abrasive band, and using the guide and support shown. Worktable has been removed for clarity.

The technique is a rapid way to cut ferrous metals under 1" (25 mm) in thickness. Thicker stock can be cut if a rocking movement is employed. Hardness is not a deterring factor.

Wheels on machines used for friction sawing are usually large in diameter and are carefully balanced for smooth vibration-free operation. Large wheels are necessary to reduce fatigue due to the high operating speed of the band.

Band machines designed for friction sawing cannot be used for conventional band machining unless they are equipped with variable speed drives. Most band machines are not adaptable for friction sawing.

Friction sawing requires a full face shield, leather gloves, and a transparent shield fitted around the cutting area.

20.7.4 Other band tools

Tooth-type bands are most commonly used on the vertical band machine. However, other types of blades have been developed for special work, **Figure 20-37.**

The *knife-edge blade* is employed to cut material that would tear or fray when machined by a conventional blade. For example, sponge rubber, cork, cloth, corrugated cardboard, and rubber would tear easily.

The *diamond-edge band* is specially designed to cut material that is difficult or impossible to cut with a conventional toothed blade. The diamonds are only on the front edge of the band where the cutting is accomplished. On a *wire band,* diamonds are fused around the circumference on the band permitting it to cut in any direction, **Figure 20-38.**

Friable materials (those that are easily crumbled or reduced to powder) of extreme hardness, or with abrasive qualities, can be cut economically with special bands.

In addition to the bands mentioned, blades with unusual characteristics are available to meet almost any band machining requirement.

20.7.5 Specialized vertical band machines

The vertical band machine is manufactured in a wide range of sizes, and has been adapted to do many kinds of band machining. Some large

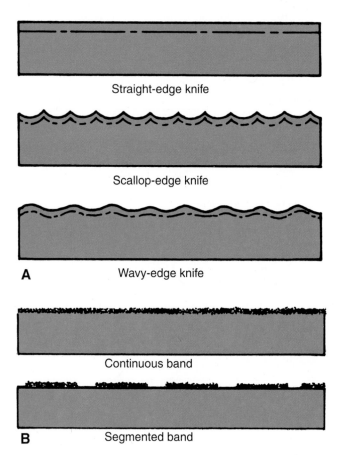

Figure 20-37. Other band tools. A—Types of knife edge blades. B— Cutting edge of diamond band is impregnated with diamond dust.

Figure 20-38. Diamond-impregnated abrasive wire is being used to cut an extrusion die opening. Note guides used to allow unidirectional cutting. Guides have been raised to show blade more clearly. (DoALL Co.)

machines have been fitted with closed circuit TV and a remote control console to permit the operator to contour-machine large sections of material or to perform hazardous or dangerous work. For example, this type of machine is employed to machine toxic and/or radioactive materials.

Band machines with computer numerical control (CNC) are now available. They have "X" and "Y" table movement capability plus circular interpolation (allowing them to cut circular and elliptical contours).

20.8 TROUBLESHOOTING BAND MACHINES

There are a number of problems that can ocur during band maching. Use the table shown in **Figure 20-39** to identify various problems and the methods that can be used to correct each of them.

Troubleshooting Band Machines	
Problem	**Correction**
1. Teeth dull prematurely.	a. Use slower cutting speed. b. Replace blade with a finer pitch band. c. Be sure proper type cutting fluid is used. d. Increase feed pressure. e. Check to be sure band is installed with teeth pointing down.
2. Band teeth breaking out.	a. Reduce feed pressure. b. Use finer pitch band if thin material is being cut. c. Be sure work is held solidly as it is fed into band. d. Use a heavier-duty cutting fluid.
3. Band breaks.	a. Change to a heavier band. b. Reduce cutting speed. c. Check wheels for damage. d. If blade breaks at weld, use longer annealing time. Reduce heat gradually. e. Use finer pitch balde. f. Reduce feed pressure. g. Decrease band tension. h. Check blade guides for proper adjustment. i. Use cutting fluid.
4. Cutting rate too slow.	a. Increase band speed. b. Use coarser pitch blade. c. Increase feed presssure. d. Use cutting fluid.
5. Band makes "belly-shaped" cut.	a. Increase blade tension. b. Adjust guides close to work. c. Use coarser pitch band. d. Increase feed pressure.
6. Band does not run true against saw guide backup bearing.	a. Remove burr on back of band where joined. b. If hunting back and forth against backup bearing on guide, reweld blade with back of band in true alignment. c. Check alignment of wheels. d. Check backup bearing. Replace if worn.
7. Premature loss of set.	a. Use narrower band. b. Reduce cutting speed. c. Apply cutting fluid.

Figure 20-39. Troubleshooting band machines.

20.9 BAND MACHINING SAFETY

- Do not attempt to operate a band machine until you have received instructions on its safe operation!
- Never attempt to operate the machine while your senses are impaired by medication or other substances. Be sure all guards are in place before starting to operate a band machine.
- Do not start a cut until the guides have been set properly! Guides should be positioned as close to the work as the job will permit.
- Other than changing blade speeds (some machines require band to be running when speed changes are made), make no adjustments until the blade has come to a complete stop!
- Wear eye protection and leather gloves when handling band blades or blade material.
- Get help when handling heavy material.
- Remove burrs and sharp edges from the work as soon as possible. They can cause serious cuts. Have cuts and bruises treated immediately. Report all injuries!
- Do not clean chips from the machine with your hands. Use a brush. Stop the machine before attempting to clean it!
- Keep your hands away from the moving blade! Use a push stick or piece of metal for additional safety. Never have your hands in line with the cutting edge of the band.

TEST YOUR KNOWLEDGE

Please do not write in the text. Write your answers on a separate sheet of paper.

1. Band machining makes use of a _____ blade.

2. The cutting tool must be installed with the teeth facing _____.

3. List three advantages band machining has over other machining techniques.

4. What two points must be considered when selecting a blade for a specific job?

5. Blade pitch refers to:
   a. Width of the blade in inches or millimeters.
   b. Thickness of the blade in inches or millimeters.
   c. Number of teeth per inch of blade or tooth spacing in millimeters.
   d. All of the above.
   e. None of the above.

6. Tooth form is the:
   a. Shape of the tooth.
   b. Thickness of the blade.
   c. Number of teeth on the blade.
   d. All of the above.
   e. None of the above.

7. When making straight cuts, use the _____ blade the machine can accommodate.

8. The joint of a properly welded blade should be as _____ as the blade itself.

9. The blade must be _____ after welding because the joint is extremely _____ and cannot be used in this condition.

10. A _____ welder is used to weld band machine blades.

11. Blade tension is the pressure put on the saw band to:
    a. Cut the metal more rapidly.
    b. Keep it taut and tracking properly.
    c. Reduce the power needed to do the cutting.
    d. All of the above.
    e. None of the above.

12. Cutting fluids are applied on the band machine in the form of:
    a. _____ is recommended for heavy-duty sawing.
    b. _____ is used for high-speed sawing of free machining nonferrous metals.
    c. _____ are applied when the machine is not fitted with a built-in coolant system.

13. What is the simplest form of band machining?

14. The worktable on many vertical band machines can be tilted to make _____ cuts.

15. How can internal cuts be made on a band machine?

16. Smooth, uniformly finished surfaces are possible when the machine is fitted for _____.

17. Describe friction sawing.

18. Of what use is a knife-edge blade on a band machine?

19. When are diamond-edge bands used?

20. What is unique about a diamond impregnated wire band?

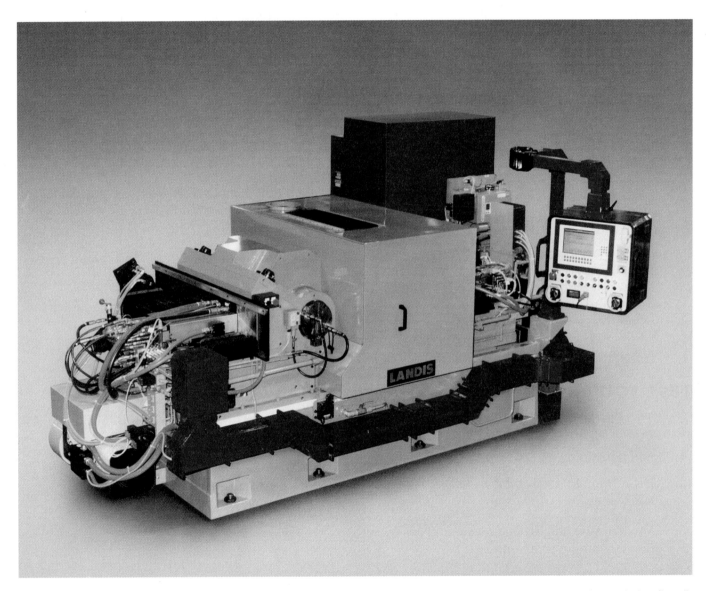

Computer numerical control is used to produce automotive crankshafts accurately and efficiently on this centerless grinder. (Landis Div. of Western Atlas)

Chapter 21

Computer Numerical Control

series of coded instructions, **Figure 21-2.** The code consists of *alphanumeric data* (numbers, letters of the alphabet, and other symbols) that are translated

LEARNING OBJECTIVES

After studying this chapter, you will be able to:
- ❍ Define the term "numerical control."
- ❍ Describe the difference between the incremental and absolute positioning methods.
- ❍ Explain the operation of NC (numerical control), CNC (computer numerical control), and DNC (direct or distributed numerical control) systems.
- ❍ Point out how manual and computer-aided programming is done.

IMPORTANT TERMS

absolute positioning
Cartesian Coordinate
   System
circular interpolation
closed loop system
continuous path system

incremental positioning
machine control unit
   (MCU)
open loop system
point-to-point system
straight-cut system

21.1 COMPUTER-AIDED MACHINING TECHNOLOGY

Note: *Because of the complexity of this area of machining technology, only a basic description of it can be included in a textbook of this nature.*

In *manual machining*, where cutting operations are performed on conventional machine tools, the machinist first studies the print. After determining the machining sequence, the cutter is installed and the work mounted and positioned on the machine.

Machining is done by moving one or more of the machine's lead and feed screws. After selecting the cutting speed and feed rate, the cutter is fed into the material and manually guided through the various machining operations that will produce the part specified on the print, **Figure 21-1.**

Numerical control (NC) is not a machining *process*, but the *operation* of the machine tool by a

Figure 21-1. *This machinist is manually operating a vertical milling machine. (U.S. Army)*

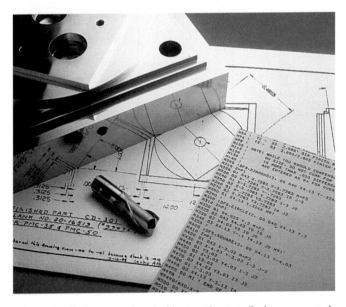

Figure 21-2. *A series of coded instructions, called a program, is used to control machine tool movements and operations in numerical control (NC) machining. (Giddings & Lewis, Inc.)*

into pulses of electric current. These pulses of current activate motors and other devices to run the machine through the specified machining cycles.

The motors, called *servos*, are connected to the machine's lead and feed screws. They provide the power to position the work and feed the work into the cutter. The coded instructions may be punched into paper or plastic tape, encoded on magnetic tape or a floppy disk, or sent directly from a computer. These instructions control the electronic impulses that tell the servos when to start, in what direction to move, and how far to move. Feed rate, cutter speed, coolant flow, and (on some machines) tool changes are controlled by the same set of instructions or *program*. See **Figure 21-3**.

Numerical control (NC) has evolved into *CNC* (*computer numerical control*) and *DNC* (an abbreviation used for both direct numerical control and distributed numerical control).

The term "CNC," is usually used to describe a self-contained NC system for a single machine tool utilizing a computer controlled by stored information (*part program*) to perform basic NC functions. See **Figure 21-4**.

Figure 21-4. *This CNC machine tool includes a computer that, through the part program, controls all of machining functions. The computer console at right is used to monitor operation and to enter and modify part programs. (Sharnoa Corp.)*

The NC program can be entered by direct electronic transfer from a *CAD* (*computer-aided drafting*) drawing or manually written and entered into the machine tool's on-board computer.

DNC is an abbreviation used to identify two different numerical control methods: *direct* numerical control and *distributed* numerical control. In both cases, the system consists of a group of CNC machine tools (and sometimes other devices, such as robots and inspection stations) connected to a central or *mainframe* computer that has substantial memory for storage of many parts programs. See **Figure 21-5**. The machine tools can be located in widely dispersed locations. DNC systems do not require the use of punched tapes or disks.

Direct numerical control systems connect the central computer directly to the CNC machine tools. NC data is fed to each machine tool in the system segment by segment to meet the machine's peak rate demands (for example, when machining an undulating profile). The machine tool's *MCU* (*machine control unit*) is constantly on-line with the central or host computer.

Distributed numerical control is typically used in large manufacturing situations. A number of intermediate computers are located between the central computer and the machine tools. Each of these intermediate computers distributes the necessary machining instructions to a separate group of CNC machine tools. This control method provides greater flexibility than direct numerical control and permits more rapid response to changing conditions.

Figure 21-3. *The program also controls the starting and stopping of coolant flow and may be used to perform automatic tool changes. (Giddings & Lewis, Inc.)*

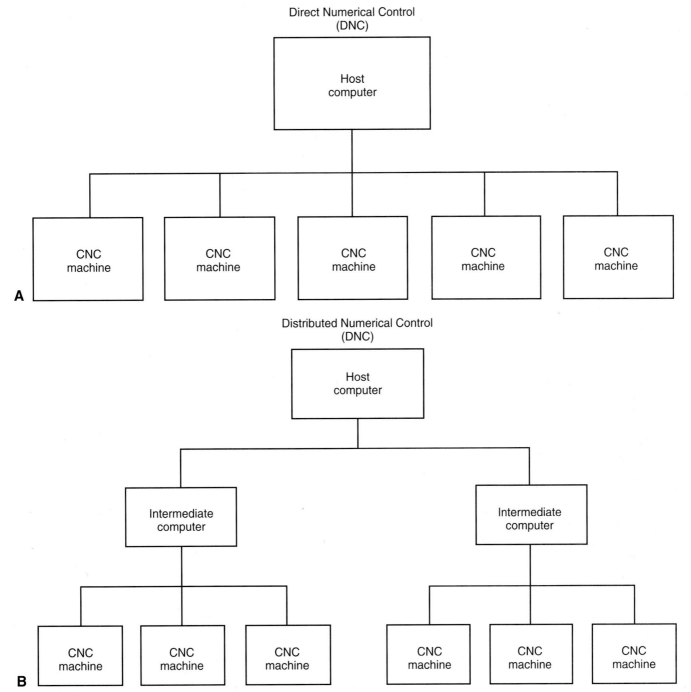

Figure 21-5. The term "DNC" has two different meanings. A—Diagram of a Direct Numerical Control (DNC) system. B—Diagram of a Distributed Numerical Control (DNC) system.

21.2 POSITIONING WITH NUMERICAL CONTROL

The Massachusetts Institute of Technology (MIT) is given credit for coining the term *numerical control* in the early 1950s. Numerical control permits the precise positioning of a tool's cutting point in space through the use of numbers to express distance and direction of movement. Movement may be up/down, side-to-side, or back-to-front (and vice versa). Depending upon the type of machine being used, the movement may involve the tool, workpiece, or both.

More sophisticated NC machines are fitted with a **closed loop system** to control the positioning of the tool or work. This system uses an electronic feedback device **(transducer)** to continually monitor tool position. It "tells" the servo or servos to stop when the instructed distance (desired tool position) has been reached. The feedback unit also serves as a check on the accuracy of the desired tool movement.

Other NC machine tools are fitted with an *open loop system* to control positioning. It has no feedback for monitoring or comparing purposes. The system relies on the integrity of the control unit itself for positioning accuracy.

21.2.1 Cartesian Coordinate System

Since the NC program must instruct the servos on the direction and distance the work and/or tool must move, there must be some way to define this movement.

The *Cartesian Coordinate System,* shown in **Figure 21-6,** is the basis of NC programming. Programs, written in either inch or metric units, specify the destination of a particular movement. With the destination established, the *axis of movement* (X, Y, or Z) and the direction of movement (+ or –) can be identified. To determine whether the movement is positive (+) or negative (–), the program is written as though the tool, rather than the work, is doing the moving.

Spindle motion is assigned the Z axis. This means that for a drill press or vertical milling machine, the Z axis is vertical. For machines such as a lathe or horizontal milling machine, however, the Z axis is horizontal. See **Figure 21-7.** The system of coordinates used for machine axis designation is specified according to the *right-hand rule* of Cartesian coordinates, **Figure 21-8.**

Most NC machine tools manufactured today are of the CNC (computer numerical control) type and are equipped with an on-board computer control

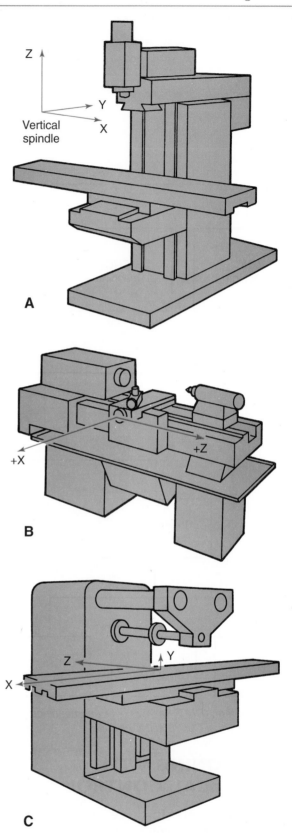

A

B

C

Figure 21-7. *Axes of machine tool movements. A—Vertical milling machine. B—Lathe. C—Horizontal milling machine. Spindle motion is assigned Z axis. Note how Z axis differs between machines with vertical spindle and machines with horizontal spindle.*

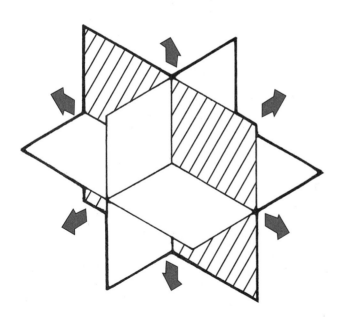

Figure 21-6. *The Cartesian Coordinate System is the basis of all NC programming.*

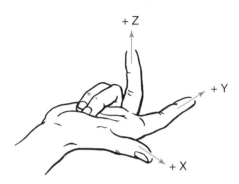

Figure 21-8. *Machine tool axes are specified according to right-hand system of Cartesian coordinates. When right hand is held as shown, the thumb, forefinger, and next finger point in positive (+) directions of X, Y, and Z axes. (IBM)*

system, **Figure 21-9**. Machining instructions can be programmed directly into computer memory and do not require input via punched or magnetic tapes. However, since many tape-controlled machine tools are still in use, tape information will be included in this chapter.

21.2.2 NC Tool Positioning Methods

NC tool positioning may be *incremental* or *absolute*. With **incremental positioning**, each tool movement is made with reference to the prior (last)

Figure 21-9. *CNC machine tools are equipped with onboard computers that permit computer-aided or manual programming. A CRT screen displays all important machining information, such as the tool path shown in yellow. (Giddings & Lewis, Inc.)*

tool position, **Figure 21-10**. *Absolute positioning* measures all tool movement from a fixed point of origin, or *zero point*. See **Figure 21-11**.

Whenever possible, the *absolute positioning* method should be employed, because a mistake in dimensioning an individual point does not affect remaining dimensions. It is also easier to check for errors.

A *machine control unit (MCU)* is a microcomputer or electronic circuit for controlling the machine. Most will accept program information in either the incremental or absolute mode. It is also

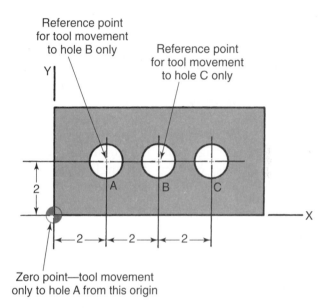

Figure 21-10. *Incremental positioning system. Each set of coordinates has its point of origin from last point established.*

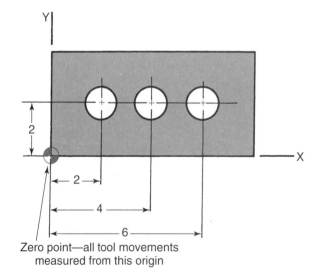

Figure 21-11. *Absolute positioning system. In this system, all coordinates are measured from fixed point (zero point) of origin.*

possible to switch back and forth from one mode to the other in the same program. Either inch or metric dimensions may be used. The choice can be made by a switch or by a specific code in the part program.

21.3 NC MOVEMENT SYSTEMS

There are three basic NC movement systems: point-to-point, straight-cut, and contour or continuous path, **Figure 21-12**.

21.3.1 Point-to-Point System

The *point-to-point system* is typically used for drilling, punching, spot welding, or any operation performed at a fixed location in terms of a two-axis coordinate position. While moving from one work position to the next, the cutting tool is not in contact with the work, so a rapid traverse feed can be used. Tool movement from one point to the next does not have to follow a specific path, **Figure 21-13**.

Some point-to-point NC machines have the capability of straight-line milling along either the X-axis or Y-axis, or at a 45° angle between two points, with no change in depth along the line that the tool travels.

Point-to-point systems generally do not require computers for program preparation. Anyone with a basic knowledge of machining practices and the ability to read engineering drawings can prepare a simple program.

21.3.2 Straight-cut System

A *straight-cut system* permits controlled tool travel along one axis at a time. Length of cut and

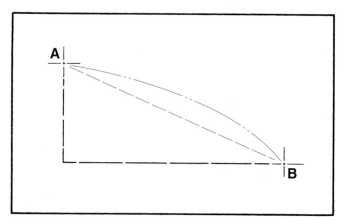

Figure 21-13. In point-to-point NC systems, there is no concern what path is taken when tool moves from Point A to Point B, except when performing straight-line milling.

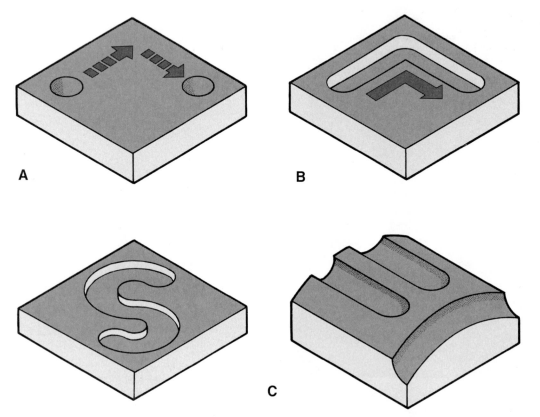

Figure 21-12. The basic NC systems. A—Point-to-point. B—Straight cut. C—Contour or continuous path (in either two or three dimensions).

feed rate are determined by the program. Straight-cut systems are used on simple milling machines and lathes, **Figure 21-14**.

21.3.3 Contour (Continuous Path) System

A *contour* or *continuous path system* precisely controls machine and tool movement at all times, in all planes, as the cutter moves along the programmed path, **Figure 21-15**. Cutting is continuous and can be on six axes simultaneously, **Figure 21-16**.

Cutter location is monitored continuously through a feedback mechanism to the machine control unit (MCU). This is necessary to maintain the correct direction of cutter movement and proper cutting speed and feed. Cutter size and other variables must be considered when programming, **Figures 21-17** and **21-18**.

Geometrical complexity makes a computer mandatory when preparing programs to machine two- and three-dimensional shapes, **Figure 21-19**. This is especially necessary for making circular, elliptical, or similar complex cuts, **Figure 21-20**, since the cutting tool must be fed a constantly changing series of instructions.

In a way, continuous path machining might be described as a very sophisticated point-to-point system, because tool movement is only a series of straight lines. However, the straight line movements are so small (0.0001" or 0.003 mm), and blend so well, they appear to be a continuous smooth curved cut, **Figure 21-21**.

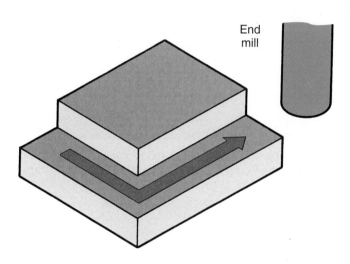

Figure 21-14. *Straight-cut NC systems are used on simple milling machines and lathes.*

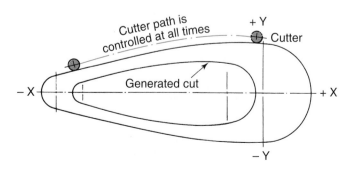

Figure 21-15. *With contour or continuous path system, cutter path is controlled at all times. By adding a Z axis, three-dimensional machining is possible.*

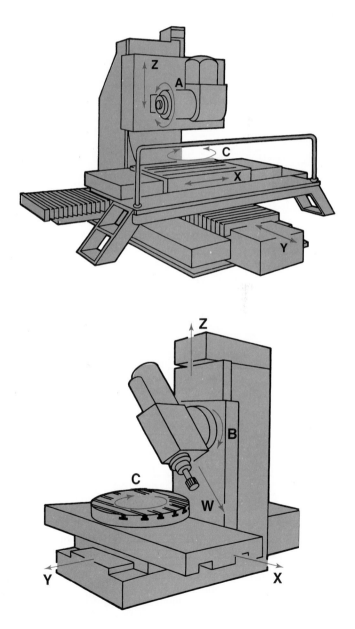

Figure 21-16. *Multiaxis machining. Top—5-axis NC milling machine. Bottom—6-axis machining center.*

To reduce the number of calculations required to program a curved surface, most NC machines have a MCU encoded for *circular interpolation*, **Figure 21-22**. Information needed for the program includes:

- The coordinate locations of the end points of the arc component.
- The radius of the circle.
- The coordinate location of the circle.
- The direction the cutter is to proceed.

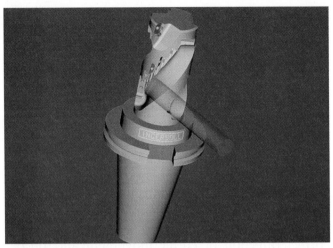

Figure 21-19. *Complex part geometry makes a computer vital to preparing a program to machine two- and three-dimensional shapes. (CG Tech)*

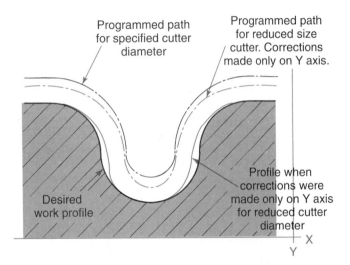

Figure 21-17. *In a contour program, adjustments must be made if a cutter is smaller or larger than one specified; otherwise, contour will deviate from required part outline.*

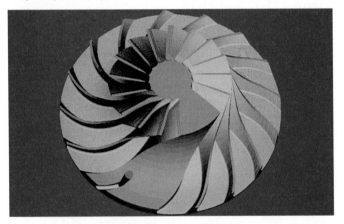

Figure 21-20. *The program required to machine this impeller is far too complex to be calculated manually. (CG-Tech)*

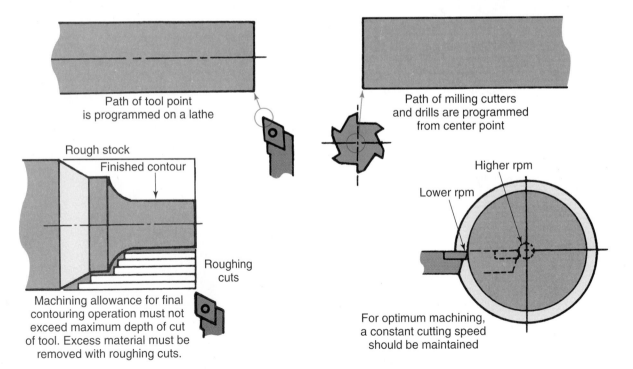

Figure 21-18. *Variables that must be considered when designing an NC program.*

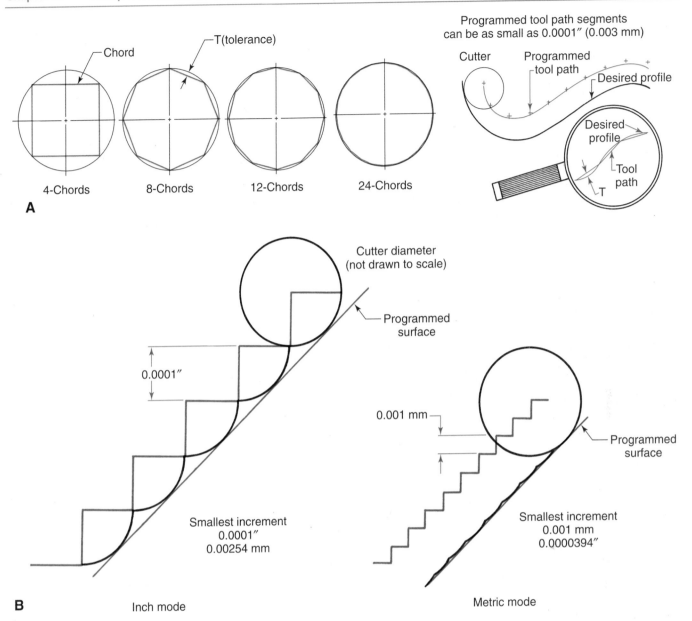

Figure 21-21. *A—Contours obtained from contour or continuous path machining are result of a series of straight-line movements. The degree to which a contour corresponds with specified curve depends upon how many movements or chords are used. Note how, as number of chords increase, the closer the contour is to a perfect circle. The actual number of lines or points needed is determined by the tolerance allowed between design of the curved surface and one actually machined. B—This exaggerated illustration shows why metric machine movement increments are often preferred when contour machining. The benefit has to do with the least input increment allowed in the metric mode. In the inch mode, the least input increment is 0.0001", which means you can input program coordinates and tool offsets down to 0.0001". In the metric mode, the least input increment is 0.001 mm, which is less than one-half the least input increment when using the inch mode. The coordinates going into the program will then will be much closer to what is desired for accurately machined parts.*

The circular interpolator in the MCU will automatically compute the necessary number of intermediate points to describe the circular cut. It also generates the electronic signals that will run the servos and guide the cutting tool in making the cut.

Using the continuous path system, **Figure 21-23**, very complex parts can be produced economically in small numbers, without the need for expensive jigs, fixtures, or templates. On most NC machine tools, the reversal of plus and minus values along an axis will allow the machining of a *mirror image part*. That is, a left-handed part can be machined using a right-handed part program, or vice versa. See **Figure 21-24**. This technique is known as *axis inversion*.

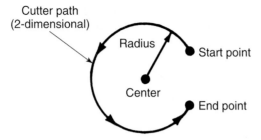

Cutter path
(2-dimensional)

Radius • Start point

Center

End point

Figure 21-22. On NC units encoded with circular interpolation, only the coordinate location of the start and end points of the arc, radius of the circle, coordinate location of the circle's center, and the direction of cut need to be programmed. The intermediate points required to describe circular cut are then computed by the MCU's circular interpolator. On some machines, the circle is divided into four 90° segments and must be programmed accordingly.

Figure 21-23. Complex parts, like this impeller, are produced quickly and economically by a continuous path NC system. (Courtesy of SURFCAM by Surfware)

21.4 PROGRAMMING NC MACHINES

A *machining program* is a sequence of instructions that tells the machine what operations to perform, and where on the material they are to be done. A program is also referred to as *software*.

Each line of a program is called a *block*. It consists of a sequence number, preparatory codes for setting up the machine, coordinates for the destination of the tool, and additional commands (feed rate, coolant on, coolant off, etc.). Some typical codes are shown in **Figure 21-25**.

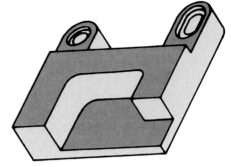

Right-hand unit

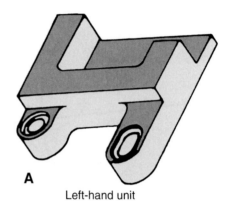

A

Left-hand unit

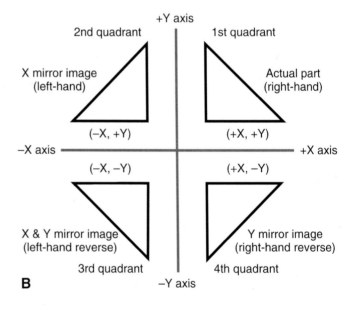

B

Figure 21-24. Mirror image machining. A—On many NC machine tools, mirror image parts can be machined using the same program. B— Mirror image to produce a reverse duplicate part. The technique is known as axis inversion.

Preparatory Functions (G-codes)*	
Code	Function
G00	Rapid traverse (slides move only at rapid traverse speed).
G01	Linear interpolation (slides move at right angles and/or at programmed angles).
G02	Circular interpolation CW (tool follows a quarter part of circumference in a clockwise direction.)
G03	Circular interpolation CCW (tool follows a quarter part of circumference in a counterclockwise direction).
G04	Dwell (timed delay of established duration. Length is expressed in X or F word).
G33	Thread cutting.
G70	Inch programming.
G71	Metric programming.
G81	Drill.
G90	Absolute coordinates.
G91	Incremental coordinates.

*G-codes may vary on different N/C machines.

Miscellaneous Functions (M-codes)*	
Code	Function
M00	Stop machine until operator restart.
M02	End of program.
M03	Start spindle–CW.
M04	Start spindle–CCW.
M05	Stop spindle.
M06	Tool change.
M07	Coolant on.
M09	Coolant off.
M30	End program and rewind tape.
M52	Advance spindle.
M53	Retract spindle.
M56	Tool inhibit.

*M-codes may vary on different N/C machines.

Figure 21-25. Examples of NC programming codes for preparatory functions and miscellaneous functions.

NC equipment can be programmed manually (drilling, straight-line cutting, spot welding, etc.) or with the aid of a computer (complex two- and three-dimensional shapes).

21.4.1 Manual Programming

As noted earlier, *manual programming* may be done if the part is not too complex. It can be accomplished by anyone who can interpret engineering drawings and has a working knowledge of machine tool operations, **Figure 21-26**.

A program is developed by converting each machining sequence and machine function into a coded block of information that the MCU can understand, **Figure 21-27**. The code consists of alphanumeric data (letters, numbers, punctuation marks, and special characters). Each code identifies a different machine function.

Each block of information is a line on the *program sheet* or *program script* and is identified by a sequence number, **Figure 21-28**. Included in this information are the coordinate dimensions (X, Y, and Z movement) or the location where the operation (drilling, punching, spot welding, etc.) is to take place, along with miscellaneous functions (spindle on, spindle off, tool change, etc.).

Blocks of information must be separated with *end of block (EOB) codes*. An *end of program code* completes the program. If punched tape is used, a final EOB rewinds the tape. It will then be ready to repeat the machining cycle.

The coded information may be recorded on magnetic tape, or punched into a paper or Mylar® tape on a *tape punch/reader unit*. See **Figure 21-29**. Today, however, it is more common for program information to be entered directly into the machine's onboard computer as electronic data. The input information is proofed for omissions and/or errors and corrected, if necessary, before being released to the production area.

21.4.2 Computer-aided Programming

Computer-aided programming reduces and simplifies the numerical calculations that the programmer must perform when programming the machining of more complex parts, **Figure 21-30**.

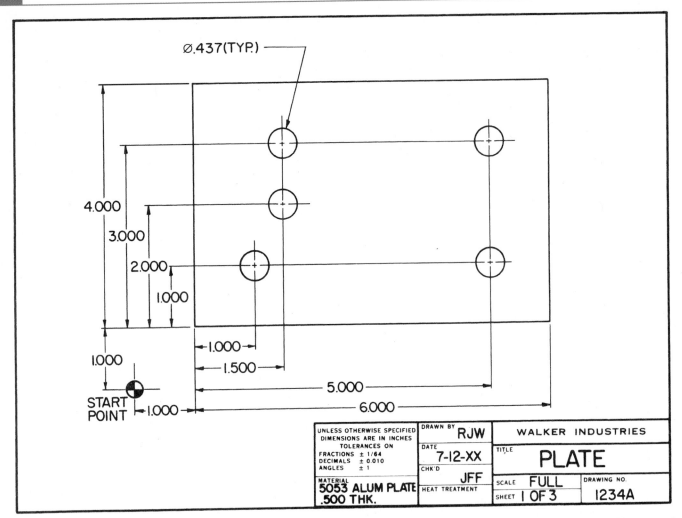

Figure 21-26. Point-to-point programming can be done manually by anyone who can interpret engineering drawings and has a working knowledge of machine tool operations.

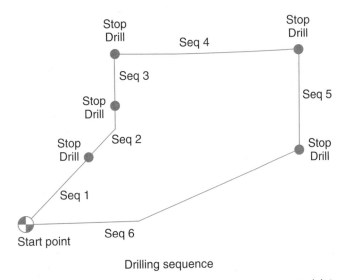

Drilling sequence

Figure 21-27. Machining operations must be converted into individual coded blocks of information that the MCU can understand. This is the planned drilling sequence for the part shown in Figure 21-26.

PART NO. 1234A	MACHINE 2-AXIS DRILLING WITH FIXED FEED RATE	REMARKS SET POINT (0,0) M56 IS TOOL INHIBIT
PART NAME PLATE		
TAPE NO. 1234A	TOOLING 7/16 DIA. DRILL	
DATE 7-12-XX	PAGE 1 OF 1	PROCESSOR JFF / APPROVED LJ

N	G	X	Y	Z	F	EOB	M	INSTRUCTIONS
000	90					EOB		7/16 DIA. DRILL
00						EOB		SET DEPTH STOP
0		0	0			EOB	03	
1		2000	2000			EOB	07	
2		2500	3000			EOB		
3		2500	4000			EOB		
4		6000	4000			EOB		
5		6000	2000			EOB		
6		0	0			EOB	0956	
7						EOB	05	
8						EOB	30	CHANGE PART

Figure 21-28. Information given on print shown in Figure 21-26 was developed into this program.

Machine tools manufactured today have control units of the CNC type. That is, the machines have a self-contained NC system with an onboard computer dedicated exclusively to that particular machine, **Figure 21-31**. They can also be connected to, and controlled by, a more powerful central host computer. A *host computer* is a mainframe computer that can simultaneously monitor and control a number of CNC machines.

Programming, in one of the computer languages that describe the geometry of the part to be made, can be entered into the system in a number of ways. The display screen (cathode ray tube or CRT) can show the operational data as the part is being

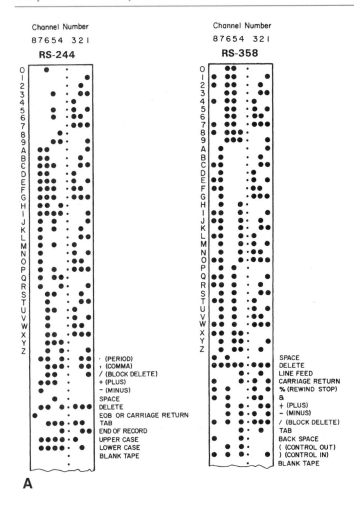

A

Figure 21-30. Computer-aided programming is common today, especially when machining complex parts. The computer simplifies the calculations a programmer must perform in such cases. (Giddings & Lewis, Inc.)

B

Figure 21-29. Punched tape. A—Tape code. Note that every level of RS-244 tape has an odd number of punches for parity check. RS-358 tape has an even number. Parity check is a method of automatically checking to reduce possibility of tape errors caused by malfunctioning tape punch. Both tapes are in current use. B—Tape punch/reader unit is used to transfer program information to tape. It can also be used to read tape prepared elsewhere. (Facit)

Figure 21-31. Most machine tools manufactured today have control units of the CNC type. Shown is a lathe capable of high-volume production. (Mazak Corp.)

machined, **Figure 21-32**. Graphic capability may include a picture of the part and tools, and a simulation of the cutter path, often in vivid colors. See **Figure 21-33**.

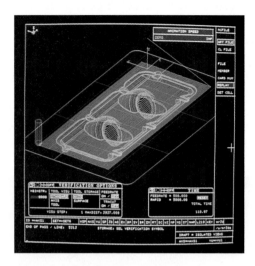

Figure 21-32. The computer's display screen (a cathode ray tube, or CRT) can show full operational data as part is being machined. (Cincinnati-Milacron)

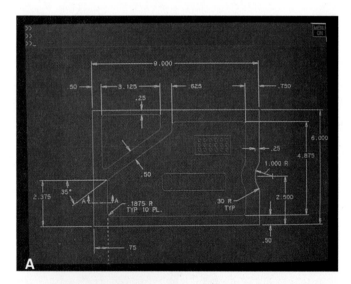

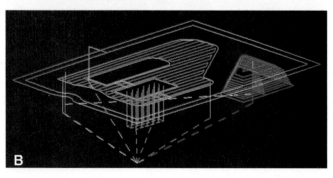

Figure 21-33. Computer graphics display. A—Top view of job with dimensions. B—An isometric view of tool paths relative to part design. (Computervision)

Computer memory capacity on CNC machines is often given an equivalent tape capacity in feet and/or meters. After use, the program can be moved to another larger computer, or stored on magnetic tape or disk.

21.5 COMPUTER LANGUAGES

There are many different computer-aided programming languages, such as APT (*Automatic Programmed Tools*), ADAPT (*Adaptation of APT*), APT III (provides for five axes of machine tool motion), COMPACT II, and AUTO MAP (*Automatic Machining Programming*).

Some are very powerful and allow programming for such tasks as milling complex curved three-dimensional surfaces that require five and sometimes six axes of motion by machine tools, **Figure 21-34**. Other programming languages are designed to be used by a specific machine or for a particular machining application.

Programming languages consist of a vocabulary of words, numbers, and other symbols. When combined according to certain rules, called *syntax*, the languages are capable of producing sets of instructions for machining parts.

APT is the most powerful of the computer-aided programming languages in common usage. It is a system that defines, in a series of statements, part geometry, cutter operations, and machine tool capabilities.

Figure 21-34. A controllable pitch propeller blade for a ship is being machined on a 5-axis milling machine. Reflections from overhead lights show how the geometry of machined surface changes from hub to tip. The blade is about 8′ or 2.5 m long. (Bird-Johnson Company)

An APT word used in a programming statement must be spelled exactly as it is spelled in the **APT system dictionary**. The dictionary specifies the only spelling the computer recognizes. For example:

- CØØLNT = Coolant. CØØLNT/ØN means turn the coolant on. It will continue on until getting a CØØLNT/ØFF or STØP command.
- FEDRAT = Feed rate. FEDRAT/6, IPM means feed rate in all directions will be 6″ per minute.
- SPINDL = Spindle. SPINDL/1000 RPM, CLW means the spindle is to rotate 1000 revolutions per minute in a clockwise direction. Spindle stays on until SPINDL/ØFF or STØP command.
- GØRGT = Tool movement is to the right. See **Figure 21-35**.
- GØLFT = Tool movement is to the left.
- FINI = Part program is completed.

APT programming, as with most computer assisted programming languages, is done in three parts:

- *Geometry:* part features are defined in geometric figures.
- *Machining statements:* used to direct the cutter around the geometry of the part in a predetermined sequence.
- *Auxiliary function statements:* include tool changes, cutting speeds, feed rate, coolant on, coolant off, etc.

21.6 ADAPTIVE CONTROL

Normally, programmed cutting speeds and feeds are on the conservative side. The programmer must include a safety factor to account for variables such as thickness of cut, tool wear, and other elements that can occur during a machining operation.

CNC machines are often fitted with **adaptive control (AC)**. A machine control for which fixed cutting speeds and feeds are not programmed, AC automatically optimizes speeds and feeds by judging machining conditions through sensors. See **Figure 21-36**. These sensors monitor such factors as cutting torque, heat, tool deflection, vibration, cutter wear, adherence to programmed work dimensions, and quality of surface finish, **Figure 21-37**.

Many types of AC are available. They range from simple automatic tool wear compensation devices to sophisticated systems which monitor and control a multitude of machining variables. Adaptive control benefits include increased productivity, reduced machining costs, extended cutter life, reduced scrap, improved work quality, and greater machine utilization.

21.7 NC AND THE FUTURE

Computer technology has reached the stage where an engineer can use computer graphics to design and control the manufacture of a part. See **Figure 21-38**. The computer is used to place the **design elements** (lines, curves, circles, etc.) and dimensions on a display screen, **Figure 21-39**.

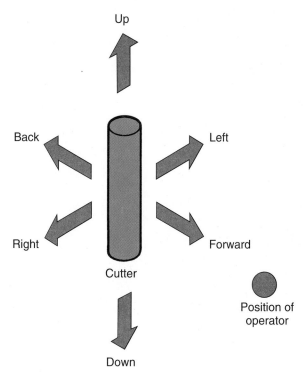

Figure 21-35. Tool motion commands used when programming in APT.

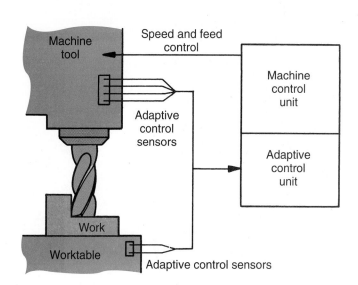

Figure 21-36. Diagram of one type of adaptive control (AC). Adaptive control automatically optimizes cutting speeds and feeds by judging machining conditions through sensors that monitor various operating factors of the machine.

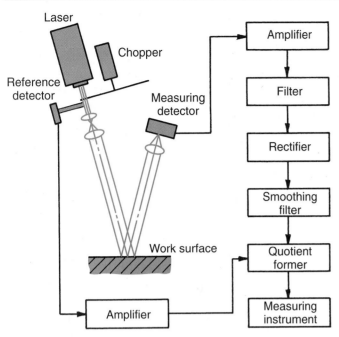

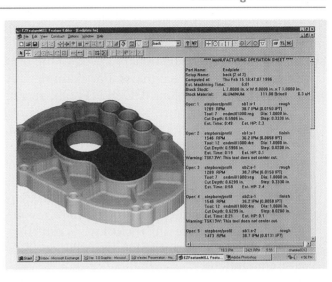

Figure 21-37. Measurement of surface finish is being done by detecting the intensity of a laser beam reflected from the work.

Figure 21-39. Display screen can show necessary machining information and picture of part being machined. (Bridgeport Machines, Inc.)

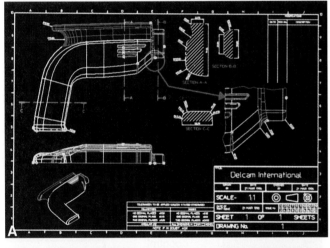

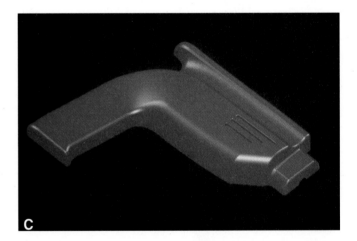

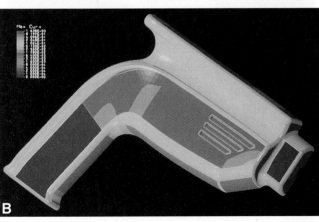

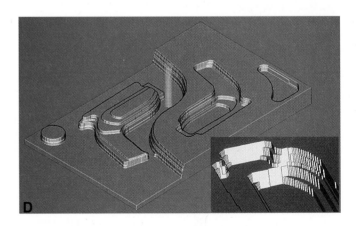

Figure 21-38. The use of computer graphics to design a product and the die to mold it. A—First step in designing a product, a tool housing. B—Curvature analysis of the surface of the design. The model is shaded in a range of colors, from blue for flat surfaces to dark orange for those with maximum curvature. C—Corrected design of the tool housing. Its geometry will provide the information to develop the program to machine the dies to mold it. D—Developing the program to machine the dies needed to mold the tool housing. (DELCAM International Inc., Ontario, Canada)

Design changes can be made easily, using the computer keyboard or other input devices. When the design has been edited and proofed, the computer can be instructed to analyze the geometry of the part and calculate the tool paths that will be required when machining the part.

The tool path is translated into a detailed sequence of machine axis movement commands that will enable the appropriate CNC machine to produce the part, **Figure 21-40**. No engineering drawing or program script is required. However, the data can be used to produce *hard copy* (detail and assembly drawings) for the part and determine how it fits into the overall product assembly. The computer-generated data can be stored in a central master computer for direct transfer to a CNC machine tool for parts manufacture. See the discussion of *direct numerical control* earlier in this chapter. The data can also be stored on punched tape (being phased out), magnetic tape (similar to an audio tape cassette), or a disk.

Before being released for production, the program is often verified by machining a sample part from plastic, wax, or a similar inexpensive material, **Figure 21-41**. Newer prototype techniques like

Figure 21-41. NC and CNC programs are usually checked out by using them to produce a part in an inexpensive material, such as plastic or wax. (Maho Machine Tool Corp.)

stereolithography and *Laminated Object Manufacturing* are also used to proof a part design. These techniques are described in Chapter 22.

The system that makes all of this possible is called *CAD/CAM* (computer-aided design/computer-aided manufacturing). See **Figure 21-42**.

New machine tools are being introduced where the cutter moves over the work on all axes rather than the work moving under the cutter, **Figure 21-43**.

CNC machine tools can be upgraded to take advantage of pinpoint accuracy and high speed of the laser-based digitizing technology that has been developed. See **Figure 21-44.** The laser digitizing system optically scans and converts surface details into digital information at speeds up to 70 points per second, 110″ per minute. Noncontact, nondeforming technology preserves shape integrity, even if the model is soft clay or rubber. Detail is not limited by shape and size of a physical contact probe, so detail is more accurate and complete. True surface data is saved as a separate file, and can be modified and used to generate NC code repeatedly without having to generate the original surface code.

21.8 ADVANTAGES AND DISADVANTAGES OF NC

NC can offer many advantages over traditional machining techniques. A few of the more important advantages include:

- Increased production.
- Reduced tooling costs.

Figure 21-40. Tool path data developed on a computer can be translated into a detailed sequence of machine axis movement commands that will enable an appropriate CNC machine to produce the part. (Courtesy of SURFCAM by Surfware)

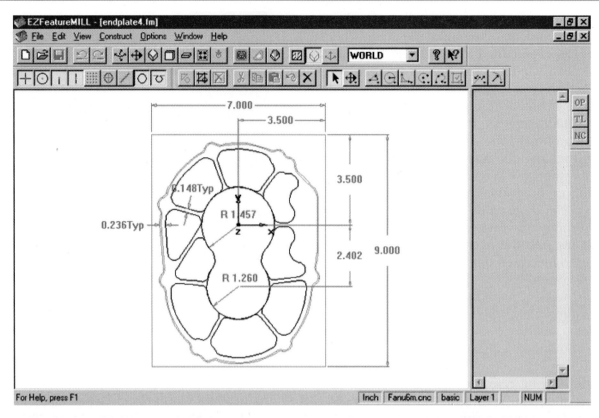

A

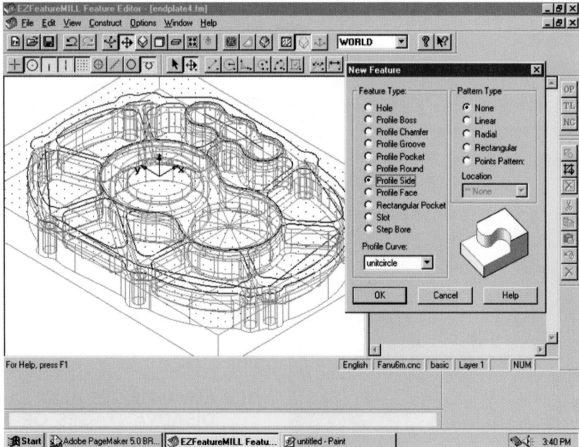

B

Figure 21-42. *CAD/CAM. A—With some software, the geometry of the part to be machined can be created at the machine tool or imported from files. B—The software defines the features of the part. C—The software automatically produces ready-to-use Operations Sheets and Tool Lists for the machinist. D—The software generates NC code that is ready to run and machine the part. (EZFeatureMILL—Engineering Geometry Systems)*

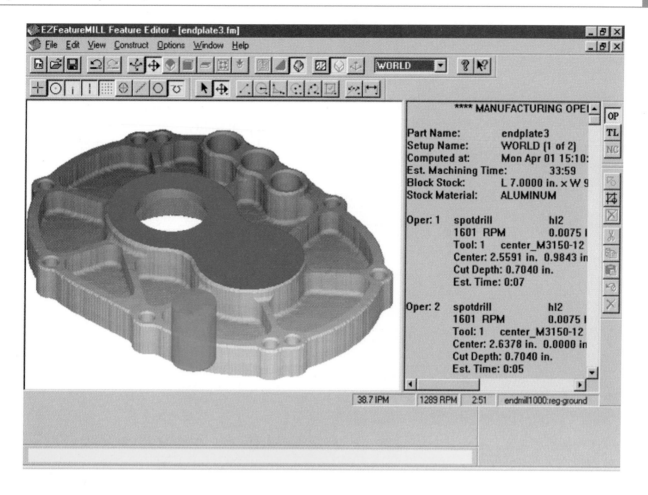

C

D

Figure 21-42. CAD/CAM (continued).

Figure 21-43. The CNC milling machine of the future has six degrees of freedom, provided by three pairs of variable-length legs. The legs move the multiaxis cutting head through three-dimensional space for rapid and precise positioning. (Giddings & Lewis, Inc.)

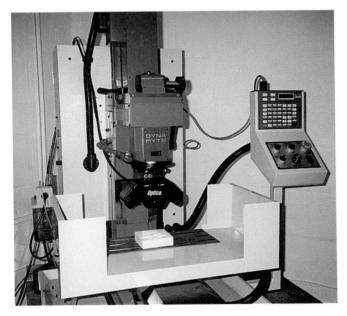

Figure 21-44. A laser digitizing system optically scans and digitizes any complex 3D part or model. The data obtained can be manipulated for male/female conversions, reductions, and enlargements, and then be used to duplicate the part itself, or to machine a die, punch, mold, or electrode. It is a noncontact operation. (Optica)

- Improved quality control.
- Economical production of difficult-to-manufacture parts.
- Versatility, since NC machines can be programmed to produce a single piece or a large scale production run.
- Repeatability.
- Elimination of many expensive jigs and fixtures.
- All, or nearly all, the machining operations can be performed on a single machine set-up.
- Direct savings in labor.
- Easy inch/metric conversion.
- Increased machining capabilities.

NC does have some disadvantages. Most can be overcome by careful planning.

- High initial cost of equipment.
- Shortage of skilled technicians to service equipment.
- Increased maintenance costs over traditional machine tools.
- Machine capabilities must be fully utilized.

21.8.1 NC for Training Programs

Many NC and CNC machine tools have been developed for both production and training programs, **Figure 21-45.** The complexity of the work they are capable of performing is limited by machine size and onboard computer capacity.

When using these machines, students and trainees are able to develop programs, prepare the tooling, program the machine, and see the resulting finished part within a reasonable time. **Figure 21-46** shows a trainee using this type CNC machine tool.

21.8.2 Other NC Applications

NC and CNC systems have been adapted to a broad range of metalworking-related equipment, **Figure 21-47.** Spot welding, riveting, and punching were among the first NC systems because they are basically point-to-point operations. Today, virtually every type of machining operation has been adapted to NC.

Multiple-operation equipment, like that employed to machine automobile engine blocks, is typically computer controlled. The machine shown in **Figure 21-48** has rough castings loaded into one end. The castings are then transferred from station to station, through several machining operations. The system unloads the finished blocks automatically at the end of the machining cycle.

Figure 21-46. Industry and many schools use relatively inexpensive CNC machine tools for educational purposes. They simulate actual industrial practices on how programs are developed. Students and trainees can see resulting finished part within a reasonable time. (Millersville University, Dept. of Industry and Technology)

Figure 21-45. Machines for training. A—Tabletop vertical machining center. Its size and cost make it ideal for CNC training programs as well as for industrial use. The complexity of the work it is capable of performing is limited only by machine size and onboard computer capacity. (Defiance Machine and Tool Co., Inc.) B—A small horizontal CNC turning center controlled by a separate computer unit. Excellent for training purposes and small size production work. (Light Machines Corporation)

Coordinates for machining auto body dies (shaped steel blocks that form sheet metal body sections) can be scanned or "lifted" from carefully made clay and wooden models, **Figure 21-49.** (Some die shapes are of such complexity that they cannot be fully described by CAD.) The coordinates mathematically define die geometry and are stored in computer memory. The information is then used to control machine tools when the three-dimensional body dies are machined, **Figure 21-50.**

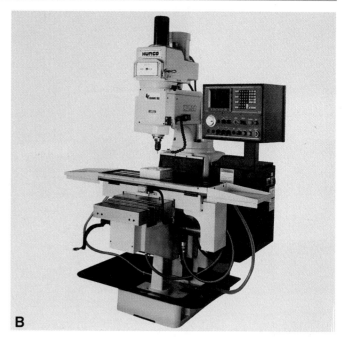

Figure 21-47. *CNC applications. A—Precise positioning control is a must for this laser cutting tool as it makes hexagonal openings in tubing. (Rofin-Sinar) B—The versatility of a CNC knee-type vertical milling center makes it a piece of equipment found in many machine shops. (Hurco Manufacturing Company) C— A CNC machine used to cut, punch, and bend metal sheet and plate. (U.S. Amada)*

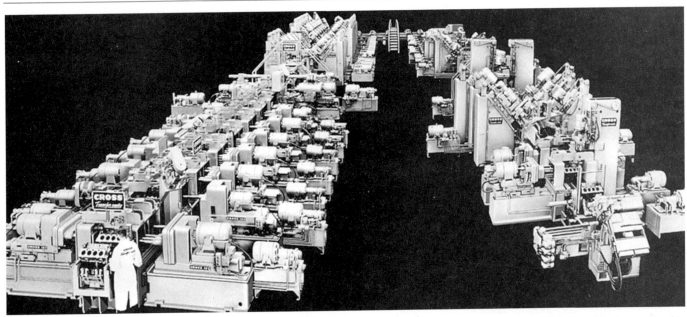

Figure 21-48. *A large machining complex employed to machine automobile engine blocks. At any one time, 104 blocks are having some machining or inspection operation performed on them. The machine is almost two city blocks long and performs 555 operations, including: 265 drilling, 6 milling, 56 reaming, 101 countersinking, 106 tapping, and 133 inspection operations. It produces 100 pieces an hour. (Cross Co.)*

Figure 21-49. *Coordinates for machining auto body dies are scanned from clay and/or wood models. The coordinates mathematically define body geometry and can be used to control cutting tools when three-dimensional body dies are machined. (Advanced Concept Center, General Motors Corp.)*

Figure 21-50. *Body dies machined using technique shown in Figure 21-49. (Buick Div., General Motors Corp.)*

TEST YOUR KNOWLEDGE

Please do not write in the text. Write your answers on a separate sheet of paper.

1. Describe the differences between manual machining techniques and NC methods.

2. Prepare a sketch showing the Cartesian Coordinate System.

3. Prepare two similar sketches. Show incremental dimensioning on the first sketch and absolute dimensioning on the second sketch.

4. What is an NC program?

5. List the three basic NC systems.

6. Using sketches, show how the three NC systems differ.

7. Which of the three NC methods require the use of a computer? Why is a computer required?

8. What do the following terms mean?
   a. MCU.
   b. Alphanumeric data.
   c. Program sheet.

9. How is an NC program often verified?

10. Briefly explain adaptive control (AC).

This large machining center, with its five-axis head, was developed to perform cutting and machining operations of large panels used in aircraft production. It can handle workpieces as large as 14 m × 4 m × 2 m (46′ × 13′ × 7′) and can process materials ranging from wood, plastics, and composites to steel, aluminum, and most other metals. (AsquithButler)

Chapter 22

Automated Manufacturing

LEARNING OBJECTIVES

After studying this chapter, you will be able to:
- ○ Define the term "automation."
- ○ Describe several automated production systems.
- ○ Define the term "industrial robot."
- ○ Discuss the use of robotics in automated production systems.

IMPORTANT TERMS

computer integrated
  manufacturing (CIM)
flexible manufacturing
  system (FMS)
Fused Disposition
  Modeling (FDM)
JIT

Laminated Object
  Manufacturing (LOM)
manipulator
robot
smart tooling
stereolithography
work envelope

Automation is a term coined in 1947 by a Ford Motor Company engineer. It is not a revolutionary new form of manufacturing, but a *manufacturing procedure* that has evolved over many years. Automation is concerned with the continuous automatic production of a product, **Figure 22-1**. It involves a machine or group of machines, **Figure 22-2**, activated electronically, hydraulically, mechanically, or pneumatically (or a combination of these means) to automatically perform one or more of the five basic manufacturing processes:

- Making
- Inspecting
- Assembling
- Testing
- Packaging

The principles of automation have been known for many years. An automated flour mill was in operation in the late 1700s near Philadelphia. It featured the continuous milling of grain into flour. The mill used many of the elements found in modern automated operations.

Figure 22-1. A robotic system used to assemble the components of automobile engines. (Hirata Corporation of America)

Figure 22-2. This integrated manufacturing system is used to automate the production of motorcycle parts. (Westech Automation Systems)

The integration of the computer, **Figure 22-3**, with specially designed machine tools and equipment has revolutionized production technology while improving product quality and reducing manufacturing costs. Human involvement is also reduced to an absolute minimum.

22.1 FLEXIBLE MANUFACTURING SYSTEM

One of the more recent automated production techniques is known as the *flexible manufacturing system (FMS)*. This general category of machining/manufacturing technology is also widely referred to as *computer integrated manufacturing (CIM)*; the grouping of machines used to perform multiple operations automatically is often called a *flexible machining cell (FMC)*.

The FMS brings together workstations (machine tools), automated handling and/or transfer systems, and computer control in an integrated manner, **Figure 22-4**. It is capable of producing a selected range of work configurations randomly and simultaneously. Work is transferred to and from member machines by automated fixture carts, conveyors, or specially designed loaders, **Figure 22-5**. Robots may also be employed in some operations.

Machine tools typically used in FMS are CNC (computer-aided numerical control) vertical and horizontal machining centers. Other CNC machine tools (such as grinders) and automatic gaging equipment also may be included in such systems.

Each step of the manufacturing process is computer-controlled and linked with the succeeding one, **Figure 22-6**. Measuring sensors, often utilizing lasers, are sensitive enough to detect tool wear as it occurs and automatically compensate for it. Every part is inspected. Problems such as tool malfunction, tool breakage or damage, etc., are immediately identified so corrections can be made before additional parts are produced that do not meet specifications.

FMS also includes the use of *smart tooling*, *JIT* (just-in-time arrival of materials, parts, subassemblies, etc.), and *robots*.

Smart tooling uses cutting tools and work-holding devices that can be readily reconfigured to produce a variety of shapes and sizes within a given part family. This makes it economically feasible to manufacture products in smaller lot sizes.

JIT is a system that eliminates the need for large inventories of materials and parts. They are scheduled for arrival at the time needed and not before. While JIT lessens (and often eliminates) storage costs, production can be reduced or stopped if the delivery of the needed items is hampered by weather delays and strikes.

22.2 ROBOTICS

The Robotic Industries Association (RIA) has adopted the following definition of the industrial robot, **Figure 22-7**.

"A *robot* is a programmable, multifunctional manipulator designed to move material, parts, tools, or specialized devices through variable programmed motions for the performance of a variety of tasks."

22.2.1 Robotics in Automation

There are two implied requirements in the RIA definition of a robot:

- The robot must be a multiaxis machine that is capable of moving parts or tools to a specific location as well as positioning them at any attitude. The work envelope of a robot is the volume of space defined by the reach of the robot arm in three-dimensional space. **See Figure 22-8**.

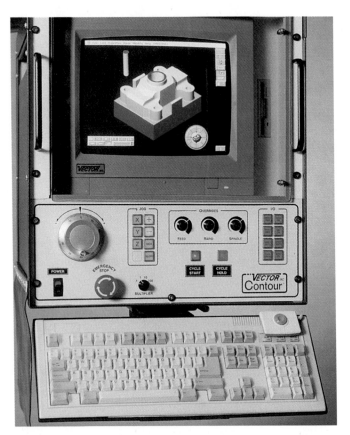

Figure 22-3. *The computer, integrated with specially designed machine tools and equipment, has revolutionized automation technology while improving product quality and reducing manufacturing costs. (Vector, a Division of Artran, Inc.)*

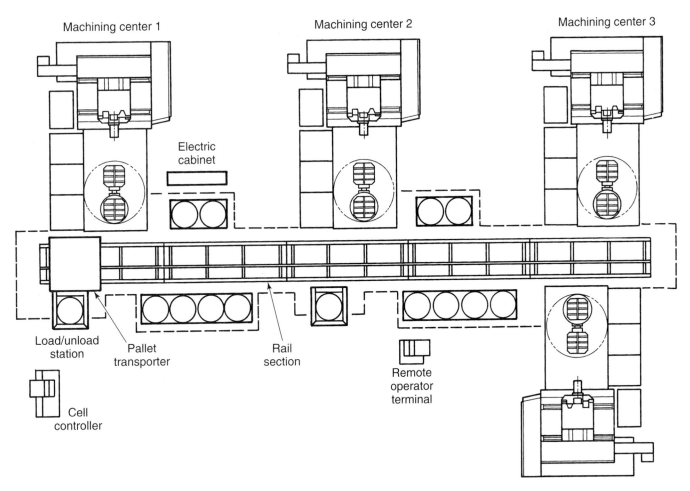

Figure 22-4. *Flexible manufacturing cell that uses a pallet transporter to link the machines. A cell controller automatically queues work for immediate delivery to the next machine available. (Cincinnati Milacron)*

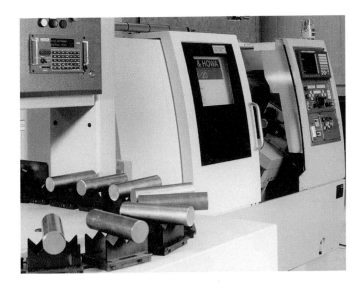

Figure 22-5. *Gantry loader transfers workpieces from conveyor (foreground) to an unattended horizontal machining center in this manufacturing cell. It is capable of handling parts weighing up to 70 lbs. (Westech Products Group/Gantrex Machine Tool Loaders)*

Figure 22-6. *In a flexible manufacturing system, each step in the manufacturing process is linked with the succeeding one. There is an automated flow of raw material, total machining of parts across machines, and removal and storage of finished parts. (Mazak Corporation)*

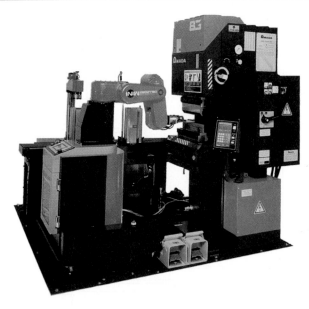

Figure 22-7. Robots can be programmed to do many types of jobs that are hazardous or very monotonous for human operators. Spot welding, paint spraying, and loading parts onto machines are among jobs performed by robots. This robot is paired with a press brake to load and remove sheet metal parts. (U.S. Amada, Ltd.)

- There must be a control system that can be programmed to drive the manipulator (device that grasps the part or tool) through a series of specified motions and be capable of interacting with other machines and equipment.

22.2.2 Design of a Robot

As shown in **Figure 22-9,** a robot consists of four basic components.

- *Controller.* Performs computations for controlling the movement of the arm and wrist to the proper location.

- *Power Supply.* May be hydraulic, pneumatic, or electric. Most modern robots have electric drive.
- *Manipulator.* The articulated "arm" of the robot. The end of the arm is fitted with a "wrist" capable of angular and/or rotational motion.
- *End-of-arm Tooling.* A device attached to the robot wrist for specific applications, such as a gripper, welding head, spray gun, etc.

22.2.3 Robot Applications

Industrial robots have a number of applications in which they are widely used:

- *Hazardous and harsh environments.* Fumes produced by spray painting and welding, heat in foundry or forging operations, feeding material into punch presses, etc. See **Figure 22-10.**
- *Tedious operations.* Repetitive operations such as parts feeding, loading and unloading parts from machines, some assembly operations, etc. See **Figure 22-11**.
- *Precision operations.* Precisely repeating positioning operations, maintaining consistent tool speed, follow complex welding and cutting paths without patterns, quality control using lasers, clean room operations, etc., **Figure 22-12.**
- *Handling heavy materials.* Lifting material onto or from a stack, moving material beyond the normal reach of a human, mounting heavy workpieces on machine tools, etc.

Many types of robots have been developed, but almost all can be classified into one of the four basic geometric configurations shown in **Figure 22-13.**

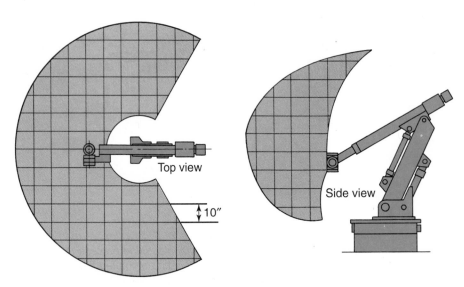

Figure 22-8. The work envelope of a robot is the volume of space defined by the reach of the robot arm in three dimensions.

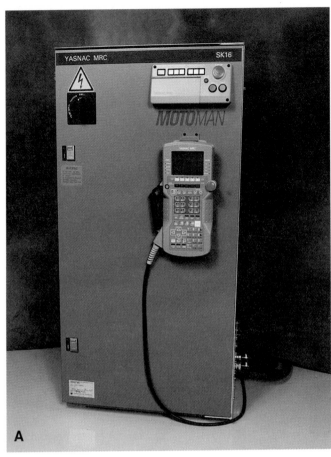

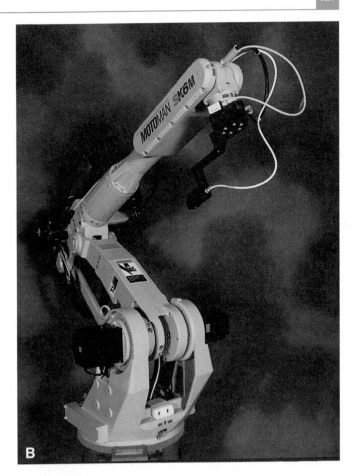

Figure 22-9. A robot consists of four basic parts: controller, power supply, manipulator, and end-of-arm tooling. A—The controller and power supply provide signals to the robot's servomotors to perform necessary operations and movements. The teach pendant hung on the front of the cabinet is used to program robot movements. B—The manipulator, or articulated robot arm, is moved in various directions by servomotors at each joint. The end-of-arm tooling on this robot is a vacuum gripper used to pick up workpieces. (Motoman).

Figure 22-10. A robot ladles molten aluminum into a wheel mold. (Fanuc Robotics North America, Inc.)

Figure 22-11. This robot is being used to load and unload castings from a drilling machine. (Fanuc Robotics North America, Inc.)

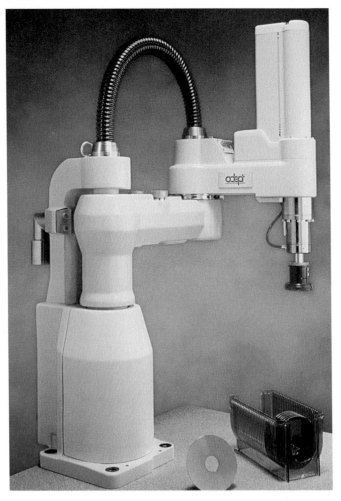

Figure 22-12. *Specially equipped robots are used for handling tasks in clean room environments where computer chips, precision disks for computer hard drives, and similar products are produced. (Adept Technology, Inc.)*

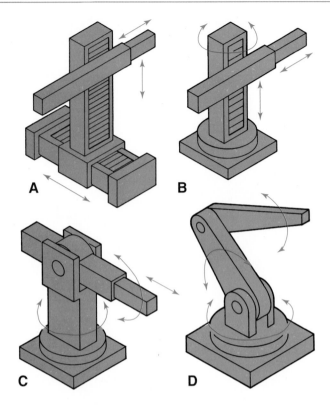

Figure 22-13. *Basic geometric configurations of robots. All provide three articulations (specific arm movements). A—Cartesian coordinates. B—Cylindrical coordinates. C—Polar coordinates. D—Revolute coordinates.*

Advanced technology robots can be programmed to serve a wide range of automated manufacturing applications. They can be interfaced with testing devices to weigh, gage, or measure.

Some robots are capable of selecting and positioning complex shaped parts for machining or storage. A laser is utilized to "see" and define the part outline so the correct item will be selected and positioned for machining.

22.3 SAFETY IN AUTOMATED MANUFACTURING

- Observe the same safe operating procedures that are used with traditional machine tools and machining.
- Wear approved eye protection and a snug-fitting apron or shop coat.
- Be sure work-holding devices are positioned correctly and are securely fastened to the worktable.

- Carefully check tool clearance to be sure the cutter will clear the workpiece, work-holding devices, clamps, etc., when manually postioning the work and during rapid traverse. Make a "dry run" for a safety check of tool positioning for each machining operation.
- Never enter the work envelope of a robot until you are sure power is turned off.
- Establish that the machine tool and control unit are functioning correctly, and continually monitor machining operation to be sure each tool is cutting properly. Know how to safely stop machining operations in case of an emergency.
- To avoid cuts, remove burrs and sharp edges before inspecting finished parts for accuracy or finish of machined surfaces.

22.4 RAPID PROTOTYPING TECHNIQUES

A number of new techniques have been developed to allow designers to quickly generate three-dimensional models or prototypes of parts in relatively inexpensive materials. These techniques help to improve design and identify potential machining problems.

22.4.1 Laminated Object Manufacturing (LOM)

Laminated Object Manufacturing (LOM)™ is a new technology for the rapid generation of models, prototypes, accurate patterns, and molds. See **Figure 22-14.** This state-of-the-art three-dimensional modeling method was developed in 1991 by Helisys, Inc. Since then, many automotive, aerospace, appliance, and medical product manufacturers have adopted the process, **Figure 22-15.** The equipment is compact and can operate 24 hours a day.

LOM uses inexpensive solid sheet material, such as paper, plastic, and composites, to form the desired designs. Molds made from the composite materials can be used in direct tooling applications for such processes as injection molding, vacuum forming, and sand casting. See **Figure 22-16.** LOM parts are accurate enough for form-and-fit verification applications. The finished part has a composition similar to wood and can be easily machined or modified to obtain the exact fit required. Modifications and corrections can be incorporated into the CAD design before manufacturing.

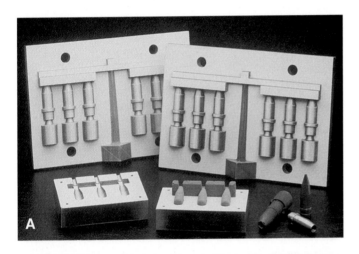

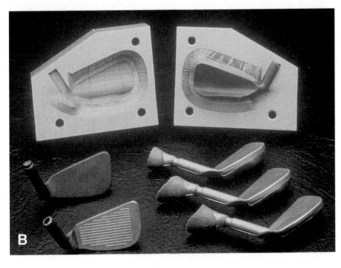

Figure 22-14. Laminated Object Manufacturing. A—LOM was used to produce these extremely accurate, full-size matchplates (patterns) and core box to be used in casting prototype metal projectiles. B—Rapid prototyping of the molds to make the wax patterns for the investment casting of oversized golf club irons was done with LOM. Design changes could be carried out before making the expensive molds necessary for quantity production. (Helisys, Inc.)

Figure 22-15. Aluminum sand casting of the prototype of a large rear-engine power takeoff housing for a diesel truck engine. The pattern was made by the LOM process. Using this technique cut development time in half. (Helisys, Inc.)

Figure 22-16. Multiple vacuum formed plastic prototypes and sand cast production tools were produced directly off the same LOM pattern. (Helisys, Inc.)

Operation of the LOM System

CAD data is programmed into the LOM's process controller. A cross-sectional slice is generated and a laser cuts the outline of the cross section, then cross hatches the excess material for later removal, **Figure 22-17**. A new layer of material is bonded to the top of the previously cut layer. The next cross section is prepared and cut. This automatic process continues until all layers are laminated and cut. Excess material is then removed to expose the finished part. The completed object's surface can then be sanded, polished, or painted as desired.

22.4.2 Stereolithography

Another new technology, called *stereolithography*, can also produce complex design prototypes of castings and other objects in hours, instead of the days or weeks previously required, **Figure 22-18**. The three-dimensional hard plastic models can be studied to determine whether they are the best solution to a design problem. Since new models can be made quickly, design changes and modifications can be evaluated without the expense of making new patterns or molds.

The stereolithography process uses a computer-guided low-power laser beam to harden a liquid photocurable polymer plastic into the programmed shape, **Figure 22-19**.

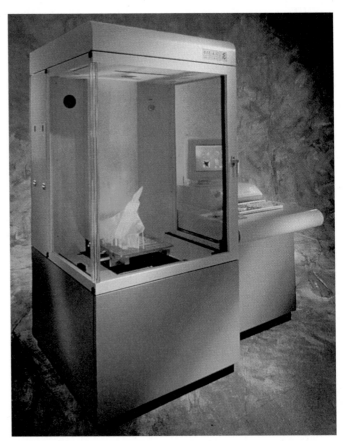

Figure 22-18. *Stereolithography equipment used to produce durable, finely detailed patterns extensively used in modeling, functional prototyping, and tooling applications. (3D Systems)*

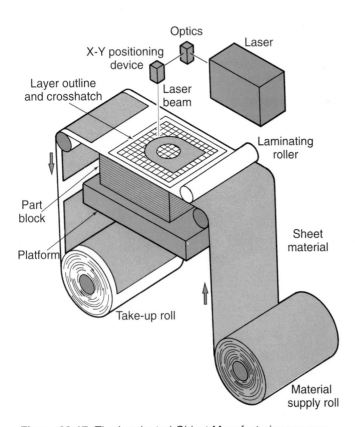

Figure 22-17. *The Laminated Object Manufacturing process.*

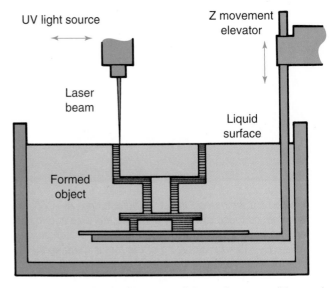

Figure 22-19. *Basic diagram of how the stereolithography process works.*

The process starts by creating the required design on a CAD system and orienting it in three-dimensional space, **Figure 22-20**. A support structure is added to hold the various elements in place while the model is built up. The design is then sliced into cross sections of 0.005" to 0.020" (0.12 mm to 0.50 mm) in thickness.

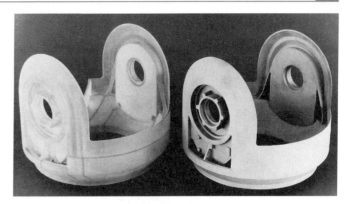

Figure 22-21. The casting design as made by the stereolithography process and the casting that was used in the night vision system. The part design was proofed before the dies to produce the castings were made. (Hughes Electro-Optical & Data Systems Group)

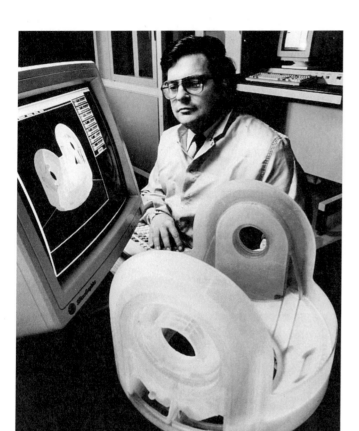

Figure 22-20. The stereolithography process starts by creating the required design on a CAD (Computer-Aided Design) system. The CAD-designed part, in this case a yoke housing casting for a helicopter night vision system, is shown on the computer screen. The prototype part made by the technique is shown in the foreground. (Hughes Electro-Optical & Data Systems Group)

The design data is downloaded into the stereolithography machine, which operates in a way similar to a CNC machine tool. The machine's control unit guides a fine laser beam onto the surface of a vat containing the liquid photocurable polymer. The liquid solidifies wherever it is struck by the laser beam.

The model is created from the bottom up, on a platform located just below the surface of the liquid plastic. After each "slice" is formed, the platform drops a programmed distance. The sequence is repeated until the entire model is formed, **Figure 22-21**.

The finished part requires final ultraviolet curing. It is removed from the platform and the support structure is clipped away. It can be finished by filing, sanding, and polishing. Paint or dye can be applied. The entire stereolithography process is shown in **Figure 22-22**.

22.4.3 Other Rapid Modeling Techniques

Rapid modeling techniques that operate on a system similar to stereolithography can produce fully functional prototypes made of ABS (acrylonitrile butadiene styrene), medical ABS, or investment casting wax. When made of investment casting wax, the three-dimensional models can be used as the master for a cast part.

Called *Fused Disposition Modeling (FDM)*™, the process can produce fully functioning prototypes, **Figure 22-23**. When made of ABS, the parts can be installed and run, for the best proof that a design works. The parts can be made in color, **Figure 22-24**.

The FDM process forms three-dimensional objects from CAD-generated solid or surface models. A temperature-controlled head extrudes thermoplastic material layer by layer. The designed object emerges as a three-dimensional part without tooling.

A variation of the technology has been developed that produces parts made of ceramic material, instead of plastic. Layers of a special ceramic powder are built up, using an inkjet style printer to spray a quick-hardening binder to solidify each layer. The technology has been used to quickly produce shell molds for casting such metals as inconel and aluminum. The moldmaking process is called *Direct Shell Production Casting (DSPC)*.

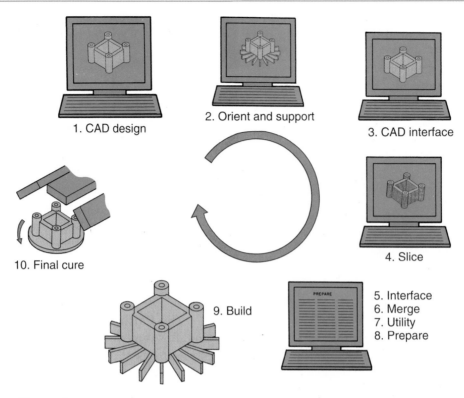

1. CAD design

2. Orient and support

3. CAD interface

4. Slice

5. Interface
6. Merge
7. Utility
8. Prepare

9. Build

10. Final cure

Figure 22-22. The stereolithography process.

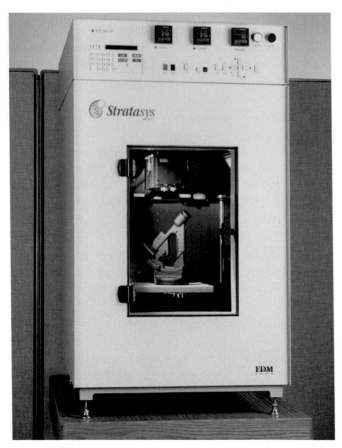

Figure 22-23. Fused Disposition Modeling equipment can produce fully functional prototypes from ABS, medical ABS, or investment casting wax. When made from ABS, the parts can be installed and run to proof the design. (Stratasys, Inc.)

Figure 22-24. ABS parts can be made in color. (Stratasys, Inc.)

22.5 THE FUTURE OF AUTOMATED MANUFACTURING

With technological advances being made at such a rapid pace, only time will tell what effect the computer and new automated manufacturing techniques will have on our society.

Some workers, mostly unskilled and semi-skilled, will lose their jobs, much like the home artisans did during the Industrial Revolution. The same thing happened to the carriagemaker, blacksmith, buggy whip manufacturer, and feed dealer when Henry Ford started to mass-produce the automobile, **Figure 22-25**.

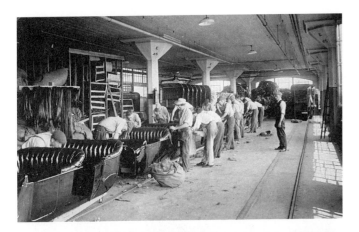

Figure 22-25. *Then and now in the manufacture of motor cars. When Henry Ford opened the first production line, he paid the workers $5 per day, a very good salary for the time. (Ford Motor Company)*

Such developments were condemned then, just as automated manufacturing and robots seem to be condemned today. However, machines eventually helped employ, directly or indirectly, many more people than the number they originally displaced. Better jobs, with higher pay and improved working conditions, were created. These changes also demanded that workers and technicians become better educated. The key to the future, then, will be men and women who are well versed in the various areas of industrial technology.

TEST YOUR KNOWLEDGE

Please do not write in the text. Write your answers on a separate sheet of paper.

1. What is automation?

2. How are the machines in automation activated?

3. List the five basic manufacturing processes that automated machines can perform.

4. What do the following acronyms stand for in their relation to automation? (An acronym is a word formed using the initial letters of words in a phrase. For example: RPM stands for revolutions per minute.)
   a. FMS.
   b. CIM.
   c. FMC.
   d. CNC.
   e. CAD/CAM.
   f. LOM.

5. In a flexible machining cell, specially designed _____ are often used to place workpieces in a machine.

6. Explain the following terms:
   a. Smart tooling.
   b. JIT.
   c. Robot.

7. In robotics, what is meant by the term "work envelope"?

8. List three current uses of the robot.

9. List three techniques that are now available for the rapid prototyping of a CAD design.

10. In several of the rapid prototyping systems, a _____ is used to harden a thin layer of material as the object is built up.

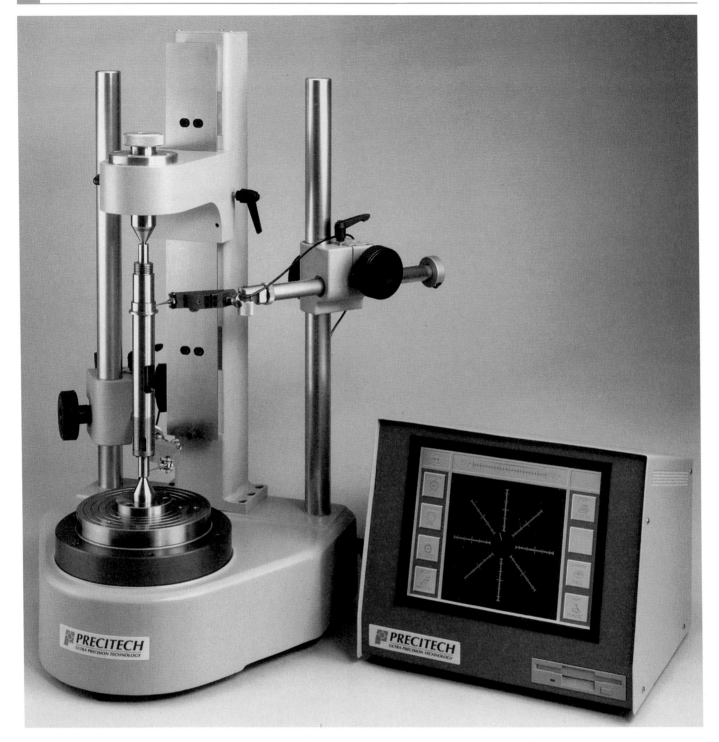

Circular geometry gage is designed to make exact measurement of out-of-roundness, eccentricity, flatness, and perpendicularity. (Precitech, Inc.)

Chapter 23

Quality Control

LEARNING OBJECTIVES

After studying this chapter, you will be able to:
○ Explain the need for quality control.
○ Point out the difference between the two basic quality control techniques.
○ Describe how some nondestructive testing methods operate.
○ Explain advanced methods for assuring quality control during and after machining.

IMPORTANT TERMS

coordinate measuring
   machine (CMM)
eddy-current test
fluorescent penetrant
   inspection
magnetic particle
   inspection

megahertz (MHz)
optical comparator
profilometer
statistical process control
   (SPC)
ultrasonic testing
ultraviolet light

The primary purpose of *quality control* is to seek out and prevent potential product defects in the manufacturing process before they can cause injuries or damage and substandard products. The eventual goal of quality control is *not* to detect imperfect products after manufacture, but to *prevent* them from ever being made.

Quality control is one of the most important segments of industry. It plays a vital role in improving the competitive position of a manufacturer.

23.1 THE HISTORY OF QUALITY CONTROL

The development of the science of quality control closely parallels the development of the airplane. Early aircraft were made "by-guess-and-by-golly," with little concern for quality control, **Figure 23-1**.

As the theory of flight became more refined, more care was taken in selecting materials that went into the planes, **Figure 23-2**. Tests were made to determine the strength of materials. Engines were made to specific standards and were checked many times during their manufacture. As a result, aircraft dependability increased greatly. In the automotive field, Henry Ford began the mass production of the Model-T using similar quality control methods.

Figure 23-1. *There was little or no quality control in the manufacture of early aircraft, like this one on display at an aeronautical museum. If a wood strip was straight with no knots, and looked strong enough, it was used. To build such an aircraft, a full-size drawing was usually drawn in chalk on the shop floor. The plane was built over the chalk drawing.*

Figure 23-2. *The favorite type aircraft of World War I's "Red Baron." This Fokker DRI triplane was constructed more-or-less to specifications. Some of the materials were inspected and tested before use. However, some aircraft components had tolerances of ± 25 mm or 1". (Bill Hannan)*

The volume production of all-metal aircraft in the early 1930s introduced many new quality control techniques, **Figure 23-3.** *Jigs* and *fixtures*, which held the parts while they were machined or fabricated, were aligned with optical tools. The use of iron filings and a magnetic current, called *magnaflux,* helped to find defects and flaws in ferrous metals. Measurements of 0.0001″ (0.0025 mm) were common in engine components. Inspectors made up a larger percentage of the workforce than ever before. Many other industries established quality control programs.

The need for thousands of high performance aircraft during World War II, **Figure 23-4,** led to the introduction of many of the quality control techniques in use today. The inspector became a vital part of the manufacturing team.

Aerospace vehicles, such as the Space Shuttle, **Figure 23-5,** require quality control programs of great scope and magnitude. Because they are subjected to pressures, stresses, and temperatures seldom found on earth, a breakdown or malfunction of any of the thousands of critical parts would be disastrous.

Following the lead of the aerospace industry, other industries began placing increasing emphasis on quality control techniques. As products became more complex and sophisticated, and demand for reliability continued to grow, an increasing percentage of industry's budget had to be spent on quality control.

Figure 23-5. *The Space Shuttle is one of the most complex products ever constructed. Its manufacture required a quality control program second to none to assure its integrity. Before each flight, 488 nondestructive and visual inspections must be made. (NASA)*

Figure 23-3. *The manufacture in quantity of all-metal aircraft during the 1930s brought about the development of quality control techniques that are still in use. This was one of the first PBY Catalina amphibian aircraft constructed. It was built in 1935 and is still in flying condition. (John Winter)*

23.2 CLASSIFICATIONS OF QUALITY CONTROL

Quality control techniques fall into two basic classifications:
- *Destructive testing* results in the part being destroyed during the quality control testing program.
- *Nondestructive testing* is done in such a manner that the usefulness of the product is *not* impaired.

23.2.1 Destructive Testing

Destructive testing is a costly and time-consuming quality control technique. A specimen to use for testing is selected at random from a great number of pieces, **Figure 23-6.** Statistically, at least, it indicates the characteristics of the untested (and

Figure 23-4. *The B-17 Flying Fortress was the backbone of the Air Force during the 1940s. Mass production was made possible only because of modern quality control procedures. (Bob Walker)*

Figure 23-6. *Automobile bodies, selected at random, are cut apart so that quality control inspectors can check placement and size of welds made on the plant's highly-automated framing line. Destructive testing is done to assure that structural integrity of bodies is maintained.*

undestroyed) remaining pieces. It gives no assurance of perfection, however, since many of the untested parts could be defective and be used in the manufacture of a product.

23.2.2 Nondestructive Testing

Nondestructive testing, **Figure 23-7**, is a basic tool of industry. It is well-suited for testing electronic and aerospace products where the performance of *each* part is critical. Each piece can be tested individually *and* as part of a completed assembly.

23.3 NONDESTRUCTIVE TESTING TECHNIQUES

You are familiar with several methods of nondestructive testing: measuring, weighing, and visual

Figure 23-7. *This quality control technician is checking the concentricity of a tire mold tread ring using dial indicators. This is a nondestructive quality control technique, because the part is not damaged in any way. (Aluminum Company of America)*

inspection. While satisfactory for many products, they leave much to be desired for others. To assure effective quality control in all areas of manufacturing, industry has developed more sophisticated testing techniques.

23.3.1 Measuring

The use of micrometers, vernier tools, dial indicators, gages, and similar devices falls into the *measuring* category of quality control testing. See **Figure 23-8**. To guarantee their accuracy, these tools must be checked at frequent intervals against known standards. The calibrating is done in a *precision tool calibration laboratory*, **Figure 23-9**.

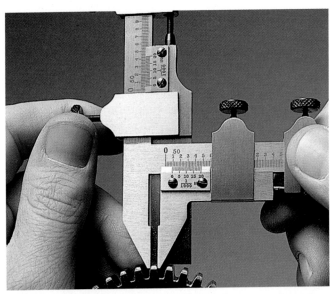

Figure 23-8. *Some quality control techniques are based on measurement. To guarantee accuracy of measuring tools, they must be carefully checked against known standards at frequent intervals. (L.S. Starrett Co.)*

Figure 23-9. *Personnel in the precision tool calibration laboratory test and keep an accurate set of records on all production measuring tools to assure their accuracy. This technician is using a dial indicator to check the accuracy of a plate lamination gage used by automatic press operators at a production site. (Master Lock Co.)*

The shape of some products, **Figure 23-10**, prevent accurate measurements from being made by conventional measuring tools. The *coordinate measuring machine (CMM)*, **Figure 23-11**, has been devised to make measurements electronically. A CMM can be programmed to check thousands of individual reference points on the object against specifications.

Measuring can be done visually on an *optical comparator*, which is a gaging system for inspection and precise measurement of small parts and sections of larger parts. See **Figure 23-12**. A part can be magnified up to 500x without distortion to permit accurate measurement. An enlarged image of the part being inspected is projected onto a screen, where it is superimposed upon an accurate drawing overlay of the part. Variations as small as 0.0005″ (0.013 mm) can be noted by a skilled operator, **Figure 23-13**.

An *optical gaging system*, **Figure 23-14**, automates the process of optical comparator inspection. The software builds a graphical model of the measurements and remembers each step as the user measures a part. The resulting procedure is saved on a disk. When measuring additional samples, the software prompts the user by highlighting each measurement in the model. The user simply follows these graphical instructions to finish the inspection.

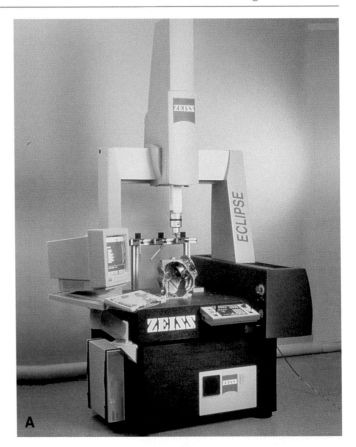

Figure 23-10. Complex parts, such as these turbine wheels, cannot be checked for accuracy with conventional measuring tools. Why do you think those measuring tools cannot be used? (Howmet Corp.)

Figure 23-11. Coordinate measuring machines. A—This CNC 3-axis coordinate measuring machine, with a weight capacity of 1000 lbs., can achieve an accuracy of 0.0002″ (0.005 mm). Because of ceramic material used in parts of its construction, accuracy is guaranteed between 64°F and 78°F. The probe system can scan the work at a velocity of 17″ (425 mm) per second. (Carl Zeiss, Inc.) B—This probe is being used to verify the accuracy of the holes machined into a complex part. A CMM can achieve accuracy far higher than any manual measuring method. (Renishaw, Inc.)

Figure 23-12. Parts can be magnified up to 500× on an optical comparator. (Bridgeport Machines, Inc.)

Figure 23-14. Optical gaging system. It automates optical comparator inspection. (Optical Gaging Products, Inc.)

Figure 23-13. A quality control engineer is checking hole location on a plate lamination of an automatic lock body assembly, using an optical comparator with a part overlay transparency. (Master Lock Co.)

Figure 23-15. Performing frequency distribution (random inspection of parts to determine that they are within established standards) on a part for a lock assembly, using a desk top computer interfaced with a snap gage and digital readout to collect data and assure that the manufacturing process is within statistical control. (Master Lock Co.)

Quality control can also be assured by a *statistical method* that involves measuring a number of parts in a production run, **Figure 23-15**. This is carefully worked out in a mathematical approach to quality measurement known as **statistical process control (SPC).** A wide variation in the accuracy of the parts being inspected will make it necessary to increase the inspection rate to include more or all of the units manufactured.

Special gaging and inspection tools can be designed for almost any application, **Figure 23-16**. Combining precision tools with electronic devices permits accurate inspection to be made by semiskilled workers.

23.3.2 Radiography (X-ray) Inspection

Inspection by **radiography** involves passing gamma rays through a part and onto light sensitive film to detect flaws (crack, pores, etc.). It has become routine in the acceptance or inspection of critical parts and materials, **Figure 23-17**. The technique involves the use of X-rays and **gamma radiation**

Figure 23-16. *A technician employing a radial measuring system to assure roundness of the taper at the cutting unit gageline of an element used in a quick-change tooling system. (Kennametal, Inc.)*

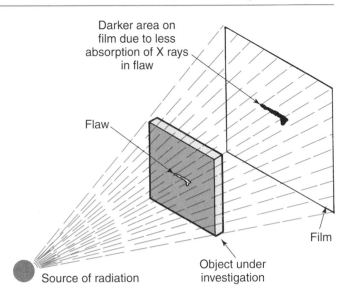

Figure 23-18. *How radiographic inspection works.*

Figure 23-17. *A laser alignment device is being used with an X-ray generator, to precisely locate X-ray radiation on a titanium structure that has been electron-beam welded. Highly accurate alignment is required because of the very narrow heat-affected zone characteristic of EB welds. (Northrop-Grumman Corp.)*

Figure 23-19. *X-ray examination is being used to check a number of similar parts. Hidden flaws are easily detected. (Westinghouse Electric Corp.)*

(highly energetic, penetrating radiation found in certain radioactive elements) projected through the object under inspection and onto a section of photographic film, **Figure 23-18**. The developed film has an image of the internal structure of the part or assembly, **Figure 23-19**.

Many kinds of *peripheral* (outside circumference) inspection operations can be made because of the *omnidirectional* (all directions) characteristics of the rays, **Figure 23-20**.

Film

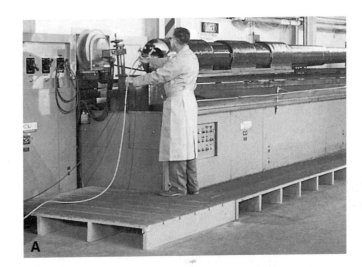

Figure 23-20. *Technique employed to inspect cylindrical objects by radiographic (X-ray) means. A flaw causes more exposure of the film, so an image of the flaw is shown on the film when it is developed.*

The radiographic inspection process offers many advantages:

- Inspection sensitivity is high.
- The projected image is geometrically correct.
- A permanent record is produced.
- Image interpretation is highly accurate.

In addition, the internal and hidden parts of complex assemblies can be inspected without fear of damage. Objects can also be inspected without taking them out of use. For example, wheels and axles on locomotives and railway cars are inspected at regular intervals without having to remove them and take them to special facilities.

If you make radiographic tests, you must observe strict safety precautions when handling radioactive materials. Exposure to X-rays beyond established limits can be harmful.

23.3.3 Magnetic Particle Inspection

Magnetic particle inspection, commonly known as *"magnafluxing"* (a trademark of Magnaflux Corp.) is a nondestructive testing technique employed to detect flaws on or near the surface of ferromagnetic (iron-based) materials. The technique is rapid, but shows only serious defects. It does not show scratches or minor visual defects, **Figure 23-21.**

The magnetic particle inspection technique is based on the theory that every conductor of electricity is surrounded by a circular magnetic field. If the part is made of ferromagnetic material, these lines of force will, to a large extent, be contained within the piece. A circular magnetic field, if *not* interrupted, has no poles. However, because a flaw or other imperfection present in the piece is oriented (positioned) to cut through these magnetic lines of force, poles will be formed at each edge of the flaw.

These poles will hold the finely divided magnetic particles, thus outlining the flaw. The limitations of this technique are apparent when the flaw is parallel to the line of magnetic force. The flaw will not interrupt the force, so no indication of it will appear when magnetic particles are applied, **Figure 23-22.**

In practice, a magnetic field is introduced into the part and fine particles of iron are blown (dry method) or flowed in liquid suspension (wet method) over it. As noted, a flaw will disturb or distort the magnetic field, giving it magnetic properties different from the surrounding metal. Many of the

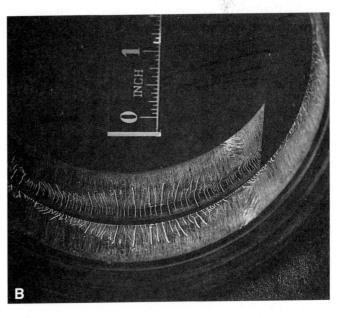

Figure 23-21. *Magnetic particle inspection. A—A large magnetic particle inspection machine being used to check a huge machined section. B—Highlighted flaws in a fitting that were found by magnetic particle inspection. (U.S. Army)*

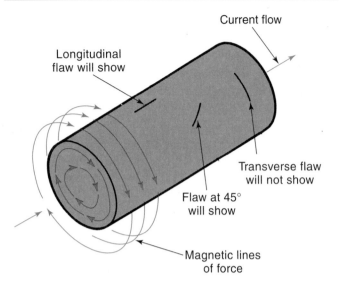

Figure 23-22. Theory, scope, and limitations of magnetic particle inspection technique. (Magnaflux Corp.)

Figure 23-24. Three ways to inspect a kingpin for a truck front axle. A—Under visual or "eyeball" inspection, the part appears to be sound and safe for service. B—Examination with magnetic particle inspection shows that excess heat generated during grinding has caused dangerous cracks. C—The same kingpin after being treated with a fluorescent penetrant testing material and photographed under ultraviolet light. The cracks show much more clearly. (Magnaflux Corp.)

iron particles will be attracted to the area and form a definite indication of the flaw (its exact location, shape, and extent), **Figure 23-23**.

23.3.4 Fluorescent Penetrant Inspection

The theory of *fluorescent penetrant inspection* is based on capillary action to show flaws in parts, **Figure 23-24**. A penetrant solution is applied to the part's surface by dipping, spraying, or brushing. Capillary action literally pulls the solution into the defect. The surface is rinsed clean, and a developer is applied. This acts as a blotter drawing the penetrant back to the surface.

When the part is inspected under *ultraviolet light* ("black light"), any defects will glow with fluorescent brilliance, making them easily visible. This is because dye penetrant has flowed into and remains in the flaw.

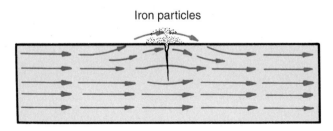

Figure 23-23. Crack in steel bar generates a magnetic field outside the part to hold iron particles. Buildup of iron particles makes even tiny flaws visible.

23.3.5 Spotcheck

Another of the penetrant inspection tools, *Spotcheck* is easy to use, accurate, economical, and does not require a black light to bring out the flaws. Application is similar to that described for the fluorescent type penetrant. The specimen is coated with a red liquid dye which soaks into the surface crack or flaw. The liquid is then washed off and the part dried. A developer is dusted or sprayed on the part, and flaws and cracks show up red against the white background of the developer. See **Figure 23-25**.

23.3.6 Ultrasonic Inspection

Ultrasonic testing techniques make use of sound waves above the audible range. They can be used to detect cracks and flaws in almost any kind of material that is capable of conducting sound. Sound waves may also be employed to measure the thickness of the same materials from one side. Ultrasonic test equipment is shown in **Figure 23-26**.

The human ear can hear sound waves whose frequencies range from about 20 to 20,000 *hertz* (cycles per second, abbreviated *Hz*). Sound waves that oscillate (vibrate) with a frequency greater than 20,000 Hz are *inaudible* (cannot be heard) and are known as *ultrasound.* Ultrasonic testing equipment utilizes waves measured in *millions* of cycles per second, or *megahertz (MHz).*

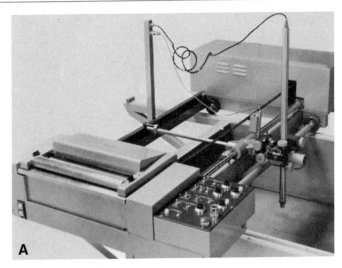

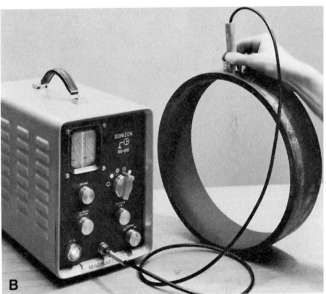

Figure 23-25. *Spotcheck penetrant inspection technique. A—Spray on penetrant, then apply developer. B—Inspect for flaws. They will stand out sufficiently to be seen easily. (Magnaflux Corp.)*

Figure 23-26. *Ultrasonic test equipment. A—The part being inspected is located in the tank below the scanning bridge on this automatic ultrasonic immersion testing system. Note the vertical search tube where the sound waves are emitted. B—A portable contact-type ultrasonic testing device. (Magnaflux Corp.)*

Sound waves are employed to obtain information about the interior structure of a material. By observing the echoes that are reflected from within the material, it is possible to judge distances by the length of time required to receive an echo from an obstruction (flaw). See **Figure 23-27.** The ultrasonic testing equipment is partly a timing device for measuring the relative length of elapsed time between the sending of the sound waves and the return of the various echoes.

High-frequency sound is produced by a piezoelectric transducer (crystal) which is electrically pulsed and then vibrates at its own natural frequency, **Figure 23-28.**

In order to transmit the sound waves from the transducer to the metal and return the echoes to the transducer, it is necessary to provide a liquid coupling. This is accomplished with a film of oil, glycerine, or water between the transducer and the test piece. The same results can be achieved by immersing both the test piece and transducer in water, **Figure 23-29.** Immersion-type testing is ideal for production testing, since there is no contact between the transducer and the work; thus, no transducer wear occurs.

The transducer vibrates for about two-millionths of a second. The result is a very short burst of sound waves that travel through the liquid to the surface of the test material. A portion of the sound is immediately reflected from the surface of the metal as a very large echo. Part of the sound will *not* be reflected and will continue into the test material.

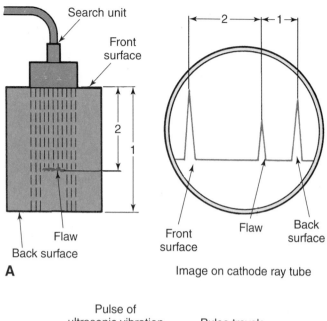

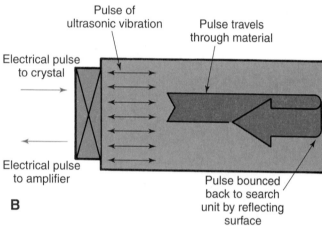

Figure 23-27. *Ultrasonic waves. A—How ultrasonic sound waves are used to detect and locate a flaw in a test piece. B—How sound waves travel through a part and bounce back to locate flaws in the material.*

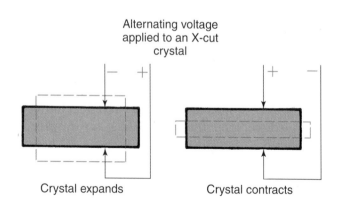

Figure 23-28. *A piezoelectric transducer is a device that receives energy from one system and retransmits it to another system, often in a different form. For inspection, it is pulsed electrically to generate ultrasonic sound waves.*

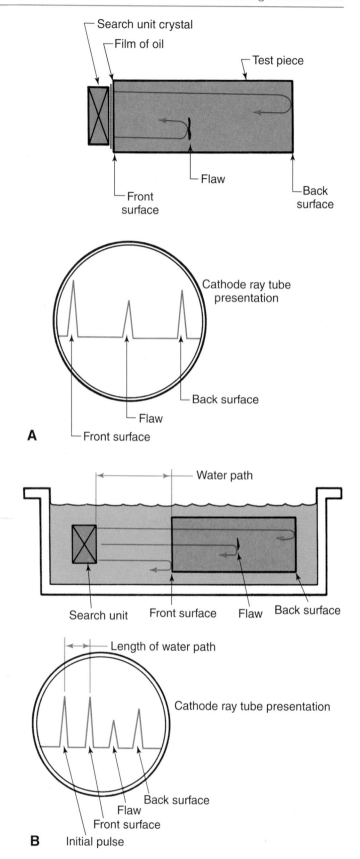

Figure 23-29. *Liquid coupling. A—Action of a contact-type ultrasonic inspection device. A film of oil, water, or glycerine is employed to make a positive contact between the transducer and the test piece. B—Immersion type ultrasonic testing. Note the extra spike on the CRT, indicating the path through water.*

If this portion of the sound encounters no interference, in the form of a discontinuity (flaw) in the material, it will continue until it is partially reflected from the back surface as a second echo or "back reflection." If there is a flaw in the interior, a portion of the sound will be reflected from the flaw and will return to the receiver as a separate echo between the echoes received from the front and back surfaces.

After the transducer has given off its short burst of sound waves, it stops vibrating and "listens" for the returning echoes. When the echoes are received, they cause the transducer to vibrate and to generate an electric current which can be visually displayed on a trace of a *cathode ray tube (CRT)*, which is similar to a television picture tube, **Figure 23-30**. Information transmitted to the CRT can be expanded or condensed to improve readability. This is shown in **Figure 23-31**.

There are two basic categories of ultrasonic testing, as shown in **Figure 23-32**:
- *Pulse echo* uses sound waves, generated by a transducer, that travel through the test piece. The reflected sound waves (echoes) locate the flaws. One crystal is used to both transmit the sound and receive the echoes.
- *Through inspection* has one crystal that transmits the wave through the piece, and another crystal that picks up the signal on the opposite end of the piece.

The beam is partially blocked by a flaw. The reduced intensity of the beam activates a signaling device (flashes a light, rings a bell, etc.) to alert the operator to the flaw. The technique is frequently employed to check the integrity of helicopter rotor blades, for example.

Apparently, there is no size limitation on work that can be tested by ultrasonic techniques. **Figure 23-33** shows one application.

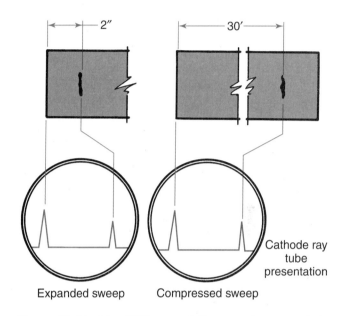

Figure 23-31. *The CRT presentation can be expanded or compressed for easier reading or to compensate for different material thicknesses.*

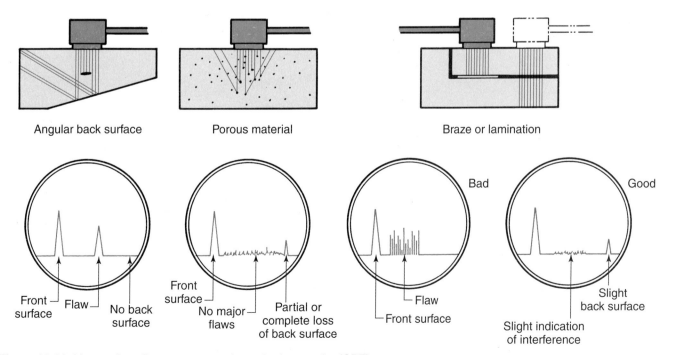

Figure 23-30. *How various flaws appear on the cathode ray tube (CRT).*

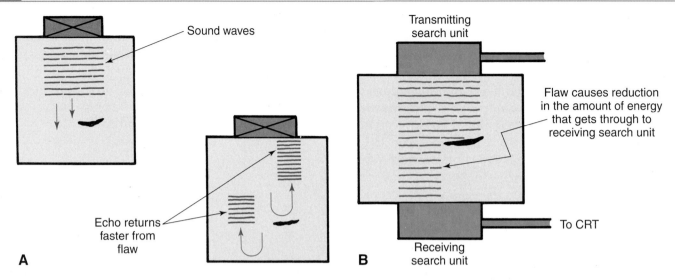

Figure 23-32. *Types of ultrasonic inspection. A—With pulse echo ultrasonic inspection, reflected wave will return sooner when it bounces off a flaw. B—Through-type ultrasonic inspection uses sensor to detect waves on opposite side of the piece.*

Figure 23-33. *This aircraft wing section is the longest adhesively bonded panel ever produced. Ultrasonic test equipment is being used to check the integrity of the bonding. (AVCO Aerostructures Div., AVCO Corp.)*

23.3.7 Inspection by Laser

The laser has also been adapted for quality control, **Figure 23-34**. In addition to being able to check the fit of components on an assembly line, the computer-controlled laser can be made sensitive enough to detect tool wear and automatically compensate for it, **Figure 23-35**. This helps prevent imperfect parts from ever being made. Laser measuring

devices can also be used to evaluate the condition of products that are in use. One such application is shown in **Figure 23-36**.

23.3.8 Eddy-current Inspection

The *eddy-current test* is based on the fact that flaws in a metal product will cause impedance (resistance) changes in a coil brought near it.

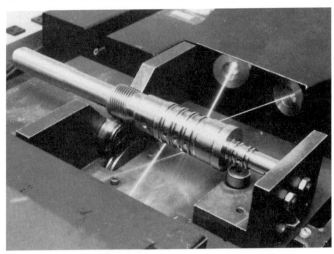

Figure 23-34. A laser being used to inspect a part used in an automatic transmission. Parts inspection gives an immediate indication whether machine adjustments are required. (Ford Motor Company)

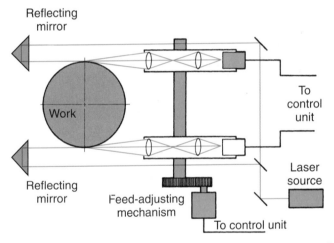

Figure 23-35. One type of laser inspection device. With it, tool wear can be continuously monitored. Note how laser beams can detect work that is oversize or undersize.

Figure 23-36. This technician is using a laser step and gap tool to measure the heights and distances between thermal protection tiles on the Space Shuttle Columbia. The measurements are necessary to ensure that the tiles are precisely aligned. A rough surface or unevenly spaced tiles would increase friction and keep the thermal protection system from working properly. (NASA)

Slightly different eddy-currents will result in test coils placed next to metal parts *with* and *without* flaws. This difference will show which parts pass or fail inspection.

Eddy-current testing methods can be divided into two general categories:

- The *eddy-current differential system* is used to detect cracks, seams, holes, or other flaws in metal parts (such as wire, tubing, and bar stock) as they move off the production line. The test equipment must be sensitive to rapid change.

- The *eddy-current absolute system* is used to detect variations in dimension, composition, and other physical properties of a metal product. Instrumentation for this type operation must be responsive to comparatively small changes.

Eddy-current instrumentation is designed to detect changes and convert them into a form that allows them to be monitored by the operator. See **Figure 23-37.**

Work being tested can be passed through encircling detection coils, **Figure 23-38,** at speeds up to 500 fpm (150 mpm). When a flaw in the test piece is

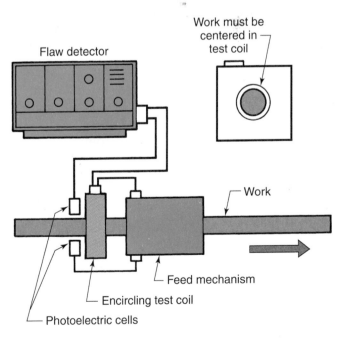

Figure 23-37. Operation of the eddy current flaw detection system. Photoelectric cells turn off alarm system when the end of a test piece passes into the test coil. Any flaw would cause small current changes in the test coil.

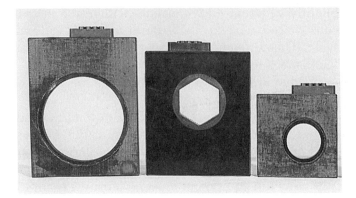

Figure 23-38. *Various sizes and types of encircling test coils that are used to check tubing, bar stock, or wire with the eddy-current system.*

Figure 23-39. *Methods of checking surface finishes. A—Machinist is using a surface finish comparative-type checking panel. B—Surface roughness can also be checked electronically with a profilometer. (Mitotoyo/MTI Corp.)*

detected, the eddy-current unit does one or more of the following:

- Flashes a warning light to alert the operator.
- Sounds a tone alarm.
- Activates a rejection device to eject the part that does not meet standards.
- Marks the section that contains the flaw so it can be removed.
- Provides an electronic record of the section under test.

23.4 OTHER QUALITY CONTROL TECHNIQUES

In addition to the quality control techniques described, industry makes wide use of highly specialized testing devices. For example, the quality of a machined surface may be highly critical. The surface finish can be inspected *visually* with a *comparative checking unit* or *electronically* with a *profilometer*. See **Figure 23-39**.

Because of the versatility of the computer, many new quality control techniques are being developed as the need arises. Computers will also make 100% inspection of parts or entire products possible, **Figure 23-40**.

Remember! The goal of quality control is *not* to detect and discard imperfect parts, but to *prevent* the defective parts from being manufactured in the first place.

Figure 23-40. *New quality control devices make 100% inspection possible. Here lasers scan front end, door frames, and other assemblies to an accuracy of 3.0 mm or about 1/8″. (Pontiac Div., General Motors Corp.)*

TEST YOUR KNOWLEDGE

Please do not write in the text. Write your answers on a separate sheet of paper.

1. Quality control is an important segment of industry. Its purpose is to:
    a. Improve product quality.
    b. Maintain quality.
    c. Help to reduce costs.
    d. All of the above.
    e. None of the above.

2. Quality control falls into two basic classifications. Name and explain each.

3. Precision measuring tools, such as micrometers, vernier tools, or dial indicators, are inspected and calibrated in a _____ laboratory.

4. Inspection by radiography involves the use of _____ and gamma radiation.

5. List four advantages of the radiographic inspection process.

6. Magnetic particle inspection is commonly known as _____.

7. Describe the fluorescent penetrant inspection process.

8. Quality control is an important industrial tool because:
    a. It can be done easily.
    b. It guarantees that the parts being produced meet standards and specifications.
    c. It can be done by unskilled labor.
    d. All of the above.
    e. None of the above.

9. The _____ is an optical gaging instrument designed for the inspection of small parts and sections of larger parts.

10. Explain how the magnetic particle inspection technique operates.

11. Only _____ metals can be inspected by the magnetic particle technique.

12. Ultrasonic inspection makes use of:
    a. Accurately made measuring fixtures.
    b. A high-frequency sound beam.
    c. X-rays.
    d. All of the above.
    e. None of the above.

13. Make a sketch showing the two methods of liquid coupling used for ultrasonic testing.

14. List the two basic categories of ultrasonic testing. Briefly describe each.

15. Make a sketch showing how ultrasonic testing is done.

Tool selection, cutting speeds and feeds, and other machining considerations are affected by the characteristics of the metal being processed. This carbide-coated cutter insert is being used for semi-finished turning on a high-carbon steel workpiece. (Valenite, Inc.)

Metal Characteristics

IMPORTANT TERMS

alloy	ferrous
Aluminum Association Designation System	honeycomb
	nonferrous
base metal	red hardness
carbon content	tungsten carbide
ductility	

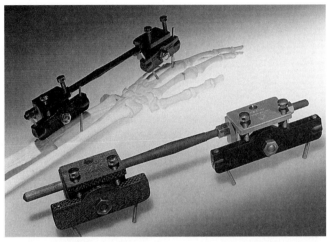

Figure 24-1. An example of the revolution in materials technology is this splint used to immobilize severe fractures. The original metal version, shown at top, has been replaced by the high-strength carbon fiber composite at bottom. The composite version is lighter, stiffer, and does not interfere with X-rays or other diagnostic tools. (Polygon Company).

More than a thousand different metals and alloys are used by the metalworking industry. Most of them are worked in the machine shop. However, it must be noted that there is a "new materials revolution" taking place, especially in the aerospace and medical equipment industries. Many nonmetals and composites of metals and nonmetals are rapidly emerging in applications that were once the exclusive domain of metals. See **Figure 24-1**.

Can you tell by looking at a piece of metal whether it is *ferrous* (containing iron), or *nonferrous* (containing no iron)? Is it an *alloy* (a mixture of two or more metals)? Could it be a *base metal*, like tin, copper, or zinc? From a practical point of view, it is almost *impossible* to find out much about a piece of metal by just looking at it.

A machinist is not expected to have a complete understanding of all the technicalities of metals. However, a working knowledge of the various materials, and the common terms associated with them, is essential.

In addition to the conventional metals worked in a modern machine shop, several of the newer materials will be described in this chapter. Plastics are described in Chapter 29.

24.1 CLASSIFYING METALS

Modern industrial metals may be classified as:
• Ferrous metals.
• Nonferrous metals.
• High-temperature metals.
• Rare metals.

Because of space exploration and military research, great strides have been made in the development of the latter two groups in recent years.

Note: For tables that show physical properties of metals, dimensional tolerances, and feeds and speeds for machining, see the Reference Section of this text.

24.2 FERROUS METALS

The irons and steels, and their alloys, make up the family of ferrous metals. For simplicity and easier understanding, ferrous metals can be further subdivided into various categories.

24.2.1 Cast Irons

Cast irons are iron alloys that contain 2.0% to 5.0% carbon with small quantities of silicon and manganese. There may also be traces of other elements in the alloy.

Of the cast irons, *gray iron* and *malleable iron* are most widely used in industry. They can be found in great quantities in automotive, railroad, farm equipment, and machine tool bodies. See **Figure 24-2**.

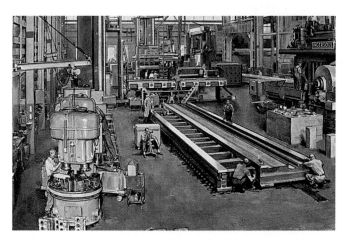

Figure 24-2. Many quality gray and malleable iron castings are used in modern machine tool construction, because of iron's rigidity and stability with wide changes in temperature, and its ease of machining to close tolerances. Shown is the assembly of a rail-type multiaxis milling machine. (Ingersoll Rand)

Malleable iron can be hammered into shape without fracturing. Most cast irons can be readily machined once the hard surface scale has been penetrated. Carbide cutting tools are recommended because of the abrasive nature of the iron scale. No cutting fluids should be used on cast iron. Compressed air is recommended if a coolant is needed. Use extreme care when employing compressed air as a coolant. To prevent injury, chips and dust be contained.

24.2.2 Steels

Steel is often considered the "backbone" of the metalworking industry, **Figure 24-3**. *Carbon steel* is very common. It is an alloy of iron and carbon and/or other alloying elements, but carbon is the

Figure 24-3. Molten iron from a blast furnace is being charged into a basic oxygen furnace vessel at a Bethlehem Steel plant. After charge has been completed, vessel will return to its upright position for oxygen "blow." Blast furnace iron, mixed with scrap and selected additives, will then be refined into steel.

major alloying agent. The alloying elements impart to iron the desired characteristics needed to perform a specific job.

The physical properties of steel are unique. Steel can be made soft enough to be easily machined. By careful heat treatment, the soft steel can be transformed into a "glass-hard" material. By slightly varying the *heat treatment* procedure, steel can be given a hard, wear-resistant surface while retaining a soft, tough core to resist breaking. Its magnetic qualities also make steel ideal for many electrical applications.

Carbon steels are classified according to the amount of carbon they contain. The *carbon content* is measured in percentage or in points (100 points equal 1%). They are available in all standard mill forms, **Figure 24-4**.

Low-carbon steels do not contain enough carbon (less than 0.30% or 30 points) to be hardened. They are easy to work and can be case-hardened. Low-carbon steel is often called *mild steel* or *machine steel.* It is used for nuts, bolts, screws, gun parts, precision shafting, tie rods, tool cylinders, and similar applications.

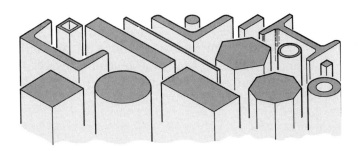

Figure 24-4. A small portion of the hundreds of shapes and sizes of metals that are available.

Hot rolled steel is steel that has been rolled to finished size while hot. It is easily identified by its black oxide surface scale, **Figure 24-5**.

Cold finished steel is steel that has been *"pickled"* or treated with a dilute acid solution to remove the oxide coating. After pickling, the steel is drawn or rolled to finished size and shape while cold. Cold finished steel is characterized by a smooth bright finish. The process improves the machinability of the steel.

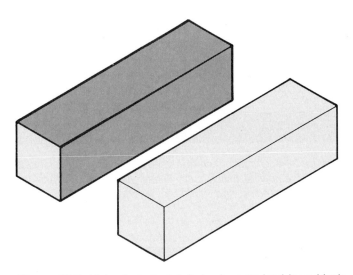

Figure 24-5. Hot rolled steel, left, is characterized by a black oxide coating. Cold finished steel, at right, has a smooth shiny surface.

Medium-carbon steel contains 0.30% to 0.60% (30 to 60 points) carbon. The carbon content is sufficient to allow partial hardening with proper heat treatment. The heat-treating process improves the strength of the steel. Available in all standard mill forms, medium-carbon steels are used for machine parts, automotive gears, camshafts, crankshafts, cap screws, precision shafting, etc.

High-carbon steels contain 0.60% to 1.50% (60 to 150 points) carbon. They are available in hot rolled form. However, some high-carbon steel shapes may

be purchased with ground surfaces. Drill rod and ground flat stock are examples.

High-carbon steels are found in products that must be heat-treated, **Figure 24-6**. Applications include heavy machinery parts, control rods, wrenches, hammers, screwdrivers, pliers, springs, and a large variety of agricultural equipment.

Figure 24-6. The teeth of this giant helical gear are being hardened by the induction method. Only the faces of the teeth are hardened to minimize their wear. (Philadelphia Gear Corp.)

Adding controlled amounts of sulfur and/or lead to carbon steels will result in improved machinability without greatly affecting the metal's mechanical properties. Usually, machining involves the removal of considerable metal. The machine tool must be capable of increased cutting speeds before the machining of high-carbon steels will prove economical.

24.2.3 Alloy Steels

Alloy steels have other metal elements added to change their characteristics. Alloy steels are more costly to produce than carbon steels because of the increased number of special operations that must be performed in their manufacture, **Figure 24-7**.

Elements such as nickel, chromium, molybdenum, vanadium, manganese, and tungsten are used to make alloy steels harder, stronger, or tougher. A combination of two or more of the above elements usually imparts some of the characteristic properties of each.

Figure 24-7. Preparing to lift an engine into the number one position on the Discovery Space Shuttle. Many alloy and high-temperature steels were used in the manufacture of this huge engine (for size comparison, note technician at bottom center). The engine has been designed to safely handle the super cold of liquid oxygen at one end and the blast furnace temperature of exhaust gases at the other, while operating in the near absolute zero temperature of space. (NASA)

Chromium-nickel steels, for example, develop good hardening properties with good *ductility* (a property of metal that permits permanent deformation by hammering, rolling, and drawing without breaking or fracturing). Chromium-molybdenum combinations develop excellent hardenability with satisfactory ductility and a certain amount of heat resistance.

24.2.4 Alloy Metallic Elements

Metallic elements that may be added to alloy steels, and properties they impart to the steel, include:
- *Nickel* imparts toughness and strength, particularly at low temperatures. Nickel steels permit more economical heat treatment and have improved resistance to corrosion. They are especially suitable for the case-hardening process and are used for applications such as armor plate, roller bearings, and aircraft engine parts.
- *Chromium* is added when toughness, hardness, and wear resistance are desired. It is the basis of *stainless steel*. Chromium steel is found extensively in automotive and aircraft parts, **Figure 24-8**.
- *Molybdenum* is employed as an alloying agent when the steel must remain tough at high temperatures.
- *Vanadium,* when added as an alloying element, produces a steel that has a fine grain

structure and increased toughness at high temperatures.
- *Manganese* purifies steel and adds strength and toughness. Manganese steel is used for parts that must withstand shock and hard wear, **Figure 24-9**.

Figure 24-8. Chromium steel is used extensively in the landing gear assembly of this aircraft. It has ability to withstand the shock of landing the multi-ton aircraft. (U.S. Air Force Thunderbirds)

Figure 24-9. Many components of earthmoving vehicles are made from manganese steel. This type of steel is strong and tough enough to withstand constant abrasive wear.

- *Tungsten,* when added in the proper amount, makes steel that has a fine, dense structure with improved heat treatment qualities. It is one of the principal alloying agents in many tool steels. Tools made with these steels retain their strength and hardness at high temperatures.
- *Cobalt* is the chief alloying element in high-speed steels because it improves the *red hardness* (quality of remaining hard when red hot) of cutting tool materials. Wear resistance is also improved.

Some alloy steels possess relatively high strength at moderately elevated temperatures. They are finding many applications in airframe structures of aerospace vehicles.

24.2.5 Tool Steel

Tool steel is the term usually applied to steels found in devices that are used to cut, shear, or form materials. They may be either carbon or alloy steels. Steels in the lower carbon content range (0.70% to 0.90% or 70 to 90 points) are used for tools subject to *shock.* Higher-carbon-content tool steels (1.10% to 1.30% or 110 to 130 points) are utilized when tools with *keen cutting edges* are required.

Drills, reamers, milling cutters, punches, and dies are made from alloy tool steels. Although several tool steels can be hardened using water as a quenching medium, most must be hardened in oil or in air. The latter are often known as *oil-hardened* or *air-hardened steels.*

Some alloy steels are also classified as *high-speed steels,* because they are capable of making deeper cuts at higher cutting speeds than regular tool steels. They possess red hardness, or the ability to retain their hardness at high temperatures, **Figure 24-10.** They also possess high abrasion resistance. In spite of the development and wide spread use of cemented carbides and ceramics, high-speed steels remain a major cutting tool material.

24.2.6 Tungsten Carbide

Tungsten carbide is the hardest human-made metal. It is almost as hard as a diamond. The metal is shaped by molding tungsten, carbon, and cobalt powders under heat and pressure in a process known as *sintering.* The metals fuse together without melting.

Tungsten carbide, while not a true steel, is usually classified with the steels. Tools made from this family of materials can cut many times faster than high-speed cutting tools, **Figure 24-11.**

Figure 24-10. *Two high-speed steel (HSS) milling cutters are being used to mill an aluminum plate into the specified shape.*

Figure 24-11. *Cutting tools made from tungsten carbide can cut many times faster than high-speed steel cutters. (Valenite, Inc.)*

Applying a 0.0001″ (0.002 mm) thick coating of *titanium nitride (TiN)* and *titanium carbide (TiC)* to the surface of carbide tools extends their life 3 to 8 times longer than uncoated tools. Uncoated tools acquire a buildup of material on the cutting edge, which produces a ragged surface finish on the work. Coated tools resist buildup from the chips. They also run cooler, last longer, hold tolerances better, and produce a better surface finish. See **Figure 24-12.**

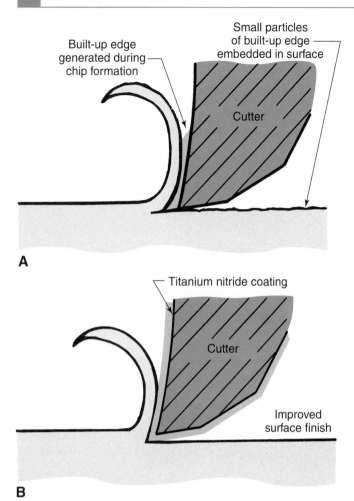

Figure 24-12. *Tool coating benefits. A—Uncoated cutting tools acquire a buildup of material on the cutting edge, which produces a "ragged" surface finish as parts of the buildup flake off. B—Tools coated with titanium nitride (TiN) and titanium carbide (TiC) resist the buildup from the chips. The coated cutting tools run cooler, stay sharp longer, and produce a better surface finish.*

Recommended carbide grade applications and how the various manufacturers list their corresponding products are shown in chart form in **Figure 24-13.**

24.2.7 Stainless Steels

There are more than a hundred different stainless steels. However, one characteristic common to all of them is that they contain enough chromium to render them corrosion resistant. *Stainless steels* may be divided into three basic groups:

- The *austenitic* classification includes the chromium-nickel and chromium-nickel-manganese stainless steels. Generally, they are hardenable only by cold working. The American Iron and Steel Institute (AISI) 300 series of stainless steels are in this category.
- The *martensitic* stainless steel alloys of iron, carbon, and chromium are characteristically magnetic in nature and obtain their hardness through normal heat-treating processes.
- The *ferritic* stainless steels have more than 18% chromium. They are nonhardenable and all of them are magnetic.

Stainless steels may be machined with techniques normal for mild steels. However, some precautions must be observed with stainless steel:

- Feeds must be high enough to ensure that the cutting edge(s) get under the previous cuts and thus avoid the hardened portions.
- Tools must be as large as possible, because the life of the cutting edge(s) depends on good heat dissipation into the body of the cutting tool.
- Finishing cuts should be used when working to close tolerances.
- The machine should be adjusted so there is minimum play. Otherwise, the cutting tool may "ride" the work and glaze and/or harden the surface.

24.2.8 Identifying Steels

Because different kinds of steels look alike, several methods of identification have been devised. They include identification by chemical composition, mechanical properties, the ability to meet a standard specification or industry accepted practice, or the ability to be fabricated. The shape of the *mill form* (rod, bar, structural shape, etc.), and the intended use for the metal, can also determine the method of identification.

AISI/SAE codes

The American Iron and Steel Institute (AISI) and the Society of Automotive Engineers (SAE) have devised almost identical standards that are widely used for identifying steel. Both systems use an identical *four-number code* (some steels require a fifth digit) that describes the physical characteristics of the steels. The AISI system also makes use of a *prefix letter* (A, B, C, etc.) that indicates the steel manufacturing process used.

Recommended Carbide Insert Grade Applications	
Grade C2 uncoated	For cast iron, nonferrous materials, and general-purpose use.
Grade C4 uncoated	Light finishing cast iron, nonferrous, and general-purpose.
Grade C6 uncoated	For steel, steel casting, malleable cast iron, stainless steels, and free cutting steels.
Grade C5–C6 Titanium nitride coated (TiN)	For carbon steels, tool steels, alloy steels, steel castings, malleable cast iron, austenitic and martensitic stainless steels, and free cutting steels.
Grade C6–C7 Titanium carbide coated (TiC)	For steel, steel castings, malleable cast iron, nodular iron, and martensitic stainless steels.
Grade C2–C4–C6–C8 Aluminum oxide, ceramic coated	For carbon steels, tool steels, stainless steels, alloy steels, steel castings, gray cast iron, malleable cast iron, and nodular iron.

A

Carbide Comparison Chart												
Class	RTC	Newcomer	Carboloy	Iscar	Kennametal	Mitsubishi	Sandvik	Seco	Sumitomo	Valenite	V.R. Wesson	RTW
C-2	RTC2	N21	833	IC20	K-68	UTi20T HTi10	CG-20	HX	G10E	VC-1 VC-2/VC-28	2A5	CQ-2
C-6	RTC5	N60	395 78	IC54	K420	STi20 UTi20T	S-2 S-35	S-2 S-4	ST20E	VC-6	VR-75	Cy-5
TiN	RTC052	NN60	516	IC656	KC810 KC950	–	GC-425 GC-315 GC-225	TP15	AC720 AC815 AC15	VN-5 VC-7 VC-88 VO1	663 650 660	TRW-755
TiN & AlO₂ & TiN	RTC054	1000	550 560	IC635	KC850	U610	GC4025 GC235	TP10	–	VN2	653	718

B

Figure 24-13. Recommendations for various tungsten insert type cutting tools. A—Recommended applications. B—Carbide insert comparisons.

The four-numeral code works as follows: The *first digit* classifies the steel. The *second digit* indicates the approximate percentage of the alloying element in the steel. The *last two digits* show the approximate carbon content of the steel in points or hundredths of one percent. For example, a steel designated SAE 1020 is a carbon steel with approximately 20 points or 0.20% carbon. The AISI and SAE four-digit code applies primarily to bar, rod, and wire products. See the Reference Section of this text for more information on this subject.

Color coding

Color coding is another method of identifying the many kinds of steel, **Figure 24-14**. Each

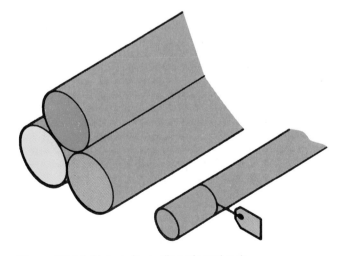

Figure 24-14. Note color coding of steel rods.

commonly used steel is designated by a specific color. The color coding is painted on the ends of bars that are 1″ (25 mm) or larger in diameter. On bars smaller than 1″ (25 mm), the color code may be applied to the end of the bar or on an attached tag.

Spark test

The *spark test* is also employed at times to determine grades of steel. The metal is touched to the grinding wheel lightly and the resulting sparks are carefully observed. See **Figure 24-15**.

When performing or observing a spark test, wear approved eye protection. The grinder eyeshield should be clean and in position. The tool rest must also be properly adjusted.

24.3 NONFERROUS METALS

There are many metals that do not have iron as their basic ingredient. Known as *nonferrous metals*, they offer specific properties, or combinations of properties, that make them ideal for tasks where ferrous metals are not suitable, **Figure 24-16**.

24.3.1 Aluminum

Aluminum has come to mean a large family of aluminum alloys, not just a single metal. As first

Figure 24-16. *Aluminum alloys are utilized in construction of this high-speed train designed to travel at 200 mph (320 kph). The metal is light, strong, and corrosion resistant. (French National Railroads)*

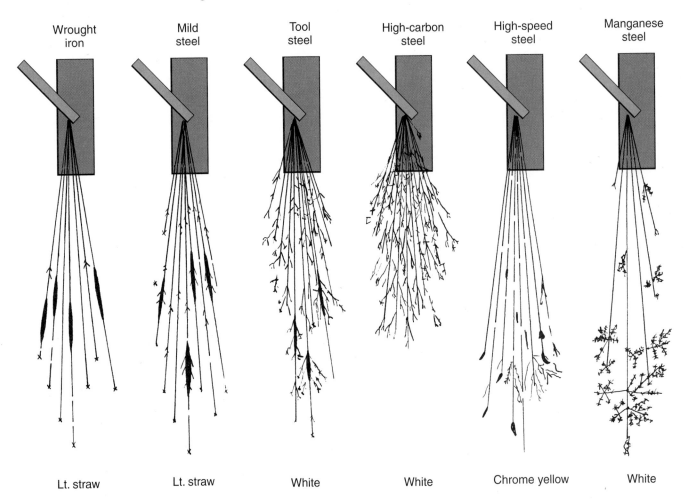

Wrought iron	Mild steel	Tool steel	High-carbon steel	High-speed steel	Manganese steel
Lt. straw	Lt. straw	White	White	Chrome yellow	White

Figure 24-15. *A spark test is sometimes employed to determine the grade of steel. Touch steel to the grinding wheel lightly, and observe the color and form of the resulting sparks.*

produced, aluminum is 99.5% to 99.76% pure. It is somewhat soft and not very strong.

The strength of aluminum can be greatly increased by adding small amounts of alloying elements, by heat-treating, or by cold working. A combination of the three techniques has produced aluminum alloys that, pound for pound, are stronger than structural steel. In addition to increasing strength, alloying elements can be selected to improve welding characteristics, corrosion resistance, machinability, etc.

There are two main classes of aluminum alloys: wrought alloys and cast alloys. The shape of *wrought alloys* is changed by mechanically working them by forging, rolling, extruding, hammering, or other techniques. *Cast alloys* are shaped by pouring metal into a mold and allowing it to solidify, **Figure 24-17**.

Figure 24-17. *This die cast aluminum wheel is emerging from die (mold) of a casting machine. The wheel is lighter, stronger, and more attractive than conventional pressed steel wheels. (Kelsey-Hayes)*

Each alloy is given an identifying number. Known as the *Aluminum Association Designation System*, it is a four-digit code, plus a temper designation. *Temper designation* indicates the degree of

hardness of the alloy. It follows the alloy identification number and is separated from it by a dash.

Aluminum alloys possess many desirable qualities. They are extremely strong and corrosion-resistant under most conditions. The alloys are lighter than most commercially available metals. They can be shaped and formed easily, and are readily available in a multitude of sizes, shapes, and alloys.

Machining aluminum

Most of the wrought aluminum alloys possess excellent machining characteristics. They are capable of being machined to intricate shapes at high cutting speeds. However, the makeup of an aluminum alloy is a factor that can affect machinability. Some aluminum alloys of a *nonabrasive nature* (those containing copper, magnesium, or zinc), have improved machinability. Other alloys with *abrasive constituents* (such as silicon) reduce tool life and machined surfaces may have a slightly gray finish with little luster.

Most aluminum alloys are easier to machine to a good finish when in full hard temper than when in an annealed state. Machining characteristics of more commonly used aluminum alloys are:

- *Number 1100* and *3003 alloys* have good machinability but are gummy in nature. Turnings are long and stringy, causing difficulty in chip disposal. Good results can be obtained if the cutting tools have large top and side rake angles, with keen, smooth cutting edges.
- *Number 5052 alloy* has turnings that are long and stringy, and the machined surface is not as good as on 3003. Machinability is good, however.
- *Number 5056 alloy* has good machinability with the advantage of fairly easy chip disposal.
- *Numbers 2017-T4* and *2014-T6 alloys* machine to an excellent finish. Of the two, 2014-T6 has better machinability because of the heat-treating method employed. It causes greater tool wear.
- *Number 2024-T3 alloy* has good machining characteristics with properly sharpened and honed tools. Surface finishes are excellent.
- *Number 6061-T6 alloy* contains silicon and magnesium. It is more difficult to machine than the 2000 series alloys. Properly sharpened cutting tools and coolants with good lubricating qualities are essential. Fine finishes are obtainable with moderately heavy cuts.

• *Number 7075-T6 alloy* is the highest strength aluminum alloy that is commercially available. Machining qualities are good.

High-speed cutting tools will produce satisfactory results when machining most aluminum alloys. However, optimum results will require the use of carbide or ceramic tools. Recommended tool geometry for aluminum is shown in **Figure 24-18**.

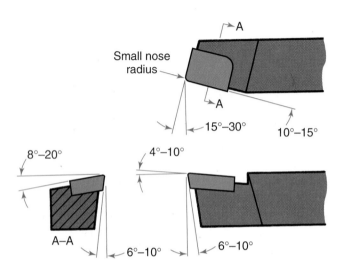

Figure 24-18. *Configuration of carbide lathe tool for machining aluminum.*

24.3.2 Magnesium

Magnesium alloys are the lightest of the structural metals. They have a high strength-to-weight ratio. These alloys have excellent machining properties, and can be machined by all common metalworking techniques. The lathe tool recommended for magnesium is shown in **Figure 24-19**.

Despite their many advantages, magnesium alloys must be worked with extreme care. Several of the alloys developed for use at elevated temperatures in aerospace vehicles contain thorium, a low-level radioactive material. They must be handled according to strict safety precautions for radioactive materials.

Another concern when machining magnesium is that extreme care must be taken because the chips or particles are highly flammable. (Because of the relatively high thermal conductivity of magnesium, there is normally no fire hazard when a solid section of the metal is exposed to fire, however.) Burning magnesium chips are so intensely hot (600°F or 315°C) they cannot be extinguished by conventional firefighting techniques. Water or commercial extinguishing agents will actually intensify the fire. A special (Class D) extinguishing agent is made for flammable metals fires.

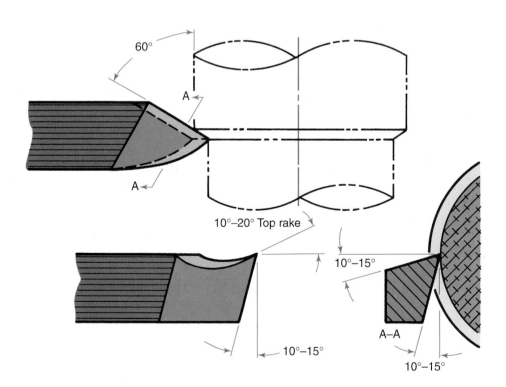

Figure 24-19. *Configuration of HSS lathe tool recommended for machining magnesium.*

To guard against magnesium fires, do not allow magnesium chips to accumulate on or around the machine, and use a straight mineral oil cutting fluid in adequate quantities to flood the work. Avoid water-based coolants, since they react with the chips and actually intensify a magnesium fire once it gets started.

24.3.3 Titanium

Titanium is a metal as strong as steel but only half as heavy. It bridges the gap between aluminum and steel. It is silvery in appearance, and extremely resistant to corrosion. Most titanium alloys are capable of continuous use at temperatures up to about 800°F (427°C). This makes titanium ideal for use in high-speed aircraft components. Aluminum fails rapidly at temperatures above 250°F (121°C).

Machining titanium

Titanium can be machined with conventional tools if the following practices are observed:
- Tool and work setups must be rigid.
- Tools must be kept sharp.
- Good coolants must be applied in adequate quantities.
- Cutting speeds must be slower, with heavier feeds than those used for steel.

Turning titanium that is commercially pure is very similar to turning 18-8 stainless steel. The alloys are somewhat more difficult to machine. Tungsten carbide and some types of ceramic tools produce the best results.

Milling titanium is more difficult than turning because the chips tend to weld to the cutter teeth. Climb milling may alleviate the problem to a great extent. Cast alloy tools often prove more economical to use than carbide tools. A water-based coolant is recommended.

Drilling titanium with conventional high-speed steel drills will produce satisfactory work. Drills should be no longer than necessary to produce the required depth hole and still allow the chips to flow unhampered.

Tapping titanium is one of the more difficult machining operations. The tap has a tendency to freeze or bind in the hole. Careful selection of cutting fluid will minimize this problem.

Sawing titanium requires a slow speed of about 50 fpm (15 mpm) with heavy, constant pressure. The tooth geometry of the blade must be designed for sawing titanium.

24.4 COPPER-BASED ALLOYS

Brass and bronze are the most familiar of the *copper-based alloys.* However, lesser-known heat-treatable alloys are available. The newer alloys include copper and exotic metals like zirconium and beryllium. Most copper-based alloys are available in rod, bar, tube, wire, strip, and sheet forms.

24.4.1 Copper

Copper is a *base metal;* that is, a pure metallic element. It is probably the oldest known metal. It can be shaped easily, but becomes hard when worked and must be annealed or softened. Copper is difficult to machine because of its toughness and softness.

With copper, keep tools honed sharp and make as deep a cut as possible. Cutting fluids are *not* ordinarily needed. except when tapping.

24.4.2 Brass

Brass is an alloy of copper and zinc. It ranges in color from reddish yellow to a silvery yellow, with the color determined by the percentage of zinc it contains. Most brasses can be readily machined.

24.4.3 Bronze

Bronze, **Figure 24-20,** is an alloy of copper and tin. It is harder than brass and is much more expensive. Many special bronze alloys include additional alloying elements, such as aluminum, nickel, silicon, and phosphorous. Most bronzes are relatively easy to machine with sharp tools, **Figure 24-21**

24.4.4 Beryllium Copper

Beryllium is one of the newer copper-based alloys. Its machining qualities are similar to those of copper.

Machining beryllium copper can pose a definite health hazard if precautions are not observed. The fine dust generated by machining and filing can cause severe respiratory damage. A respirator-type face mask must be worn. Special procedures must also be followed when cleaning machines used to machine beryllium copper.

A vacuum system should be employed to remove the beryllium copper dust, or the work should be liberally flooded with cutting fluid. Do not permit cutting tools to become dull; since dull tools generate more dust than sharp tools.

Figure 24-20. Bronze. A—Bronze is an excellent metal for ship propellers. The one shown is 31 feet in diameter. (Bethlehem Steel Co.) B—Ship propellers are machined on large CNC multiaxis machine tools. (Bird-Johnson Company)

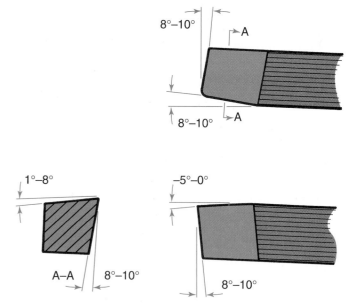

Figure 24-21. Configuration of HSS lathe tool for turning brass and bronze.

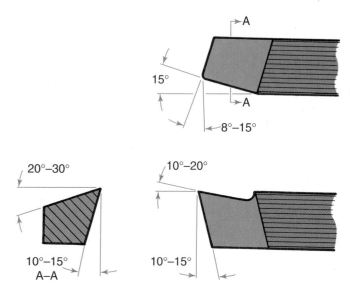

Figure 24-22. Configuration of HSS lathe tool for turning beryllium copper.

Beryllium copper can be heat-treated. It should *not* be machined in the annealed state. Recommended tool geometry for lathe tools used with beryllium copper is shown in **Figure 24-22**. When machining beryllium, employ a mineral oil-based cutting fluid. The fluid should be selected for its cooling properties, rather than for lubrication.

24.5 HIGH-TEMPERATURE METALS

The nuclear and aerospace industries are chiefly responsible for the development of a number of *high-temperature metals*. These metals have the unique properties of high strength for extended periods at elevated temperatures, **Figure 24-23**. They are sometimes called *superalloys*.

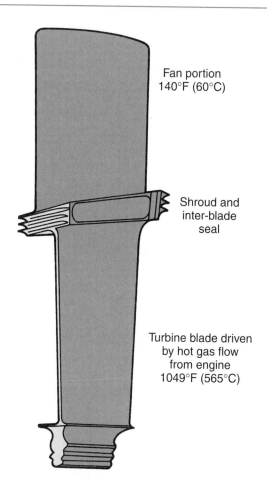

Fan portion
140°F (60°C)

Shroud and
inter-blade
seal

Turbine blade driven
by hot gas flow
from engine
1049°F (565°C)

Figure 24-23. Metal used for the manufacture of this fan jet turbine blade must be able to withstand high temperatures for long periods of time without failing.

24.5.1 Nickel-based Alloys

Nickel-based alloys are known commercially as *Iconel-X, Hastealloy-X, Rene 41,* etc. They have many uses in jet engines, rocket engines, and electric heat-treating furnaces. These applications require metals that can operate at temperatures of 1200°F to 1900°F (649°C to 1038°C). These metals are not easy to machine by conventional methods.

24.5.2 Molybdenum

Molybdenum has excellent elevated temperature strength in the 1900°F to 2500°F (1038°C to 1372°C) range, and has found many applications in modern technology. It also has great resistance to corrosion by acids, molten glass, and metals.

Molybdenum machines similar to cast iron if both the work and cutting tool are mounted rigidly. For most work, tungsten carbide and ceramic tools are preferred over high-speed steel tools.

24.5.3 Tantalum

Tantalum alloys are specified where dependability at temperatures above 2000°F (1094°C) is required. Tantalum is used for rocket nozzles, heat exchangers in nuclear reactors, and in some space structures.

Tantalum is not an easy metal to machine. It is gummy and has a tendency to tear. High-speed steel tools are usually recommended. Extreme cutting angles are used to keep the tool and chip clear of the work. The tool should be well supported, with little overhang.

24.5.4 Tungsten

Tungsten melts at a higher temperature —6200°F (3429°C) — than any other known metal. However, tungsten is not resistant to *oxidation* at high temperatures (above 930°F or 499°C) and must be protected with a suitable coating, such as one of the silicides. It has many uses in rocket engines, welding electrodes, and high-temperature furnaces. Tungsten is an ideal metal for breaker points in electrical devices.

Machining is quite difficult, but can be done with carbide and ceramic tools if the work is preheated to about 400°F (204°C). The final shaping of a tungsten part is frequently done by grinding. Adequate cooling of the grinding wheel with an oil-based compound is recommended.

24.6 RARE METALS

Name just about any rare metal, and the odds are that someone is attempting to find a way to use it for aerospace applications. Most metals in this category are available only in small quantities for experimental purposes. Many of them cost considerably more than gold.

Included in the rare metals group are such elements as scandium, yttrium, cerium, europium, lanthanum, and holmium. While they may seem strange and almost unknown at the present time, it was not too long ago that uranium, titanium, and beryllium were in the same category. In fact, it has only been slightly more than 100 years since *aluminum* was considered a rare metal worth many times more than gold!

24.6.1 Other Materials

In addition to conventional metals and plastics, the modern machine shop is also expected to work other types of materials.

Honeycomb

Many ways have been devised to give existing metals greater strength and rigidity while reducing weight. *Honeycomb* sandwich structures, **Figure 24-24,** are an example. Sections of thin material (aluminum, stainless steel, titanium, and nonmetals like Nomex® fabric) are bonded together to form a structure that is similar in appearance to the wax comb that bees create to store honey.

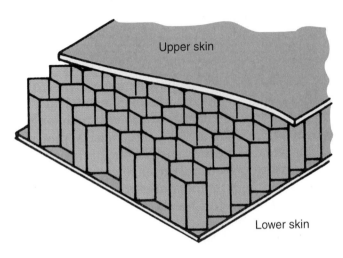

Figure 24-24. Honeycomb has great strength and rigidity for its weight. It has many applications in aerospace industries.

Figure 24-25. Machining honeycomb can be a delicate operation. A three-dimensional milling machine is being used to carve aluminum honeycomb to airfoil shape for the tail section of an aircraft. (Hexcel Corp.)

When honeycomb is rigidly bonded between two metal or composite sheets to form a sandwich panel, it becomes a strong structure. It has a very high strength-to-weight ratio and rigidity-to-weight ratio. The bonding is done with an adhesive, or the materials fused by brazing or resistance welding.

Because the material is fragile (before being shaped and bonded into rigid units), it can cause problems in machining, **Figure 24-25.** In addition to special tools that literally pare the material off, *electrolytic grinding* is the most rapid method for machining honeycomb. It does not leave a burr that would create problems in its removal.

The space shuttle and other late-model aerospace vehicles utilize large quantities of aluminum, stainless steel, and titanium honeycomb in their structures. See **Figure 24-26.**

Composites

Composites are a relatively new development that utilize fibers of conventional materials (and some *not* so common materials), in both pure and alloy forms. Fibers such as pure iron, graphite, boron, and fiberglass are bonded together in a special *plastic matrix* (binding substance, such as an

Figure 24-26. An orbiter (space shuttle) being mated to the back of its 747 carrier aircraft. The metals and composites used in the manufacture of each vehicle were carefully selected to provide maximum strength while keeping weight at a minimum. This was no easy task, given their operating requirements. (NASA)

epoxy) under heat and pressure. These materials are generally lighter, stronger, and more rigid than many conventional metals. Some are being used in commercial jet engines for parts like bushings and washers. Some are capable of withstanding temperatures of 600°F (315°C) for long periods of time, and temperatures of 1000°F (540°C) for short periods of time.

Present uses for composites are concentrated in the aerospace and automotive industries, **Figure 24-27.** However, composites are finding applications in such things as fishing poles, skis, golf clubs, tennis rackets, bicycle frames, and other products, **Figure 24-28.** Much research is being done

Figure 24-27. *The F-117 "Stealth" aircraft utilizes large quantities of composites in its manufacture. Many new manufacturing techniques had to be developed to construct an aircraft of this type. (Lockheed Martin Corp.)*

Figure 24-28. *One of the many types of composite materials is glass or carbon fiber bound together with a resin. Here, glass fiber is being wound around a large metal mandrel before resin is applied. Once the resin cures (hardens), grinding and other machining techniques can be used to achieve final dimensions. Note the dust mask being worn by the technician to prevent inhaling tiny airborne fibers. (Compositek Corporation)*

to reduce the cost of composites so they can be employed to make lighter and safer automobile bodies and other components.

TEST YOUR KNOWLEDGE

Please do not write in the text. Write your answers on a separate sheet of paper.

1. What are the four main categories of metals?

2. Iron and steel are classified as _____ metals.

3. Carbide cutting tools are recommended for machining cast iron because:
   a. Cast iron is hard and brittle.
   b. Cast iron has a hard surface scale.
   c. Cast iron is difficult to machine.
   d. All of the above.
   e. None of the above.

4. Carbon steel is an alloy of _____ and _____.

5. How are carbon steels classified?

6. Hot rolled steel is characterized by the _____ on its surface.

7. The machinability of carbon steel is improved if _____ or _____ is added as an alloying element.

8. Nickel, chromium, molybdenum, vanadium, and tungsten are used to make steel _____, _____, and _____.

9. Drills, reamers, some milling cutters, and similar tools are usually made from _____ steel.

10. The chief characteristic of stainless steel is its resistance to _____.

11. List the three basic groups of stainless steel.

12. Aluminum, magnesium, and titanium are _____ metals.

13. Magnesium is the _____ of the structural metals.

14. Titanium is a metal that is as _____ as steel but only _____ as _____.

15. Brass and bronze are _____-based alloys.

16. Brass is an alloy of copper and _____.

17. Bronze is an alloy of copper and _____.

18. Why must a machinist take special precautions when working beryllium copper?

19. Nickel-based alloys, molybdenum, tantalum, and tungsten are classified as _____ metals.

20. What is the structural material known as honeycomb?

21. What are composites?

This technician is adjusting the controls of a multizone carbonitriding furnace used to case-harden steel. (Master Lock Co.)

Chapter 25

Heat Treatment of Metals

LEARNING OBJECTIVES

After studying this chapter, you will be able to:
○ Explain why some metals are heat-treated.
○ List some of the metals that can be heat-treated.
○ Describe some types of heat-treating techniques and how they are performed.
○ Case harden low-carbon steel.
○ Harden and temper some carbon steels.
○ Compare hardness testing techniques.
○ Point out the safety precautions that must be observed when heat-treating metals.

IMPORTANT TERMS

annealing
Brinell hardness tester
case hardening
hardness number
normalizing

Rockwell hardness tester
scleroscope
stress-relieving
tempering
Webster hardness tester

Since many parts produced in the machine shop must be heat-treated before use, it is important that the machinist be familiar with the basic science of heat-treating metals.

Heat treatment involves the controlled heating and cooling of a metal or alloy to obtain certain desirable changes in its physical characteristics, **Figure 25-1**. These changes include improving resistance to shock, developing toughness, and increasing wear resistance and hardness, **Figure 25-2**.

Heat treatment is done by heating the metal to a predetermined temperature, then quenching it (cooling it rapidly) in water, brine, oil, blasts of cold air, or liquid nitrogen. Refer to **Figure 25-3**.

Desired qualities do not always prevail after quenching. Stresses can develop that, under certain conditions, may cause some steels to shatter. Therefore, the metal may have to be reheated to a

Figure 25-1. *A computerized material testing system being used to determine tensile strength of metals. Metal is subjected to a slowly applied force that pulls it apart. The metal fractures when the ultimate tensile strength of the specimen is exceeded. (MTS Systems Corp.)*

Figure 25-2. *Many parts of this huge ore-carrying truck are heat-treated. Without heat-treated parts (wheels, drive shaft, gears, axles, etc.), the vehicle would not be able to maintain its grueling workload for long without part wear and failure. (Euclid)*

lower temperature, followed by another cooling cycle to develop the proper degree of hardness and toughness.

Figure 25-3. A cryogenic quenching area in a modern heat-treating facility. Characteristics of several aluminum alloys and some space-age metals are greatly improved by heating them to a predetermined temperature and quickly quenching (cooling) them in liquid nitrogen at about –300°F (–185°C). (Grumman Aerospace Corp.)

Heat treatment involves a number of processes, which are described in later sections of this chapter. Similar techniques can be employed to *anneal* (soften) metals to make them easier to machine, or to *case harden* (produce a hard exterior surface) steel for better resistance to wear.

25.1 HEAT-TREATABLE METALS

Steel and most of its alloys are hardenable. However, when heat-treating carbon steel, it must be remembered that the carbon content of the metal is an important consideration. Carbon steels are classified by the percentage of carbon they contain in "points" or hundredths of 1%. For example: 60 point carbon steel would contain 60/100 (0.60) of 1% carbon. Steel with less than 50 points carbon cannot be hardened.

Magnesium, copper, beryllium, titanium, and many aluminum alloys are also capable of being heat-treated.

25.2 TYPES OF HEAT TREATMENT

The heat treatment of metals may be divided into two major categories. One deals with *ferrous*

metals, the other with *nonferrous* metals. Because each area is so broad, it is beyond the scope of this text to include more than basic information on the heat treatment of metals.

Changes in the physical characteristics of steel and its alloys can be affected by these basic types of heat treatment.

25.2.1 Stress-relieving

Stress-relieving is done to remove internal stresses that have developed in parts that have been cold worked, machined, or welded. See **Figure 25-4**. To stress-relieve, steel parts are heated to 1000°F to 1200°F (547°C to 660°C), held at this temperature one hour or more per inch of thickness, and then slowly air- or furnace-cooled. The technique is sometimes called *process annealing.*

25.2.2 Annealing

Annealing is a process that reduces the hardness of a metal to make it easier to machine or work, **Figure 25-5**. It involves heating the metal to slightly above its critical temperature, but never more than 50°F to 75°F (28°C to 40°C) above this point, **Figure 25-6**. The time that it is held at this temperature depends upon the shape and thickness

Figure 25-4. The stresses that develop in metals when they are machined or welded are removed by stress-relieving.

of the part. After the holding period, the piece is allowed to cool slowly in the furnace or other insulated enclosure.

For some steels, it may be necessary to use the box annealing method or a controlled atmosphere furnace to prevent the work from scaling or *decarbonizing* (loss of carbon on surface). With *box annealing*, the part is placed in a metal box and the entire unit is heated, then allowed to cool slowly in the sealed furnace.

Figure 25-5. *Annealing reduces hardness of metal and improves its machinability. When annealing is done in a controlled atmosphere, oxidation does not take place and the part remains bright. (Lindberg Steel Treating Co.)*

In reducing the hardness of the metal, its machinability is improved. Many nonferrous metals can also be softened by annealing.

25.2.3 Normalizing

Normalizing is a process where the metal is heated to slightly above its upper critical temperature and allowed to cool to room temperature. Normalizing is employed to refine the grain structure of some steels and thereby improve machinability. It is a process closely related to annealing.

25.2.4 Hardening

Hardening is a technique normally employed to obtain optimum physical qualities in steel, **Figure 25-7**. This is accomplished by heating the metal to a predetermined temperature for a specified period of time.

The temperature at which steel will harden is called its **critical temperature**, and ranges from 1400°F to 2400°F (760°C to 1316°C), depending upon the alloy and carbon content. For a "rule-of-thumb" range of hardening temperatures for carbon steel, see **Figure 25-8**.

After heating, the part is quenched in water, brine, oil, liquid nitrogen, or blasts of cold air. Water or brine is employed to quench plain carbon steel. Oil is usually used to quench alloy steels. Blasts of cold air or liquid nitrogen are used for high-alloy steels.

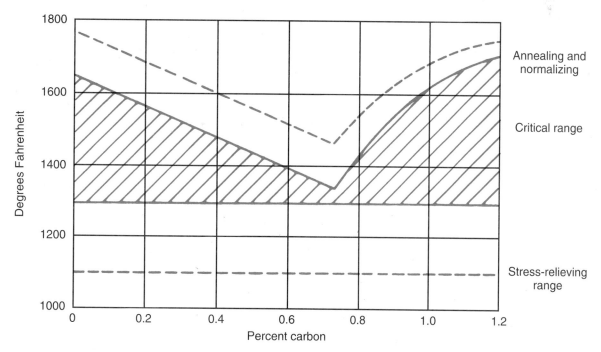

Figure 25-6. *Critical range diagram for plain carbon steel.*

Figure 25-7. Molds for plastic bowling pins are being placed in a furnace for hardening. (Lindberg Steel Treating Co.)

Carbon Content	Hardening Temperature Range
0.65% to 0.80%	1450°F to 1550°F (788°C to 843°C)
0.80% to 0.95%	1410°F to 1460°F (766°C to 793°C)
0.95% to 1.10%	1400°F to 1450°F (760°C to 788°C)
Over 1.10%	1380°F to 1430°F (749°C to 777°C)

Figure 25-8. This chart shows the "rule-of-thumb" (rough practical method) way to determine a range of hardening temperatures for carbon steel. Heated metal must be quenched in water, brine, light oil, or blasts of cold air.

Quenching leaves the steel hard and brittle. It may fracture if exposed to sudden changes in temperature. For most purposes, this brittleness and hardness must be reduced by a tempering or drawing operation.

25.2.5 Surface Hardening

Surface hardening is often used when only a medium-hard surface is required on high-carbon or alloy steels, **Figure 25-9**. The internal structure of the metal is not affected. Flame hardening, induction hardening, and laser hardening are used to attain these characteristics.

Flame hardening involves the rapid heating of the surface with an acetylene torch, and immediate quenching of the heated surface, **Figure 25-10**. The flame must be moved constantly to prevent burning or hardening the metal too deeply.

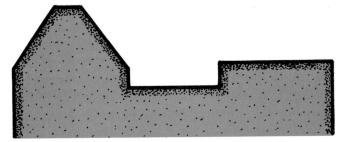

Figure 25-9. Lathe ways are frequently surface-hardened for improved wear resistance.

Figure 25-10. A lathe bed being flame hardened.

Induction hardening makes use of a high-frequency electrical induction current to heat the metal, **Figure 25-11**. The quenching medium follows the induction coil. The technique is rapid and tends to minimize distortion in the piece being heat-treated. Induction hardening is ideal for production hardening operations.

Laser hardening works on the same principle as the previously described hardening techniques. A laser beam 1/8″ to 5/8″ (3.2 mm to 15.9 mm) wide is focused on the area to be hardened, **Figure 25-12**. The light energy emitted by the laser is converted into heat energy and is absorbed by the metal.

The surface heats rapidly, so the part must be moved under the beam or the area will be heated to the melting temperature. Since there is very little heat input into the part, the hardened area cools rapidly (self-quenches) enough to touch within a few seconds.

With the laser technique, the area being hardened can be carefully controlled. It may be as small as 1/4″ (6.5 mm) square.

Figure 25-11. A stand-alone vertical-scanning induction system for heat treating. It is equipped with a touch screen and program recall from program/file base. The furnace may be loaded/ unloaded manually or by a robotic manipulator. (Radyne)

Figure 25-12. A laser beam being used to heat-treat the bearing area of a shaft. Black paint on the bearing area prevents the beam from being reflected back from the surface without heating the metal. The small flame rising from the spot being treated is caused by the black paint burning off.
(Light Beam Technology, Inc.)

There is very little chance of part warping or distortion with laser hardening. The process produces a fine grain structure that has a tougher wearing surface than other techniques.

25.2.6 Case Hardening

Low-carbon steel cannot be hardened to any great degree by *conventional* heat treatment. However, a hard shell can be put on the surface, while the inner portion remains relatively soft and tough, by the process called case hardening, **Figure 25-13**.

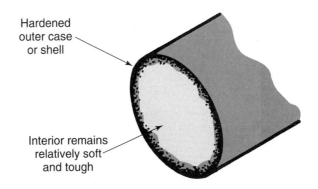

Figure 25-13. Cross section of a case-hardened part shows that interior remains relatively soft and tough, while a hard "shell" is formed on the exterior.

Case hardening is accomplished by heating the piece to a red heat and introducing small quantities of carbon or nitrogen to the metal's surface. This can be done by one of the following methods: *carburizing, cyaniding,* or *nitriding.*

- During *carburizing,* sometimes termed the *pack method,* the steel is buried in a dry **carbonaceous material** (material rich in carbon) and heated to just above its transformation range, **Figure 25-14**. In the *transformation range* (1350°F-1650°F or 730°C-900°C), steels undergo internal atomic changes that radically affect their properties. The part is

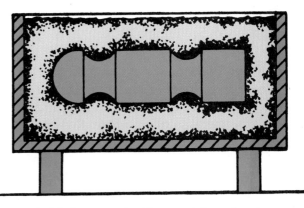

Figure 25-14. A part packed in a container of carbonaceous material, ready to be case hardened.

held at this temperature for 15 minutes to one hour, until the desired case thickness is attained. The part is removed from the furnace and quenched. Deep (very thick) cases can be obtained by this method.

- During the *liquid salt method* of case hardening, also known as *cyaniding*, the part is heated in a molten cyanide salt bath, then quenched, **Figure 25-15**. The immersion period is usually less than one hour. A high hardness is imparted to the work, and the parts treated have good wear resistance.

Figure 25-16. *Nitriding operation develops high hardness without quenching, and distortion is virtually nonexistent. (Lindberg Steel Treating Co.)*

Figure 25-15. *This operator is carefully removing a die block from a cyanide salt pot. (Master Lock Co.)*

- In the *nitriding method* or *gas method* of case hardening, parts are placed in a special airtight heating chamber where ammonia gas is introduced at high temperature. The ammonia decomposes into nitrogen and hydrogen. The nitrogen enters the steel to form nitrides which give an extreme hardness to the metal's surface. Wear-resistance and high-temperature hardness are greatly increased. See **Figure 25-16**.

25.2.7 Tempering

Tempering or *drawing* is used to lower a metal's brittleness or hardness. It involves heating the steel to below the metal's critical range. Refer to **Figure 25-6**. The exact temperature will depend upon the type of steel used and its application. This information can be found in steelmakers' catalogs and the various machinist's handbooks. Hold the temperature until complete penetration is achieved, then quench.

With the internal stresses released, the toughness and impact resistance increase. As the temperature is raised, ductility is improved, but there is a decrease in hardness and strength.

25.3 HEAT TREATMENT OF OTHER METALS

In addition to steel, many other metals and their alloys are potentially heat-treatable. You should have a basic understanding of these processes.

25.3.1 Heat-treating Aluminum

Aluminum is a general term applied to the base metal and its many alloys. When heat-treating aluminum, it is imperative that the *exact alloy* be known or the part being heat-treated may be ruined.

Aluminum alloys are heat-treatable in much the same manner as steel. That is, the metal is heated to a predetermined temperature, quenched, and then reheated to a lower temperature. However, some aluminum alloys *age-harden* at room temperature. These alloys must be kept refrigerated to remain soft and *ductile* (property of metal which permits it to be drawn out or hammered thin) while they are being worked.

Heat-treating temperatures for several of the more common aluminum alloys are shown in **Figure 25-17**. Information on other alloys, each of which requires special treatment to bring out optimum physical qualities, can be obtained from handbooks available from the various producers of aluminum.

Heat-treatable Aluminum Alloys						
Solution				Precipitation (aging)		
Aluminum alloy	Heat to °F (°C)	Quench	Resulting temper	Heat to °F (C°)	Hold for hours	Resulting temper
2024-0	910°F–930°F (488°F–499°C)	In cold water as quickly as possible after removal from furnace	—	Room	48–96	2024–T4
6061-0	960°F–980°F (515°C–527°C)		6061-W	315°F–325°F (157°C–162°C)	16–20	6061–T6
				345°F–355°F (174°C–179°C)	6–10	
7075-0	860°F–930°F (460°C–499°C)		7075-W	245°F–255°F (118°C–124°C)	20–26	7075–T6
Note: When heat treating clad 2024 and clad 7075 aluminum, hold temperature for shortest possible time.						

Figure 25-17. *Heat-treating temperatures for several common aluminum alloys.*

25.3.2 Heat-treating Brass

Brass can be annealed after cold working by heating to 1100°F (593°C) and cooling. The rate of cooling has no appreciable effect on the metal.

25.3.3 Heat-treating Copper

Copper is annealed in much the same manner as brass. The metal may be quenched or allowed to cool slowly at room temperature.

25.3.4 Heat-treating Titanium

Most titanium alloys are heat-treatable. However, special facilities are necessary because titanium is a *reactive metal:* it readily absorbs oxygen, carbon, and nitrogen. These elements greatly affect the strength of titanium and its resistance to fatigue and corrosion.

25.4 HEAT-TREATING EQUIPMENT

Heat-treating involves three distinct steps: high temperature heating, rapid cooling or quenching to harden, and tempering for final hardness and physical properties. Each step is significant in its effect on final results.

25.4.1 Quenching Media

The main problem in heat-treating is to cool the metal at a uniform rate over its entire area. Water, oil, air, and liquid nitrogen are standard *quenching media* used to draw heat from the part being treated. Quench tanks for oil and water are available with temperature controlling systems, **Figure 25-18**.

Water has the most severe cooling effects. It is employed mainly when treating plain carbon steel for maximum hardness. Water has the disadvantage

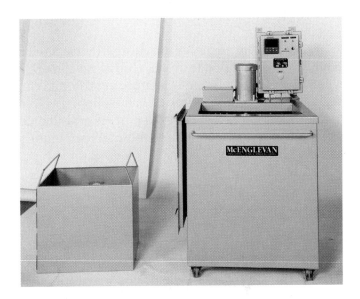

Figure 25-18. *A circulating oil-quench system. The shop rule for a circulating quench tank is one gallon of quench medium for each pound of steel quenched at 1500°F per hour. Standard quench oils give best results when heated to 120°F to 140°F (50°C to 60°C).*
(MIFCO McEnglevan Industrial Furnace Co., Inc.)

of forming gases when the hot metal is immersed in it. The gas bubbles adhere to the metal, retard cooling, and cause soft spots in the treated piece. The use of *brine* (5% to 10% salt in the quench water) prevents the formation of gases and gives better cooling results.

Mineral oils cool more slowly and produce less distortion in the treated part than water. Special quenching oils have been developed. They have a high *flash point* (the lowest temperature at which the oil vapor will ignite in air) and do not have a disagreeable odor. In production heat treatment using oil, the quenching bath must be filtered and cooled down to room temperature. Oils are used to harden alloy steels.

Quenching heated metal in oil should be done only in a well-ventilated area. Avoid inhaling any of the fumes.

Freely circulated *air* is used to cool some highly alloyed steels. The air used as a cooling medium must be dry, because any moisture may cause the steel to fracture.

Liquid nitrogen, a supercold medium for quenching, is used with several aluminum and space-age alloys. Special facilities are required.

25.4.2 Furnaces

The *heat-treating furnace* must be capable of reaching and maintaining the temperatures needed for heat-treating. They are heated by electricity, gas, or oil.

Most smaller furnaces are heated by electricity. They are safe to operate, quiet, require no elaborate venting systems, reach temperature quickly, and can be controlled with accuracy. When equipped with a microprocessor-based controller, it is possible to program the furnace for precise time/temperature cycles, **Figure 25-19**.

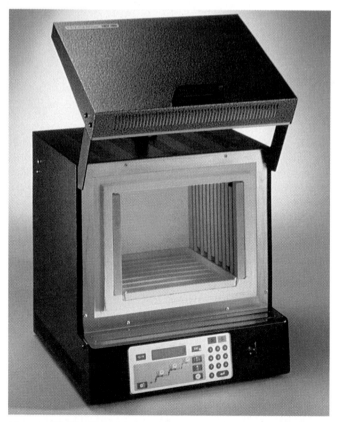

Figure 25-19. Bench-top muffle type furnace with a programmable controller for precise time/temperature management. This electric-powered furnace has a temperature range of 90°F to 2012°F (32°C to 1100°C). (NEYTECH)

Some models are equipped with two chambers, **Figure 25-20**. The upper high-temperature chamber can be equipped with atmosphere control for heat-treating with inert gases. The lower chamber is used for tempering and drawing.

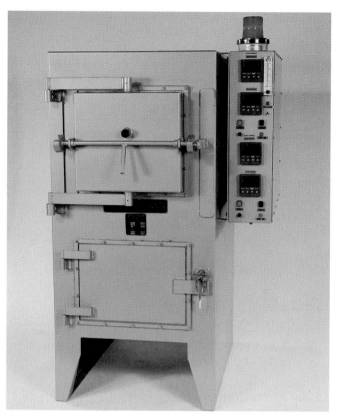

Figure 25-20. Dual electric furnace with digital readout and microprocessor-based controller. Top unit features atmospheric control for hardening with inert gases; the lower unit is used for drawing and tempering.
(MIFCO McEnglevan Industrial Furnace Co., Inc.)

When fitted with atmospheric control, the furnace can be sealed and a vacuum drawn to remove atmospheric gases that might contaminate the metal being heat-treated. The chamber is then flooded with an inert gas (one that will not oxidize or be absorbed by the metal's surface) during the heat-treating operation.

Modern electric furnaces are fitted with numerous safety devices. Avoid using a furnace until you are thoroughly versed in their safe operation.

Gas-fired furnaces are also widely employed for heat-treating. For safe operation of a gas-fired furnace, refer to the following section of this chapter. Gas-fired furnaces are noisy. Hearing protectors must be worn when working near them.

Industry employs many types of heat-treating furnaces. Most are automated and continuous in operation. See **Figure 25-21.**

Figure 25-21. Heat-treating furnaces. A—Two automatic furnaces. The one on the right has a water tank quench with a conveyor. Furnace on the far left incorporates an automatic oil quench cycle. B—A production type, controlled atmosphere heat-treating furnace. (Lindberg Steel Treating Co.)

25.5 HARDENING CARBON STEEL

Heat-treating any metal requires maintaining accurate temperatures. A *pyrometer,* **Figure 25-22,** is an instrument that accurately measures furnace temperature. On electric furnaces, the pyrometer can be set to the desired temperature and used to maintain that temperature once it is reached.

If the furnace is *not* equipped with a pyrometer, it will be necessary to judge the temperature by the color of the metal as it heats. Color charts are available from the steel companies. See **Figure 25-23.**

The following procedure is recommended when hardening carbon steels. Oxidation of the metal during heat-treating can be avoided by wrapping the part in stainless steel foil, **Figure 25-24.**

Figure 25-22. Pyrometer is used to measure temperatures in a furnace. This pyrometer is used on gas-fired furnaces. (Johnson Gas Appliance Co.)

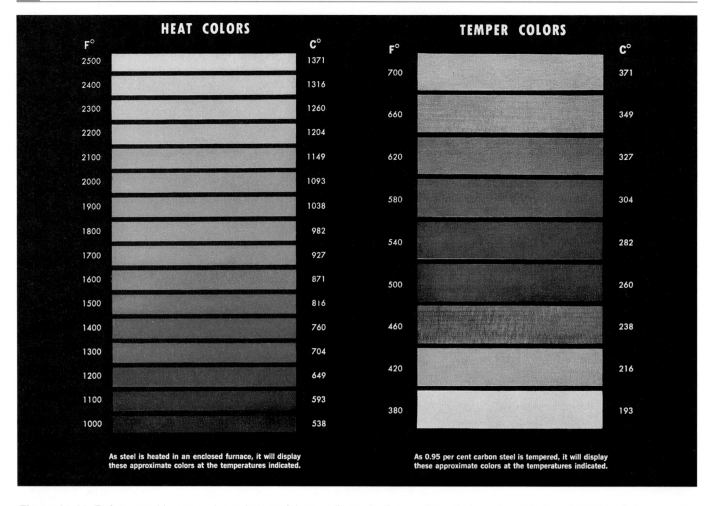

Figure 25-23. *Before metal becomes incandescent (glows red), steel will pass through the colors listed on this table. Colors are also useful for tempering steel if no pyrometer is fitted to the furnace. (Bethlehem Steel Co.)*

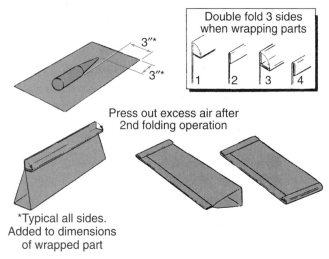

Figure 25-24. *Procedure for wrapping parts in stainless steel foil as protection against oxidation and decarburization during heat treatment.*

1. Place the metal in the furnace and set controls (if applicable) for the desired temperature. If the furnace is gas-fired, light it according to the manufacturer's instructions. When lighting a gas furnace, stand to one side and do not look into the fire box. After the gas has ignited, adjust the air and gas valves for the best operation.

2. Heat the metal to its critical temperature (1300°F to 1600°F or 705°C to 870°C). Avoid placing the part being heat-treated directly in the gas flames. If it has not been wrapped in stainless steel foil, position it in a piece of iron pipe, as shown in **Figure 25-25**. Allow the part to "soak" in the furnace until it is heated evenly throughout.

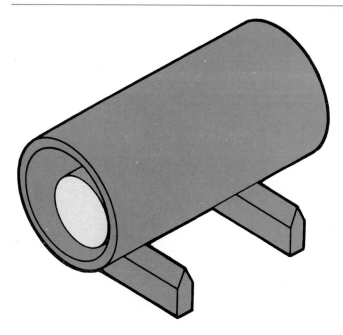

Figure 25-25. Work being heat-treated must be protected from direct flames in a gas furnace by inserting it in a section of pipe. Elevate unit from furnace floor to permit uniform heating.

3. Preheat the tong jaws, then remove the piece from the furnace. Dress properly for any operation involving the furnace, **Figure 25-26**. The metal is very hot and serious burns can result from relatively minor accidents.

Figure 25-26. Heat-treating temperatures are very hot. Dress properly for job and keep the area around furnace clean so there is no danger of slipping or stumbling. Also, preheat tongs before grasping the heated part.
(MIFCO McEnglevan Industrial Furnace Co., Inc.)

4. Quench the piece in water, brine, or oil (depending upon the type of steel being treated). Except for some alloy steels, steels are usually classified as *water-hardening* or *oil-hardening* types, according to the quenching medium to be employed on them. The quenching technique is critical.

5. To secure even hardness throughout the piece, dip long slender sections straight down into the quenching fluid with an up-and-down motion. Avoid a circular motion, since this may cause the piece to warp. Parts with other shapes should be moved around in a manner that will permit them to cool quickly and evenly.

Steel that has been hardened properly will be "glass-hard" and too brittle for most purposes. Hardness may be checked by trying to file the work surface. A file will not cut the surface if the piece has been hardened properly. Do not use a *new file* for testing hardness—it will be damaged!

25.6 TEMPERING CARBON STEEL

As mentioned earlier, tempering or drawing is employed to relieve the stresses and strains that develop in the metal during hardening. Until tempering is done, the brittle hardened steel may crack or shatter from shock (dropping, striking, etc.) or from sudden changes in temperature.

Tempering is done as follows:

1. Polish the hardened piece with abrasive cloth.

2. Heat the piece to the correct tempering temperature. This is determined by the type of steel and the job the finished part is to do. It will range from about 380°F or 193°C (the metal will turn a silvery yellow or straw color) to 700°F or 371°C (the metal will turn light gray or blue color). Color charts available from steel companies can be employed as a guide for determining when the desired temperature is attained if no pyrometer is available on the furnace. Refer to **Figure 25-23**. Quench immediately upon reaching the required temperature.

3. Small tools are best tempered by placing them on a steel plate that has been heated red-hot. Have the point of the tool extend beyond the edge of the plate, as shown in **Figure 25-27**. Watch the temper color as it heats up. Quench when the proper color has reached the tool point.

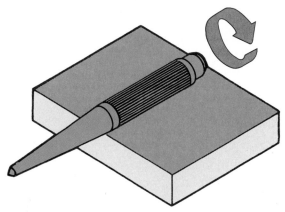

Figure 25-27. *Use a heated steel plate when tempering small tools. Have point of tool extend beyond edge of plate.*

Hot liquid baths of oil, molten salts, or lead are often employed in place of a furnace for heating parts to their proper tempering temperature. The pieces are held in the bath until the heat permeates them. They are then removed from the bath and allowed to cool in still air.

25.7 CASE HARDENING LOW-CARBON STEEL

Of the several case hardening techniques, the simplest is *carburizing*, which requires a minimum of equipment. It uses a nonpoisonous commercial compound, such as *Kasenit*™.

The hardening technique known as *cyaniding* is not recommended, because cyanide is a deadly poison and *very dangerous* to use under any but ideal conditions.

There are two methods recommended for using Kasenit to harden low-carbon steel. The first method is as follows:

1. Bring the furnace to temperature.

2. Bring the workpiece to a bright red (1650°F - 1700°F or 900°C - 930°C). Use a pyrometer or thermocouple to monitor the temperature, **Figure 25-28**.

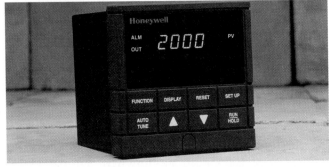

Figure 25-28. *A modern microprocessor-based digital controller for electric heat-treating furnaces. It can accept ten different thermocouple types.*
(Honeywell, Inc., Industrial Automation and Control)

3. Dip, roll, or sprinkle Kasenit on the piece, **Figure 25-29**. The powder will melt and adhere to the surface, forming a shell.

4. Reheat the workpiece to a bright red and hold at that temperature for a few minutes.

5. Quench in cold water.

Roll

Dip

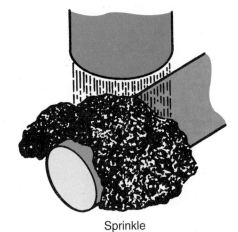

Sprinkle

Figure 25-29. *Dip, roll, or sprinkle Kasenit, or another safe commercial case hardening compound, on the work until a shell of compound has been formed.*

The second method for case hardening low-carbon steel is to:

1. Secure a container large enough to hold the work. A tin can will do if care is taken to burn off the tin coating before use.

2. Completely cover the job with Kasenit. Refer to **Figure 25-14**.

3. Place the entire unit in the furnace and heat it to a red heat. Hold the temperature for 5 to 30 minutes, depending upon the depth of case required.

4. Quench the job in clean, cool water. Be sure to use *dry* tongs to remove the piece from the container.

With either method of using Kasenit, work in a well-ventilated area and wear full face protection, leather apron, and heat-resistant gloves.

25.8 HARDNESS TESTING

Hardness testing will make sure that the metal has been given the proper degree of heat treatment, cold working, or a combination of the two, for its intended use. With hardness testing, it is also possible to establish standards for hardness that can be cited on drawings and specifications.

The most commonly utilized technique involves determining the distance a steel ball or special shaped diamond penetrates into the metal under a specific load.

Brinell and Rockwell testing machines, known as *indention hardness testers,* are used. The *hardness number* indicates the degree of material hardness. See **Figure 25-30**.

25.8.1 Brinell Hardness Tester

The *Brinell hardness tester* is employed extensively in both laboratory and production situations. The Brinell test is a measure of resistance of the material. It is an excellent index of such factors as machinability, uniformity of grade, temper after heat treatment, and body hardness of the metal.

The Brinell test for determining the hardness of a metallic material consists of applying a known load to the surface of that material through a hardened steel ball of known diameter. The diameter (depth of penetration) of the resulting permanent impression in the metal is measured. The Brinell hardness is taken as the quotient of the applied load, divided by the area of the surface of the impression (which is assumed to be spherical):

$$BHN = \frac{P}{\frac{\pi D}{2}(D - \sqrt{D^2 - d^2})}$$

Where:

BHN = Brinell Hardness Number in kilograms per mm^2.
  P = Applied load in kilograms.
  D = Diameter of steel ball in millimeters.
  d = Diameter of impression in millimeters.

To perform the Brinell test on a compressed air-type hardness tester, **Figure 25-31**, then follow these instructions:

1. Gradually turn the adjustable air regulator valve in a clockwise direction until the desired load is indicated on the load gauge. Any Brinell load from 500 to 3000 kilograms can be selected by adjusting this regulator. Check the load reading when making the initial test of a series, and adjust the air regulator valve, if necessary, so that the desired load is indicated on the dial when a specimen is actually under load.

2. Place the test specimen on the anvil, then turn the handwheel until the gap allows inserting the specimen into the machine. This distance (between the surface of the specimen and the Brinell ball) should be kept to a minimum prior to applying the load.

3. Pull out the load/unload plunger on the left side of the machine. The load will be instantly released and the test completed when the plunger is pushed in.

4. Read the impression with the special microscope and obtain the Brinell hardness number from the hardness table. *Do not apply the test load when the anvil is within travel range of the ram and Brinell ball, unless a test specimen is in place. If no test specimen is in place, a Brinell impression will be made on the machine's anvil.*

5. The test specimen must be thick enough to prevent a bulge, or other marking showing the effects of the applied load, from appearing on the side opposite the impression.

6. When necessary, the surface on which the impression is to be made should be filed, ground, machined, or polished with abrasive material, so that the edge of the impression is defined clearly enough to permit measuring the diameter to the specified accuracy.

Hardness Conversion Table
(Approximate)

Values vary depending on grades and conditions of material involved. Rockwell "B" Scale should not be used over B-100. The "C" Scale should not be used under C-20.

Brinell	Rockwell		Shore scleroscope	Tensile lbs. sq. in.	Brinell	Rockwell	Shore scleroscope	Tensile lbs. sq. in.
Hard No.	B Scale	C Scale	Hard No.	In 1000 lbs.	Hard No.	B Scale	Hard No.	In 1000 lbs.
782	...	72	107	383	163	84	25	84
744	...	69	100	365	159	83	25	82
713	...	67	96	350	156	82	24	80
683	...	65	92	334	153	81	24	79
652	...	63	88	318	149	80	23	78
627	...	61	85	307	146	78	23	77
600	...	59	81	294	143	77	22	76
578	...	58	78	284	140	76	..	74
555	...	56	75	271	137	75	..	73
532	...	54	72	260	134	74	..	71
512	...	52	70	251	131	72	..	70
495	...	51	68	242	128	71	..	69
477	...	49	66	233	126	70	..	67
460	...	48	64	226	124	69	..	66
444	...	47	61	217	121	67	..	65
430	...	45	59	210	118	66	..	63
418	...	44	57	205	116	65	..	62
402	...	43	55	197	114	64	..	61
387	...	41	53	189	112	62	..	60
375	...	40	52	183	109	61	..	59
364	...	39	50	178	107	59	..	58
351	(110)	38	49	172	105	58	..	57
340	(109)	37	47	167	103	57	..	56
332	(108.5)	36	46	162	101	56	..	55
321	(108)	35	45	157	99	54	..	54
311	(107.5)	34	44	152	97	53	..	53
302	(107)	33	42	148	96	52	..	53
293	(106)	31	41	144	95	51	..	52
286	(105.5)	30	40	140	93	50	..	52
277	(104.5)	29	39	136	92	49	..	51
269	(104)	28	38	132	90	48	..	50
262	(103)	27	37	128	88	47	..	49
255	(102)	26	36	125	87	46	..	48
248	(101)	25	36	121	86	45	..	48
241	100	24	35	118	85	44	..	47
235	99	(22)	34	115	83	43	..	47
228	98	(21)	33	113	82	42	..	46
223	97	(20)	33	109	81	41	..	46
217	96	(19)	32	106	80	40	..	45
212	95	(18)	31	104	79	39	..	45
207	94	(17)	30	101	78	38	..	44
202	93	(15)	30	99	77	37	..	44
196	92	(13)	29	96	76	36	..	43
192	91	(12)	29	94	75	35	..	43
187	90	(10)	28	91	74	33	..	42
183	89	(9)	28	90	73	31	..	42
179	88	..	27	89	72	30	..	41
174	87	..	27	88	71	29	..	41
170	86	..	26	86	70	27	..	40
166	85	..	26	85	69	26	..	40

Figure 25-30. Hardness conversion table.

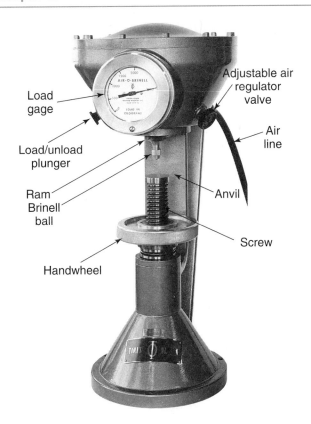

Figure 25-31. *The parts of compressed air-type Brinell hardness tester. (Tinius Olsen Testing Machine Co.)*

25.8.2 Rockwell Hardness Tester

The most widely used of all hardness testing methods is the Rockwell testing technique, **Figure 25-32.** When using the *Rockwell hardness tester,* **Figure 25-33,** either a steel ball or a specially designed diamond cone penetrator is used, depending upon the material being tested. See **Figure 25-34.**

Figure 25-32. *This technician is using state-of-the-art Rockwell hardness tester. (Wilson Instruments/Instron Corporation)*

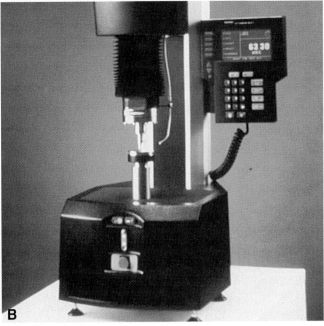

Figure 25-33. *Rockwell hardness tester. A—The basic-type tester. (Mitutoyo/MTI Corp.) B—This state-of-the-art tester offers pushbutton speed for all functions including test scales, statistical calculations, hardness scale conversions, and corrections for round parts. The machine can be set for high/low limits tolerance with a graphical icon displaying values in the go/no go mode. (Wilson Instruments/Instron Corporation)*

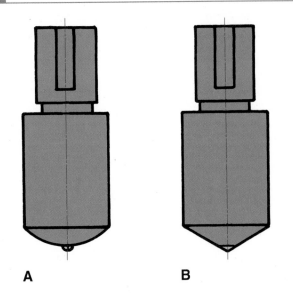

Figure 25-34. *Rockwell penetrators. A—Ball-type. B—Conical diamond-type (Brale).*

A minor load of 10 kg is first applied, then the dial gage is set to zero. The major load is then added and removed, **Figure 25-35**. The hardness number represents the *additional* depth to which the test ball

or conical diamond penetrator is driven by the major load *beyond* the depth of the previously applied light load. The hardness number is automatically indicated on the dial gage.

Typically, a 1/16″ steel ball is employed in conjunction with a 100 kg load for testing such metals as brass, bronze, and soft steel. All readings made with the 1/16″ ball and 100 kg load are **Rockwell B readings;** the letter *B* must be placed before the number. *There is no Rockwell hardness designated by a number alone.* It must always be prefixed by the proper scale letter.

The conical diamond test point, known as a **brale penetrator,** is used with a 150 kg load for testing hardened steel or any hard metals, **Figure 25-36**. All readings with the brale penetrator and 150 kg load are **Rockwell C readings**; the letter C must precede the hardness number.

Special penetrators are available for testing soft materials. The **scale designation** depends upon the ball size used to make the test. See **Figure 25-37**.

Two weights are normally supplied with a Rockwell hardness tester. One of them has a red marking and the other black. However, *three*

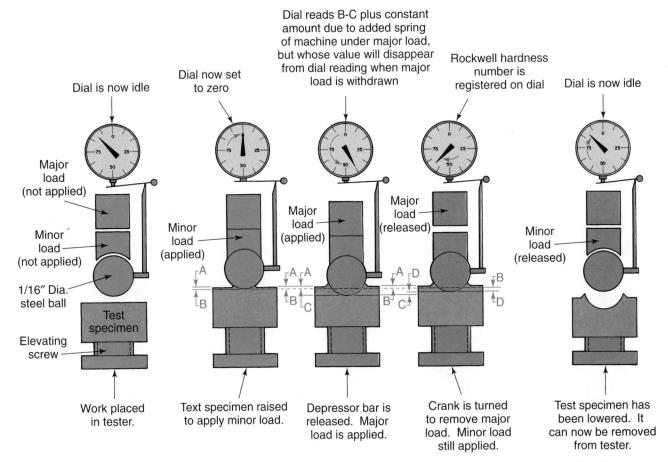

Figure 25-35. *This diagram shows how 1/16″ steel ball penetrator is used to make a Rockwell B hardness reading. Size of the ball has been greatly exaggerated for clarity.*

different loads — 60 kg, 100 kg, and 150 kg — can be applied. The weight arm, together with the link and weight pan, will apply a load of 60 kg. The weight with the red marking is placed on the weight pan for the 100 kg load. When the weight with the black marking is added, a 150 kg load is applied. The black weight is *never* used alone.

When making tests with the 100 kg load, the dial scale with red figures is used. The dial scale with black figures apply with the 150 kg load.

The 60 kg load (weight arm and weight pan alone) is employed with the brale penetrator for testing extremely hard metals, such as tungsten carbide alloys. The 1/16″ ball is extensively used with the 60 kg load for testing sheet brass.

To operate the Rockwell hardness tester, select and mount the proper penetrator point, then check that the weight for the desired test load is in position. Place the correct anvil in the elevating screw with extreme care, or the penetrator might be damaged. See **Figure 25-38.** Inspect the test specimen and remove any scale or burr that would flatten under the test and give a false reading.

Figure 25-38. Protect the penetrator with your finger when removing and replacing an anvil. Striking the hard, but brittle, diamond with the anvil may fracture the diamond. A steel ball penetrator may be deformed if it is hit by the anvil. (Wilson Mech. Inst. Div., American Chain and Cable Co., Inc.)

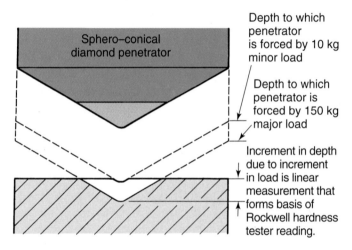

Figure 25-36. How conical diamond penetrator (brale) is employed to determine Rockwell C hardness.

Scale symbol	Penetrator	Major load (kg)	Dial figures	Typical applications of scales
B	1/16″ ball	100	Red	Copper alloys, soft steels, aluminum alloys, malleable iron, etc.
C	Diamond cone	150	Black	Steel, hard cast iron, titanium, deep case hardened steel, etc.
A	Diamond cone	60	Black	Cemented carbides, thin steel, and shallow case hardened steel.
D	Diamond cone	100	Black	Thin steel, medium case hardened steel, and pearlite malleable iron.
E	1/8″ ball	100	Red	Cast iron, aluminum and magnesium alloys, and bearing materials.
F	1/16″ ball	60	Red	Annealed copper alloys, thin soft sheet metals.
G	1/16″ ball	150	Red	Phosphor bronze, beryllium copper, malleable iron, etc.
H	1/8″ ball	60	Red	Aluminum, zinc, lead.

Figure 25-37. A letter is used as a prefix to the hardness value read from the Rockwell tester dial. The letter depends upon load, type of penetrator, and scale from which dial readings are taken.

The following procedures are considered to give the most precise hardness readings. Note that the numbers in **Figure 25-39** correspond to the following sequence numbers.

1. Place the test specimen on the anvil.

2. Gently raise the specimen until it comes into contact with the penetrator. Continue to turn the capstan handwheel slowly until the small pointer (8) is nearly vertical and slightly to the right of the dot. Continue raising the work until the long pointer is approximately upright (within five divisions plus or minus). The minor load (10 kg) has now been applied.

3. Set the dial to zero (line marked "set") by turning the knurled ring that is located below the handwheel, **Figure 25-40.**

4. Carefully push down on the depressor bar to apply the major load. The penetrator is forced into the work. The depth to which it penetrates depends upon the metal's hardness.

5. Watch the pointer until it comes to rest.

6. Pull the crank handle forward to lift the major load, but leave the minor load still applied.

7. Read the Rockwell Hardness Number. If the test has been made with the 1/16″ ball, and the

Figure 25-40. *A thumb is used to "zero-in" the dial pointer before applying the major load. Same thumb is used to push down the depressor bar to apply major load.*
(Wilson Instruments/Instron Corporation)

load is 100 kg, the reading is taken from the red scale and the letter B is prefixed to the number to signify the condition of the test. The letter C is prefixed to the number if the brale penetrator and 150 kg load were employed. The reading is then made from the black scale.

After completing the test, lower the work away from the penetrator and remove it from the testing machine.

Like other precision tools, the Rockwell Tester must be handled with care if it is to maintain its accuracy. There are a few precautions that must be observed:

- When moving the tester, grasp it only by the cast iron base, never by any parts that are attached to the base.
- The tester should be leveled on a solid bench, in a location free from grit and vibration. Keep the machine covered when not in use!
- False readings will result if the shoulder on the brale penetrator is not kept clean, **Figure 25-41.** Carefully wipe it with a clean, soft cloth before installing it in the tester.

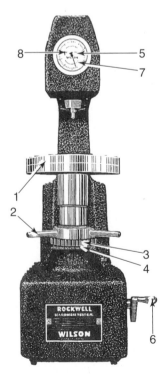

Figure 25-39. *Basic Rockwell hardness tester operating procedures.*

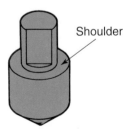

Figure 25-41. *Shoulder of the brale (penetrator) must be free of dirt and burrs before it is mounted in tester. Otherwise, incorrect readings will result.*

- Use only lubricants specified by the manufacturer.
- Support long work properly.
- Clean the work with an abrasive cloth to remove scale and roughness; a smooth surface is necessary for accurate readings. Castings and forgings should have a spot ground or machined where the test is to be made, so that the penetrator will test the true metal underneath.
- The specimen must be thick enough so that the undersurface does not show the slightest indication of the test.
- Corrections must be added to readings made on round stock, if it is not possible to file or grind a flat spot in the test area.
- Prevent damage to the penetrator or the anvil by being careful to avoid forcing them together when a test piece is not in the machine.
- If the tester is used on case hardened steel, accurate readings cannot be made unless the "case" is several times thicker than the indentation depth.

25.8.3 Webster Hardness Tester

The *Webster hardness tester*, **Figure 25-42**, is a portable test device for checking materials such as aluminum, brass, copper, and mild steel. It can be used on assemblies that cannot be brought into the laboratory, or to test a variety of shapes that other testers cannot check, such as extrusions, tubing, or flat stock.

The tester's dial indicator reading is converted to the Rockwell hardness scale by referring to a conversion chart furnished with the tool.

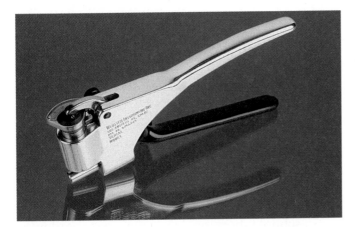

Figure 25-42. The Webster hardness tester is a portable device used with materials such as aluminum, brass, copper, or mild steel. By referring to a conversion chart, the tester's dial indicator reading can be converted to the appropriate Rockwell scale. (Webster Instrument, Inc.)

25.8.4 Scleroscope Hardness Testing Machine

A *scleroscope* is a testing device that drops a hammer onto the test piece, with the resulting bounce or rebound of the hammer used to determine hardness.

Two styles of scleroscope are in use. One is fitted with a vertical scale, the other is a dial recording instrument. See **Figure 25-43**. They may be

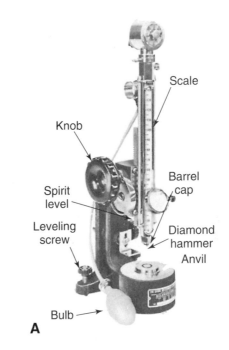

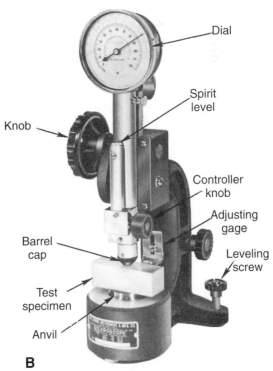

Figure 25-43. Two types of Shore scleroscope. A—A vertical-scale model. B—A scleroscope with a direct reading dial. (Shore Instrument and Manufacturing Co., Inc.)

employed for testing the hardness of all metals, ferrous and nonferrous, polished or unpolished, with virtually no limitation in size or shape. Hardness testing with the scleroscope is essentially a nonmarring test. No craters are produced that would require refinishing of the test area.

The scleroscope hardness test involves dropping a diamond hammer from a fixed height to make a *minute* (very tiny) indentation in the metal. The hammer rebounds, but not to its original height, because some of the energy in the falling hammer is dissipated in producing the tiny indentation. The rebound of the hammer varies in proportion to the hardness of the metal—the harder the metal, the higher the rebound.

The tester scale consists of units determined by dividing into 100 parts the average rebound of the hammer from quenched tool steel of ultimate hardness. These rebounds will range from 95 to 105. The scale is carried higher than 100 to cover super-hard metals. See **Figure 25-44**.

The scleroscope is capable of yielding accurate hardness readings on the softest or the hardest metals without changing the scale or diamond hammer.

When testing objects within the capacity of the clamping stand of the vertical scale scleroscope, the specimen must be mounted on the anvil. See **Figure 25-43**. The unit should be leveled. Do so by turning the leveling screws while observing the built-in spirit level. The instrument is operated pneumatically by means of a rubber squeeze bulb.

To perform a test, revolve the knob to bring the barrel cap firmly into contact with the test specimen. It is essential that you maintain a firm pressure on the specimen during the test. Squeeze and release the rubber bulb to draw the hammer to the up position. While maintaining torque on the knob, again squeeze and release the rubber bulb to release the hammer. Observe the reading on the scale.

The height to which the hammer rebounds on the first bounce indicates the hardness of the specimen. The correct hardness of the piece, however, is the average of several tests. *Do not make more than one test at a given spot or false readings will result.* The dial recording scleroscope works on the same rebound principle, but is direct reading.

While the method may sound unorthodox, the results are very close to those obtained with Brinell and Rockwell testers.

25.9 HEAT-TREATING SAFETY

- Never attempt to heat-treat metals while your senses are impaired by medication or other substances.
- Make sure that the furnace is in good operating condition before attempting to use it. Avoid lighting a furnace until you have been instructed in its safe operation. Never stand in front of a gas furnace while igniting it.
- Never look into the furnaces unless you are wearing tinted goggles or glasses under your face shield.
- Heat-treating involves raising metal to very high temperatures. Handle the hot metal with appropriate tools, and always wear an approved full face safety shield and the proper protective clothing. Wear heat resistant gloves and a leather apron (never a cloth apron, especially one that is greasy or oil-soaked.)
- Work only in areas that are well ventilated.
- Do not stand over the quenching bath when immersing hot work.
- Potassium cyanide should never be used in a school shop or lab as a case hardening medium. If you work in a situation that permits the use of potassium cyanide, never breathe the toxic fumes. Wash thoroughly after completing the heat-treating operations

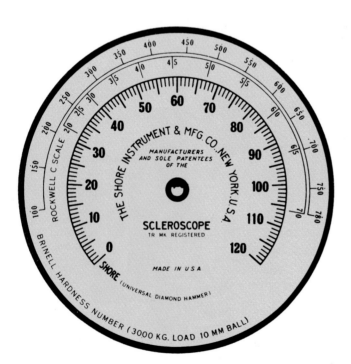

Figure 25-44. *Scales on scleroscope dial. Note that the dial includes scales for the equivalent Brinell and Rockwell C hardness values.*

TEST YOUR KNOWLEDGE

Please do not write in the text. Write your answers on a separate sheet of paper.

1. Heat-treating is done to:
    a. Obtain certain desirable changes in the metal's physical characteristics.
    b. Increase the hardness of the metal.
    c. Soften (anneal) the metal.
    d. All of the above.
    e. None of the above.

2. What does heat-treating involve?

3. Carbon steels are classified by the percentage of carbon they contain, expressed in "points" or hundredths of 1%. If this statement is true, briefly explain what it means.

4. Other than steel, list four other metals that are capable of being heat-treated.

5. In addition to water, _____, _____, blasts of cold _____ or liquid _____ may be used as a quenching medium.

The following are matching questions. Each word in the numbered list matches one of the sentences that follow. On your paper, write the letter next to the appropriate number to match the words and statements.

6. _____ Stress-relieving.

7. _____ Annealing.

8. _____ Normalizing.

9. _____ Case hardening.

10. _____ Surface hardening.

11. _____ Hardening.
    a. Involves heating metal to slightly above its upper critical temperature and then permitting it to cool slowly in insulating material. Hardness of the metal is reduced.
    b. Used to refine grain structure of steel and to improve its machinability.
    c. Done to reduce stress that has developed in parts that have been welded, machined, or cold worked during processing.
    d. Used when only a medium-hard surface is required on high-carbon or alloy steels.
    e. Only the outer surface of low-carbon steel is hardened while the inner portion remains relatively soft and tough.
    f. Accomplished by heating metal to its critical range and cooling rapidly.

12. Tempering a piece of hardened steel makes it:
    a. Brittle.
    b. Soft.
    c. Tough.
    d. All of the above.
    e. None of the above.

13. What advantages does an electric heat-treating furnace have over a gas fired heat-treating furnace?

14. The _____ is used to measure and monitor the high temperatures needed in heat-treating.

15. What is hardness testing?

16. List three types of commonly used hardness testers.

17. What safety precautions must be observed when lighting a gas-fired heat-treating furnace?

18. List five safety precautions that must be observed when heat-treating metal.

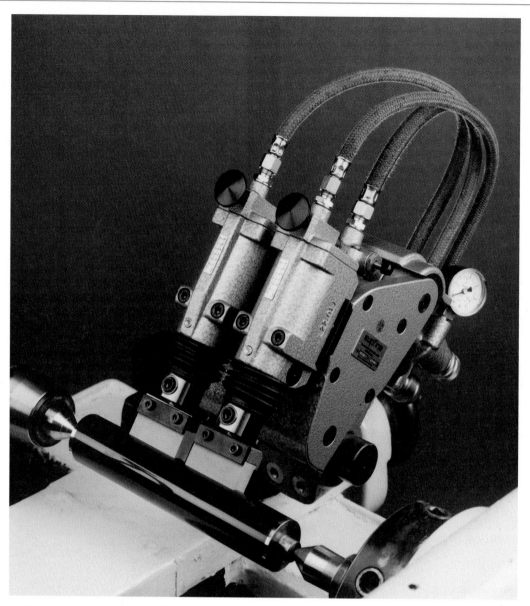

Superfinishing provides an extremely smooth finish on parts used in automotive and other applications. This superfinishing attachment, mounted on a lathe, uses fine abrasive stones and a special lubricant. An oscillating motion is employed to eliminate chatter and surface waviness. (Sunnen Products Co.)

Chapter 26

Metal Finishing

LEARNING OBJECTIVES

After studying this chapter, you will be able to:
○ Describe how the quality of a machined surface is determined.
○ Explain why the quality of a machined surface has a direct bearing on production costs.
○ Describe some metal finishing techniques.

IMPORTANT TERMS

anodizing
electroplating
lay
metal spraying
microinches
micrometers

roller burnishing
surface roughness
standards
vitreous enamel
waviness

The term *metal finish* refers to the degree of smoothness or roughness remaining on the surface of a part after it has been machined. A machined surface has geometric irregularities that are produced by the cutting action of the tool. Each type of cutting tool leaves its own characteristic surface marking, **Figure 26-1**.

26.1 QUALITY OF MACHINED SURFACES

At one time, the quality of a machined surface was noted by the symbol "*f*." This was not based on specific standards. Therefore, the engineer or drafter included explanatory notes such as *rough grind, smooth turn, surface grind*, etc., on the drawing. These notes indicated the desired quality of general surface finish.

The technique left much to be desired, since each machinist interpreted the specifications differently. Often, the piece was better finished than it had to be, increasing its production cost.

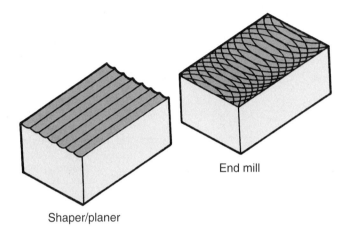

Shaper/planer

End mill

Figure 26-1. *Each type cutting tool leaves characteristic markings.*

The problem reached such proportions that in the early 1940s, the Standards Associations of Canada, Great Britain, and the United States developed tentative surface roughness or texture standards.

The terms and ratings of **surface roughness standards** or **texture standards** relate to surfaces produced by machining, grinding, casting, molding, forging, or similar processes. These standards are not concerned with luster, appearance, color, corrosion resistance, wear resistance, hardness, and the many other characteristics that may be governing considerations in specific applications.

The standards also do not define the different degrees of surface roughness and waviness suitable for specific purposes, nor do they specify the means by which any degree of such irregularities may be obtained or produced. The standards deal only with the *height, width*, and *direction* of surface irregularities, since these are of practical importance in specifications.

The present surface finish system arrives at roughness values by averaging, arithmetically, the irregular contours on a surface. The values are given in *microinches* (millionths of an inch, shown as XX µin.) or *micrometers* (millionths of a meter, shown as XX μm). With established standards, a universal set of numbers and symbols indicating surface roughness or texture are now used on drawings and in specifications. See **Figures 26-2** and **26-3.**

Symbol	Description
✓	*Basic surface roughness/texture symbol.* Surface may be produced by any method.
⊽	*Material removal by machining required.* Horizontal bar indicates that material removal by machining is required. Material must be provided for that purpose.
.187 ⊽	*Material removal allowance.* Number indicates amount of material that must be removed in inches/millimeters. Tolerances may be added.
⊘	*Material removal prohibited.* Circle in vee indicates that surface must be produced by processes such as casting, forging, hot finishing, cold finishing, powder metallurgy, or injection molding without subsequent removal of material.
√̄	*Surface texture symbol.* To be used when any surface characteristics are specified above horizontal line or to right of symbol. Surface may be produced by any method.

Figure 26-2. Surface roughness or texture symbols.

Roughness Height Rating		Surface Description	Process
Microinch	**Micrometer**		
1000	25.2	Very rough	Saw and torch cutting, forging, or sand casting.
500	12.5	Rough machining	Heavy cuts, coarse feeds in turning, milling, and boring.
250	6.3	Coarse	Very coarse surface grind, rapid feeds in turning, planing, milling, boring, and filing.
125	3.2	Medium	Machining operations with sharp tools, high speeds, fine feeds, and light cuts.
63	1.6	Good machine finish	Sharp tools, high speeds, extra fine feeds, and cuts.
32	0.8	High grade machine finish	Extremely fine feeds and cuts on lathe, mill, and shaper required. Easily produced by centerless, cylindrical, and surface grinder.
16	0.4	High quality machine finish	Very smooth reaming or fine cylindrical or surface grinding, or coarse hone or lapping of surface.
8	0.2	Very fine machine finish	Fine honing and lapping of surface.
2–4	0.05 0.1	Extremely smooth machine finish	Extra fine honing and lapping of surface.

Figure 26-3. Roughness values. When used on a drawing, a number indicates roughest surface in microinches/micrometers that is acceptable for that specific application.

In addition to surface roughness, other surface conditions are considered and values given for them.

Waviness ordinarily takes the form of smoothly rounded peaks and valleys caused by tool and machine vibration and chatter, **Figure 26-4**. Waviness is of *greater magnitude* than roughness. It is measured with reference to a nominal or geometrically perfect surface.

Waviness is specified in inches (millimeters) as the maximum allowable peak-to-valley height. It is measured using a sensitive dial indicator with a ball contact 0.06″ (1.5 mm) in diameter. **Figure 26-5** shows how acceptable waviness tolerances are specified on drawings, in reports, and as specifications.

Lay is the term that describes the direction of the predominant tool marks, grain, or pattern of surface roughness. See **Figure 26-6**.

26.1.1 Degrees of Surface Roughness

Milling and turning can produce surface finishes in the order of 125 μin. to 8 μin. (3.2 μm to 0.2 μm). Grinding, depending on the coarseness of the wheel and feed rate, has a range of 63 μin. to 4 μin. (1.6 μm to 0.1 μm).

Lapping produces the *smoothest* finish on a production basis. It is used by automotive manufacturers to produce mating surfaces that are flat and smooth enough to form a gasketless oil-tight seal in automatic transmissions and other applications. Surfaces are as fine as 2 μin. to 3 μin. (0.05 μm to 0.07 μm). See **Figures 26-7** and **26-8**.

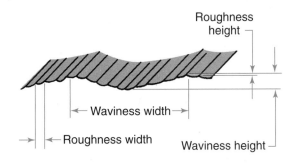

Figure 26-4. *How surface waviness is measured. Note the difference in magnitude between waviness and roughness.*

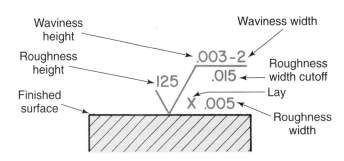

Figure 26-5. *On drawings, symbols and numbers show roughness, waviness, and lay. They specify finishes required on a surface.*

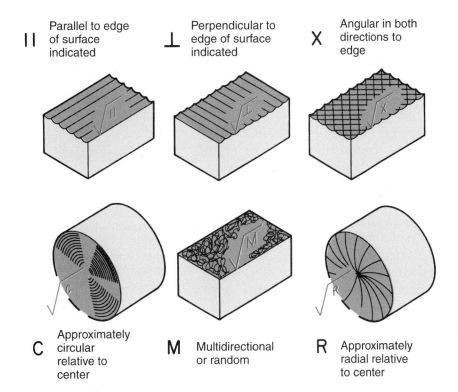

Figure 26-6. *The lay symbols and their meanings. Lay symbols are located beneath the horizontal bar on a surface texture symbol.*

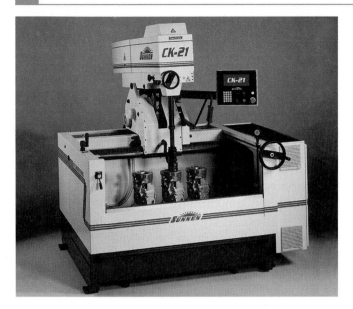

Figure 26-7. *Computerized vertical honing machine. It is designed for precision honing of bores in such applications as small engine blocks, air compressor cylinders, valve bodies, and aircraft cylinders. The unit has a microprocessor-based control system that enables the operator to control all aspects of the production cycle. (Sunnen Products Company)*

Several tools have been developed to measure surface quality. The most accurate is the ***profilometer***, which measures and amplifies surface roughness electronically, **Figure 26-9**. Surface roughness is usually read on a meter. A hard copy graphic readout can be provided by some models.

The ***surface roughness gage***, **Figure 26-10**, is a visual or comparison tool. It contains sample specimens of the various degrees of surfaces that conform to values established by the American National Standards Institute (ANSI Y14.36-1978).

26.1.2 Economics of Machined Surfaces

The quality of a machined surface has a direct bearing on production costs: the *finer* the finish, the *higher* the cost. If costs are to be kept within acceptable limits, care must be taken to meet, but not exceed, the required specifications.

The chart in **Figure 26-11** illustrates the range of surface quality normally attainable by the various machining processes. Values are relative and will vary depending upon the condition of the machine, sharpness of the cutting tool, and the material being machined.

26.2 OTHER METAL FINISHING TECHNIQUES

While the quality of the machined surface is of paramount importance in the machining of metal,

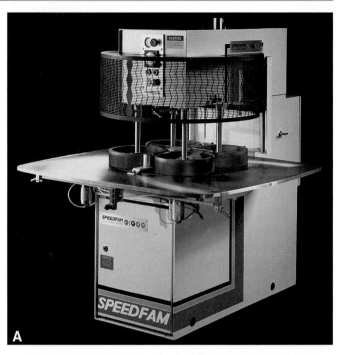

Figure 26-8. *Lapping machines. A—Free abrasive lapping machines provide exceptional accuracy with part thickness accuracies of ±0.00015″, flatness of 0.00001″, and a fine surface finish. (SpeedFam) B—A twin-plate diamond lapping system. It laps and/or polishes both sides of a workpiece simultaneously and produces parallel parts with uniform edge-to-edge flatness. The unit uses a menu-driven microprocessor controller. (Engis Corporation)*

other finishing techniques may be employed for one or more of the following reasons:

- *Appearance* affects product salability and is more important than often realized. A product is much more attractive with a proper finish than when the metal is left unfinished.

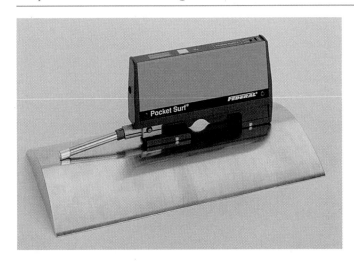

Figure 26-9. Handheld surface roughness gage (profilometer) with a digital display that presents the measured readout value in microinches or micrometers. It has a variety of interchangeable probes for different applications. (Federal Products Co.)

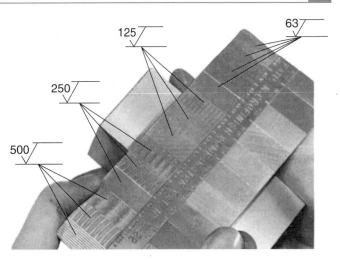

Figure 26-10. A surface roughness gage is being used to visually compare surface of a milled aluminum block. Since it is often difficult to check machined surfaces visually, a test by feel is also used. More obvious roughness standards are identified by appropriate symbols in microinches.

Machine process	Machining finishes/microinches								
	500	250	125	63	43	16	8	4	2
Abrasive cutoff									
Automatic screw machine									
Bore									
Broach									
Counterbore									
Countersink									
Drill									
Drill (center)									
Face									
File									
Grind, cylindrical									
Grind, surface									
Hone, cylindrical									
Hone, flat									
Lap									
Mill, finish									
MIll, rough									
Ream									
Saw									
Shape									
Spotface									
Super finish									
Turn, smooth									
Turn, diamond									
Turn, rough									

Figure 26-11. Typical surface finishes resulting from various machine processes. The finer the finish (the lower the roughness value in microinches), the higher the cost of obtaining it.

A car would be drab if left unfinished (unpainted), **Figure 26-12**.

- *Protection* is important because all metals are affected to some degree by contaminants in the atmosphere and by abrasion. For example, metal sheet is often textured as a means of protection. The textured surface will reduce reflected light so that surface defects (scratches and small dents that occur during product use) are "hidden" within the irregularities of the surface treatment.

- *Identification* makes the product stand out from its competition. Finishes are also applied to make the product blend into the surroundings. See **Figure 26-13**.

- A form of *cost reduction* can result by use of a surface finish. For example, an expensive metal can be coated onto a less costly metal, or other material. The finished part will be less costly than if it were made from the more costly metal. The part will retain the properties of the higher priced metal. Silver and

Figure 26-13. *A computer-designed camouflage pattern for an aircraft that will be employed in desert areas. It was developed to make aircraft blend into its surroundings, both in the air and on the ground. (Evans & Sutherland)*

gold, for example, are often applied to steel to improve electrical conduction, heat distribution properties, solderability, and appearance.

Figure 26-12. *An automobile would be quite drab if a color finish were not applied. Automotive finishes are planned in the design stage. (Ford Motor Co.)*

Figure 26-14 shows the many finishes that can be applied to aluminum. Regardless of the finish that is to be employed, the surfaces must be cleaned of all contaminants before the finish can be applied, **Figure 26-15.** Oxidation can be removed mechanically (sandblasting or burnishing), or chemically (etching with an acid, etc.). Solvents are often employed to remove oil and grease.

Special methods have been devised to clean complex machined castings. They are needed to remove loose casting sand, metal chips, and other foreign matter trapped in recesses and passages during manufacturing operations, **Figure 26-16.**

26.2.1 Organic Coatings

A wide range of finishes fall into the category of *organic coatings:* paints, varnishes, lacquers, enamels, various plastic-base materials, and epoxies in both clear and pigmented (color) formulas. With the exception of the epoxies, they are set by the evaporation of their solvents. This may be accomplished by air drying or baking. Epoxies require the addition of a *catalyst* or hardener to set. Organic finishes are seldom applied to machined surfaces.

A primer is often required to secure satisfactory bonding between the metal and the finish. Some types of castings may need a filler to smooth out the rough cast surface, **Figure 26-17.**

Organic coatings are applied in the following ways:

- *Brushing*, at one time, was the only way of applying finishes.
- *Spraying* atomizes the finishing material and carries it to the work surface by air pressure, **Figure 26-18.** Small pressurized spray cans, offered in a wide range of colors, are

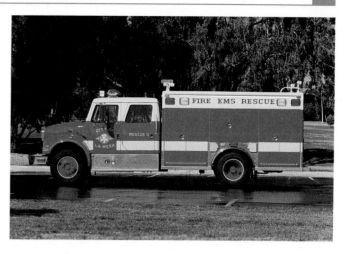

Figure 26-15. *Before the colorful protective finish can be applied to the body of this fire/rescue vehicle, metal surfaces must be cleaned of all dirt, oil, and grease. Otherwise, the finish will not adhere properly. (Emergency Vehicles, Inc.)*

Figure 26-16. *A truck engine block being cleaned after machining operations. Chemical flow and agitation of the part will remove foreign matter from deep cavities, holes, and recesses of the casting. The final operation of the cleaning cycle is application of rust-inhibiting chemicals. (Magnus Chemical)*

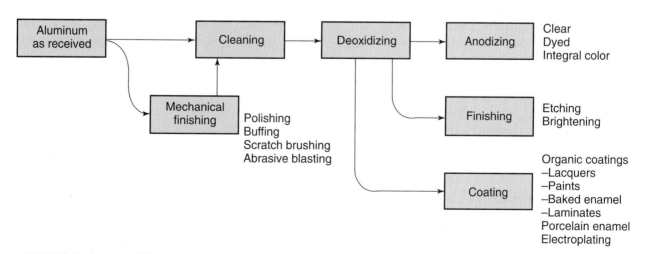

Figure 26-14. *Various types of finishes that can be applied to aluminum.*

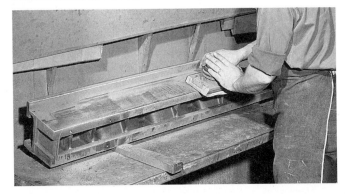

Figure 26-17. *A filler is used to smooth surface irregularities on a lathe bed casting. The filler is being smoothed with an air sander before a finish is applied. (Clausing Industrial, Inc.)*

Figure 26-19. *Robotic painting is commonly used in the automotive industry. Robot movement is controlled by computer. (Chrysler Corp.)*

Figure 26-18. *With spraying, the finishing material is atomized and carried to the work surface by air pressure. Shown is a CAD illustration used by a truck manufacturer to develop the robotic spray patterns that will completely cover the vehicle cab. (Deneb Robotics, Inc.)*

Figure 26-20. *Electrically charged particles of paint will evenly coat and protect all bare surfaces of this van body as it is submerged in tank a containing tens of thousands of gallons of primer. (Chrysler Corp.)*

available. Spraying is easily adapted to mass-production techniques, **Figure 26-19**.

- *Roller coating* can be used only on flat surfaces. It is a low-cost technique that can be mechanized.
- With *dipping*, the part is submersed into the finish, removed, and allowed to dry, **Figure 26-20**. Dipping is widely used today by the automotive industry to apply body primer and rust proofing. The coating is dried by warm air.
- During *flow coating*, the part is flooded with the finish and allowed to drain while held in an atmosphere saturated with solvent vapor. Drying is thus delayed until draining is complete.

26.2.2 Inorganic Coatings

Several well-known finishing techniques fall into the *inorganic coatings* category: anodizing, glass coating, chemical blackening, etc.

Anodizing is the best known of this type finish. This process forms a protective layer of aluminum oxide on aluminum parts.

There are three classes of anodizing: *ordinary anodizing, hard coat anodizing,* and *electrobrightening.* The basic procedure for producing all three is the same, **Figure 26-21**.

The anodized coating of *aluminum oxide* forms by reaction of the aluminum with an electrolyte when the aluminum is used as the anode. Oxygen is liberated at the surface of the aluminum and an oxide forms. Ordinary anodizing leaves a layer

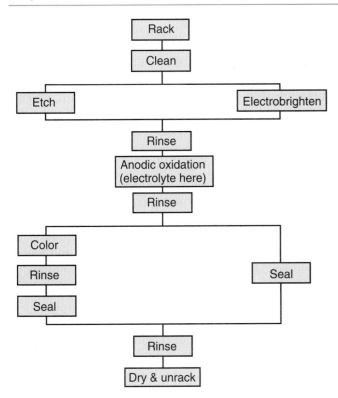

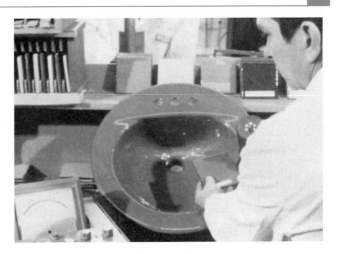

Figure 26-22. *Vitreous or porcelain enamel is a glass coating fused to steel or cast surfaces. It is an extremely hard coating that is smooth and easy to clean. Here, a quality control inspector checks color part with a standard color patch. (Eljer Plumbingware)*

Figure 26-21. *Sequence of operations when anodizing aluminum.*

0.0001″ to 0.0006″ (0.003 mm to 0.015 mm) in thickness. Hard coating produces a layer about 10 times thicker. It has superior abrasion-, erosion-, and corrosion-resistance; however, the strength of the material is slightly reduced.

Electrobrightening occurs when the aluminum is the anode in an electrolyte that dissolves the oxide film at about the same rate that it is formed. The process leaves a smooth, bright, mirror-like finish. The anodized coating can be dyed in a wide range of colors. The color becomes part of the surface of the metal.

Vitreous enamel or *porcelain* is a glass coating that has been fused to sheet or cast iron surfaces. They form an extremely hard coating that is smooth and easy to clean, **Figure 26-22**. Available in many colors, they can be applied to metals that remain solid at the firing (melting) temperatures of the coating material. The finish is applied as a powder (*frit*) or as a thin slurry known as "*slip*." After drying, the material is fired at about 1500°F (815°C) until it fuses to the metal surface, **Figure 26-23**.

Chemical blackening is a finishing technique that produces a black oxide coating on the machined surfaces of ferrous metals, **Figure 26-24**. It chemically bonds with the metal surface, instead of merely adhering to it like paint or other applied finishes.

Figure 26-23. *Porcelain application and firing. A—Porcelain is applied as a powder, called "frit," or as a thin slurry, known as "slip." B—After frit has been applied, the unit is fired at 1500°F (815°C) until porcelain has fused to the metal. (Eljer Plumbingware)*

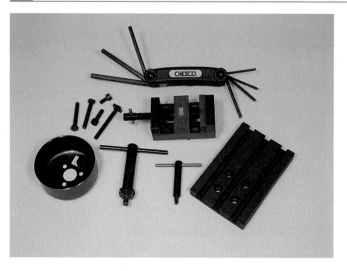

Figure 26-24. Chemical blackening is a surface finishing technique that can be applied to steel parts and products like those shown. It bonds with the metal's surface, rather than coating it; there is no surface buildup.

Blackening is employed as a finish on precision parts that cannot tolerate the added dimensional thickness of paint or plating (the blackening is 0.00003″ or 0.0008 mm thick). It will not chip or peel.

The process can be carried out at room temperature and takes about 15 minutes to complete. Operators must wear eye protection and gloves that provide chemical protection. Wear a plastic apron to protect clothing.

Chemical blackening offers several advantages:
- The black finish enhances the appearance of many items. Because the cost is so low, it is possible to finish parts that were previously left unfinished.
- It aids in protecting the machined surface against humidity and corrosion.
- Light glare is greatly reduced. When employed on moving tool and machine parts, safety factors are improved by reducing eye fatigue.
- Abrasion resistance is improved.
- Adhesion qualities are improved. Paint and other finishes applied over blackening take hold faster, adhere more strongly, and last longer.

There are two drawbacks to the process, both involving environmental treatment and disposal laws:
- The water used in the process must be filtered and treated before it can be discharged as wastewater.
- Chemical residue from the process must be handled as toxic waste.

26.2.3 Metal Coatings

With the exception of electroplating, which can be used on a number of materials, *metal coatings* are applied primarily to steel. They adhere to the surface tightly enough to offer protection from corrosion, as well as improving wear resistance and/or the appearance of the item. For example, cutting tools are now available with a titanium oxide coating that extends tool life up to 800%.

The more commonly employed metal coatings include electroplating, metal spraying, and detonation gun coating.

Electroplating is a process in which a layer of metal is deposited on a metal surface by the use of an electrical current. Practically any metal can be used as a coating. Chrome plating is one common example of electroplating.

Coating thickness can be controlled closely. Unlike many other metal coating processes, electroplating can deposit coatings that improve both wear-resistance and appearance. When a product is machined, allowance must be made for the additional metal thickness that will be deposited by the plating process.

In *metal spraying,* a metal wire or powder is heated to its melting point and sprayed by air pressure to produce the desired coating on the work surface. See **Figure 26-25**.

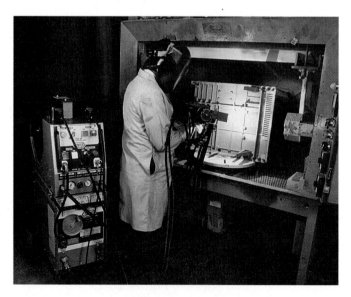

Figure 26-25. The plasma flame process is one of several metal spraying techniques used in industry. It is capable of temperatures up to 30,000°F (16 649°C) and can be used to spray any material that will melt without decomposing. Note how the operator is dressed for protection against the high heat generated by the process. (Metco, Inc.)

Most inorganic materials that can be melted without decomposition can be applied by spraying. Flame-sprayed coatings can be applied to build up worn or scored surfaces so they can be remachined to the required size, **Figure 26-26**. Superhard coatings can be sprayed when abrasion-resistant surfaces are needed.

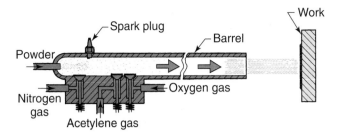

Figure 26-27. Cross section of a D-Gun or detonation gun used to apply metal coatings. (Union Carbide Corp.)

Figure 26-26. This metallizing gun is applying stainless steel onto a roller surface. (Metco, Inc.)

The *detonation gun coating process,* or D-Gun® process is another technique for depositing a metallic coating on a workpiece. The process was invented and developed by the Union Carbide Corporation. The D-Gun is essentially a water-cooled barrel several feet long and about 1″ (25 mm) in diameter. It is fitted with valving for introducing gases and material to be sprayed. See **Figure 26-27**.

A carefully measured mixture of gases (usually oxygen and acetylene) is fed into the barrel, along with a charge of coating material in powder form. The powder has a particle size of less than 100 microns. A spark ignites the gases. This heats the powder and creates a high-temperature, high-velocity gas stream. The stream of gas carries the molten material to the surface to be coated. A pulse of nitrogen purges the barrel after each detonation; the process is repeated many times a second.

Each detonation, called a "pop," results in a circle of coating material a few microns thick and about an inch in diameter. The molten or semi-

molten coating droplets quickly solidify on the workpiece surface. The complete coating consists of many overlapping "pops" that build up to the required thickness. The application process can be fully automated.

The process can be used to apply coatings with very high melting points to fully heat-treated parts without danger of changing the metallurgical properties or strength of the part and without danger of thermal distortion.

Almost any material that can be melted without decomposing can be sprayed. Coatings include pure metals and metallic alloys such as nickel and nichrome, tungsten carbide, and ceramics. These coatings are used in many applications, especially where there is need to combat wear (abrasive, erosive, or adhesive) often in very corrosive environments. Refer to **Figure 26-28**.

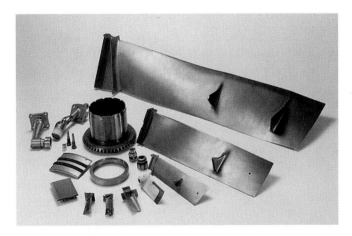

Figure 26-28. Some examples of metal and ceramic coating application on gas turbine engines by detonation gun coating process. (Union Carbide Corp.)

26.2.4 Mechanical Finishes

Buffing is a power polishing operation. Buffing wheels are attached to a buffing lathe, **Figure 26-29**.

The wheels are charged with different grades of abrasives that remove scratches and polish the metal's surface to a high luster. Unless the buffing lathe is fitted with a high efficiency dust-collecting system, an approved filter mask must be worn.

Diamond dust and air-powered, hand-held polishing units are used to polish the hardened steel dies used for die casting or the injection molding of plastics. See **Figure 26-30**.

Roller burnishing, **Figure 26-31**, is a cold-working operation. It removes no metal, but rather compresses or "irons out" the peaks of a metal's surface into the valleys. With the process, no honing or grinding is necessary. Both ID and OD burnishing is possible and can be done at speeds and feeds comparable to cutting tools.

A *roller-type burnishing tool* is also available for finishing holes, **Figure 26-32**. The tool can size, finish, and work-harden metal parts to tolerances as close as 0.0002" (0.005 mm), and surface finishes as fine as two to four microinches, **Figure 26-33**.

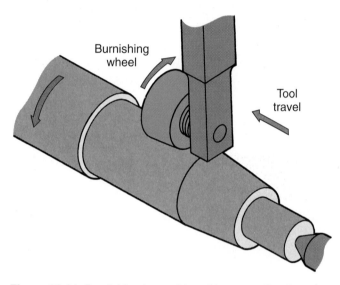

Figure 26-31. Burnishing is a cold-working operation that gives a high quality finish to the metal. It "irons out" the peaks of a metal's surface into the valleys.

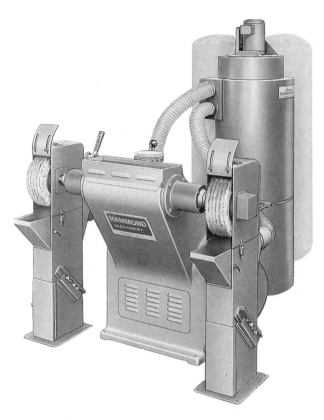

Figure 26-29. This buffing lathe is equipped with wheel guards and a dust collection system. (Hammond Machinery Inc.)

Figure 26-32 Roller burnishing tool for finishing holes. It sizes, finishes, and work-hardens the metal. The process removes no metal. Each hole size requires a different roller burnishing tool. (Cogsdill Tool Products, Inc.)

26.2.5 Deburring Techniques

Powerbrushing is frequently employed to remove burrs from machined surfaces, **Figure 26-34**. Wire or fiber brushes replace the buffing wheels on the buffing lathe. In removing burrs, the wheels produce a satin sheen on the brushed surface, **Figure 26-35**.

Figure 26-30. Polishing a steel die with a diamond dust to produce a mirror-like finish. (NSK America)

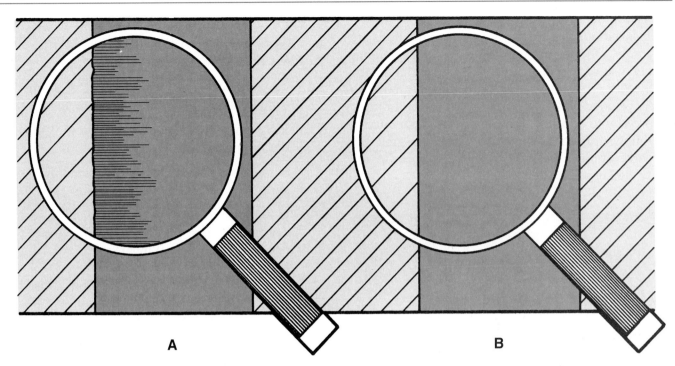

Figure 26-33 Roller burnishing. A—Closeup of reamed hole. Resulting microscopic ridges and valleys are exaggerated in the drawing. B—Finish in hole after roller burnishing.

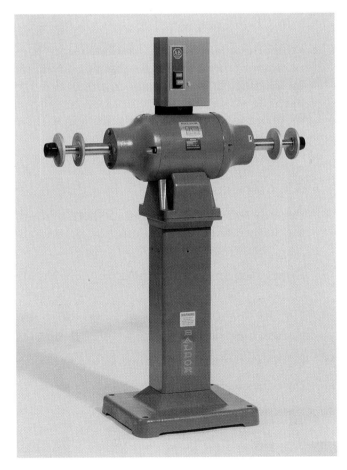

Figure 26-34 Buffing lathe that can be fitted with wire or fiber brush wheels to remove burrs, and flannel wheels to polish. Always wear a dust mask when using a buffing lathe. Guards and wheels have been removed for clarity. (Baldor)

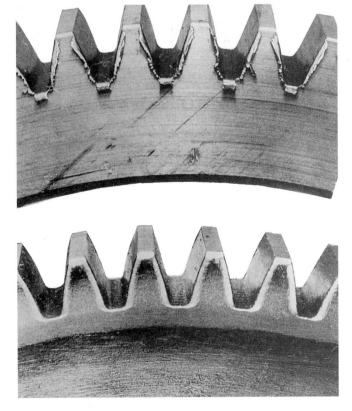

Figure 26-35. Top gear has burrs. Lower gear has had burrs removed using wire brushes. (Osborn Mfg. Co.)

Hand deburring of small holes and intricate parts is tedious and expensive so other methods are desirable. One technique, known as *extrude-hone*, makes use of a *silicon putty* (very similar to the "Silly Putty®" found in toy stores) permeated with finely divided abrasive particles. The silicon plastic is forced into and through the part to be deburred. As it flows through the openings to be deburred, the abrasive grains remove any machine burrs in the part. See **Figure 26-36**.

Figure 26-36. *Extrude hone technique. A—This block has 80 intersecting holes; a few of which are shown in cut-away sections. Access to them for hand deburring was virtually impossible. Abrasive-charged silicon putty removed all burrs rapidly and economically. B—Several difficult jobs deburred on a production basis by this process. (South Shore Tool and Development Corp.)*

TEST YOUR KNOWLEDGE

Please do not write in the text. Write your answers on a separate sheet of paper.

1. The symbol _____ was used at one time on drawings to designate a machined surface.

2. Why was the above symbol's use discontinued?

3. Surface roughness is now measured in _____ and _____. What does each equal?

4. In addition to surface roughness, other surface conditions were given values. Waviness was one such condition. It means:
   a. Very rough surfaces.
   b. Smoothly rounded undulations caused by tool and machine conditions.
   c. Scratches on the machined surface.
   d. All of the above.
   e. None of the above.

5. Lay is another surface finish condition. What does it mean?

6. A 500/12.5 surface finish is _____ than a 125/3.2 surface finish.

7. While the quality of a machined surface is of paramount importance in the machining of metal, other finishing methods are used in the machine shop. They are employed for one or more of the following reasons. Explain each.
   a. Appearance.
   b. Protection.
   c. Identification.
   d. Cost reduction.

8. Regardless of the finishing method utilized, the surface to be finished must be thoroughly _____ of all contaminants.

9. Paints, lacquers, and enamels are in the family of _____ coatings.

10. List the five ways employed to apply the finishes in Question 9.

11. List three types of anodizing.

12. Describe electroplating.

Electromachining Processes

IMPORTANT TERMS

arc
dielectric fluid
electrical discharge
   machining (EDM)
electrical discharge wire
   cutting (EDWC)
electrochemical machining
   (ECM)

electrode
electrolysis
electromachining
erosion
servomechanism

The most notable advantage of *electromachining* is that mechanical forces have no influence on the processes. Material is not removed by "brute force," as with conventional machining techniques. Rather, electrical energy is applied directly to remove metal by erosion.

Electrical discharge machining and electrochemical machining are two electromachining processes that have made a great impact on the field of metalworking and machining. Notably, neither process produces a chip as metal is removed. Instead, the particles are disposed of completely by vaporization or are reduced to microscopic particles. In order to be machined by either of these processes, the metals must conduct electricity.

Never attempt to operate electromachining equipment while your senses are impaired by medication or other substances.

27.1 ELECTRICAL DISCHARGE MACHINING (EDM)

Electrical discharge machining (EDM) is a process where hard, tough, fragile, and/or heat-sensitive metals difficult to machine by conventional "chip-making" techniques can be worked to close tolerances.

Die blanks, for example, can be worked after heat-treating, **Figure 27-1**. This eliminates warping and distortion that frequently occur when a finished die is heat-treated. Superhard metals are readily worked to tolerances as close as 0.0002" (0.005 mm). Surface finishes can be varied from very rough to almost mirror-smooth. "Washed-out" (worn) dies can also be reworked in the heat-treated state.

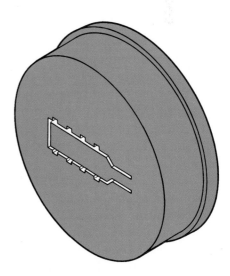

Figure 27-1. *Die blanks, like this extrusion die used to produce storm window frames, can be machined by EDM while in the heat-treating stage. The job can be done quickly, with no possibility of die warping, and at a considerable cost savings over traditional machining techniques.*

27.1.1 EDM Principle

In a gasoline engine, sparking (or arcing) takes place at the spark plug gap when the ignition coil fires to ignite the fuel mixture. After prolonged operation, the spark plug electrodes will be eroded by the action of the electric arcs. This is the basis of EDM. See **Figure 27-2**.

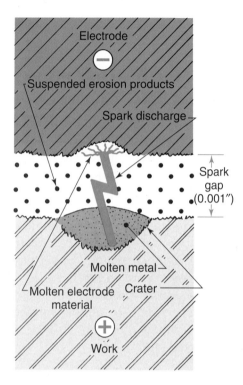

Figure 27-2. How the EDM process works. A spark (or arc) from an electrode causes the work to erode.

An *electrical discharge machine,* **Figure 27-3,** is composed of the following parts.

- The *power supply* is needed to provide direct current and to control voltage and frequency.
- The *electrode,* **Figure 27-4,** can be compared to the cutting tool on a conventional machine tool. Graphite has become the most-used material for EDM electrodes. Because of the dust produced when machining graphite electrodes, special machine tools with enclosed cutting zones and powerful suction systems have been developed, **Figure 27-5**.
- The *servomechanism* (drive unit) is used to accurately control electrode movement and to maintain the correct distance between the work and the electrode as machining progresses.
- The *dielectric fluid,* usually a light mineral oil, is used to form a nonconductive barrier between the electrode and the work at the arc gap.

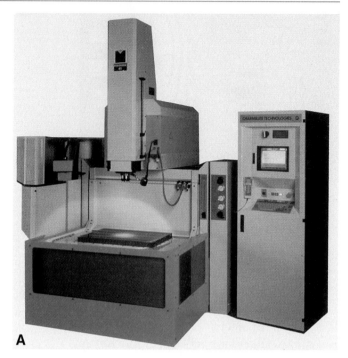

A

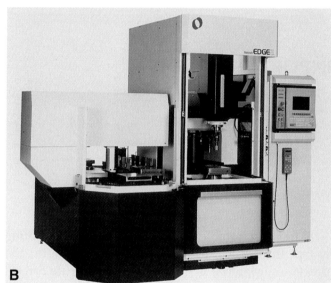

B

Figure 27-3. Electrical discharge machines. A—A CNC ram-type EDM machine. (Charmilles Technologies) B—A CNC EDM machine which has automated loading and unloading of electrodes and workpieces. (LeBlond Makino Machine Tool Co.)

The servomechanism maintains a thin gap of about 0.001″ (0.025 mm) between the electrode and the work. The electrode and the work are submerged in the fluid, **Figure 27-6**. When the voltage across the gap reaches a point sufficient to cause the dielectric fluid to break down, a spark occurs. Each spark erodes only a tiny particle of metal, but since the sparking occurs 20,000 to 30,000 times per second, appreciable quantities of metal are removed.

Besides providing a nonconductive barrier, the dielectric fluid also serves to flush particles from the gap, keep the electrode and the work cool, and

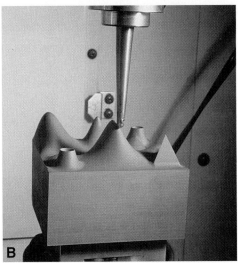

Figure 27-4. *Graphite electrodes. A—Note how thin the cutting areas are on this graphite electrode. It would be difficult to machine them by conventional means. B—A more complex shaped graphite electrode. A probe is being used to check whether the electrode's shape meets specifications. (LeBlond Makino Machine Tool Co.)*

Figure 27-5. *Machines for milling graphite electrodes must be environmentally clean. This CNC graphite mill has a totally enclosed cutting zone and a powerful suction system that contains, captures, and removes chips and dust. (LeBlond Makino Machine Tool Co.)*

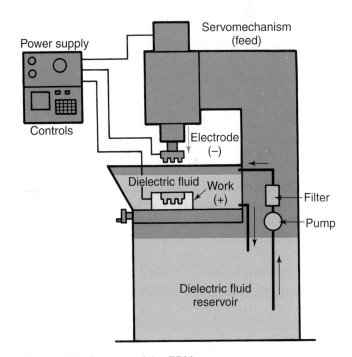

Figure 27-6. *Diagram of the EDM process.*

prevent fusion of the electrode with the workpiece. A filter removes particles from the fluid.

Roughing cuts are made at low voltage and low frequency, with high amperage and high capacitance (opposition to any change in voltage). Finishing cuts require high voltage and high frequency, with low amperage and low capacitance.

Hard metals erode at a much slower rate than soft metals. Since the electrode is also consumed, but at a much slower rate than the work, considerable savings can be effected by making interchangeable electrodes for roughing, sizing, and finishing. Long runs may require several sets of electrodes.

27.1.2 EDM Applications

The EDM process is employed to:
- Shape carbide tools and dies.
- Machine complex shapes in hard, tough metals. See **Figure 27-7**.
- Machine applications where the physical characteristics of the metal or its use make it impractical or very expensive to machine by conventional methods, **Figure 27-8**.

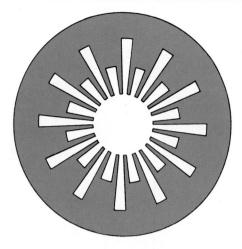

Figure 27-7. *EDM makes it possible to machine complex shapes to close tolerances in hard, tough materials. This die is employed to make electronic parts.*

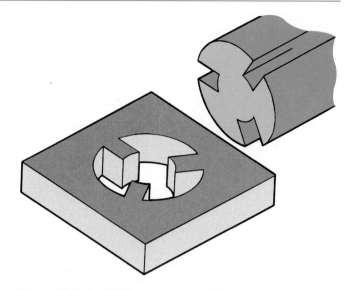

Figure 27-9. *An EDM electrode must be an exact reversal or mirror image of the cut to be made. Many electrodes are made from graphite.*

Inserts

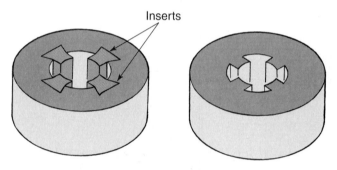

Figure 27-8. *This guide would be expensive to manufacture using conventional machining techniques.*

- Eliminate tedious and expensive handwork in die making because the cavity produced in the metal is a "mirror image" of the electrode, **Figure 27-9**.
- Drill holes ranging in size from 0.0012″ to 0.120″ (0.3 mm to 3 mm) in diameter. They can be square, rectangular, triangular, or round. Multiple holes can be produced at the same time. Hardness is not a factor as long as the material will conduct electricity.

27.2 ELECTRICAL DISCHARGE WIRE CUTTING (EDWC)

A process similar to band machining, *electrical discharge wire cutting (EDWC)* makes use of a small-diameter wire electrode in place of a saw. Like EDM, this process removes material by spark erosion as the wire electrode is fed through the workpiece, **Figure 27-10**. The wire electrode is fed from a spool over sapphire or diamond guides. A starter or threading hole is required. A steady stream of

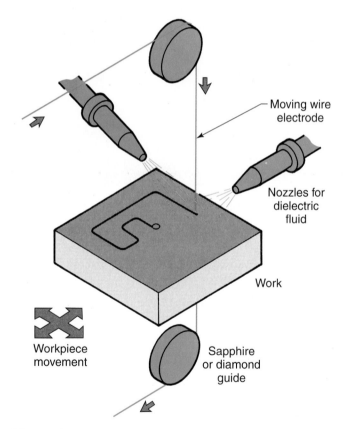

Moving wire electrode

Nozzles for dielectric fluid

Work

Workpiece movement

Sapphire or diamond guide

Figure 27-10. *How electrical discharge wire cutting (EDWC) works. It differs from EDM in that a fine, moving wire electrode is used for cutting instead of a solid electrode. This technique is ideal for CNC operations.*

dielectric fluid cools the electrode and the work. The wire electrode is used only once because it becomes warped or distorted after just one pass through the work.

EDWC is well-suited for CNC applications, **Figure 27-11**. This makes it possible to produce dies, shaped carbide cutting tools, and punches in less than one-third the time required by conventional methods. When employed for other types of work, layers of sheet metal stacked up to 6″ (150 mm) thick can be gang-cut to produce a number of parts in one pass. See **Figure 27-12** for examples of EDWC work.

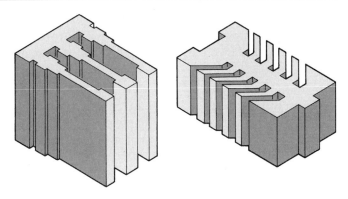

Figure 27-12. Examples of work produced using EDWC.

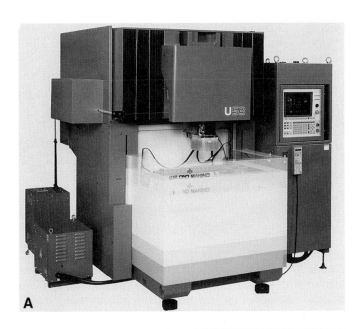

Figure 27-11. EDWC equipment. A—Submerged wire EDM machine. Submerged processing allows thicker cuts and can produce ultrafine finishes. The machine is fitted with an automatic wire threader and is self-cleaning. (LeBlond Makino Machine Tool Co.) B—Automated wire EDM cell with robotic loading and unloading.
(Charmilles Technologies/Mecatool USA Ltd.)

27.3 ELECTROCHEMICAL MACHINING (ECM)

Electrochemical machining, more commonly known as ECM, might be classified as electroplating in reverse. Like electroplating, the process requires DC electricity and a suitable *electrolyte* (an electrically conductive fluid). However, with ECM, the metal is removed from the work rather than deposited onto it, **Figure 27-13**.

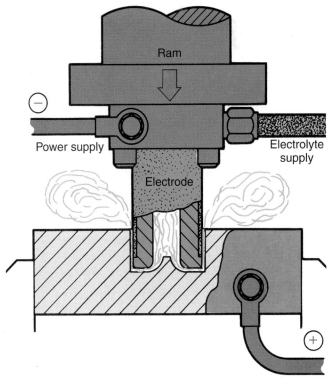

Figure 27-13. How electrochemical machining (ECM) works.

The electrolyte for ECM is usually common salt (NaCl) mixed with water. A stream of electrolyte is pumped at high pressure through a gap between the positively (+) charged work and the negatively (–) charged tool (electrode). The current passing through the gap removes material from the work by electrolysis, duplicating the shape of the electrode tool as it advances into the metal. In some applications, tolerances as close as 0.0004″ (0.010 mm) can be maintained.

The work is not touched by the tool; hence, no friction, heat, sparking, or tool wear occur. The machined surface is free of burrs and, in some instances, is highly polished. The operation of the machine is unique because the only sound heard is the rush of liquid. See **Figure 27-14**.

Figure 27-14. *The rush of fluid is the only noise heard when an electrical discharge machine is in operation.*

27.3.1 ECM Advantages

ECM offers many advantages:
- Metal is removed rapidly — up to 1 cu. in. per minute for each 10,000 amps of machining current.
- The kind of metal or its hardness does not affect the speed of material removal. Cast iron is about the only metal that offers problems, and it is machined by other processes.
- ECM is accurate. Difficult shapes can be machined easily, **Figure 27-15**.
- The machined metal is stress-free and will not warp or spring out of shape when removed from the machine.
- There is no tool wear.
- Several operations (milling, grinding, deburring, and polishing) often can be eliminated with ECM. See **Figure 27-16**.

Figure 27-15. *The blades of this titanium turbine wheel were machined in a metal blank. One blade was machined in two wheels simultaneously at a feed rate of approximately 0.250″ (6.5 mm) per minute.*

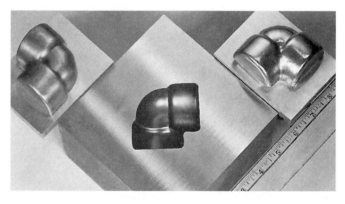

Figure 27-16. *This pipe elbow forging die was machined into hardened steel to the shape and finish shown in 20 minutes. Conventional machining and polishing would require several hours to produce the same form. Danger of warping would also be present when the machined die was heat-treated.*

- Advances continually are being made in ECM. The ability to produce highly complex shapes with simple tooling widens the range of application for this machining process.

TEST YOUR KNOWLEDGE

Please do not write in the text. Write your answers on a separate sheet of paper.

1. EDM stands for _____.

2. EDWC stands for _____.

3. ECM stands for _____.

4. Metals to be machined by EDM and ECM must conduct _____; otherwise, neither process can be used.

5. Neither of the processes in Question 4 produces a _____ as metal is removed. Particles are disposed of completely by _____ or reduced to _____.

6. EDM is a process that permits metals that are:
   a. Hard or tough to be machined.
   b. Difficult or impossible to machine by conventional means to be worked to close tolerances.
   c. Fragile or heat-sensitive to be machined.
   d. All of the above.
   e. None of the above.

7. Explain the functions of the dielectric fluid used in EDM.

8. The EDM machine _____ maintains a very thin gap of about 0.001″ (0.025 mm) between the electrode and the work.

9. In EDM, the cavity produced in the work is an exact _____ of the electrode.

10. How does EDWC differ from EDM?

11. ECM might well be classified as _____ in reverse.

12. In ECM:
    a. The work is not touched by the tool.
    b. There is no friction or heat generated.
    c. There is no tool wear.
    d. All of the above.
    e. None of the above.

13. In ECM, metal is removed _____ .

14. What are five advantages that ECM offers?

Operating under computer numerical control, this five-axis laser beam machining system can produce some complex parts in minutes, rather than the hours required with conventional machining. (Lumonics Corp.)

Chapter 28

Nontraditional Machining Techniques

LEARNING OBJECTIVES

After studying this chapter, you will be able to:
○ Describe several nontraditional machining techniques.
○ Explain how nontraditional machining techniques differ from traditional machining processes.
○ Summarize how to perform several nontraditional machining techniques.
○ List the advantages and disadvantages of several of the nontraditional machining techniques.

IMPORTANT TERMS

chemical blanking
chemical machining
chemical milling
electron beam machining
etchant

hydrodynamic machining
impact (slurry) machining
laser beam machining
ultrasonic machining
water-jet cutting

The metalworking industry is responsible for cutting, shaping, and fabricating both metals and nonmetals. Given the broad range of materials and rapid development of new ones, new machining techniques have been devised to keep pace. This chapter will describe several of the new techniques that differ from the traditional chip-removal method of the lathe, drill press, milling machine, grinder, and saw.

28.1 CHEMICAL MACHINING

Chemical machining is a refinement of the process once used by photoengravers to prepare printing cuts and plates. Chemicals, usually in an *aqueous* (with water) solution, are employed to etch away selected portions of the metal to produce an accurately contoured part.

In general, chemical machining falls into two categories: chemical milling and chemical blanking.

28.1.1 Chemical Milling

Chemical milling, also called *chem-milling* or *contour etching*, is a recognized and accepted technique for machining metal to exacting tolerances through chemical action. The process makes possible the selective removal of metal from relatively large surface areas, **Figure 28-1**. For example, chem-milling may be employed to reduce the weight of sheet metal parts, critical to aerospace vehicle performance.

Basically, chem-milling is a process in which the prepared part is immersed in an *etchant* (usually a strong alkaline solution) where the resulting chemical action removes the desired metal. The

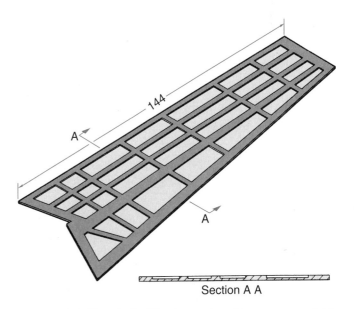

Figure 28-1. *Chemical milling is employed to remove metal to close tolerances. This aircraft wing panel has been chemically reduced in sections where spars are not attached. Considerable weight reduction is accomplished with no sacrifice of structural strength.*

immersion time must be carefully controlled. Areas that are not subject to metal removal are protected by *masks* (special coating materials) that do not react to the etching solution.

Chem-milling and conventional milling are complementary processes. Refinements in chem-milling make it possible to remove metal to form shapes or microscopic parts that would be difficult or impossible to do by conventional machining techniques. Tapers and multiple-depth cuts also can be produced. See **Figures 28-2** and **28-3**.

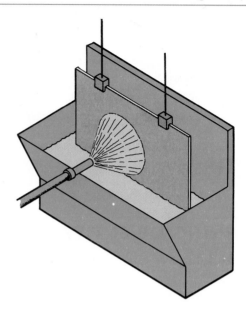

Figure 28-4. Cleaning removes all grease and dirt that might affect the etching process.

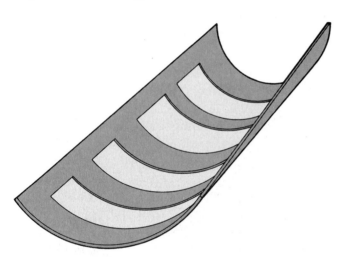

Figure 28-2. The outer skin of an aircraft engine housing is integrally stiffened. Aluminum sheet is approximately 48″ × 72″ (1219 mm × 1829 mm) and is chem-milled after being formed.

Figure 28-3. Multiple depth cuts are possible with chem-milling by masking shallower sections when the correct depth has been reached. Tapers are produced by withdrawing metal from an etchant at a predetermined rate.

Steps in chemical milling

There are six major steps in the chemical milling process:

- *Cleaning.* Removes all grease and dirt that might affect the etching process. See **Figure 28-4**.
- *Masking.* The entire part is coated with a masking material, applied by brushing, dipping, spraying, or roller coating. The masked metal sheet is then baked to remove all solvents.
- *Scribing and stripping.* A template is placed over the entire part. Then, the areas to be exposed are circumscribed, and the masking material is stripped away. See **Figure 28-5**.
- *Etching.* Parts are racked and lowered into an etchant for milling operation. See **Figure 28-6**.

Figure 28-5. A template is placed over the entire part; then, areas to be exposed are circumscribed and masking material is stripped away. (Northrop-Grumman Corp.)

- *Rinsing and solvent stripping.* After rinsing, parts are lowered into a solvent tank which releases the maskant bond. Maskant residue is stripped from the part.
- *Inspection.* The accuracy of the chemical milling etch is measured with the aid of an ultrasonic thickness gage.

Advantages of chemical milling

Chemical milling offers many advantages:
- Tooling costs are low.
- Tolerances of +/−0.003″ (0.07 mm) are obtainable on cuts up to 0.50″ (12.5 mm) deep.

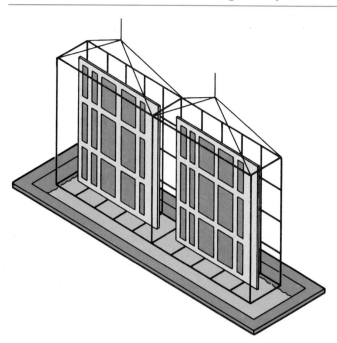

Figure 28-6. Parts are racked and lowered into an etchant for milling operation.

- The size of the workpiece size is limited only by the size of the immersion tank.
- Warping and distortion of formed sections is negligible.
- Contoured or shaped parts can be chem-milled after they are formed, **Figure 28-7**.
- Many parts can be produced simultaneously.
- Unsupported pieces that are as thin as 0.015″ (0.4 mm) can be machined without danger of buckling.
- Both sides of the metal can be milled at the same time.
- Any metal, regardless of its state of heat treatment, can be machined chemically.
- No burrs are produced in the area machined.

Disadvantages of chemical milling

The chemical milling process does have some disadvantages:

- The process is slow; it takes considerable time to remove large quantities of metal.
- All surface imperfections must be removed before etching; otherwise, these areas would etch at a faster rate and be amplified on the finished surface.
- Chem-milling is not recommended for etching holes.
- Surface finishes on deep etches are not comparable to conventionally machined surfaces.
- Lateral dimensions are difficult to hold because the etchant works sideways as well

as in depth. Typical lateral tolerances work on an etch factor of 3:1. This means that for every 0.003″ (0.08 mm) of etched depth, 0.001″ (0.03 mm) of undercut will occur. The top edges of the cavity will be sharp; however, the inside edges and corners will have a radius approximately equal to the cut depth, **Figure 28-8**.

28.1.2 Chemical Blanking

Chemical blanking, also called *chem-blanking*, *photoforming*, or *photoetching*, involves total removal

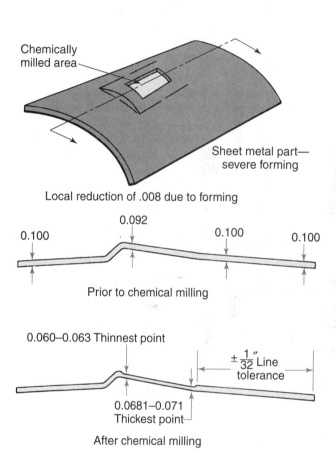

Figure 28-7. This aircraft part was chem-milled after it was formed. Pieces even more severely formed than this part can be chem-milled economically. (Northrop-Grumman Corp.)

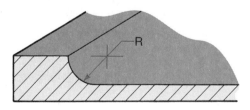

Figure 28-8. The inside edges of a chem-milled section will have a radius equal to the depth of the etch, an advantage in many applications.

of metal from certain areas by chemical action. It is a variation of chemical milling. Chem-blanking is employed by the aerospace and electronics industries to produce small, intricate, ultrathin parts. See **Figure 28-9**.

Metal foil as thin as 0.00008" (0.002 mm) can be worked by chem-blanking. (By comparison, a cigarette paper is about 0.001" or 0.025 mm thick.) This ultrathin metal is laminated to a plastic backing to protect it from damage during shipping, **Figure 28-10**.

The process is not recommended for metals thicker than 0.09" (2.3 mm). However, almost any metal can be chem-blanked.

Steps in chemical blanking

There are five major steps in the chemical blanking process:

1. A master drawing of the part is made, **Figure 28-11**. Depending upon the tolerances that must be met, it may be drawn as large as 50 times the size of the required part.

2. The drawing is reduced photographically. This produces a film or glass master the exact size of the required part. Since many identical parts usually must be made, a negative with multiple images often is often produced on a photo-repeating machine, **Figure 28-12**.

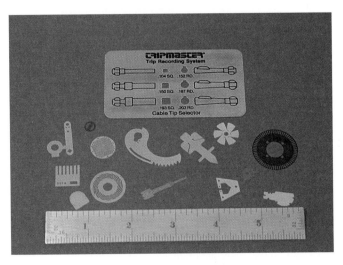

Figure 28-9. Made by chemical blanking, these electronic parts range in thickness from 0.0001" to 0.002" (0.0025 mm to 0.05 mm). (Microphoto, Inc.)

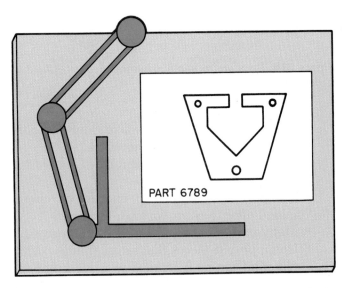

Figure 28-11. A master drawing may be up to 50 times the actual size of the required part, depending upon tolerances that must be met. (Microphoto, Inc.)

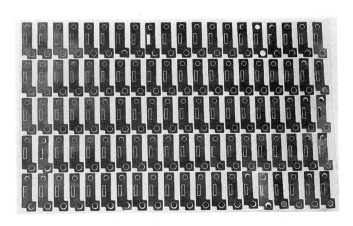

Figure 28-10. A plastic sheet holds 0.001" (0.025 mm) thick computer components. Computer parts are laminated to plastic to assure a flat, crease-free condition.

Figure 28-12. A master drawing is photographically reduced to the required size. Since many identical parts must be made, it is most economical to etch several at a time. (Microphoto, Inc.)

3. A metal sheet is thoroughly cleaned and coated with photosensitive resist. The multiple-image negative is contact-printed on the surface of the metal blank, **Figure 28-13**. After exposure, the images are developed. The resist in unexposed areas is dissolved during the developing process, exposing bare metal.

Figure 28-14. After designs have been developed on metal, strips are placed in a spray etcher. These are stainless steel computer parts. (Hamilton Watch Co.)

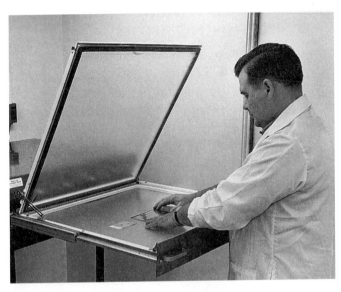

Figure 28-13. The technician is using a double-sided printing unit. Metal is placed between two identical, perfectly aligned photographic negatives. A light source deposits the part's image onto the metal blank. (Hamilton Watch Co.)

4. The processed metal blank is placed in a spray etcher, **Figure 28-14**. Spraying is preferred over immersion or splashing because it offers a higher etching rate and better tolerance control. **Figure 28-15** shows one type of horizontal conveyorized spray etcher used for chem-blanking. The spray nozzles move back and forth while the holding tray oscillates to assure maximum exposure of the work to the etchant. Etching time varies from just over three minutes for metal 0.0001″ (0.002 mm) thick, to as long as an hour for 0.010″ (0.25 mm) material. Etching removes all of the metal not protected by the resist coating.

5. After etching, the photoresist is removed with solvents, and the metal is flushed with warm water and dried. Visual inspection of each chem-blanked unit assures strict adherence to specifications, **Figure 28-16**.

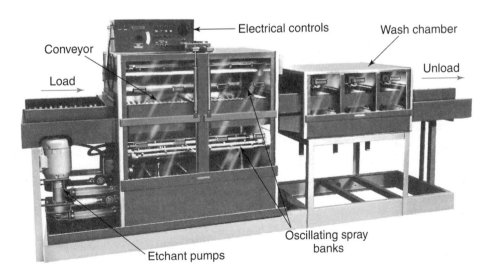

Figure 28-15. This continuous spray etching machine was developed for a conveyorized operation. It also includes a rinsing compartment at the exit end. (Chemcut)

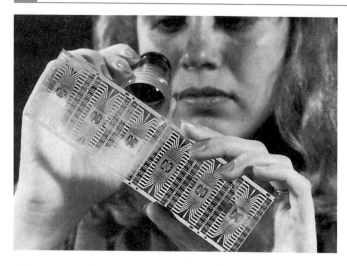

Figure 28-16. An inspector visually checks a microchip section that was chem-blanked. It will become part of the electronic control module of an automotive emission/fuel control system. (Delco Electronics Div. of General Motors Corp.)

Advantages of chemical blanking

Like all nontraditional machining techniques, chemical blanking offers certain advantages over conventional machining techniques:

- Tooling costs are lower. The technique utilizes an inked drawing on dimensionally stable plastic film, so almost any pattern that can be drawn can be produced. Small parts can be drawn oversized. A good drafter can render these large drawings within +/−0.005″ (0.13 mm) on Mylar® film. Any appreciable error diminishes when the drawing is reduced to actual size.
- There are no burrs on the milled part.
- Initial quantities of newly designed parts often can be produced within hours.
- Design changes are made easily by revising the existing artwork.
- Ultrathin metal foils can be worked without fear of distortion.
- Accuracy increases as the metal thickness decreases.
- Metal characteristics (brittleness, hardness, etc.) have no significant effect on the process.

Disadvantages of chemical blanking

Chemical blanking has some disadvantages:

- The process is relatively slow. Metal removal seldom exceeds 0.001″ (0.02 mm) per minute.
- Highly skilled workers are required.
- Chemicals are extremely corrosive, so etching equipment must be isolated from other plant equipment. Care must be employed when

disposing of expended etchant and rinse water to prevent environmental damage.
- Good photographic facilities are required.
- The maximum thickness of metal that can be worked is limited. For production purposes, the practical limit is 0.09″ (2.3 mm).
- Tolerances increase with metal thickness.

28.2 HYDRODYNAMIC MACHINING (HDM)

Hydrodynamic machining (HDM) or *water-jet cutting*, **Figure 28-17**, is a relatively new cutting process. It was developed to shape *composites*—layers of a tough fabric-like material bonded together into three-dimensional shapes called *layups*. The texture of composites quickly dulls conventional cutting tools.

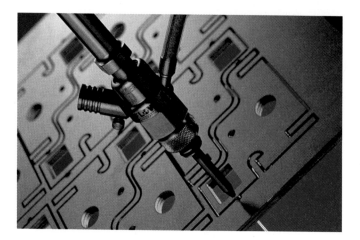

Figure 28-17. How hydrodynamic machining or abrasive water-jet cutting operates. (Flow International Corp.)

Hydrodynamic machining can be accomplished with or without abrasives. The addition of abrasives to the water jet is preferred when shaping metals and harder nonmetallic materials, **Figure 28-18**.

The computer-controlled technique uses a 55,000 psi (379 000 kPa) water jet to cut complex shapes with minimal waste, **Figure 28-19**. Depending upon the material being cut, tolerances can be held to +/−0.004″ (0.1 mm). No heat is generated to damage the material, being cut, nor is airborne dust produced; *particulates* (fragments) are carried away by the water jet.

Since the speed of the water jet leaving the nozzle is approximately three times the speed of sound, it can cause serious injury. Hands must be kept clear of the work area.

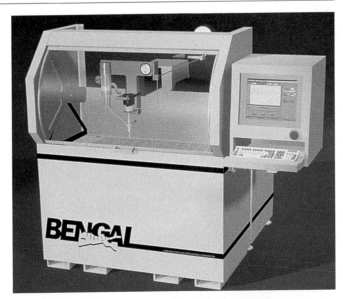

Figure 28-18. A CNC hydrodynamic machine programmed to cut circular saw blades. Abrasives were added to the water jet because the blades are cut from heat-treated, high-speed steel (HSS). The process is powerful enough to cut through two inches of tool steel at the rate of 1.5″ per minute. (Ingersoll-Rand Co., Waterjet Systems)

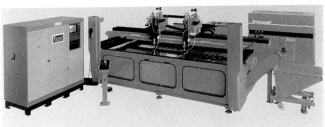

Figure 28-19. Water-jet machines. A—A two-axis (X-Y), abrasive water-jet machine with a built-in CAD/CAM system. It has a work envelope of 39.3″ × 19.6″ (1 m × 0.5 m) and can cut virtually any material up to 4″. (100 mm) thick. Note that the machine is enclosed to protect the operator. (Flow International Corp.) B—This CNC abrasive water-jet cutting system has two separate cutting heads that travel over the cutting area. It is also fitted with an automatic drill to pre-drill a piercing hole. The drill is useful when cutting stacked material because the water jet, which is normally used to pierce, could separate laminated materials of varying hardness. Parts are programmable on the built-in computer. (Bystronic, Inc.)

28.3 ULTRASONIC MACHINING

The science of *ultrasonics* (sound waves at frequencies higher than the human ear can detect) has found applications in many areas—machining, welding, quality control, and cleaning, to name a few. Considerable research is being done to develop new uses and to improve existing techniques.

The average person can hear sounds that vibrate between 20 to 20,000 cycles (times) per second. Below 20 cycles per second, sound waves are called *infrasonic*. Above 20,000 cycles per second, sound waves are known as *ultrasonic*. Industrial applications make use of energy up to 100,000 cycles per second.

Ultrasonic waves are created by passing an electric current (usually 60-cycle ac) into a suitable generator (transducer) to produce the desired frequency. The sound waves may be used in conjunction with a fluid (as in quality control and cleaning applications) or applied directly to the cutting tool or metal as it is being machined, welded, or formed. See **Figure 28-20**.

28.3.1 Ultrasonic-assist Machining

Ultrasonic machining applies sound waves to the tool or metal as it is cutting or being cut. A fairly recent development is an *ultrasonic assist* for conventional metalcutting processes. The transducer is fitted to a standard machine tool to make the tool vibrate at a high frequency. The ultrasonic assist makes possible a 10% to 50% reduction in tool forces with almost complete elimination of tool chatter. Chatter reduction is a distinct advantage when the boring operations require the use of long, slender boring tools. Tool wear is reduced and more cutting can be done between sharpenings. Surface finishes are also improved.

Similar applications have been made in grinding. Ultrasonic waves are passed through the grinding wheel. Particles and chips that normally become embedded in the wheel are vibrated loose and washed away by the coolant. Grinding temperatures are reduced, and wheel life is extended. Material can be removed more rapidly and surface

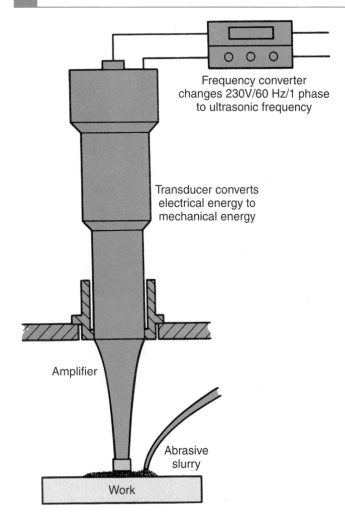

Figure 28-20. *Ultrasonic machining setup.*

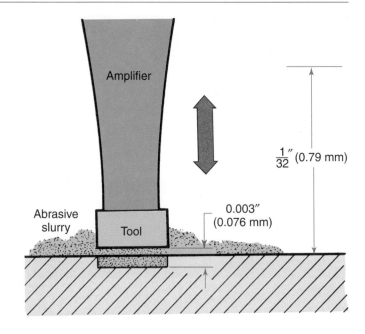

Figure 28-21. Tool motion in ultrasonic (impact) machining is slight, only 0.003" (0.076 mm). The 1/32" (0.79 mm) measurement is used to indicate scale.

finishes improved for the same expenditure of power. Ultrasonic vibrations imparted to grinder coolant have been found to produce similar results.

Drilling, reaming, honing, milling, and EDM (electrical discharge machining) techniques all employ ultrasonic assist.

28.3.2 Impact Machining

Impact or *slurry machining* uses ultrasonics and a special tool to force an abrasive against the work. With the exception of diamond tools, it is the only commercially feasible way to machine extremely hard, brittle, and frangible (breakable) materials. The technique works best on hard, brittle materials (glass, quartz, silicon, and carbides); it is ineffective on soft materials, such as aluminum and copper.

Machining is done by a shaped cutting tool oscillating about 25,000 times per second. This rapid oscillation pounds a slurry of fine, abrasive particles against the work. The tool stroke, at the vibrating end, is only 0.003" (0.076 mm), **Figure 28-21.**

A solid funnel-shaped horn must be employed to amplify and transmit the vibrations. Ultrasonic vibrations directly from the transducer do not produce enough motion to produce the required tool movement. The tool does not touch the work, and no heat is generated, thereby eliminating possible distortion of the work. Microscopic portions of the work are chipped away to produce the desired shape. The machined section is a mirror image of the cutting tool, **Figure 28-22.**

Industrial applications of impact machining include:

- Slicing and cutting germanium and silicon wafers into tiny chips for transistors, diodes, and rectifiers for the electronics industry.
- Machining complex shapes in nonconductive and semiconductive materials that cannot be satisfactorily handled by EDM and ECM.
- Shaping virtually unmachinable space-age materials.

Impact machining is slow and the surface finish is dependent on the size of the abrasive grit used. One inch (25 mm) is about the deepest cut possible with existing equipment. On the other hand, tolerances of 0.001" (0.025 mm) can be maintained on hole size and geometry in most materials. Equipment cost is moderate. The training period for machine operators is short and does not require special skills.

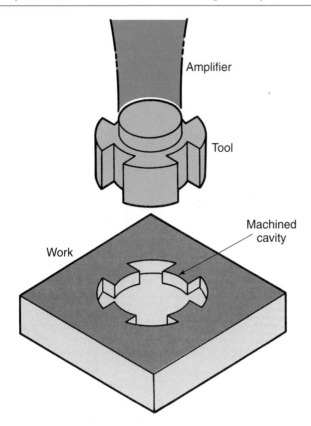

Figure 28-22. The machined section is a mirror image of the tool.

28.3.3 Other Ultrasonic Applications

Industrial applications for ultrasonics vary. A principal use is to improve the cleaning power of chemical solvents. As the waves pass through the cleaning solution, microscopic vapor pockets are created in the fluid. These "bubbles" (high vacuum areas) form and collapse about 20,000 times per second. This creates local pressure as much as 10,000 psi, as well as heat. The bubbles smash against the work and literally tear away the dirt, oil, grease, chips, flux, and other contaminants on the surface.

The cleaning action is so thorough that the technique is employed to decontaminate work that has been exposed to radioactive solutions and gases.

Nondestructive testing using ultrasonics has been developed to the point where the testing procedure is fully automated. Flaws as small as 0.001″ × 0.250″ (0.025 mm × 6.25 mm) can be detected on a continuous basis.

Sound waves are also used to weld metals to nonmetals. Precision welding of dissimilar materials to form a solid bond has been developed successfully. For example, ultrasonic waves can weld aluminum wires to special glass on some electronic circuits.

28.4 ELECTRON BEAM MACHINING (EBM)

The *electron beam microcutter-welder* uses a high-energy, highly focused beam of electrons to weld or cut materials. Electron beam machining (EBM) is the direct result of the special needs of the atomic energy, electronics, and aerospace industries. In some aspects, it is the most precise and versatile of the new machining techniques. See **Figure 28-23**.

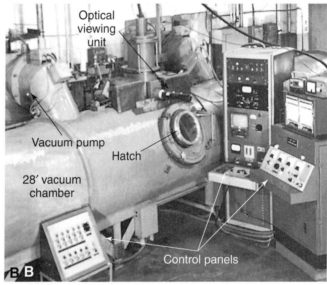

Figure 28-23. Electron beam machines. A—An electron beam microcutter-welder. (Hamilton Standard Div., United Technologies) B—A partial view of a 28′ (8.5 m) electron beam cutting and welding machine. Lead shielding is required to protect the operator from X-rays produced by the electron beam. (Westinghouse Electric Corp.)

Because the electron beam will cut any known metal or nonmetal that can exist in a high vacuum, its microcutting capabilities are almost unlimited.

The development of refined focusing systems has made it possible to control the cutting action with a high degree of precision.

Electron beam cutting action can be controlled so precisely that it is possible to drill holes as small as 0.0002" (0.005 mm) in diameter and mill slots having widths of 0.0005" (0.0125 mm). The finish of the completed work is similar to a very fine machined edge. See **Figure 28-24**.

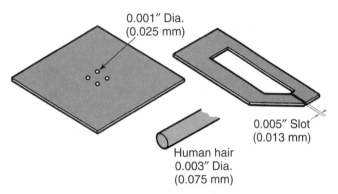

Figure 28-24. *Note the size of the work done using the electron beam technique. The parts and the shaft of human hair are drawn to the same scale.*

The electron beam machine is similar in some respects to a television set. In the television set, a heated tungsten filament emits a stream of electrons. The electron stream is concentrated into a small-diameter beam by an electro-optical system. The beam is moved by a deflection system so rapidly that a glowing picture is produced on a fluorescent screen (television picture tube).

The electron beam machine tool also makes use of a beam of electrons. However, the resulting electron beam is several times more intense than the one used to produce a television picture. Control dials are used to focus and adjust the electron beam for cutting or welding, **Figure 28-25**.

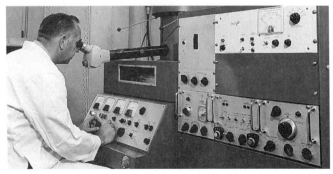

Figure 28-25. *Adjustment and position of the electron beam are controlled by dials similar in function to those found on some audio equipment.*
(Hamilton Standard Div., United Technologies)

The electron beam machine is basically a source of thermal energy, **Figure 28-26**. The beam of electrons can be focused to a very sharp point. Cutting is achieved by alternately heating and cooling the area to be cut.

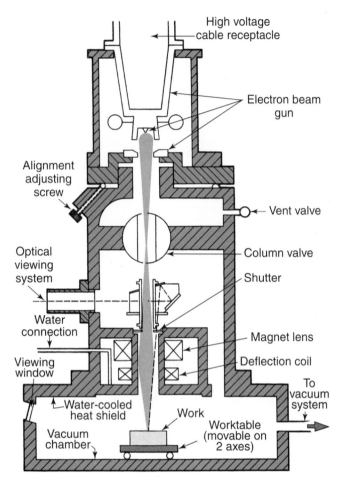

Figure 28-26. *Cross-sectional view of an electron beam microcutter-welder.*

The heating and cooling must be controlled carefully so the material at the point of focus is heated to a temperature high enough to vaporize it, yet not enough to cause the surrounding area to melt. This is accomplished by employing a pulsing technique. The beam is on for only a few milliseconds and is off for a considerably longer period of time. A temperature of 12,000°F (6649°C) can be achieved at the focal point of the electron beam.

Cut geometry (the shape of the cut) is controlled by movement of the worktable in the vacuum chamber and by employing the deflection coil to bend the beam of electrons to the desired cutting path. Initial hole diameter or cut width is controlled by the amount of power applied and the duration of the cutting time. See **Figure 28-27**.

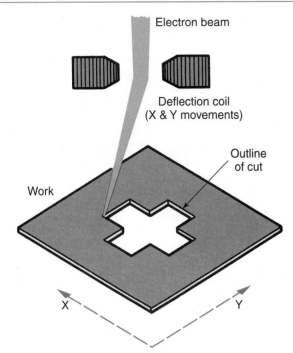

Figure 28-27. The path of the cut is controlled by a deflecting electron beam or by movement of the worktable. Cutting must be done in a vacuum.

28.5 LASER BEAM MACHINING

The term *laser* is an acronym for **L**ight **A**mplification by **S**timulated **E**mission of **R**adiation. The laser produces a narrow and intense beam of light that can be focused optically onto an area only a few microns in diameter. See **Figure 28-28**.

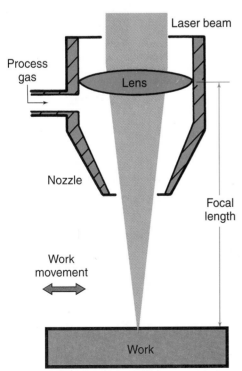

Figure 28-28. A laser produces a narrow and intense beam of light. The beam can be concentrated to a point as small as 0.0002". (0.05 mm) diameter.

Depending upon the initial energy source used to activate the laser, it is possible to instantaneously create temperatures up to 75,000°F (41 650°C) at the point of focus. This is almost seven times the average temperature of the sun. No known material can withstand such heat, **Figure 28-29**.

There has been a dramatic increase in the use of lasers in parts manufacturing, **Figure 28-30**. Lasers

Figure 28-29. At its focal point, the laser beam is almost seven times the average temperature of the sun.

Figure 28-30. A multiaxis laser system that can handle parts up to 13' (4 m) wide. Its design permits multiple setups. Parts can be loaded and unloaded while other parts are being machined. (Lumonics Corp.)

are used for cutting, drilling, slotting, scribing, heat-treating, and welding. See **Figures 28-31**.

The energy output of a laser is usually not continuous. It lasts only a fraction of a second. When employed for machining and cutting, it operates at 1 to 5 cycles per second. The cycle can be controlled manually or electronically. The laser operates on the principle described in **Figure 28-32**.

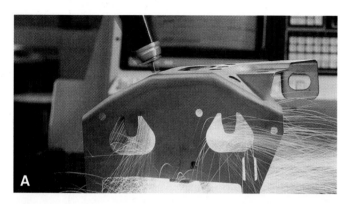

Figure 28-31. Laser cutting. A—When used with CNC, a laser is capable of cutting 3-D workpieces. (Lumonics Corp.) B—A CNC CO$_2$ laser cutting specially shaped slots in a stainless steel plate. The high velocity gas jet aids in material removal by blowing out molten metal through the backside of the work. It also protects the lens from spatter ejected from the cut zone, especially during the piercing operation. (Rofin-Sinar, Inc.)

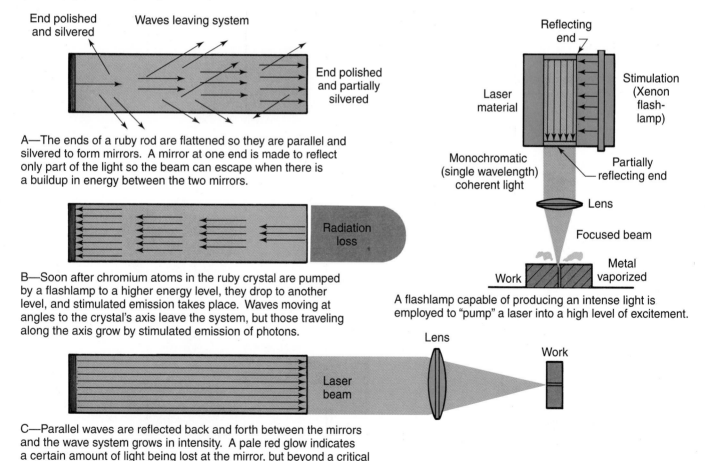

A—The ends of a ruby rod are flattened so they are parallel and silvered to form mirrors. A mirror at one end is made to reflect only part of the light so the beam can escape when there is a buildup in energy between the two mirrors.

B—Soon after chromium atoms in the ruby crystal are pumped by a flashlamp to a higher energy level, they drop to another level, and stimulated emission takes place. Waves moving at angles to the crystal's axis leave the system, but those traveling along the axis grow by stimulated emission of photons.

C—Parallel waves are reflected back and forth between the mirrors and the wave system grows in intensity. A pale red glow indicates a certain amount of light being lost at the mirror, but beyond a critical point, the waves intensify enough to overcome this loss. An intense red beam flashes out of the partially silvered end of the crystal.

A flashlamp capable of producing an intense light is employed to "pump" a laser into a high level of excitement.

Figure 28-32. Operation of a laser.

The laser unit in **Figure 28-33** can operate in either continuous or pulsed modes. With most materials, the edge quality of the cut is better with a continuous beam. Cutting speed is also much greater. The disadvantage of pulsing is that starting, stopping, and turning corners concentrates heat in localized areas on the work. This overheating causes the part to burn away from the programmed path, affecting cut quality and dimensional accuracy.

Figure 28-33. A fully automatic CNC laser cutting cell. It is designed to process flat material up to 80" × 160" (2032 mm × 4064 mm). (Bystronic, Inc.)

Aluminum, stainless steel, and titanium parts that were cut to shape by a laser used to require a time-consuming secondary operation to remove the oxide and dross that formed on the edge and surface of the cut. Titanium had a tendency to become embrittled by the oxygen that was absorbed during the cutting operation.

A recently developed process eliminates these problems by flooding the work surface with argon or nitrogen during the cutting operation. Since no oxygen can contaminate the metal, there is no oxide formation or embrittlement. Parts can be welded immediately after cutting because they require no cleaning.

TEST YOUR KNOWLEDGE

Please do not write in the text. Write your answers on a separate sheet of paper.

1. Chemical machining falls into two categories. Briefly describe each of them.

2. Chemical milling is also known as _____ or _____ .

3. List the six major steps in chemical milling.

4. A mask protects the portion of a chemically-milled job that is:
   a. Not to be etched.
   b. To be etched.
   c. To be cleaned.
   d. All of the above.
   e. None of the above.

5. List the five major steps in chemical blanking.

6. Briefly describe water-jet machining.

7. What machining processes use ultrasonics?

8. Sound waves below 20 cycles per second are called _____ .

9. Sound waves above 20,000 cycles per second are called _____ .

10. Impact machining makes use of a _____ tool that forces _____ against the work to do the cutting.

11. Impact machining is one of the very few commercially feasible methods for machining which types of materials?
    a. Hard.
    b. Brittle.
    c. Frangible.
    d. All of the above.
    e. None of the above.

12. What are three disadvantages of impact machining?

13. With impact machining, tolerances of _____ can be maintained on hole size and geometry in most materials.

14. List five areas where the science of ultrasonics has found industrial applications.

15. The development of the electron beam machine was the direct result of the special needs of what industry?
    a. Electronics.
    b. Atomic energy.
    c. Aerospace.
    d. All of the above.
    e. None of the above.

16. Holes as small as _____ in diameter can be drilled using the electron beam technique.

17. The electron beam machine is basically a source of what type of energy?
    a. Thermal.
    b. Sonic.
    c. Fluid.
    d. All of the above.
    e. None of the above.

18. The electron beam technique cuts material by:
    a. Alternately heating and cooling the area to be cut.
    b. Vaporizing the material.
    c. Making use of a pulsing technique.
    d. All of the above.
    e. None of the above.

19. List two methods employed to control the shape of the cut with EBM.

20. What does *LASER* stand for?

21. Describe how a laser operates.

Chapter 29

Other Processes

LEARNING OBJECTIVES

After studying this chapter, you will be able to:
- ○ Discuss the general machining characteristics of various plastics.
- ○ Describe the hazards associated with machining plastics.
- ○ Sharpen cutting tools to machine plastics.
- ○ Describe the five basic operations of chipless machining and their variations.
- ○ Explain how the Intraform process differs from other chipless machining techniques.
- ○ Describe how powder metallurgy parts are produced.
- ○ Relate how powder metallurgy parts can be machined.
- ○ Compare the advantages and disadvantages of various HERF techniques.
- ○ Explain how the science of cryogenics is used in industry, and list some applications.

IMPORTANT TERMS

briquetting
chipless machining
cold heading
cryogenic
electrohydraulic forming

explosive forming
high-energy-rate forming
magnetic forming
powder metallurgy
sintering

New materials and new processes are continually being introduced to the field of machining and its related areas. In addition to the nontraditional machining processes described in Chapter 28, such processes as chipless machining, high-energy-rate forming, and cryogenic treatment are being used today. The machining of various types of plastics and the fabrication of parts from powdered metals have become important. This chapter will provide you with a basic introduction to these emerging technologies.

29.1 MACHINING PLASTICS

Plastics are no longer rarities in the machine shop and are being machined in increasingly larger quantities each year, **Figure 29-1**. For this reason, it is important that the machinist be familiar with the machining characteristics of the common plastics, such as nylon, Teflon®, Delrin®, acrylics, and laminated plastics.

Figure 29-1. *More and more plastics are being machined. These fixtures hold both plastic and metal parts for machining. With CNC, machine tools automatically change cutting speeds and feeds when different materials come into position for machining. (Chick Machine Tool, Inc.)*

Special care must be taken when working with plastics. The dust and fumes given off by some plastics may be irritating to the skin, eyes, and respiratory system. Other plastics have fillers such as asbestos or glass fibers that can be harmful to your health. Be sure you are aware of the safety precautions that must be observed before you attempt to machine any plastic.

29.1.1 Nylon

Nylon is the name for a group of polyamide resins. Nylon is tough and exhibits high tensile (pull), impact, and flexural (bending) strengths. This material is highly resistant to abrasion and is not affected by most common chemicals, greases, and solvents.

This plastic is excellent for bearings, gears, cams, and rollers. Another application for parts machined from nylon is in the preparation of experimental components. Research and development projects also make considerable use of machined nylon.

General machining precautions for nylon

Most types of nylon can be machined using the same techniques employed for machining soft brass. Coolants will permit higher cutting speeds, but are not necessary to produce good-quality machined surfaces. When used, coolants should be of the soluble-oil type.

Since nylon is not as rigid as metal, it must be well supported during machining operations. Otherwise, deflection of the unsupported stock will result in inaccurate dimensions and distortions. To assure accuracy, bring parts machined from nylon to room temperature before checking dimensions.

Turning nylon

Nylon can be turned on a standard metalworking lathe. While tool bits sharpened to machine soft brass will prove satisfactory, best results are obtained using the tool bit shapes shown in **Figure 29-2**. Tools must be kept very sharp! The best finish on nylon is obtained with a high speed and fine feed.

Milling nylon

Conventional milling cutters, providing they are kept very sharp, can be used for milling nylon, **Figure 29-3**. Climb milling will minimize the formation of burrs. Surface speeds in excess of 100 fpm (30 mpm) can be employed.

Vertical milling is practical using fly cutters or two-lip end mills. Cutters must be kept very sharp to prevent plastic from melting or becoming gummy. Feeds of 10″ (25 mm) per minute or higher have proven satisfactory. Smoother surface finishes can be achieved using lighter feeds.

Figure 29-3. *Nylon and most other plastics can be readily milled with very sharp cutters.*

Drilling nylon

Since drilling produces considerable heat, it requires extra care. Standard twist drills, when sharpened as shown in **Figure 29-4**, will produce acceptable results. However, best results are obtained by using drills designed specifically for plastics. Drills for plastics have flutes that are highly polished and have a long lead.

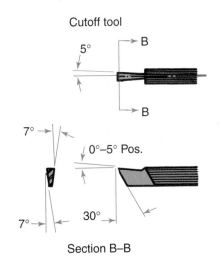

Figure 29-2. *Typical cutting tools for machining nylon on the lathe. (DuPont Co.)*

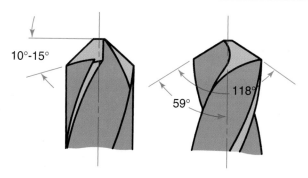

Figure 29-4. *Drill point recommended for drilling nylon. (DuPont Co.)*

To prevent the excessive heat that results when the drill *scrapes* the plastic rather than cutting it, use heavy feeds. If a coolant is not employed, the drill must be withdrawn from the hole frequently to clean out chips and prevent overheating. If the drill is kept cool, holes will be drilled to size.

Nylon can be reamed using an expansion reamer that is adjusted to a few thousandths oversize. Holes that are finished with solid reamers tend to be undersize.

Threading and tapping nylon

Nylon can be threaded and tapped with conventional equipment. But before tapping, the hole should be chamfered to reduce the chance of the first few threads tearing. Production tapping requires a tap that is 0.005″ (0.125 mm) oversize unless a self-locking thread is desired. The tap can be made oversize by chrome plating.

Threading can be done in nylon with a regular single point cutting tool, **Figure 29-5.** Use the same procedure as with metal. However, because of nylon's resilience, the finish cut should not be less than 0.005″ (0.125 mm). Support long work with a ballbearing follower rest.

Figure 29-5. *Plastics can be threaded on a lathe using same procedures recommended for metal. However, the cutting tool must be kept very sharp and have plenty of clearance.*

Sawing nylon

For sawing nylon and other plastics, good results can be obtained by employing a band saw. A band saw blade quickly dissipates the heat. Dry cutting is best accomplished with a skip tooth metal-cutting blade that has 4 to 6 teeth per inch. However, the blade must be sharp to prevent gumming of the nylon. If gumming occurs, it will usually freeze the blade in the cut. Hollow-ground circular saw blades are also used extensively in cutting plastics. Blades with a slight set are available for sawing quantities of both extra-thick or thin plastics.

Annealing nylon

Like metal, machined nylon parts require *annealing* to prevent dimensional changes. Annealing of plastics should be carried out in the absence of air, preferably by immersion in a suitable liquid. High-temperature boiling hydrocarbons, such as waxes and oils, are recommended for annealing nylon. A temperature of 300°F (148°C) is often employed for general annealing.

An annealing time of 15 minutes per 0.125″ (3.175 mm) of cross section is normally required. Allow the part to cool slowly in a draft-free area. Placing the heated piece in a cardboard container is a simple way to ensure slow, even cooling.

29.1.2 Delrin

Parts manufactured from **Delrin** (the DuPont trade name for acetal resin) have an unusual combination of physical properties that bridge the gap between metals and plastics. These properties include excellent dimensional stability, high strength, and rigidity. The plastic is used extensively for parts in business machines (gears, cams, bearings, and printing wheels). Delrin is also replacing brass and zinc for many applications in the automotive and plumbing industries. The material has low friction, and thus requires a minimum use of lubricants. It is very quiet in operation.

In machining characteristics, Delrin is very similar to nylon. Recommended machining, cutting, and finishing operations are shown in **Figure 29-6.**

29.1.3 Teflon

Teflon (the DuPont trade name for its fluorocarbon resins) is filling a wide range of needs in the electronic, electrical, chemical, and processing industries. It has a very low friction coefficient: rubbing two flat pieces of Teflon together generates about the same friction as rubbing together two ice

Machining, Cutting, and Finishing
with Delrin Acetal Resin

	Equipment		Cutting Speed			
Operation	Machines	Tools	RPM/FPM	Feed	Coolant use	Remarks
Sawing	Std.	Std. 14 T.P.I. Slight set	100 – 300 FPM (30.5–91.5 M/min)	Med.	—	Coolant improves finish of cut
Drilling	Std.	Std. twist drills 118°	1500 RPM for 0.500" (1.27 cm) drill	Med.	At med. and high speeds	On-size holes drilled without coolant
Turning	Std.	Std.	690–840 FPM (210–256 M/min)	.002"–.005" (0.051–0.13 mm)	At high speeds	Depth of cut— .016"–.200" (0.41–5.08mm); support long lengths
Milling	Std.	Std. cutters; single fluted end mills	Similar to brass	Similar to brass	Not required	—
Shaping	Std.	Std.	Max.	Similar to brass	Not required	—
Reaming	Std.	Expansion type preferred	Similar to brass	Med.	At med. and high speeds	—
Threading and tapping	Std.	Std.	Similar to brass	Similar to brass	At med. and high speeds	Coolant facilitates cutting to dimensions
Blanking and punching	Std.	Std.	—	—	Not required	Primarily for 1/16" (0.16 cm) thick stock
Filing and sanding	Std.	Std. file, std. abrasive paper and discs	—	—	Wet sanding	Finish will vary with type of file
Finishing	Std.	6"–12" (15.2–30.4 cm) dia. muslin pumice and water polishing compound	1000–2000 RPM	—	Not required	Use light pressure and rotate part

Figure 29-6. Recommended machining, cutting, and finishing operations for Delrin acetal resins. (DuPont Co.)

cubes. This makes Teflon ideal for use as bearings and seals in food processing equipment, where lubricating oil would contaminate the food.

Teflon works well in both high-temperature and very-low-temperature (*cryogenic*) applications. It is an expensive material, but will do things that no other material can do as well.

Teflon general machining characteristics

Teflon has a high thermal expansion rate. It expands at a rate about 10 times that of steel. When tolerances are critical, measurements should be made at room temperature, or (where applicable) at the temperature at which the part will be used. It is recommended that the plastic be stored at a temperature of 74°F (21°C) or higher for at least 48 hours before the machining operations.

Teflon has a tendency to pick up metal shavings and chips. For this reason, no machining should be attempted until the equipment has been thoroughly cleaned of all metal particles.

Turning Teflon

Teflon is more flexible than many other machinable plastics. To prevent deflection of the material away from the cutting tool, take care to support the work properly.

Tools must be sharp and have generous clearances so that the cutting edge of the tool will not rub, **Figure 29-7**. Chips must not be allowed to accumulate around the work because they prevent the heat from dissipating. See **Figure 29-8**.

Cutting fluids are needed when tolerances are critical. Large amounts of water-based coolant will deter thermal expansion in Teflon. Best surface finishes are achieved at cutting speeds of 200-500 fpm (60-150 mpm) with feeds of 0.0002″ to 0.010″ (0.005 mm to 0.25 mm).

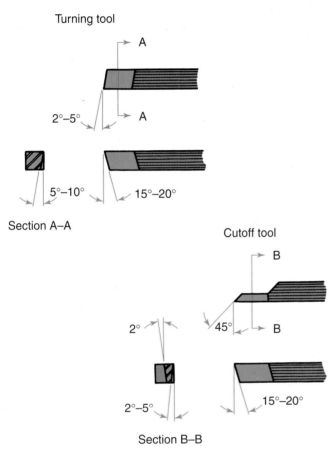

Figure 29-7. *Typical lathe cutting tools for machining Teflon. (DuPont Co.)*

Figure 29-8. *When turning plastics, do not allow chips to accumulate around work. Chips prevent heat from being dissipated and work may become distorted.*

Drilling Teflon

Drills sharpened as shown in **Figure 29-9** will provide satisfactory cutting action in Teflon. Teflon tends to swell slightly during the drilling operation, which results in a hole smaller than the drill. To compensate for this swelling, the machinist must use a drill that is slightly oversize. You should test drill a piece of scrap material to determine the exact drill size needed. For close-tolerance drilling in Teflon, feeds of 0.004″ to 0.006″ (0.10 mm to 0.15 mm) are suggested.

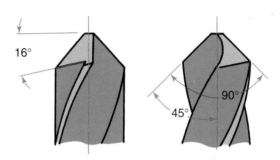

Figure 29-9. *Drill point is recommended for Teflon. (DuPont*

Milling Teflon

Teflon is milled in much the same manner as the other plastics. Only newly sharpened and honed cutters should be employed. The work must be solidly supported.

With shell mills, the milling head of the machine should be tilted slightly into the cut, as shown in **Figure 29-10**. This will eliminate cutter drag marks on the material.

Reaming Teflon

Reaming of Teflon is not advised. If close tolerances are specified, holes should be bored with a single-point tool.

Threading and tapping Teflon

Teflon is threaded and tapped with the same general techniques as those suggested for nylon.

Sawing Teflon

If a square cut is to be made in Teflon, a rigid machine with first-class saw guides is essential no coolant is needed. Maximum machine speeds can be used with a skip tooth blade having 4 to 6 teeth per inch.

Annealing Teflon

To maintain dimensional stability, Teflon should be annealed. Normally, the plastic is heated to a

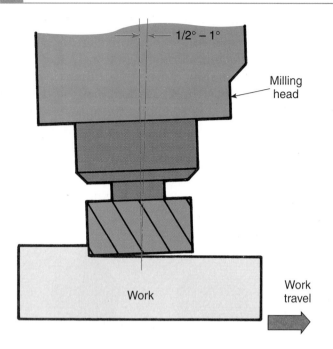

Figure 29-10. When shell milling, it is recommended that the cutter be tilted slightly (1/2° to 1°) into the cut to perform machining with the periphery of the cutter. This will eliminate cutter marks on machined surface.

temperature above that to which the finished part will be exposed, but below 621°F (333°C). Above 621°F, Teflon turns to a gel. One hour of annealing for each 1″ (25 mm) of thickness is adequate. Allow the part to cool slowly. Heating is usually done in an oven.

The following oven annealing procedure for Teflon is recommended:

1. Anneal the rod or tubing.

2. Rough machine the part to within 0.06″ (1.5 mm) of finished size.

3. Anneal again, but at temperatures slightly lower than that of the initial annealing.

4. Finish machine after the Teflon has returned to room temperature.

29.1.4 Acrylics

The *acrylic plastics* Lucite® (the DuPont trade name) and Plexiglas® (the Rohm & Haas Co. trade name) have an unusual combination of desirable characteristics. They have good dimensional stability and high impact strength to temperatures as high as 200°F (94°C). Acrylics are easy to machine, form, and polish.

Because of their unusual "light piping" capability (they transmit light as a hose carries water) and edge lighting qualities, acrylics find considerable use in light control and optical applications.

General machining characteristics for acrylics

Acrylics are machined in much the same manner as other plastics, **Figure 29-11**. Generally, little difficulty will be encountered if sharp tools with adequate clearance are used. Drills should be sharpened as shown in **Figure 29-12**.

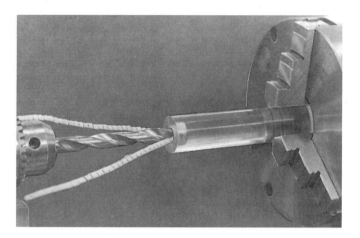

Figure 29-11. Transparent acrylics present a unique machining experience. You can see the tool cutting inside the work. Poor cutting action can be spotted immediately and corrections made.

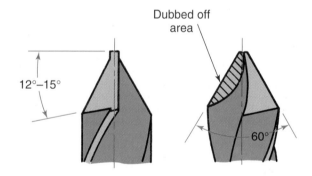

Figure 29-12. Drill point recommended for making through holes in acrylics. The tip angle should be increased to 118° for blind holes. Smoothly finished holes can be produced by first drilling a pilot hole and filling it with wax. Wax will allow chips to move up flutes without sticking to drill. (Rohm and Haas)

29.1.5 Laminated Plastics

Laminated plastics consist of layers of reinforcing materials (cotton fabric, paper, asbestos, glass fiber, etc.) that have been impregnated with synthetic resins. The resins are cured under heat and pressure.

Machining of laminated plastics can be accomplished with conventional machine tools. Some laminated plastics, like those containing glass fiber and asbestos, are highly abrasive. Carbide tools are recommended.

A dust collector system and filtered dust mask or respirator must be used for operator safety when machining plastics containing glass fiber or asbestos.

Turning laminated plastics

A round-nose lathe tool produces the best surface finish. Speeds up to 4000 fpm (1220 mpm) can be used. Lathe work is usually done dry. However, internal threading may sometimes require use of a lubricant.

Drilling laminated plastics

Drilling operations are similar to those used with nylon. However, drilling parallel with the laminations should be avoided whenever possible, because the material may split along the laminations. See **Figure 29-13**.

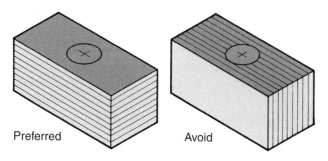

Preferred   Avoid

Figure 29-13. Whenever possible, holes should be drilled at a right angle to the laminations. Holes drilled parallel to laminations tend to split some types of material.

Milling laminated plastics

Speeds up to 1000 fpm (305 mpm) are possible with good results. Feeds up to 20″ (500 mm) per minute have been used. For cotton fabric-base laminates, best results are obtained by using the highest spindle speed the cutter will stand, with the maximum feed that will produce an acceptable surface finish.

Climb milling is recommended to keep the work held tightly in the holding device and to prevent an edge from being raised.

Sawing laminated plastics

Band sawing is advised for curved or straight cuts in laminated plastics when smooth edges and close tolerances are not specified.

Blades with 5 to 8 teeth per inch and medium to high set can be operated at speeds up to 8000 fpm (2400 mpm). Feed the work as fast as it will cut without forcing the blade.

Threading and tapping laminated plastics

Hand threading of laminated plastics can be done with standard taps and dies. It is normally done dry. High-speed steel taps that are oversize by 0.002″ to 0.005″ (0.05 to 0.13 mm) should be used if available.

A slight chamfer on the hole to be tapped, or rod to be threaded, will improve the work quality by preventing the first few threads from tearing.

29.2 CHIPLESS MACHINING

Chipless machining forms a metal wire or rod into the desired shape using a series of dies. This metalworking technique will not replace conventional machining, but it does make substantial cost savings possible for some jobs. It can reduce the amount of scrap metal that results and can increase production speed. The process is sometimes called *cold heading* or *cold forming*.

In chipless machining, a series of dies replaces the usual cutting tools on the lathe, drill press, or milling machine. See **Figure 29-14**. Material used in chipless machining is usually in coil form and is referred to as wire. This material is turned into needed, and often complex, shapes. See **Figure 29-15**.

Figure 29-14. A series of dies is used in chipless machining to replace the usual cutting tools. Work is transferred from station-to-station where various forming operations are carried out. (National Machinery Co.)

Accuracy typically can be held to tolerances of 0.002″ (0.05 mm), and closer tolerances can be achieved, if required. However, costs increase in proportion to the precision wanted. In most cases, waste is totally eliminated.

Figure 29-15. Parts can be manufactured quickly and inexpensively using chipless machining techniques. No scrap was produced in making this part, and tolerances were held to within 0.002" (0.05 mm). (National Machinery Co.)

There are five basic operations done by machines using this process, **Figure 29-16**. Combinations and variations of these operations make possible a wide range of applications.

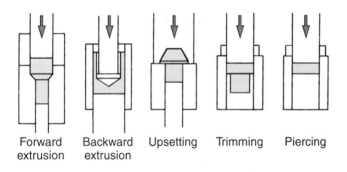

| Forward extrusion | Backward extrusion | Upsetting | Trimming | Piercing |

Figure 29-16. The five basic operations performed by machines designed for cold forming (chipless machining).

The cold heading technique is an economical and efficient way to make bolts, nuts, screws, and other fasteners, **Figure 29-17**. Another example is spark plug bases, almost all of which are made by chipless machining, **Figure 29-18**.

Metals ranging from aluminum alloys to medium-carbon steel can be shaped using this technique. Stainless steel, copper, and nickel alloys can be cold formed, but the ease with which they are shaped depends upon the part design. Material up to 1.5" (37.5 mm) in diameter can be formed on some machines.

29.2.1 Intraform Machining

Intraform® is a development of chipless machining that makes it possible to form profiles on the inside diameter of a cylindrical workpiece. See **Figure 29-19**. Forming inside profiles would be extremely difficult and expensive to do by conventional machining techniques.

In this process, a section of hollow cylindrical stock is placed over a steel mandrel, **Figure 29-20**. It is them squeezed by rapidly pulsating dies, **Figure 29-21**. At the completion of the operation, the mandrel's profile is produced on the inside diameter of the part, **Figure 29-22**.

Figure 29-23 shows how fixed rolls cause the four dies to pulsate rapidly around the outside diameter of the work. The tops of the cams are shaped to permit a smooth continuous squeezing action of the dies. Even though the work is being squeezed by the dies more than 1000 times per minute, noise and vibration are not a problem.

The technique has proven to be a practical way to produce rifle barrels for example, **Figure 29-24**. Predrilled steel blanks are fed into the machine which forms the chamber and rifling. In addition to improving the surface finish of the bore, the operation also improves the physical characteristics of the metal.

29.3 POWDER METALLURGY

Powder metallurgy, abbreviated *P/M,* is a technique used to shape parts from metal powders. These parts can be quite complex, **Figure 29-25**. Sometimes called *sintering,* the process was developed in the late 1920s to make self-lubricating electric motor bearings for the automotive industry. The steps involved in fabricating powder metallurgy products are shown in **Figure 29-26**.

29.3.1 Powder Metallurgy Applications

The P/M process is widely employed by industry for such applications as:

- Self-lubricating bearings and bearing materials.
- Precision finished machine parts, such as gears, cams, or ratchets, with tolerances as close as ± 0.0005" (0.0127 mm).
- Permanent metal filters (sintered bronze fuel filters, for example).

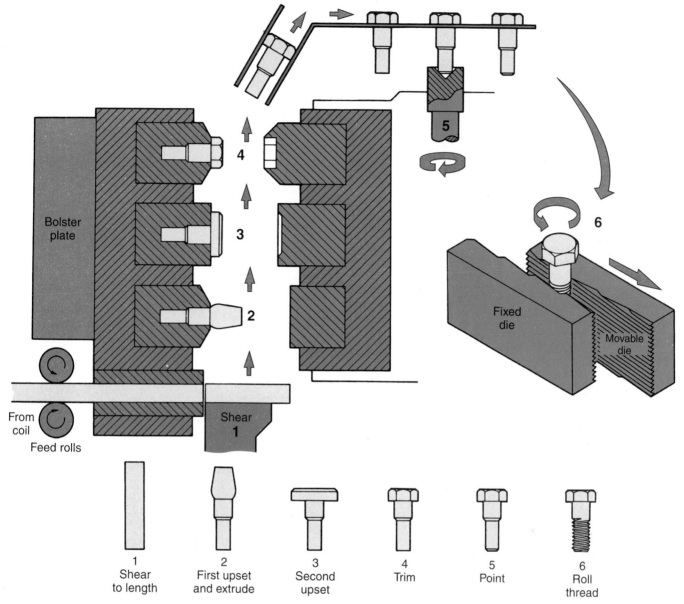

Figure 29-17. *Arrows and numbers indicate sequence involved in producing bolts by chipless machining. Trace part flow through the sequence of operations.*

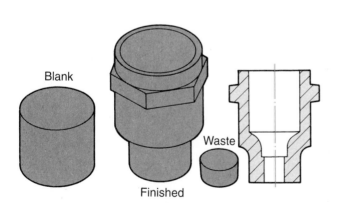

Figure 29-18. *Almost all spark plug bases are made by chipless machining. Resulting scrap was less than 2%.*

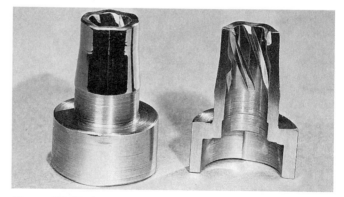

Figure 29-19. *An auto starter clutch housing produced by the Intraform process. The cutaway view shows details. The housing was formed in one operation at the rate of 220 parts per hour. (Cincinnati Milacron)*

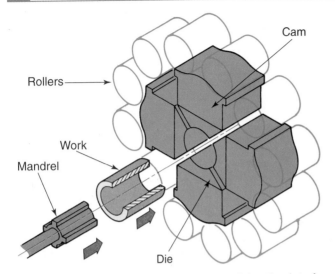

Figure 29-20. A part ready to be shaped by the Intraform process.

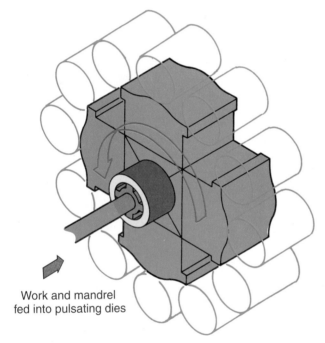

Work and mandrel fed into pulsating dies

Figure 29-21. Work and mandrel are placed in the dies. Contact with the rotating dies causes the work and mandrel to rotate at about 80% of die rpm. Work feeds over the mandrel.

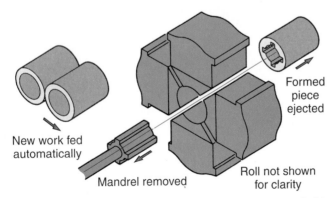

New work fed automatically

Mandrel removed

Formed piece ejected

Roll not shown for clarity

Figure 29-22. When the operation is completed, the mandrel is retracted. The next piece feeds automatically into position while the formed part is ejected.

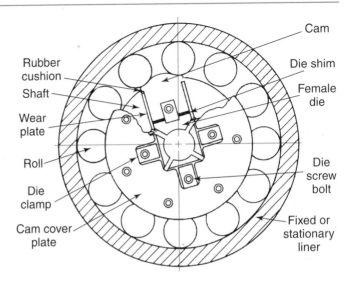

Figure 29-23. The Intraform machine die head in open position. Interaction of the cams and rollers squeezes each die more than 1000 times per minute.

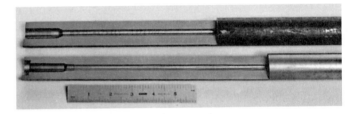

Figure 29-24. A predrilled steel rifle barrel blank (top) and one that has had the chamber and rifling finished by the Intraform process. Approximately 80% of all barrels produced by the process are of target rifle quality, compared to only 10% shaped by conventional methods. (Cincinnati Milacron)

Figure 29-25. An assortment of products made by the powder metallurgy process. (Metal Powder Industries Federation)

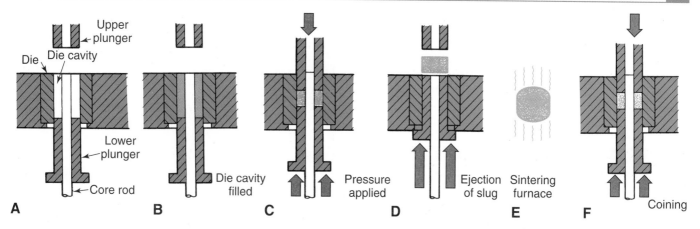

Upper plunger

Die **Die cavity**

Lower plunger

Core rod

A

B Die cavity filled

C Pressure applied

D Ejection of slug

E Sintering furnace

F Coining

Figure 29-26. Steps in fabricating a part, using the powder metallurgy process. A—Note cross section of die and die cavity. Depth of the cavity is determined by thickness of the required part, and the amount of pressure that will be applied. B—Die cavity is filled with proper metal powder mixture. C—Pressure as high as 50 tons per square inch is applied. D—Briquette or "slug" is pushed from die cavity. E—Pieces are then passed through a sintering furnace to convert them into a strong, useful product. F—Some pieces can be used as they come from the furnace. Others may require a coining or sizing operation to bring them to exact size and to improve their surface finish.

- Fabricating materials that are difficult to work:
  - Tough cutting tools (tungsten carbide), **Figure 29-27**.
  - Supermagnets (aluminum-nickel alloys or Alnico).
  - Mixtures of metals and ceramics for jet and rocket applications that require the heat resistance of ceramics as well as the heat transfer qualities of metals. This process is also used to make special cutting tools (cermets).
  - High-density counterweights for aerospace instruments that require maximum weight concentration in a minimum space.
  - Storage battery (nickel-cadmium or NiCad) elements.
- Wrought powder metallurgy tool steels are now available in mill forms (bars, rods, flats, wire, and plate). They offer improved machinability, wear resistance, increased toughness, and better dimensional stability than conventional cast/wrought tool steels. P/M tool steels are primarily used for metal-cutting and metalforming operations.

29.3.2 Powder Metallurgy Process

The first phase in the manufacture of powder metal products is the careful mixing of high-purity metal powders. The powders are carefully weighed and thoroughly mixed into a blend of correct proportions. Many materials can be used: iron, steel, stainless steel, brass, bronze, nickel, chromium, etc.

Figure 29-27. Technician examines a tray of tungsten carbide cutting tool inserts following application of a wear-resistant coating that extends tool life. Inserts were made using the powder metallurgy process. (Kennametal, Inc.)

Combinations of these metals and nonmetals can also be used.

Briquetting

The powder blend is then fed into a precision die. The die cavity has the shape of the desired part, but it is several times deeper than the thickness of the part. The powder is compressed by an upper and lower punch. Pressures applied range from 15 to 50 tons per square inch. This portion of the operation is known as *briquetting.* See **Figure 29-28.**

The piece, as ejected from the die, appears to be solid metal. However, this "green compact" is quite brittle and fragile. It will crumble if not handled carefully.

Figure 29-29. *Atmosphere within the sintering furnace is carefully controlled to prevent oxidation or contamination of the parts being processed. These finished pieces emerging from furnace may be used "as is," or may require additional operations before use.*

Figure 29-28. *In this closeup view of briquetting press, green compacts are shown moving out of the press. (Delco Moraine Div. of GMC)*

Sintering

To transform the green compact into a strong, useful unit, it must be heated to a temperature of 1500°F to 2300°F (815°C to 1260°C) for 30 minutes to 2 hours, depending upon the powder mixture. This process, known as *sintering*, is done in a controlled atmosphere furnace, **Figure 29-29.**

Forging

Many parts can be utilized as they come from the furnace. Because of shrinkage and distortion caused by the heating operation, the pieces may have to go through a *sizing, coining,* or *forging* operation, **Figure 29-30.** The part must be reheated just prior to the forging operation, which consists of pressing the sintered pieces into accurate size dies to obtain precise finished dimensions, higher densities, and smoother surface finishes, **Figure 29-31.** Powder metal parts can be drilled, tapped, plated, heat treated, machined, and ground, **Figure 29-32.**

Figure 29-30. *Hot forging takes place in closed dies with compact at red heat. After forging, part is ready for additional machining, if needed. (Burgess-Norton Mfg. Co.)*

Figure 29-31. *This quality control manager is gaging finished P/M parts to determine whether they meet design specifications. (Burgess-Norton Mfg. Co.)*

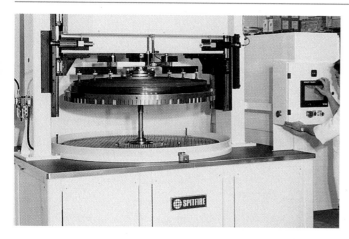

Figure 29-32. *Lapping is a grinding operation that generates a flat, smooth finish. This machine is capable of lapping both sides of a workpiece simultaneously to produce perfectly parallel surfaces. (SpeedFam-Spitfire)*

29.3.3 Powder Metallurgy Costs

The tool cost is moderate and the process is best suited to quantity production. If production quantities exceed several thousand units, the finished piece can be produced by the P/M method at less than the cost of rough sand castings.

29.4 HIGH-ENERGY-RATE FORMING (HERF)

The introduction of super-tough alloys for aerospace vehicles and the need for shaping thin, brittle metal has been responsible for the development of new ways to do the work. One of these new techniques is known as *high-energy-rate forming* or *HERF*. There is little similarity between it and conventional metalworking processes like turning, drilling, milling, etc.

Shaping metals by the use of conventional presses and drop hammers parallels HERF. However, problems develop when attempting to shape the super-tough alloys by conventional means. They exhibit *"springback,"* where the metal tends to try to regain its original shape, **Figure 29-33**. It is difficult and costly to shape these metals to acceptable tolerances by conventional means.

In HERF, the metal is shaped in microseconds with pressures generated by the sudden application of large amounts of energy. The metal, in most cases, is slammed against the die and shaped so rapidly that there is no tendency for the material to try to return to its original shape.

The great pressures are generated by detonating explosives, releasing compressed gases, discharging powerful electrical sparks, or electromagnetic energy.

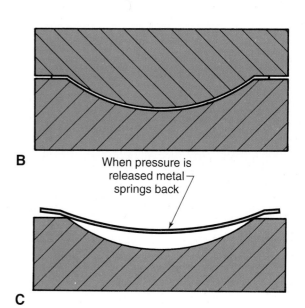

Figure 29-33. *Many metals tend to spring back to near their original shape after being formed by conventional means. A—Flat sheet metal blank ready for forming. B—Metal is formed between dies. C—Metal tries to return to its original shape when male die is removed or pressure released.*

HERF offers many advantages. Tool costs are reduced, and there is usually no need for expensive machinery. There appears to be no limitation to the size of the sections that can be formed.

29.4.1 Explosive Forming

Explosive forming uses the high-pressure wave of an explosive charge to form the metal. It is an older technique — it originated in the late 1800s to shape ornate door knobs and similar products — that has been adapted to the space age. See **Figure 29-34**.

The sheer size of many aerospace and marine components, up to 144" × 230" × 1" thick (3657 mm × 5842 mm × 25 mm), makes it impossible to form them in existing presses, **Figure 29-35**. The presses are either too small or are not powerful enough to develop the pressure required to shape the high-strength alloys.

Explosive forming makes use of the pressure wave generated by an explosion in a fluid to force the material against the walls of the die, **Figures 29-36** to **29-37**. The fluid has the effect of rounding off the pressure pulse generated by the detonation. This helps to assure that forces will be equally distributed across the surface of the material.

Figure 29-34. *High-energy-rate forming (HERF) shaped end sections of the external fuel tanks of Space Shuttle. (NASA)*

Figure 29-37. *A few of many aircraft and missile parts shaped by explosive forming. Water is used to round off the pressure pulse generated by the explosive, equalizing forces across entire surface of material being formed. (The Ryan Company)*

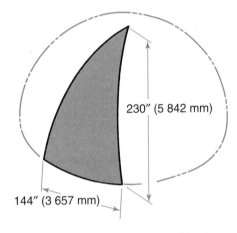

Figure 29-35. *Explosive forming was used to shape segments for top and bottom domes of giant aerospace vehicle fuel tanks. Segments were welded after forming.*

To prepare for explosive forming, the metal is cut or fabricated to a shape determined by the contours of the finished part. This *preform* is placed in the die, the die is filled with water, and an explosive charge suspended in the water.

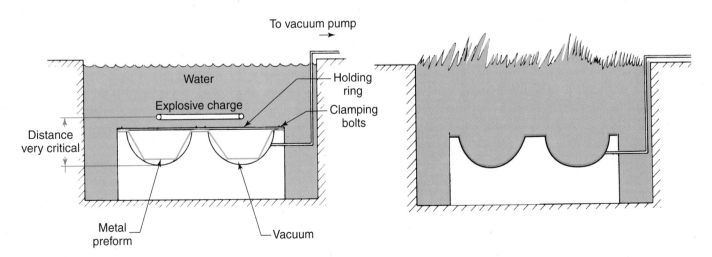

Figure 29-36. *This diagram shows the principle of explosive forming process. In some applications, a few dollars worth of explosives will do work of a press that may cost a million dollars or more.*

A large holding ring, clamped over the outer edge of the work, assures the necessary seal for drawing a vacuum in the die. A *vacuum* is necessary between the work and the die; otherwise, an air cushion would develop, preventing the metal from seating in the die and assuming its proper shape.

When the explosive is detonated, the resulting pressure slams the material against the die walls. The forming work is accomplished in microseconds. See **Figure 29-38**.

Placement and quantity of the high explosive is critical. The charge can range from a few ounces to form small parts, to the many pounds needed to form large sections of aluminum and steel up to 4" (100 mm) thick. For example, the heavy steel missile hatches on the Navy's submarines, **Figure 29-39**, are formed by this technique. Many forms of explosives are utilized: rod, sheet, granules, liquid, stick, cord, and plastic.

Depending upon placement of the explosive, HERF operations fall into two categories: *stand-off* and *contact*.

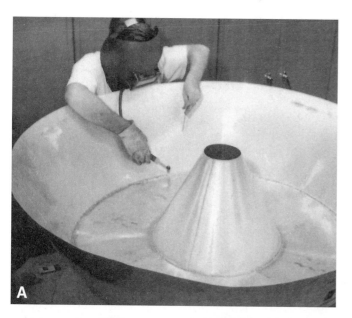

Figure 29-38. *Explosive forming. A— This worker is welding the preform for a fuel tank for explosive forming. This is first step in the process. Formed conventionally, the tank would require forty sections rather than twelve and would require 60% more welding footage. B—Once resembling a cake pan, exploded metal now is a torus, or donut shape. Engineers examine inner contour of the formed part, which has been removed from die underneath.*

Figure 29-39. *Many parts used on nuclear submarines are shaped by HERF. (U.S. Navy)*

Stand-off operations

During **stand-off HERF,** the charge is located some distance from the work. Its energy is transmitted through a fluid medium, such as water. This technique is used to form and size parts.

Contact operations

During **contact HERF,** the charge is touching the work and the explosive energy acts directly on the metal. Welding, hardening, compacting powdered metals, and controlled cutting are done with this technique.

Explosive forming—advantages and disadvantages

While explosive forming offers many advantages, there *are* some drawbacks associated with the process:

- The technique has not been developed to the stage where a part can always be formed properly on the first shot.
- Since the operation utilizes the "big-bang" principle (an explosion) to do forming, the noise can be a problem. Also, strict laws prohibit the use of explosives in populated areas. These factors usually make it necessary to locate the facility in an isolated site.

This increases transportation and handling costs. Personnel must be highly skilled in the safe handling of high explosives. Insurance rates are high.

29.4.2 Electrohydraulic Forming

Electrohydraulic forming, also called *capacitor discharge forming* or *spark forming,* is a variation of explosive forming, **Figure 29-40**. High voltage electrical energy is discharged from a *capacitor bank* (an array of devices used to store electrical energy) into a thin wire or foil suspended between two electrodes. The unit is immersed in water, as is done with explosive forming. Many of the parts on aerospace vehicles made from titanium (tough, light metal that is difficult to work) are formed using this technique.

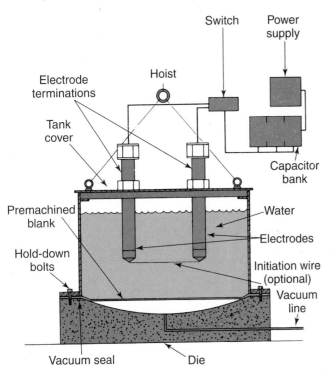

Figure 29-40. *Diagram shows setup for electrohydraulic forming, which uses electrical energy as a source of power for HERF operations. (NASA)*

As the wire or foil is vaporized by the electric current discharge, the vapor products expand, converting the electrical energy to hydraulic energy. The shock wave forms the metal against the die. Since the energy produced is less than that associated with explosives, it is usually necessary to repeat the operation several times to achieve the desired results.

A well designed electrohydraulic forming facility can be adapted to automation. The capability of generating high pressures without a "bang"

permits the unit to be employed in conventional industrial facilities.

29.4.3 Magnetic Forming

Magnetic forming, also termed *electromagnetic forming* or *magnetic pulse forming,* uses an insulated induction coil wrapped around or placed within the work, **Figure 29-41**. As very high momentary currents are passed through the coil, an intense magnetic field is developed. This causes the work to collapse, compress, shrink, or expand depending on the design of the coil. Coil location depends upon whether the metal is to be squeezed inward or bulge outward, The coil is shaped to produce the desired shape in the work. See **Figure 29-42.**

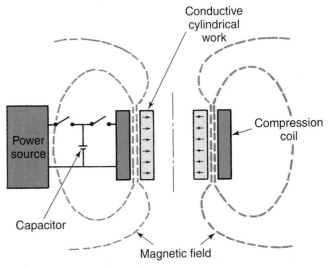

Figure 29-41. *How magnetic pulse metal forming works. A strong magnetic field is produced by discharging a capacitor through a coil. During the brief impulse, eddy currents in the work restrict the magnetic field to surface of the workpiece. This creates a uniform force to form metal. The process can be employed directly on highly conductive metals. Low conductivity metals can be shaped by using a layer of highly conductive aluminum between work and coils.*

The power source is basically the same as that used for electrohydraulic forming — a capacitor bank. The spark gap, however, is replaced by a coil. Energy is obtained by capacitor discharge through the coil. In fact, properly designed equipment can be used to perform either electrohydraulic or magnetic pulse operations.

Energy storage capacity and the ability to utilize that energy determines the size of the work that can be formed. Highly conductive metals can be formed easily by this process. Nonconductive or low-conductivity materials can be formed if they are wrapped or coated with a high-conductivity auxiliary material.

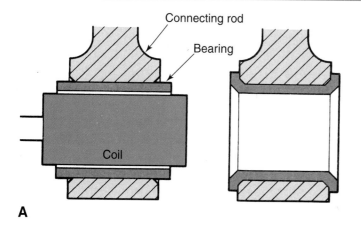

A

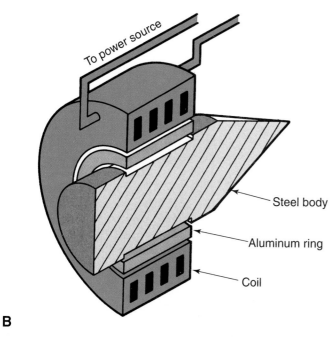

B

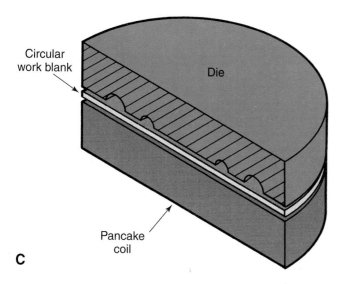

C

Figure 29-42. *Magnetic pulse forming applications. A—Magnetic pulse being used to expand a bearing sleeve into a connecting rod. B— A magnetic field being employed to shrink or squeeze parts together. C—Flat forming applications, which involve forcing sheet metal into a die, require a pancake coil to provide uniform magnetic pressure.*

29.4.4 Pneumatic-Mechanical Forming

Pneumatic-mechanical forming uses a punch and die operated by high-pressure gas. It was the first of the HERF techniques to become a standard production tool, **Figure 29-43**. The operation has much in common with conventional forging, since a punch and die are employed. However, the forces developed are many times more powerful and are sufficient to shape hard-to-work materials. The metal blank is heated prior to the forming operation. The machine requires less space than the conventional forging press.

Figure 29-43. *A pneumatic-mechanical press. Because of a unique self-reacting framework system, machine imparts no shock load to floor. This makes it possible to use it in close proximity to conventional chip-making machines. (General Dynamics Corp.)*

In pneumatic-mechanical forming, a high-pressure gas is used to accelerate a punch into a die. The punch and die are mounted on opposed rams that meet with equal force, thus taking most of the strain off the frame of the machine. Some machines make use of a recoil mechanism similar to that used on an artillery piece.

Pneumatic-mechanical forming can be accomplished by several means. One makes use of a blank cartridge to power the ram for shaping small parts. However, the most widely employed method utilizes a two-part cylinder, **Figure 29-44**. Gas is stored in one part of the cylinder at approximately 2000 psi

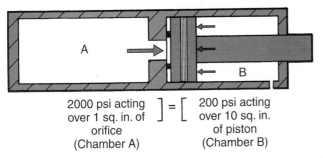

2000 psi acting over 1 sq. in. of orifice (Chamber A) = 200 psi acting over 10 sq. in. of piston (Chamber B)

Figure 29-44. High-pressure gas acts on only a small area of piston (area of the orifice) while low-pressure gas acts on entire area of the piston. This maintains a balanced situation between the two.

(13 800 kPa). Gas is also stored in the second section of the cylinder, but at a much lower pressure, about 200 psi (1380 kPa). A plate with an orifice separates the two sections of the cylinder. The piston (ram),

with a special seal, closes off the orifice. In this way, the high-pressure gas acts only on a small section of the piston (area of the orifice) while the low-pressure acts on the entire area of the piston. This maintains a stable balance between the two sections.

To make the machine operate, the pressure is increased slightly in the high-pressure section. This upsets the balance and the piston starts to move. When the seal is disengaged, the high pressure acts on the entire surface area of the piston driving it forward at tremendous speed, **Figure 29-45**.

Hard to shape metals are usually formed with one blow, **Figure 29-46**. Some space-age materials can best be shaped by this technique.

With the pneumatic-hydraulic method of forging, precise control is possible. Fewer operations are necessary (most parts can be formed with one stroke

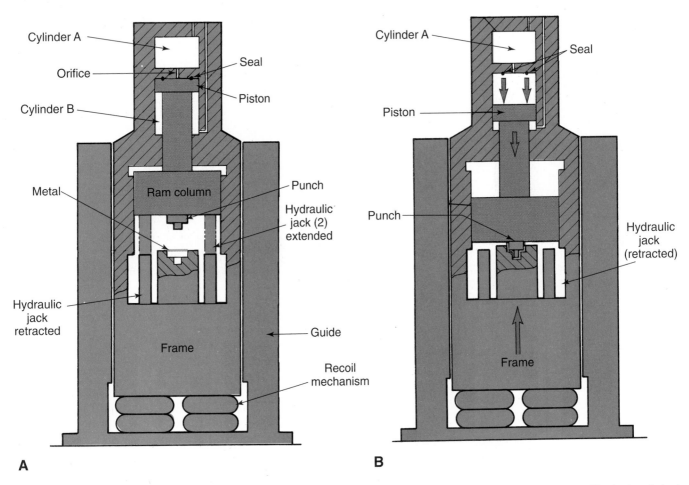

A **B**

Figure 29-45. Pneumatic-mechanical forming. A—Cross-sectional view of a pneumatic-mechanical forming press. The hydraulic jacks extend at the end of each operating cycle to lift ram column back into position for the next cycle. B—Operation of the press is triggered when pressure in Cylinder A is increased enough to break seal. This slight movement allows high-pressure gas to act instantaneously over entire area of the piston. The ram is driven downward at great speed. At the same time, frame moves upward by reaction of gas pressure over the driven piston. The frame and ram, are acted upon with equal thrust; so each has equal momentum but in opposite directions. To reset for the next cycle, the jacks lift ram column upward until it seats against the seal. (General Dynamics Corp.)

Figure 29-46. This front-wheel spindle for an automobile was formed from alloy steel in one blow on a pneumatic-hydraulic press. (General Dynamics Corp.)

of the press), so production is rapid The parts are produced to close tolerances with smoothly finished surfaces that require a minimum of machining, **Figure 29-47.** Because less material is used, fewer operations must be performed to finish the part, and higher production rates are obtained, substantial savings are possible.

Figure 29-47. Compare the finish of the wheel hub formed on a conventional press (cutaway at center) and the hub formed on a HERF machine (right). The better finish produced by HERF part means lower costs, since little or no additional machining must be done. The blank from which the wheel hub is formed is at left. (General Dynamics Corp.)

29.5 CRYOGENIC APPLICATIONS

The science of cryogenics is one of the more recent additions to the technology of machining or metalworking. Cryogenics is not a new or different way to work metal; rather, it is employed to improve or reinforce other metalworking techniques. The term *Cryogenic* means, literally, "to make icy cold;" it combines the Greek word *kryos*, meaning "icy cold," and the Latin *generatus*, which means "to make or create."

Cryogenics deals with temperatures beginning at the point where oxygen liquefies (approximately –300°F or –184°C), and goes down to just about absolute zero (–460°F or –293°C). At this point, every element but helium freezes. Oxygen and nitrogen look something like white beach sand or table salt. Metal also acts strangely. Lead coils act like steel springs, some metals increase tremendously in strength, and others become *superconductors* of electricity.

Cryogenics is widely used in annealing and heat treating metals. The characteristics of a number of aluminum alloys and some space-age metals are greatly improved by first heating them, and then quenching them in liquid nitrogen.

Another application of cryogenics is shrink-fitting metal parts together. Since most metals shrink in size as they become cold, one part of the assembly is made slightly oversize, then immersed in liquid nitrogen. The exact amount of oversize is determined by part size and type of metal. The diameter is reduced (shrunk) by the extreme temperature drop until it fits easily into its mating part. As it returns to room temperature, the cooled part expands and is thus locked in place. The parts do not become distorted as they would if they were mechanically pressed together or heated (expanded) until they could be fitted together. See **Figure 29-48.**

29.5.1 Cryogenic Treatment of Cutting Tools

Research has shown that cryogenic treatment of cutting tools has produced considerable improvement in their operating life. Such supercold treatment produces carbide inserts that last two to eight times longer than untreated inserts. Similar treatment of copper spotwelding electrodes increases their useful life by 300%. Other types of cutting tools have shown increases of up to 400% in tool life.

Liquid nitrogen enters the cooling chamber as a gas, at a carefully controlled rate. The cutting tools are *not* immersed in liquid nitrogen. Cooling occurs at a slow rate to avoid a sudden change in temperature that would cause damaging thermal shock.

The full cooling cycle takes about 40 hours. Four to six hours are required to lower the temperature to about –320°F (–160°C). The tools "soak" at this temperature for 24 hours or more. It requires another four to six hours to return to room temperature. See **Figure 29-49.** For larger or more massive items the time required to lower and raise temperature in the cooling chamber must be increased.

Figure 29-48. *An example of cryogenic shrink-fitting: mating a wing pivot for a prototype Super Sonic Transport (SST) aircraft to the wing clevis joint. The pivot has been cooled to –300°F (–184°C) to shrink it. Note the special protective gloves worn by the technicians. (Boeing Company)*

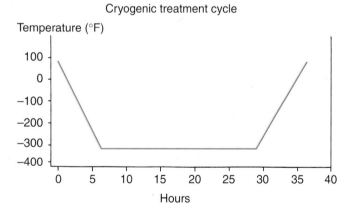

Figure 29-49. *Maximum benefit in the cryogenic treatment of cutting tools requires a gradual lowering of temperature, an extended soak period, and a slow return to room temperature.*

TEST YOUR KNOWLEDGE

Please do not write in the text. Write your answers on a separate sheet of paper.

1. Give a common trade name for each of the following types of plastics:
   a. Polyamide resins.
   b. Acetal resins.
   c. Fluorocarbon resins.
   d. Acrylic resins.

2. If plastics are to be machined with any degree of accuracy, the cutting tools must be _____.

3. When hand-threading plastics, why is it recommended that the hole or rod be chamfered?

4. Like metal, many plastics require _____ to ensure against dimensional changes.

5. When turning many plastics on a lathe, care must be taken to prevent the _____ from accumulating around the work. If this is not done, heat will build up and cause the plastic to become _____.

6. What is unique about Teflon?

7. Machining plastics can create health problems for the machinist if precautions are not taken. What are these problems and how can they best be handled?

8. What are laminated plastics?

9. When drilling laminated plastics, what should be avoided?

10. Chipless machining is also known as _____ or _____.

11. How does chipless machining make substantial savings possible?

12. In chipless machining a series of _____ replaces the usual cutting tools of the lathe, drill press, and milling machine.

13. Chipless machining is still the most economical way to make _____, _____, _____, and other types of _____.

14. List the five basic operations performed by machines making use of the chipless machining process.

15. *Intraform* is a chipless machining technique that can form profiles on the _____ of _____ pieces.

16. The Intraform technique has proven to be a practical way to produce:
    a. Socket wrenches.
    b. Rifle barrels.
    c. Automotive starter clutch housings.
    d. All of the above.
    e. None of the above.

17. Powder metallurgy, abbreviated _____, is the technique of shaping parts from _____.

18. The powder metallurgy process is used to make:
    a. Self-lubricating bearings.
    b. Precision machine parts.
    c. Permanent metal filters.
    d. All of the above.
    e. None of the above.

19. List the steps in making a part by the powder metallurgy technique.

20. Parts made from metal powder can be:
    a. Drilled.
    b. Heat treated.
    c. Turned on a lathe.
    d. All of the above.
    e. None of the above.

21. What is a briquette or "green compact?"

22. Why is it often necessary to size, coin, or forge parts made from metal powders after they have been sintered?

23. What do the above operations do to the sintered piece?

24. The abbreviation HERF means _____.

25. In HERF, metal is shaped:
    a. By the slow application of great pressure.
    b. In microseconds, with pressure generated by the sudden application of large amounts of energy.
    c. By conventional forging methods.
    d. All of the above.
    e. None of the above.

26. The pressures needed in HERF are generated by:
    a. Detonating explosives.
    b. Releasing compressed gases.
    c. Electromagnetic energy.
    d. All of the above.
    e. None of the above.

27. Many metals tend to _____ to their original shape after being formed by conventional means. This problem is greatly reduced or entirely eliminated when _____ is used to shape the metal.

28. What is explosive forming?

29. What are some of the disadvantages of explosive forming?

30. Why must a vacuum be pulled in the die when explosive forming?

31. Depending upon the placement of the explosive, most explosive forming operations fall into two categories. List them.

32. _____ is a variation of explosive forming. However, _____ is used in place of the explosive charge to generate the required energy.

33. What HERF technique employs a very high electric current passing through an induction coil shaped to produce the required configuration in the work?

34. The technique in question 33 can be used to _____ or _____ the work to produce the desired shape, depending upon placement of the coil.

35. In pneumatic-mechanical forming, _____ is used to accelerate the _____ into the _____.

36. What does the term "cryogenic" mean?

37. The science of cryogenics deals with temperature starting at _____ (_____) and goes down to temperatures near _____ (_____).

38. What does shrink-fitting mean?

39. Why is it better to use the super-low temperatures of cryogenics, rather than heat, to shrink-fit parts together?

40. Why must the cooling of treated cutting tools be done at a slow rate?

Chapter 30

Occupations in Machining Technology

LEARNING OBJECTIVES

After studying this chapter, you will be able to:
○ List the requirements for the various machining technology occupations.
○ Explain where to obtain information on occupations in machining technology.
○ State what industry expects of an employee.
○ Describe what an employee should expect from industry.
○ Summarize the information given on a resume.

IMPORTANT TERMS

all-around machinist   part programmer
apprentice programs   resume
career   semiskilled workers
engineering   skilled workers
job shops   technician

Many people want a *career* that is both challenging and interesting. Their philosophy is that a career is enjoyable; you only "work" when you dislike a job. Others are satisfied with whatever job comes along. In which of these categories are you?

If you are looking for a career that is challenging, interesting, and rewarding, the field of machining offers many opportunities. See **Figure 30-1**. Whether you choose one of the machine shop areas or select a career in a related field, you will find that the study of *Machining Fundamentals* is basic to all of them.

No matter which occupational choice you make, you should realize that you will have to keep up with technical progress. To be successful and advance in your career, a continuing program of education is usually necessary.

Figure 30-1. *The field of metal machining offers many opportunities, for semiskilled and skilled workers, technicians, and professional personnel. These machinists and engineers are discussing the best method for machining a part with CNC equipment. (Giddings & Lewis, Inc.)*

30.1 MACHINING JOB CATEGORIES

Jobs in material machining fall into four general categories:
• Semiskilled.
• Skilled.
• Technical.
• Professional.

30.1.1 Semiskilled Workers

Semiskilled workers are those who perform basic, routine operations that do not require a high degree of skill or training. Semiskilled workers may be classified into the following general groups:
• Those who are helpers for skilled workers.
• Those who operate machines and equipment used in making things. The machines are set up by skilled workers.

- Those who assemble the various manufactured parts into final products, **Figure 30-2**.

There is little chance for advancement out of semiskilled jobs without additional study and training. Most semiskilled work is found in production shops where there are great numbers of repetitive operations. In general, semiskilled workers are told what to do and how the work is to be done. They are often the first to lose their jobs when there is a downturn in the economy.

Figure 30-2. *Many semiskilled workers are employed in assembly industries, where they assemble manufactured parts into complete units. Training periods to learn these job skills are relatively short. (Saturn)*

30.1.2 Skilled Workers

Skilled workers have been trained to do more complex tasks. They are found in all areas of material machining. Many skilled workers receive their training in *apprentice programs* (on-the-job training while working with skilled machinist), **Figure 30-3**.

Four or more years of instruction under an experienced machinist is generally required. In addition to working in the shop, an apprentice usually studies related subjects, such as: math, science, English, print reading, metallurgy, safety, and production techniques. Upon completion of an apprentice program, the worker is capable of performing the precise work essential to the trade.

Figure 30-3. *The apprentice studies under an experienced machinist for a period of four or more years. The training program also includes the study of related subjects like math, English, science, etc. (William Schotta, Millersville University)*

In recent years, the number of apprentice programs being offered has declined. Most workers entering the field receive their training in the armed forces, **Figure 30-4,** or in vocational/technical

Figure 30-4. *The Army, and other branches of the Armed Forces, offer excellent opportunities for learning a trade. As a bonus, you get paid while you learn. (U.S. Army)*

programs offered in high schools and community colleges. Many community college programs are offered in conjunction with local industry.

Specialized machinists

There are several areas in which the machinist may specialize:

The *all-around machinist* is a competent person who can set up and operate most types of machine tools. He or she must be familiar with both manual and computer-controlled machine tools and how they are programmed, **Figure 30-5.** An all-around machinist is expected to plan and carry out all of the operations needed to machine a job.

Figure 30-5. *The all-around machinist can set up and operate most types of machine tools, whether manual or CNC. (Giddings & Lewis, Inc.)*

Many all-around machinists work in *job shops* (shops where special and experimental work is machined, or where production runs are very small).

A *toolmaker* is a highly skilled person who specializes in producing the tools needed for machining operations. These include:

- *Dies* (special tools for shaping, forming, stamping, or cutting metal or other materials).
- *Jigs* (devices that position work and guide cutting tools).
- *Fixtures* (devices to hold work while it is machined).

These tools are necessary for modern mass production techniques. Toolmakers must have a broader background in machining operations

and mathematics than most other skilled workers in the trade. See **Figure 30-6.**

A *diemaker* is a toolmaker who specializes in making the punches and dies needed to stamp out such parts as auto body panels, electrical components, and similar products. He or she will also produce the dies for making *extrusions* (metal shaped by being pushed through an opening in a metal disc of proper configuration) and *die castings* (parts made by forcing molten metal into a mold). Like the toolmaker, a die maker is a highly skilled machinist.

Figure 30-6. *A tool-and-die maker checking the dies used for molding a plastic pattern used to cast a jet engine component. Master tooling assures that other sections of the engine, made elsewhere in the United States, Israel, and Europe, will fit together perfectly. (Precision Castparts Corp.)*

The *layout specialist* is a machinist who interprets the drawings and uses precision measuring tools to mark off where metal must be removed by machining from castings, forgings, and metal stock. This person must be very familiar with the operation and capabilities of machine tools. He or she is well-trained in mathematics and print reading.

A *setup specialist* is a person who locates and positions ("sets up") tooling and work-holding devices on a machine tool for use by a machine tool operator. This worker may also show the machine tool operator how to do the job, and often checks the accuracy of the machined part. See **Figure 30-7.**

A *part programmer* inputs data into a computer-controlled (CNC) machine tool for machining a product. CNC machine tools are revolutionizing the field of material machining. However, computers have no inherent intelligence and cannot think or exercise judgment. They must be *programmed* (instructed) by a highly skilled part programmer who studies the drawings and determines the sequences, tools, and motions the machine tool must carry out to machine the part, **Figure 30-8.**

Figure 30-7. A setup specialist is a master machinist who prepares machine tools for operation by less highly trained personnel. After thoroughly checking the machined part to be sure it will meet specifications, the setup specialist will turn the machine tool over to a machine operator. (Hydromat, Inc.)

Figure 30-8. The programmer prepares the information (tools, tool paths, machining sequence, etc.) for a computer-controlled machine tool. This information or computer data is called a program and will direct the entire machining of a specific part. The programmer must thoroughly understand machining technology. (Tri-Tool, Inc.)

To perform this task, a part programmer must have a background that includes the following:

- Formal training in computer technology as it relates to machine tool operation.
- Experience at reading and interpreting drawings.
- A thorough grounding in machining technology and procedures.
- A working knowledge of cutting speeds and feeds for various tools and materials.
- An extensive training in mathematics.

Many community colleges and vo-tech centers offer programs in machine tool CNC programming to qualified persons, generally skilled machinists or other persons with extensive machine tool experience.

A *supervisor* or *manager* is usually a skilled machinist who has been promoted to a position of greater responsibility. This person will direct other workers in the shop and is responsible for meeting production deadlines and keeping work quality high. In many shops, the manager may also be responsible for training and other tasks, **Figure 30-9.**

Figure 30-9. The supervisor or manager of the production department works very closely with machinists, engineers, metallurgists, and other staff. He or she may also conduct training sessions, like this one, when new equipment or procedures are introduced. (LeBlond Makino Machine Tool Co.)

30.1.3 Technicians

The *technician* is a member of the production team who operates in the realm between the shop and engineering departments, **Figure 30-10.** The position is an outgrowth of today's highly technological and scientific world. The job usually requires at least two years of college, with a program of study centered on math, science, English, computer science, quality control, manufacturing, and

Figure 30-10. *These technicians are checking out and adjusting newly manufactured CNC machining centers to be sure that they meet engineering specifications before being shipped to customers. (Giddings & Lewis, Inc.)*

Figure 30-11. *Technicians are responsible for the repair and maintenance of such computer-controlled devices as welding robots. This robot is welding automotive components. (Fanuc Robotics North America, Inc.)*

production processes. Many state and community colleges offer two year programs devoted to preparing students for such technical positions.

The technician assists the engineer by testing various experimental devices and machines, compiling statistics, making cost estimates, and preparing technical reports. Many inspection and quality control programs are managed by technicians. Technicians also repair and maintain computer controlled machine tools and robots, **Figure 30-11.**

30.1.4 The Professions

The professions offer many excellent opportunities in the field of metalworking.

Teaching is one of the most satisfying of the professions; it is a field that students too often overlook, **Figure 30-12.** Teachers of industrial arts, industrial technology, industrial education, and vocational/technical education are in a fortunate position. Teaching is a challenging profession that offers a freedom not found in most other professions. It is not an overcrowded profession and it appears there will be a demand for teachers for many years to come.

To teach machining, four years of college training are usually needed, **Figure 30-13.** While industrial

Figure 30-12. *The teaching profession is a challenging one. Many skilled educators will be needed in machining technology if the United States is to maintain its position as a world leader in that industry.*

Figure 30-13. *This college student may someday teach machine technology. During four or more years of training, she will learn all phases of machine tool operation and computer programming as it relates to machining.*

experience is ordinarily not required, it will prove very helpful.

Engineering is a fast growing and challenging profession. Engineers use mathematics, science, and a knowledge of manufacturing to develop new products and processes for industry, **Figure 30-14.**

Figure 30-14. *This engineer checks out an experimental jet engine capable of sustaining flight at more than 4500 miles per hour. The liquid hydrogen used as fuel also cools the engine's combustion chamber components to keep them from melting. (NASA)*

A bachelor's degree is usually the minimum requirement for entering the engineering profession. However, some men and women have been able to enter the profession without a degree after a number of years experience as machinists, drafters, or engineering technicians. They are usually required, however, to take additional college-level training.

The *industrial engineer* is primarily concerned with the safest and most efficient use of machines, materials, and personnel, **Figure 30-15.** In some instances, he or she may be responsible for the design of special machinery and equipment to be utilized in manufacturing operations.

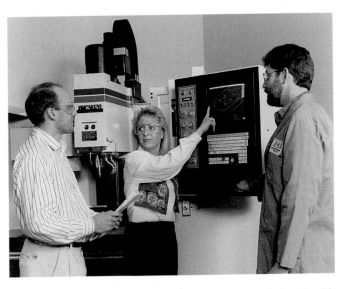

Figure 30-15. *Industrial engineers have many duties. In this case, an engineer is explaining and demonstrating the capabilities of a CNC vertical machining center (VMC) for a new job shop customer. (Giddings & Lewis, Inc.)*

A *mechanical engineer* is normally responsible for the design and development of new machines, devices, and ideas. This engineering specialty is also involved with the redesign and improvements of existing equipment. Some mechanical engineers specialize in various areas of transportation (ships, transit vehicles, robotics, etc.), **Figure 30-16.**

The *tool and manufacturing engineer* often works with the other engineers. A principal concern of the mechanical engineer is the design and development of the original or prototype model. When this model has been thoroughly tested and has met design requirements, the product is turned over to the tool and manufacturing engineer to devise methods and means required to manufacture and assemble the item, **Figure 30-17.**

A *metallurgical engineer,* **Figure 30-18,** is involved in the development and testing of metals that are used in products and manufacturing processes.

Figure 30-16. Mechanical engineers are responsible for the design and development of new machines, devices, and ideas. (Nachi Robotics)

Figure 30-17. Tool and manufacturing engineers devise new methods to manufacture complex products like the Venture Star, *which will eventually replace the present Space Shuttle. (Lockheed Martin Skunk Works)*

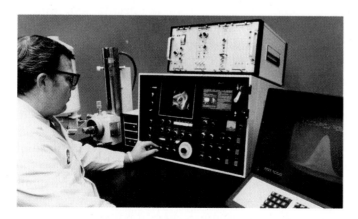

Figure 30-18. Metallurgical engineers are responsible for developing and testing metals that will eventually be used in manufacturing products. This engineer is using an electron probe to check for possible defects in a casting made from a new alloy. (Central Foundry Div. of GMC)

30.2 PREPARING TO FIND A JOB IN MACHINING TECHNOLOGY

Machining technology is a technical area with constantly developing new ideas, materials, processes, and manufacturing techniques. This means that occupational opportunities are created that were not previously available. One study reported that the average graduate will be employed in at least *five* different jobs in his or her lifetime, and three of them *do not even exist yet!*

30.2.1 Obtaining Information on Machining Occupations

There are many sources of occupational information. The most accessible for the student usually are the school's career center and industrial/technical education instructors.

State employment services are also excellent sources for getting information on local and state employment opportunities, as are the various trade unions concerned with the metalworking trades.

The field and regional offices of the Bureau of Apprenticeship, United States Department of Labor, may also be contacted for information on apprenticeship programs in your area. The *Occupational Outlook Handbook,* available at most libraries, describes many specific job categories and estimates the future demand for workers in each occupation. See **Figure 30-19.**

Information on technical occupations is also available from community colleges. Many of them offer Associate Degrees in technical areas.

Figure 30-19. Your school or public library has many sources of career information, such as the Occupational Outlook Handbook.

Keeping Your Skills Current

You will be very disappointed if you think the completion of your formal schooling means the end of training and study. To keep a job and advance in it, you will have to keep up-to-date with the knowledge and new skills that advanced technology demands. High manufacturing costs and keen competition with foreign-made products make this a real necessity.

30.2.2 Traits Employers Look for in an Employee

Industry is always on the lookout for bright young people who are not afraid to work and assume responsibility. Employers also look for the following traits in an employee, often referring to scholastic records, references, and/or to previous employers to obtain the necessary information.

- *Skills and knowledge.* Has the technical skills and knowledge necessary for an entry position. Does work neatly and accurately. Pays attention to details.
- *Integrity and honesty.* This trait is on the same level of importance as technical skills and knowledge.
- *Comprehension.* Is able to understand oral and written instructions and to read and interpret prints
- *Dependability.* Has a good attendance and punctuality record in class.
- *Teamwork.* Show the ability to work well with peers and supervisors.
- *Communication.* Is able to communicate ideas and suggestions orally and in writing.
- *Self-confidence.* Takes pride in work and will not knowingly turn out inferior or substandard material.
- *Accountability.* Is able to assume responsibility and be accountable for his/her actions.
- *Initiative.* Volunteers ideas. Demonstrates leadership.
- *Grooming and dress.* Presents a positive personal appearance.

30.2.3 Factors for Rejection for Employment

There are many factors and/or traits that can cause a person to be rejected for employment. They include:

- Poor personal appearance.
- Poor scholastic performance.
- Poor attendance record.
- Lack of maturity.
- Lack of interest or enthusiasm for the job being sought.
- Knowing little or nothing about the company where employment is sought.
- Too much interest in security and benefits.
- Unrealistic salary demands.
- Lack of ability to express himself or herself.

30.3 HOW TO GET A JOB

Securing your first job after graduation will be a very important task. To be successful, you will have to spend as much time looking for this position as you would working at a regular job. There are several other things that you can do to make this task easier.

You will have to decide what type of work you would like to do. Most schools and state employment services administer tests that will help you determine the areas of employment where you will have a good chance of succeeding.

Answering the following questions will give you additional help:

- What can I do with some degree of success?
- What have I done that others have commended me for doing well?
- What are the things I really *like* to do?
- What are the things I do *not* like to do?
- What jobs have I held? Why did I leave them?
- What skills have I acquired while in school?

You will probably have two or more areas of interest. After listing them, start gathering information on these areas of interest. Use as many different sources as time permits. This may include reading, talking with persons doing this type of work, and visiting industry.

If time permits, plan your educational program to prepare for entry into a specific job, or for advanced schooling if you are near graduation.

The next problem is, how do you go about getting that job? Jobs are always available. Workers get promoted, they retire, some quit, die, or get fired. Technological progress also creates new jobs. However, you must "track down" these jobs. There is no easy way to get a challenging job.

Concentrate on getting the job. Make your initial request for a job *in person*. Always be specific on the type of job you are seeking. Make sure you are qualified for that job. Never ask for "just any job" or inquire, "What openings do you have?"

Dress appropriately. Job hunting is not the time to wear old clothes or torn and beat-up shoes. Be clean and well-groomed.

When filling out a *job application*, avoid leaving any spaces blank. The employer may think there is something you do not want to answer. If the question is not applicable to you, write in "Not applicable," "Does not apply," or "NA." For example, there might be a question on the application that asks, "What was your highest rank in the armed forces?" If you were not in the armed forces, you would write in, "Does not apply," or "NA," rather than leaving the answer space blank.

Last, but not least, know *where* to look for a job. Check the classified advertising section of local newspapers each day. Talk with friends and relatives who are employed. They may be aware of job openings at their places of employment before the jobs are advertised.

A new office or factory building may indicate potential job openings. It would also be to your advantage to prepare a list of desirable employers in your community and visit their personnel offices. Plan these visits on a routine basis when jobs are not readily available. The personnel office will then know you are interested in working for their firm and may give you preference.

One thing you must remember: The job will *not* come to you, *You* must search for it!

30.3.1 Prepare a Resume

To speed the tedious task of filling out job applications, prepare a job resume in advance. A *resume* is a summary of your educational and employment background. It will assure uniform information with little chance for confusing responses. The resume is submitted when applying for a job. Your resume should include:

- Your full *name.*
- Your full *address,* and *telephone number.* Do not forget area and zip codes.
- *Place of birth* and *date of birth.* For some jobs, it may be necessary to include a certified copy of your birth certificate. Have a copy available in advance.
- Your *Social Security Number.*
- *Names of previous employers.* List the places you have worked, starting with the most recent. Include the items that follow (for each previous employer).
- Company name and address.
- Dates employed.
- Immediate supervisor's name.
- Salary or pay rate.
- Reason for leaving.

- *Education* and any special training. Include dates of all educational attendance.
- The types of equipment that you can safely operate.
- Names and addresses of *references.* Do not include relatives unless you have worked for them. Make sure you secure permission before using a person for a reference. (Today, many job seekers note "References available on request," instead of listing the names and addresses on the resume.)

30.3.2 What an Employee Should Expect from Industry

From the preceding sections, you have some idea of what industry expects from an employee. However, are you aware of what an employee should expect from industry? Over and above salary and fringe benefits, what should you expect from an employer?

Following are a few questions to which you might seek answers when selecting a place of employment:

Is a relatively safe and clean work area provided? Obviously, some areas can never be made as safe as others. For example, tapping a blast furnace is inherently more dangerous than working on a small lathe or drill press.

Are work areas adequately lighted, heated, and ventilated? Are noxious fumes and dust particles filtered from the air?

Is proper safety clothing and equipment available for all dangerous work? Safety items such as goggles, hearing protectors, and steel-tipped safety shoes may be provided free or at minimum cost.

Are all necessary precautions observed when hazardous materials are involved?

Is there a preventative safety program, and are safety regulations and precautions rigorously enforced?

30.3.3 Factors That Can Lead to Job Termination

The following factors can lead to failure to get a promotion, or possibly being terminated (fired) from a job. They include:

- Alcohol and/or illegal drug abuse on the job.
- Inability or refusal to perform the work required.
- Being habitually tardy or missing work repeatedly without adequate reasons.
- Inability to work with supervisors or peers.

- Fighting with or making threats to fellow workers or supervisors.
- Inability to work as a team member.

TEST YOUR KNOWLEDGE

Please do not write in the text. Write your answers on a separate sheet of paper.

1. List the four general categories into which metalworking occupations fall.

2. _____ workers are those who perform operations that do *NOT* require a high degree of skill or training.

3. The _____ worker usually starts his or her career as an apprentice.

4. Since the number of apprentice programs is on the decline, where can this training now be obtained?

5. Describe what an all-around machinist is expected to do.

6. What does a layout specialist do?

7. To perform his or her job properly, a part programmer should have the following background: (List five items.)

8. What are some of the duties of a technician?

9. List the areas of study usually included in a technician's educational program.

10. List three sources of information on metalworking occupations.

11. What does industry expect from you when you are on the job?

12. What is a job resume?

13. Why should a resume be prepared in advance?

14. Explain why you think that references are important.

15. List five traits an employer wants in a prospective employee.

16. What are three factors that can lead to job termination?

Reference Section

The following pages contain a number of tables, charts, and other materials that will be useful as reference in a variety of machining-related areas. To make locating information easier, the material in this section is listed below, along with the page number.

Common Shapes of Metals

Shapes	Length	How Measured	How Purchased
Sheet less than 1/4″ thick	Up to 144″	Thickness × width, widths to 72″	Weight, foot, or piece
Plate more than 1/4″ thick	Up to 20′	Thickness × width	Weight, foot, or piece
Band	Up to 20′	Thickness × width	Weight, or piece
Rod	12′ to 20′	Diameter	Weight, foot, or piece
Square	12′ to 20′	Width	Weight, foot, or piece
Flats	Hot rolled 20′-22′ Cold finished	Thickness × width	Weight, foot, or piece
Hexagon	12′ to 20′	Distance across flats	Weight, foot, or piece
Octagon	12′ to 20′	Distance across flats	Weight, foot, or piece
Angle	Up to 40′	Leg length × leg length × thickness of legs	Weight, foot, or piece
Channel	Up to 60′	Depth × web thickness × flange width	Weight, foot, or piece
I-beam	Up to 60′	Height × web thickness × flange width	Weight, foot, or piece

Color Codes for Marking Steels

S.A.E. Number	Code Color	S.A.E. Number	Code Color	S.A.E. Number	Code Color	S.A.E. Number	Code Color
	Carbon steels	2115	Red and bronze	T1340	Orange and green	3450	Black and bronze
1010	White	2315	Red and blue	T1345	Orange and red	4820	Green and purple
1015	White	2320	Red and blue	T1350	Orange and red		Chromium steels
X1015	White	2330	Red and white		Nickel-chromium steels	5120	Black
1020	Brown	2335	Red and white	3115	Blue and black	5140	Black and white
X1020	Brown	2340	Red and green	3120	Blue and black	5150	Black and white
1025	Red	2345	Red and green	3125	Pink	52100	Black and brown
X1025	Red	2350	Red and aluminum	3130	Blue and green		Chromium-vanadium steels
1030	Blue	2515	Red and black	3135	Blue and green	6115	White and brown
1035	Blue		Molybdenum steels	3140	Blue and white	6120	White and brown
1040	Green	4130	Green and white	X3140	Blue and white	6125	White and aluminum
X1040	Green	X4130	Green and bronze	3145	Blue and white	6130	White and yellow
1045	Orange	4135	Green and yellow	3150	Blue and brown	6135	White and yellow
X1045	Orange	4140	Green and brown	3215	Blue and purple	6140	White and bronze
1050	Bronze	4150	Green and brown	3220	Blue and purple	6145	White and orange
1095	Aluminum	4340	Green and aluminum	3230	Blue and purple	6150	White and orange
	Free cutting steels	4345	Green and aluminum	3240	Blue and aluminum	6195	White and purple
1112	Yellow	4615	Green and black	3245	Blue and aluminum		Tungsten steels
X1112	Yellow	4620	Green and black	3250	Blue and bronze	71360	Brown and orange
1120	Yellow and brown	4640	Green and pink	3312	Orange and black	71660	Brown and bronze
X1314	Yellow and blue	4815	Green and purple	3325	Orange and black	7260	Brown and aluminum
X1315	Yellow and red	X1340	Yellow and black	3335	Blue and orange		Silicon-manganese steels
X1335	Yellow and black		Manganese steels	3340	Blue and orange	9255	Bronze and aluminum
	Nickel steels	T1330	Orange and green	3415	Blue and pink	9260	Bronze and aluminum
2015	Red and brown	T1335	Orange and green	3435	Orange and aluminum		

Metal Sheet Materials Chart

Material (Sheet less than 1/4" thick)	How Measured	How Purchased	Characteristics
Copper	Gage number (Brown & Sharpe and Amer. Std.)	24" × 96" sheet or 12" or 18" by lineal feet on roll	Pure metal
Brass	Gage number (Brown & Sharpe and Amer. Std.)	24" × 76" sheet or 12" or 18" by lineal feet on roll	Alloy of copper and zinc
Aluminum	Decimal	24" × 72" sheet or 12" or 18" by lineal feet on roll	Available as commercially pure metal or alloyed for strength, hardness, and ductility
Galvanized steel	Gage number (Amer. Std.)	24" × 96" sheet	Mild steel sheet with zinc plating, also available with zinc coating that is part of sheet
Black annealed steel sheet	Gage number (Amer. Std.)	24" × 96" sheet	Mild steel with oxide coating, hot-rolled
Cold-rolled steel sheet	Gage number (Amer. Std.)	24" × 96" sheet	Oxide removed and cold-rolled to final thickness
Tin plate	Gage number (Amer. Std.)	20" × 28" sheet 56 or 112 to pkg.	Mild steel with tin coating
Nickel silver	Gage number (Brown & Sharpe)	6" or 12" wide by lineal feet on roll	Copper 50%, zinc 30%, nickel 20%
Expanded	Gage number (Amer. Std.)	36" × 96" sheet	Metal is pierced and expanded (stretched) to diamond shape; also available rolled to thickness after it has been expanded
Perforated	Gage number (Amer. Std.)	30" × 36" sheet 36" × 48" sheet	Design is cut in sheet; many designs available

Order of Ductility of Metals

1. Gold
2. Platinum
3. Silver
4. Iron
5. Copper
6. Aluminum
7. Nickel
8. Zinc
9. Tin
10. Lead

Physical Properties of Metals

Metal	Symbol	Specific Gravity	Specific Heat	Melting Point*		Lbs. per Cubic Inch
				°C	°F	
Aluminum (cast)	Al	2.56	.2185	658	1217	.0924
Aluminum (rolled)	Al	2.71	–	658	1217	.0978
Antimony	Sb	6.71	.051	630	1166	.2424
Bismuth	Bi	9.80	.031	271	520	.3540
Boron	B	2.30	.3091	2300	4172	.0831
Brass	–	8.51	.094	–	–	.3075
Cadmium	Cd	8.60	.057	321	610	.3107
Calcium	Ca	1.57	.170	810	1490	.0567
Chromium	Cr	6.80	.120	1510	2750	.2457
Cobalt	Co	8.50	.110	1490	2714	.3071
Copper	Cu	8.89	.094	1083	1982	.3212
Columbium	Cb	8.57	–	1950	3542	.3096
Gold	Au	19.32	.032	1063	1945	.6979
Iridium	Ir	22.42	.033	2300	4170	.8099
Iron	Fe	7.86	.110	1520	2768	.2634
Iron (cast)	Fe	7.218	.1298	1375	2507	.2605
Iron (wrought)	Fe	7.70	.1138	1500-1600	2732-2912	.2779
Lead	Pb	11.37	.031	327	621	.4108
Lithium	Li	.057	.941	186	367	.0213
Magnesium	Mg	1.74	.250	651	1204	.0629
Manganese	Mn	8.00	.120	1225	2237	.2890
Mercury	Hg	13.59	.032	–39	–38	.4909
Molybdenum	Mo	10.2	.0647	2620	47.48	.368
Monel metal	–	8.87	.127	1360	2480	.320
Nickel	Ni	8.80	.130	1452	2646	.319
Phosphorus	P	1.82	.177	43	111.4	.0657
Platinum	Pt	21.50	.033	1755	3191	.7767
Potassium	K	0.87	.170	62	144	.0314
Selenium	Se	4.81	.084	220	428	.174
Silicon	Si	2.40	.1762	1427	2600	.087
Silver	Ag	10.53	.056	961	1761	.3805
Sodium	Na	0.97	.290	97	207	.0350
Steel	–	7.858	.1175	1330-1378	2372-2532	.2839
Strontium	Sr	2.54	.074	769	1416	.0918
Tantalum	Ta	10.80	–	2850	5160	.3902
Tin	Sn	7.29	.056	232	450	.2634
Titanium	Ti	5.3	.130	1900	3450	.1915
Tungsten	W	19.10	.033	3000	5432	.6900
Uranium	U	18.70	–	1132	2070	.6755
Vanadium	V	5.50	–	1730	3146	.1987
Zinc	Zn	7.19	.094	419	786	.2598

*Circular of the Bureau of Standards No.35, Department of Commerce and Labor

Wire Gages in Decimal Inches

Number of Wire Gage	American or Brown & Sharpe	Washburn & Moen Mfg. Co., A.S.& W. Roebling	Imperial Wire Gage	Stubs' Steel Wire	Birmingham or Stubs' Iron Wire
0000000	. . .	.4900	.5000	. . .	. . .
000000	.5800	.4615	.4640	. . .	. . .
00000	.5165	.4305	.4320	. . .	.500
0000	.460	.3938	.4000	. . .	.454
000	.40964	.3625	.3720	. . .	.425
00	.3648	.3310	.3480	. . .	.380
0	.32486	.3065	.3240	. . .	.340
1	.2893	.2830	.3000	.227	.300
2	.25763	.2625	.2760	.219	.284
3	.22942	.2437	.2520	.212	.259
4	.20431	.2253	.2320	.207	.238
5	.18194	.2070	.2120	.204	.220
6	.16202	.1920	.1920	.201	.203
7	.14428	.1770	.1760	.199	.180
8	.12849	.1620	.1600	.197	.165
9	.11443	.1483	.1440	.194	.148
10	.10189	.1350	.1280	.191	.134
11	.090742	.1205	.1160	.188	.120
12	.080808	.1055	.1040	.185	.109
13	.071961	.0915	.0920	.182	.095
14	.064084	.0800	.0800	.180	.083
15	.057068	.0720	.0720	.178	.072
16	.05082	.0625	.0640	.175	.065
17	.045257	.0540	.0560	.172	.058
18	.040303	.0475	.0480	.168	.049
19	.03589	.0410	.0400	.164	.042
20	.031961	.0348	.0360	.161	.035
21	.028462	.0317	.0320	.157	.032
22	.025347	.0286	.0280	.155	.028
23	.022571	.0258	.0240	.153	.025
24	.0201	.0230	.0220	.151	.022
25	.0179	.0204	.0200	.148	.020
26	.01594	.0181	.0180	.146	.018
27	.014195	.0173	.0164	.143	.016
28	.012641	.0162	.0148	.139	.014
29	.011257	.0150	.0136	.134	.013
30	.010025	.0140	.0124	.127	.012
31	.008928	.0132	.0116	.120	.010
32	.00795	.0128	.0108	.115	.009
33	.00708	.0118	.0100	.112	.008
34	.006304	.0104	.0092	.110	.007
35	.005614	.0095	.0084	.108	.005
36	.005	.0090	.0076	.106	.004
37	.004453	.0085	.0068	.103	. . .
38	.003965	.0080	.0060	.101	. . .
39	.003531	.0075	.0052	.099	. . .
40	.003144	.0070	.0048	.097	. . .

Cutting Speeds for Round Stock

Diameter	Material	Roughing Cut (rpm)	Finishing Cut (rpm)	Threading (rpm)	Diameter	Material	Roughing Cut (rpm)	Finishing Cut (rpm)	Threading (rpm)
1/8″	Machine steel/bronze	2880	3200	1020	1 1/2″	Machine steel/bronze	240	270	67
	Cast iron	1920	2560	800		Tool steel	134	200	53
	Tool steel (annealed)	1600	2400	640		Brass	400	534	134
	Brass	4800	6400	1600		Aluminum	534	800	134
	Aluminum	6400	9600	1600					
3/16″	Machine steel/bronze	2880	3200	1120	1 3/4″	Machine steel/bronze	205	230	80
	Tool steel	1600	2400	640		Tool steel	115	170	50
	Brass	4800	6400	1600		Brass	340	450	115
	Aluminum	6400	9600	1600		Aluminum	456	680	115
1/4″	Machine steel/bronze	1440	1600	560	2″	Machine steel	180	200	50
	Tool steel	800	1200	320		Tool steel	100	150	40
	Brass	2400	3200	800		Brass	300	400	100
	Aluminum	3200	4800	800		Aluminum	400	600	100
3/8″	Machine steel/bronze	960	1066	270	2 1/2″	Machine steel	141	160	56
	Tool steel	540	800	220		Tool steel	80	120	32
	Brass	1700	2100	530		Brass	240	320	80
	Aluminum	2130	3200	540		Aluminum	320	480	80
1/2″	Machine steel/bronze	720	800	280	3″	Machine steel	120	140	40
	Tool steel	400	600	160		Tool steel	65	100	40
	Brass	1200	1600	400		Brass	200	270	65
	Aluminum	1600	2400	400		Aluminum	270	400	65
5/8″	Machine steel/bronze	576	640	160	3 1/2″	Machine steel	103	115	40
	Tool steel	320	480	200		Tool steel	60	85	23
	Brass	960	1280	320		Brass	171	228	57
	Aluminum	1280	1920	320		Aluminum	228	342	57
3/4″	Machine steel/bronze	500	550	176	4″	Machine steel	90	100	35
	Tool steel	266	400	106		Tool steel	50	75	20
	Brass	800	1066	266		Brass	150	200	50
	Aluminum	1066	1600	266		Aluminum	200	300	50
1″	Machine steel/bronze	360	400	140	4 1/2″	Machine steel	80	90	31
	Tool steel	200	300	80		Tool steel	45	67	18
	Brass	600	800	200		Brass	133	178	45
	Aluminum	800	1200	200		Aluminum	178	267	45
1 1/4″	Machine steel	288	320	112	5″	Machine steel	72	80	28
	Tool steel	160	240	64		Tool steel	40	58	16
	Brass	480	640	160		Brass	120	160	40
	Aluminum	640	960	160		Aluminum	160	240	40

Rules for Determining Speeds and Feeds

To Find	Having	Rule	Formula
Speed of cutter in feet per minute (FPM)	Diameter of cutter and revolutions per minute	Diameter of cutter (in inches) multiplied by 3.1416 (π) multiplied by revolutions per minute, divided by 12	$FPM = \dfrac{\pi D \times RPM}{12}$
Speed of cutter in meters per minute (MPM)	Diameter of cutter and revolutions per minute	Diameter of cutter multiplied by 3.1416 (π) multiplied by revolutions per minute, divided by 1000	$MPM = \dfrac{D(mm) \times \pi \times RPM}{1000}$
Revolutions per minute (RPM)	Feet per minute and diameter of cutter	Feet per minute, multiplied by 12, divided by circumference of cutter (πD)	$RPM = \dfrac{FPM \times 12}{\pi D}$
Revolutions per minute (RPM)	Meters per minute and diameter of cutter in millimeters (mm)	Meters per minute, multiplied by 1000, divided by the circumference of cutter (πD)	$RPM = \dfrac{MPM \times 1000}{\pi D}$
Feed per revolution (FR)	Feed per minute and revolutions per minute	Feed per minute, divided by revolutions per minute	$FR = \dfrac{F}{RPM}$
Feed per tooth per revolution (FTR)	Feed per minute and number of teeth in cutter	Feed per minute (in inches or millimeters) divided by number of teeth in cutter $\times$ revolutions per minute	$FTR = \dfrac{F}{T \times RPM}$
Feed per minute (F)	Feed per tooth per revolution, number of teeth in cutter, and RPM	Feed per tooth per revolutions multiplied by number of teeth in cutter, multiplied by revolutions per minute	$F = FTR \times T \times RPM$
Feed per minute (F)	Feed per revolution and revolutions per minute	Feed per revolution multiplied by revolutions per minute	$F = FR \times RPM$
Number of teeth per minute (TM)	Number of teeth in cutter and revolutions per minute	Number of teeth in cutter multiplied by revolutions per minute	$TM = T \times RPM$

RPM = Revolutions per minute
T = Teeth in cutter
D = Diameter of cutter
π = 3.1416 (pi)
FPM = Speed of cutter in feet per minute

TM = Teeth per minute
F = Feed per minute
FR = Feed per revolution
FTR = Feed per tooth per revolution
MPM = Speed of cutter in meters per minute

Recommended Turning Rates for Stainless Steels Using High-speed Tools

Nature of Stock	Type No.	Speed (sfpm)	Feed (inches per revolution)
Free machining grades	430 F	100 – 140	0.003-0.005
	416	90 – 135	for finish cuts and up to
	303	80 – 120	0.015 for roughing cuts
High-carbon grades that are slowed down due to their abrasive action on tools	410	75 – 115	0.003–0.008
	430	75 – 115	0.003–0.008
	420	45 – 85	0.003–0.008
	431	45 – 85	0.003–0.008
	440	30 – 60	0.003–0.008
	302		
	304		
	316	45 – 80	0.004–0.008

Feeds and Speeds for
HSS Drills, Reamers, and Taps

Material	Brinell	Drills (sfm)	Drills Point	Drills Feed	Reamers (sfm)	Reamers Feed	Taps (sfm) Threads per Inch 3–7 1/2	8–15	16–24	25–up
Aluminum	99–101	200–250	118°	M	150–160	M	50	100	150	200
Aluminum bronze	170–187	60	118°	M	40–45	M	12	25	45	60
Bakelite	. . .	80	60°-90°	M	50–60	M	50	100	150	200
Brass	192–202	200–250	118°	H	150–160	H	50	100	150	200
Bronze, common	166–183	200–250	118°	H	150–160	H	40	80	100	150
Bronze, phosphor, 1/2 hard	187–202	175–180	118°	M	130–140	M	25	40	50	80
Bronze, phosphor, soft	149–163	200–250	118°	H	150–160	H	40	80	100	150
Cast iron, soft	126	140–150	90°	H	100–110	H	30	60	90	140
Cast iron, medium soft	196	80–110	118°	M	50–65	M	25	40	50	80
Cast iron, hard	293–302	45–50	118°	L	67–75	L	10	20	30	40
Cast iron, chilled*	402	15	150°	L	8–10	L	5	5	10	10
Cast steel	286–302	40–50*	118°	L	70–75	L	20	30	40	50
Celluloid	. . .	100	90°	M	75–80	M	50	100	150	200
Copper	80–85	70	100°	L	45–55	L	40	80	100	150
Drop forgings (steel)	170–196	60	118°	M	40–45	M	12	25	45	60
Duralumin	90–104	200	118°	M	150–160	M	50	100	150	200
Everdur	179–207	60	118°	L	40–45	L	20	30	40	50
Machinery steel	170–196	110	118°	H	67–75	H	35	50	60	85
Magnet steel, soft	241–302	35–40	118°	M	20–25	M	20	40	50	75
Magnet steel, hard*	321–512	15	150°	L	10	L	5	10	15	25
Manganese steel, 7% – 13%	187–217	15	150°	L	10	L	15	20	25	30
Manganese copper, 30% Mn.*	134	15	150°	L	10–12	L	...	...	...	...
Malleable iron	112–126	85–90	118°	H	. . .	H	20	30	40	50
Mild steel, .20 –.30 C	170–202	110–120	118°	H	75–85	H	40	55	70	90
Molybdenum steel	196–235	55	125°	M	35–45	M	20	30	35	45
Monel metal	149–170	50	118°	M	35–38	M	8	10	15	20
Nickel, pure*	187–202	75	118°	L	40	L	25	40	50	80
Nickel steel, 3 1/2%	196–241	60	118°	L	40–45	L	8	10	15	20
Rubber, hard	. . .	100	60°-90°	L	70–80	L	50	100	150	200
Screw stock, C.R.	170–196	110	118°	H	75	H	20	30	40	50
Spring steel	402	20	150°	L	12–15	L	10	10	15	15
Stainless steel	146–149	50	118°	M	30	M	8	10	15	20
Stainless steel, C.R.*	460–477	20	118°	L	15	L	8	10	15	20
Steel, .40 to .50 C	170–196	80	118°	M	8–10	M	20	30	40	50
Tool, S.A.E., and forging steel	149	75	118°	H	35–40	H	25	35	45	55
Tool, S.A.E., and forging steel	241	50	125°	M	12	M	15	15	25	25
Tool, S.A.E., and forging steel*	402	15	150°	L	10	L	8	10	15	20
Zinc alloy	112–126	200–250	118°	M	150–175	M	50	100	150	200

*Use specially constructed heavy-duty drills.
Note: Carbon steel tools should be run at speeds 40% to 50% of those recommended for high speed steel.
Spiral point taps may be run at speeds 15% to 20% faster than regular taps.

Drill diameter	Cast iron		Bronze or Brass		Drop forgings Alloy steel Tool steel Annealed		Drop forgings Alloy steel Heat-treated		Steel castings		Mild steel	
Feeds and Speeds for HSS Drills in Various Metals												
	Feed	Speed	Feed	Speed	Feed	Speed	Feed	Speed	Feed	Speed	Feed	Speed
1/16″	.002	4550	.002	9150	.002	3650	.002	2750	.002	3650	.002	4250
	.004	6700	.004	12,000	.003	4550	.003	3650	.003	4550	.003	5600
1/8″	.002	2550	.002	4550	.002	1800	.002	1225	.002	1800	.002	2100
	.004	3350	.004	5600	.003	2250	.003	1800	.003	2250	.003	2800
3/16″	.004	1500	.004	3100	.003	1200	.003	900	.003	1200	.003	1400
	.006	2200	.007	5600	.004	1500	.004	1200	.005	1500	.005	1900
1/4″	.004	1150	.004	2300	.003	925	.003	750	.003	925	.003	1050
	.006	1650	.007	2750	.004	1150	.004	925	.005	1150	.005	1500
5/16″	.006	925	.007	1825	.004	725	.004	500	.004	725	.005	850
	.009	1325	.010	2200	.006	925	.005	725	.006	925	.007	1200
3/8″	.006	750	.007	1525	.004	600	.004	400	.004	600	.005	700
	.009	1100	.010	1850	.006	750	.005	600	.006	750	.007	925
7/16″	.009	650	.010	1300	.006	525	.005	350	.006	525	.006	600
	.012	950	.014	1525	.009	650	.006	525	.010	650	.010	800
1/2″	.008	575	.010	1150	.006	375	.005	300	.006	375	.006	525
	.012	850	.014	1375	.009	575	.006	375	.010	575	.010	700
9/16″	.012	500	.014	1000	.008	350	.007	275	.010	350	.010	575
	.016	750	.018	1200	.012	500	.010	350	.014	500	.014	625
5/8″	.012	450	.014	900	.008	300	.007	250	.010	300	.010	425
	.016	675	.018	1100	.012	450	.010	300	.014	450	.014	565
11/16″	.012	410	.014	800	.008	275	.007	225	.010	275	.010	375
	.016	625	.018	1000	.012	410	.010	275	.014	410	.014	525
3/4″	.012	375	.014	750	.008	250	.007	200	.010	250	.010	350
	.016	550	.018	900	.012	375	.010	250	.014	375	.014	475
13/16″	.014	350	.016	700	.010	240	.009	190	.014	240	.014	325
	.020	525	.022	850	.014	350	.012	240	.016	350	.016	450
7/8″	.014	325	.016	650	.010	225	.009	175	.014	225	.014	300
	.020	475	.022	800	.014	325	.012	225	.016	325	.016	400
15/16″	.014	300	.016	625	.010	200	.009	160	.014	200	.014	275
	.020	450	.022	725	.014	300	.012	200	.016	300	.016	375
1″	.014	280	.016	575	.010	185	.009	150	.014	185	.014	265
	.020	425	.022	675	.014	280	.012	185	.016	280	.016	350

(Chicago-Latrobe)

Speeds and feeds shown apply to average working conditions and materials. They are recommended with regard to conserving drills and avoiding excessive machine tool wear. Under many conditions, these speeds and feeds may be considerably increased; under others they must be decreased. This is dependent on judgment of operator, and performance obtained. Excessive speeds and feeds will show up by action of machine and drill. Same applies to lower speeds and feeds. Operator will notice whether he/she is getting proper performance by experience, and will advance or retard as case may justify. Feeds and speeds should be changed in proper proportions and a liberal use of cooling compound will increase life of tools.

Never dip a drill into water to cool it while grinding. This will cause tiny checks, or cracks at the cutting edge, which will cause the drill to dull quickly.

Do not leave a drill in after it shows signs of dulling or laboring; then is the time to regrind. Proper grinding is essential.

To determine feed and speed according to the above chart, proceed as follows:

You are going to drill heat-treated drop forgings. We suppose you will use a 1/2″ drill. Follow column down to where the 1/2″ drill meets it; there you will find that a feed from .005 to .006 and a speed of from 300 to 375 rpm are recommended. Start by using .005 feed and 300 rpm. If drill and machine seem to turn smoothly without strain, then both feed and speed can be advanced. Operator will soon find which is best.

Formulas for Machining Bar Stock

Surface speed—feet/minute	
Round bars	$\dfrac{\text{Diameter} \times 3.1416 \times \text{rpm}}{12}$
Hexagon bars (distance across corners).	$\dfrac{\text{Size} \times 3.1416 \times \text{rpm} \times 1.155}{12}$
	$\dfrac{\text{Distance} \times 3.1416 \times \text{rpm}}{12}$
Square bars (distance across corners).	$\dfrac{\text{Size} \times 3.1416 \times \text{rpm} \times 1.414}{12}$
	$\dfrac{\text{Distance} \times 3.1416 \times \text{rpm}}{12}$

Revolutions—number/minute	
Round bars	$\dfrac{\text{SFM} \times 12}{\text{Diameter} \times 3.1416}$
Hexagon bars	$\dfrac{\text{SFM} \times 12}{\text{Size} \times 3.1416 \times 1.155}$
	$\dfrac{\text{SFM} \times 12}{\text{Distance} \times 3.1416}$
Square bars	$\dfrac{\text{SFM} \times 12}{\text{Size} \times 3.1416 \times 1.414}$
	$\dfrac{\text{SFM} \times 12}{\text{Distance} \times 3.1416}$

Feed—inches/revolution	$\dfrac{\text{Feed inches per minute}}{\text{rpm}}$
	$\dfrac{\text{Diameter} \times 3.1416 \times \text{Feed}}{\text{SFM} \times 12}$
Feed—inches/tooth	$\dfrac{\text{Feed}}{\text{Number of teeth}}$
Time for actual machining—seconds	$\dfrac{\text{Revolutions required} \times 60 \text{ seconds}}{\text{rpm}}$
Machine time	Time for machining + idle time
Tapping or threading time—seconds	$\dfrac{\text{Number of threads} \times 60 \text{ seconds}}{\text{Actual threading speed in rpm}}$

Standard Dimensional Tolerances for Bar Stock

Hot-rolled Bars: Rounds and Squares

Specified Size (Inches)	Variations from Size (Inches)		Out-of-round (1) or Square (2) Inches
	Over	Under	
1/4 to 5/16 inclusive (3)	(4)	(4)	(4)
Over 5/16 to 7/16 inclusive (3)	0.006	0.006	0.009
Over 7/16 to 5/8 inclusive (3)	0.007	0.007	0.010
Over 5/8 to 7/8 inclusive	0.008	0.008	0.012
Over 7/8 to 1 inclusive	0.009	0.009	0.013
Over 1 to 1 1/8 inclusive	0.010	0.010	0.015
Over 1 1/8 to 1 1/4 inclusive	0.011	0.011	0.016
Over 1 1/4 to 1 3/8 inclusive	0.012	0.012	0.018
Over 1 3/8 to 1 1/2 inclusive	0.014	0.014	0.021
Over 1 1/2 to 2 inclusive	1/64	1/64	0.023
Over 2 to 2 1/2 inclusive	1/32	0	0.023
Over 2 1/2 to 3 1/2 inclusive	3/64	0	0.035
Over 3 1/2 to 4 1/2 inclusive	1/16	0	0.046
Over 4 1/2 to 5 1/2 inclusive	5/64	0	0.058
Over 5 1/2 to 6 1/2 inclusive	1/8	0	0.070
Over 6 1/2 to 8 inclusive	5/32	0	0.085

(1) Out-of-round is difference between maximum and minimum diameters of bar, measured at same cross section.
(2) Out-of-square is difference in two dimensions at same cross section of a square bar, each dimension being distance between opposite faces.
(3) Round sections in size range of 1/4″ to approximately 5/8″ diameter are commonly produced on rod mills in coils. Tolerances on product made this way have not been established; for such tolerances producer should be consulted. Variations in size of coiled product made on rod mills are greater than size tolerances for product made on bar mills.
(4) Squares in this size are not commonly produced as hot-rolled product.

Hot-rolled Bars: Hexagons and Octagons

Specified Sizes Between Opposite Sides (Inches)	Variations from Size (Inches)		Maximum Difference 3 Measurements for Hexagons Only (Inches)
	Under	Over	
1/4 to 1/2 inclusive	0.007	0.007	0.011
Over 1/2 to 1 inclusive	0.010	0.010	0.015
Over 1 to 1 1/2 inclusive	0.021	0.021	0.025
Over 1 1/2 to 2 inclusive	1/32	1/32	1/32
Over 2 to 2 1/2 inclusive	3/64	3/64	3/64
Over 2 1/2 to 3 1/2 inclusive	1/16	1/16	1/16

Cold-finished Bars: Flats

Specified Width Size or Thickness (Inches)	Variations from Width Over or Under (Inches)*		Variations from Thicknesses Over or Under (Inch)
	For Thicknesses 1/4″ and Under	For Thicknesses Over 1/4″	
Over 1/8 to 1 inclusive	—	—	0.002
Over 3/8 to 1 inclusive	0.004	0.002	0.002
Over 1 to 2 inclusive	0.006	0.003	0.003
Over 2 to 3 inclusive	0.008	0.004	0.004
Over 3 to 4 1/2 inclusive	0.010	0.005	0.005

*When it is necessary to heat treat, or heat treat and pickle after cold finishing, because of special hardness or mechanical property requirements, tolerances are double those shown in the table.

Hot-rolled Bars: Flats

Specified Widths (Inches)	Variations from Thickness for Thicknesses Given Over or Under (Inches)			Variations from Width (Inches)	
	1/8 to 1/2 Inclusive	Over 1/2 to 1 Inclusive	Over 1 to 2 Inclusive	Over	Under
To 1 inclusive	0.008	0.010	—	1/64	1/64
Over 1 to 2 inclusive	0.012	0.015	1/32	1/32	1/32
Over 2 to 4 inclusive	0.015	0.020	1/32	1/16	1/32
Over 4 to 6 inclusive	0.015	0.020	1/32	3/32	1/16
Over 6 to 8 inclusive	0.016	0.025	1/32	1/8	5/32
Over 8 to 10 inclusive	0.021	0.031	1/32	5/32	3/16

Cold-finished Bars: Hexagons, Octagons, and Squares

Specified Size (Inches)	Variations from Size (Inches)*	
	Over	Under
Over 1/2 to 1 inclusive	0	0.004
Over 1 to 2 inclusive	0	0.006
Over 2 to 3 inclusive	0	0.008
Over 3	0	0.010

*When it is necessary to heat treat, or heat treat and pickle after cold finishing, because of special hardness or mechanical property requirements, tolerances are double those shown in the table.

Cold-finished Bars: Rounds

Specified Size (Inches)	Variations from Size (Inches)*	
	Over	Under
Over 1/2 to 1 exclusive	0.002	0.002
1 to 1 1/2 exclusive	0.0025	0.0025
1 1/2 to 4 inclusive	0.003	0.003

*When it is necessary to heat treat, or heat treat and pickle after cold finishing, because of special hardness or mechanical property requirements, tolerances are double those shown in table.

Machine-cut Bars (Cut after Machine Straightening)

Specified Sizes as They Apply to Rounds, Squares, Hexagons, Octagons, and Width of Flats (Inches)	Variations from Specified Lengths (Inches)				
	To 12′ Inclusive		Over 12′ to 25′ Inclusive		
	Over	Under	Over	Under
To 3 inclusive	1/8	0	3/16	0
Over 3 to 6 inclusive	3/16	0	1/4	0
Over 6 to 9 inclusive	1/4	0	5/16	0
Over 9 to 12 inclusive	1/2	0	1/2	0

Allowances for Turning Machine-straightened Bars

When ordering bars that are to be turned, it is recommended that allowances be made for finishing from hot-rolled diameters not less than amounts shown in following table, and specify hot-rolled sizes accordingly:

Nominal Diameter of Hot-rolled Bar (Inches)	Minimum Allowance on Diameter for Turning (Inches)
1 1/2 to 3 inclusive	1/8
Over 3 to 6 inclusive	1/4
Over 6 to 8 inclusive	3/8

Hot or Cold Cutting Length Tolerances

Specified Sizes as They Apply to Rounds, Squares, Hexagons, Octagons, and Width of Flats (Inches)	Variations from Specified Lengths (Inches)			
	To 12′ Inclusive		Over 12′ to 25′ Inclusive	
	Over	Under	Over	Under
To 2 inclusive	1/2	0	3/4	0
Over 2 to 4 inclusive	3/4	0	1	0
Over 4 to 6 inclusive	1	0	1 1/4	0
Over 6 to 9 inclusive	1 1/4	0	1 1/2	0
Over 9 to 12 inclusive	1 1/2	0	2	0

Hot-finished and Cold-finished Bars for Machining

Camber is greatest deviation of a side from a straight line. Measurement is taken on concave side of bar with a straightedge. Unless otherwise specified, hot-finished and cold-finished bars for machining purposes are furnished machine-straightened to following tolerances:

Hot-finished: 1/8″ in any 5 feet; but may not exceed $1/8″ \times \dfrac{\text{No. of feet in length}}{5}$

Cold-finished: 1/16″ in any 5 feet; but may not exceed $1/16″ \times \dfrac{\text{No. of feet in length}}{5}$

Standard Dimensional Tolerances for Wire

Drawn, Centerless Ground, Centerless Ground and Polished Round and Square Wire

Specified Size (Inches)	Tolerance (Inches)	
	Over	Under
1/2	0.002	0.002
Under 1/2 to 5/16 inclusive	0.0015	0.0015
Under 5/16 to 0.050 inclusive	0.001	0.001

The maximum out-of-round tolerance for round wire is one-half of total size tolerance shown in above table.

Tolerance for wire for which final operation is a surface treatment for purpose of removing scale or drawing lubricant

Specified Size (Inches)	Tolerance (Inches)	
	Over	Under
1/2	0.004	0.004
Under 1/2 to 5/16 inclusive	0.003	0.003
Under 5/16 to 0.050 inclusive	0.002	0.002

Drawn Wire in Hexagons and Octagons

Specified Size* (Inches)	Tolerance (Inches)	
	Over	Under
1/2	0	0.004
Under 1/2 to 5/16 inclusive	0	0.003
Under 5/16 to 1/8 inclusive	0	0.002

*Distance across flats

(Carpenter Steel Co.)

Rules and Formulas for Bevel Gear Calculations
(Use with drawing on page 571)

To Find	Rule	Formula
Diametral pitch (P)	Divide the number of teeth by the pitch diameter.	$P = \dfrac{N}{P_d}$
Circular pitch (P_c)	Divide 3.1416 by the diametral pitch.	$P_c = \dfrac{3.1416}{P}$
Pitch diameter (P_d)	Divide the number of teeth by the diametral pitch.	$P_d = \dfrac{N}{P}$
Pitch angle of pinion $\tan (b_p)$	Divide the number of teeth in the pinion by the number of teeth in the gear to obtain the tangent.	$\tan b_p = \dfrac{N_p}{N_g}$
Pitch angle of gear $\tan (b_g)$	Divide the number of teeth in the gear by the number of teeth in the pinion to obtain the tangent.	$\tan b_g = \dfrac{N_g}{N_p}$
Pitch cone distance (C_r)	Divide the pitch diameter by twice the sine of the pitch angle.	$C_r = \dfrac{P_d}{2 (\sin b)}$
Addendum (a)	Divide 1.0 by the diametral pitch.	$a = \dfrac{1.0}{P}$
Addendum angle $\tan (A_1)$	Divide the addendum by the pitch cone distance to obtain the tangent.	$\tan A_1 = \dfrac{a}{C_r}$
Angular addendum (A_a)	Multiply the addendum by the cosine of the pitch angle.	$A_a = a \cos b$
Outside diameter (D_o)	Add twice the angular addendum to the pitch diameter.	$D_o = P_d + 2A_a$
Dedendum angle $\tan (c_1)$	Divide the dedendum by the pitch cone distance to obtain the tangent.	$\tan c_1 = \dfrac{a + c}{C_r}$
Addendum of small end of tooth (a_s)	Subtract the width of face from the pitch cone distance, divide the remainder by the pitch cone distance and multiply by the addendum.	$a_s = a \left(\dfrac{C_r - W}{C_r} \right)$
Thickness of tooth at pitch line (T_L)	Divide the circular pitch by 2.	$T_L = \dfrac{P_c}{2}$
Thickness of tooth at pitch line at small end of gear (T_s)	Subtract the width of face from the pitch cone distance, divide the remainder by the pitch cone distance and multiply by the thickness of the tooth at the pitch line.	$T_s = T_L \left(\dfrac{C_r - W}{C_r} \right)$
Face angle (F_a)	Face cone of blank turned parallel to root cone of mating gear.	$F_a = b + c_1$
Whole depth of tooth space (h_t)	Divide 2.157 by the diametral pitch.	$h_t = \dfrac{2.157}{P}$
Apex distance at large end of tooth (V)	Multiply one-half the outside diameter by the tangent of the face angle.	$V = \left(\dfrac{D_o}{2} \right) \tan F_a$
Apex distance at small end of tooth (v)	Subtract the width of face from the pitch cone distance, divide the remainder by the pitch cone distance and multiply by the apex distance.	$v = V \left(\dfrac{C_r - W}{C_r} \right)$
Gear ratio (m_g)	Divide the number of teeth in the gear by the number of teeth in the pinion.	$m_g = \dfrac{N_g}{N_p}$
Number of teeth in gear and/or pinion (N_g, N_p)	Multiply the pitch diameter by the diametral pitch.	$N_g = P_d P$ $N_p = P_d P$
Cutting angle (d)	Subtract the addendum plus clearance angle from the pitch angle.	$d = b - c_1$
Number of teeth of imaginary spur gear for which cutter is selected (N_o)	Divide the number of teeth in actual gear by the cosine of the pitch angle.	$N_c = \dfrac{N}{\cos b}$

Note: Use with table on page 570

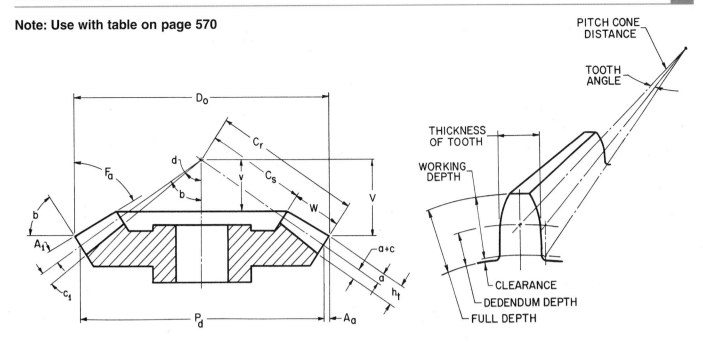

Decimal Equivalents: Number-size Drills

Drill	Size of Drill in Inches	Drill	Size of Drill in Inches	Drill	Size of Drill in Inches	Drill	Size of Drill in Inches
1	.2280	21	.1590	41	.0960	61	.0390
2	.2210	22	.1570	42	.0935	62	.0380
3	.2130	23	.1540	43	.0890	63	.0370
4	.2090	24	.1520	44	.0860	64	.0360
5	.2055	25	.1495	45	.0820	65	.0350
6	.2040	26	.1470	46	.0810	66	.0330
7	.2010	27	.1440	47	.0785	67	.0320
8	.1990	28	.1405	48	.0760	68	.0310
9	.1960	29	.1360	49	.0730	69	.0292
10	.1935	30	.1285	50	.0700	70	.0280
11	.1910	31	.1200	51	.0670	71	.0260
12	.1890	32	.1160	52	.0635	72	.0250
13	.1850	33	.1130	53	.0595	73	.0240
14	.1820	34	.1110	54	.0550	74	.0225
15	.1800	35	.1100	55	.0520	75	.0210
16	.1770	36	.1065	56	.0465	76	.0200
17	.1730	37	.1040	57	.0430	77	.0180
18	.1695	38	.1015	58	.0420	78	.0160
19	.1660	39	.0995	59	.0410	79	.0145
20	.1610	40	.0980	60	.0400	80	.0135

Decimal Equivalents: Letter-size Drills

Drill	Size of Drill in Inches	Drill	Size of Drill in Inches	Drill	Size of Drill in Inches	Drill	Size of Drill in Inches
A	0.234	H	0.266	O	0.316	V	0.377
B	0.238	I	0.272	P	0.323	W	0.386
C	0.242	J	0.277	Q	0.332	X	0.397
D	0.246	K	0.281	R	0.339	Y	0.404
E	0.250	L	0.290	S	0.348	Z	0.413
F	0.257	M	0.295	T	0.358		
G	0.261	N	0.302	U	0.368		

60° V-Type Thread Dimensions
with Sizes of Tap Drill and Clearance Drill
Fractional Sizes
(National Special Thread Series)

Nominal Size	Threads per Inch	Major Diameter Inches	Minor Diameter Inches	Pitch Diameter Inches	Tap Drill for 75% Thread †	Clearance Drill Size*
1/16″	64	.0625	.0422	.0524	3/64″	51
5/64″	60	.0781	.0563	.0673	1/16″	45
3/32″	48	.0938	.0667	.0803	49	40
7/64″	48	.1094	.0823	.0959	43	32
1/8″	32	.1250	.0844	.1047	3/32″	29
9/64″	40	.1406	.1081	.1244	32	24
5/32″	32	.1563	.1157	.1360	1/8″	19
5/32″	36	.1563	.1202	.1382	30	19
11/64″	32	.1719	.1313	.1516	9/64″	14
3/16″	24	.1875	.1334	.1604	26	8
3/16″	32	.1875	.1469	.1672	22	8
13/64″	24	.2031	.1490	.1760	20	3
7/32″	24	.2188	.1646	.1917	16	1
7/32″	32	.2188	.1782	.1985	12	1
15/64″	24	.2344	.1806	.2073	10	1/4″
1/4″	24	.2500	.1959	.2229	4	17/64″
1/4″	27	.2500	.2019	.2260	3	17/64″
1/4″	32	.2500	.2094	.2297	7/32″	17/64″
5/16″	20	.3125	.2476	.2800	17/64″	21/64″
5/16″	27	.3125	.2644	.2884	J	21/64″
5/16″	32	.3125	.2719	.2922	9/32″	21/64″
3/8″	20	.3750	.3100	.3425	21/64″	25/64″
3/8″	27	.3750	.3269	.3509	R	25/64″
7/16″	24	.4375	.3834	.4104	X	29/64″
7/16″	27	.4375	.3894	.4134	Y	29/64″
1/2″	12	.5000	.3918	.4459	27/64″	33/64″
1/2″	24	.5000	.4459	.4729	29/64″	33/64″
1/2″	27	.5000	.4519	.4759	15/32″	33/64″
9/16″	27	.5625	.5144	.5384	17/32″	37/64″
5/8″	12	.6250	.5168	.5709	35/64″	41/64″
5/8″	27	.6250	.5769	.6009	19/32″	41/64″
11/16″	11	.6875	.5694	.6285	19/32″	45/64″
11/16″	16	.6875	.6063	.6469	5/8″	45/64″
3/4″	12	.7500	.6418	.6959	43/64″	49/64″
3/4″	27	.7500	.7019	.7259	23/32″	49/64″
13/16″	10	.8125	.6826	.7476	23/32″	53/64″
7/8″	12	.8750	.7668	.8209	51/64″	57/64″
7/8″	18**	.8750	.8028	.8389	53/64″	57/64″
7/8″	27	.8750	.8269	.8509	27/32″	57/64″
15/16″	9	.9375	.7932	.8654	53/64″	61/64″
1″	12	1.0000	.8918	.9459	59/64″	1 1/64″
1″	27	1.0000	.9519	.9759	31/32″	1 1/64″
1 5/8″	5 1/2	1.6250	1.3888	1.5069	1 29/64″	1 41/64″
1 7/8″	5	1.8750	1.6152	1.7451	1 11/16″	1 57/64″
2 1/8″	4 1/2	2.1250	1.8363	1.9807	1 29/32″	2 5/32″
2 3/8″	4	2.3750	2.0502	2.2126	2 1/8″	2 13/32″

† Refer to tables "Decimal Equivalents: Number-size Drills" and "Decimal Equivalents: Letter-size Drills."
* Clearance drill makes hole with standard clearance for diameter of nominal size.
** Standard spark plug size.

60° V-Type Thread Dimensions
with Sizes of Tap Drill and Clearance Drill
International Standard (Metric)

Major Diameter (mm)	Pitch (mm)	Minor Diameter (mm)	Pitch Diameter (mm)	Top Drill for 75% Thread (mm)	Tap Drill for 75% Thread † (No. or Inches)	Clearance Drill Size*
2.0	.40	1.48	1.740	1.6	1/16″	41
2.3	.40	1.78	2.040	1.9	48	36
2.6	.45	2.02	2.308	2.1	45	31
3.0	.50	2.35	2.675	2.5	40	29
3.5	.60	2.72	3.110	2.9	33	23
4.0	.70	3.09	3.545	3.3	30	16
4.5	.75	3.53	4.013	3.75	26	10
5.0	.80	3.96	4.480	4.2	19	3
5.5	.90	4.33	4.915	4.6	14	15/64″
6.0	1.00	4.70	5.350	5.0	9	1/4″
7.0	1.00	5.70	6.350	6.0	15/64″	19/64″
8.0	1.25	6.38	7.188	6.8	H	11/32″
9.0	1.25	7.38	8.188	7.8	5/16″	3/8″
10.0	1.50	8.05	9.026	8.6	R	27/64″
11.0	1.50	9.05	10.026	9.6	V	29/64″
12.0	1.75	9.73	10.863	10.5	Z	1/2″
14.0**	1.25	12.38	13.188	13.0	33/64″	9/16″
14.0	2.00	11.40	12.701	12.0	15/32″	9/16″
16.0	2.00	13.40	14.701	14.0	35/64″	21/32″
18.0	1.50	16.05	17.026	16.5	41/64″	47/64″
18.0	2.50	14.75	16.376	15.5	39/64″	47/64″
20.0	2.50	16.75	18.376	17.5	11/16″	13/16″
22.0	2.50	18.75	20.376	19.5	49/64″	57/64″
24.0	3.00	20.10	22.051	21.0	53/64″	31/32″
27.0	3.00	23.10	25.051	24.0	15/16″	1 3/32″
30.0	3.50	25.45	27.727	26.5	1 3/64″	1 13/64″
33.0	3.50	28.45	30.727	29.5	1 11/64″	1 21/64″
36.0	4.00	30.80	33.402	32.0	1 17/64″	1 7/16″
39.0	4.00	33.80	36.402	35.0	1 3/8″	1 9/16″
42.0	4.50	36.15	39.077	37.0	1 29/64″	1 43/64″
45.0	4.50	39.15	42.077	40.0	1 37/64″	1 13/16″
48.0	5.00	41.50	44.752	43.0	1 11/16″	1 29/32″

† Refer to tables "Decimal Equivalents: Number-size Drills" and "Decimal Equivalents: Letter-size Drills."
* Clearance drill makes hole with standard clearance for diameter of nominal size.
** Standard spark plug size.

Tap Drill Sizes

Probable Percentage of Full Thread Produced in Tapped Hole
Using Stock Sizes of Drill

Tap	Tap Drill	Decimal Equivalent of Tap Drill	Theoretical % of Thread	Probable Oversize (Mean)	Probable Hole Size	Percentage of Thread	Tap	Tap Drill	Decimal Equivalent of Tap Drill	Theoretical % of Thread	Probable Oversize (Mean)	Probable Hole Size	Percentage of Thread
0–80	56	.0465	83	.0015	.0480	74	8–32	29	.1360	69	.0029	.1389	62
	3/64	.0469	81	.0015	.0484	71		28	.1405	58	.0029	.1434	51
1–64	54	.0550	89	.0015	.0565	81	8–36	29	.1360	78	.0029	.1389	70
	53	.0595	67	.0015	.0610	59		28	.1405	68	.0029	.1434	57
1–72	53	.0595	75	.0015	.0610	67		9/64	.1406	68	.0029	.1435	57
	1/16	.0625	58	.0015	.0640	50	10–24	27	.1440	85	.0032	.1472	79
2–56	51	.0670	82	.0017	.0687	74		26	.1470	79	.0032	.1502	74
	50	.0700	69	.0017	.0717	62		25	.1495	75	.0032	.1527	69
	49	.0730	56	.0017	.0747	49		24	.1520	70	.0032	.1552	64
2–64	50	.0700	79	.0017	.0717	70		23	.1540	67	.0032	.1572	61
	49	.0730	64	.0017	.0747	56		5/32	.1563	62	.0032	.1595	56
3–48	48	.0760	85	.0019	.0779	78		22	.1570	61	.0032	.1602	55
	5/64	.0781	77	.0019	.0800	70	10–32	5/32	.1563	83	.0032	.1595	75
	47	.0785	76	.0019	.0804	69		22	.1570	81	.0032	.1602	73
	46	.0810	67	.0019	.0829	60		21	.1590	76	.0032	.1622	68
	45	.0820	63	.0019	.0839	56		20	.1610	71	.0032	.1642	64
3–56	46	.0810	78	.0019	.0829	69		19	.1660	59	.0032	.1692	51
	45	.0820	73	.0019	.0839	65	12–24	11/64	.1719	82	.0035	.1754	75
	44	.0860	56	.0019	.0879	48		17	.1730	79	.0035	.1765	73
4–40	44	.0860	80	.0020	.0880	74		16	.1770	72	.0035	.1805	66
	43	.0890	71	.0020	.0910	65		15	.1800	67	.0035	.1835	60
	42	.0935	57	.0020	.0955	51		14	.1820	63	.0035	.1855	56
	3/32	.0938	56	.0020	.0958	50	12–28	16	.1770	84	.0035	.1805	77
4–48	42	.0935	68	.0020	.0955	61		15	.1800	78	.0035	.1835	70
	3/32	.0938	68	.0020	.0958	60		14	.1820	73	.0035	.1855	66
	41	.0960	59	.0020	.0980	52		13	.1850	67	.0035	.1885	59
5–40	40	.0980	83	.0023	.1003	76		3/16	.1875	61	.0035	.1910	54
	39	.0995	79	.0023	.1018	71	1/4–20	9	.1960	83	.0038	.1998	77
	38	.1015	72	.0023	.1038	65		8	.1990	79	.0038	.2028	73
	37	.1040	65	.0023	.1063	58		7	.2010	75	.0038	.2048	70
5–44	38	.1015	79	.0023	.1038	72		13/64	.2031	72	.0038	.2069	66
	37	.1040	71	.0023	.1063	63		6	.2040	71	.0038	.2078	65
	36	.1065	63	.0023	.1088	55		5	.2055	69	.0038	.2093	63
6–32	37	.1040	84	.0023	.1063	78		4	.2090	63	.0038	.2128	57
	36	.1065	78	.0026	.1091	71	1/4–28	3	.2130	80	.0038	.2168	72
	7/64	.1094	70	.0026	.1120	64		7/32	.2188	67	.0038	.2226	59
	35	.1100	69	.0026	.1126	63		2	.2210	63	.0038	.2248	55
	34	.1110	67	.0026	.1136	60	5/16–18	F	.2570	77	.0038	.2608	72
	33	.1130	62	.0026	.1156	55		G	.2610	71	.0041	.2651	66
6–40	34	.1110	83	.0026	.1136	75		17/64	.2656	65	.0041	.2697	59
	33	.1130	77	.0026	.1156	69		H	.2660	64	.0041	.2701	59
	32	.1160	68	.0026	.1186	60							

Tap Drill Sizes
(continued)

Tap	Tap Drill	Decimal Equivalent of Tap Drill	Theoretical % of Thread	Probable Oversize (Mean)	Probable Hole Size	Percentage of Thread	Tap	Tap Drill	Decimal Equivalent of Tap Drill	Theoretical % of Thread	Probable Oversize (Mean)	Probable Hole Size	Percentage of Thread
5/16–24	H	.2660	86	.0041	.2701	78	1"–14	59/64	.9219	84	.0060	.9279	78
	I	.2720	75	.0041	.2761	67		15/16	.9375	67	.0060	.9435	61
	J	.2770	66	.0041	.2811	58	1 1/8–7	31/32	.9688	84	.0062	.9750	81
3/8–16	5/16	.3125	77	.0044	.3169	72		63/64	.9844	76	.0067	.9911	72
	O	.3160	73	.0044	.3204	68		1"	1.0000	67	.0070	1.0070	64
	P	.3230	64	.0044	.3274	59		1 1/64	1.0156	59	.0070	1.0226	55
3/8–24	21/64	.3281	87	.0044	.3325	79	1 1/8–12	1 1/32	1.0313	87	.0071	1.0384	80
	Q	.3320	79	.0044	.3364	71		1 3/64	1.0469	72	.0072	1.0541	66
	R	.3390	67	.0044	.3434	58	1 1/4–7	1 3/32	1.0938	84			
7/16–14	T	.3580	86	.0046	.3626	81		1 7/64	1.1094	76			
	23/64	.3594	84	.0046	.3640	79		1 1/8	1.1250	67			
	U	.3680	75	.0046	.3726	70	1 1/4–12	1 5/32	1.1563	87			
	3/8	.3750	67	.0046	.3796	62		1 11/64	1.1719	72			
	V	.3770	65	.0046	.3816	60	1 3/8–6	1 3/16	1.1875	87			
7/16–20	W	.3860	79	.0046	.3906	72		1 13/64	1.2031	79	No		
	25/64	.3906	72	.0046	.3952	65		1 7/32	1.2188	72	test results		
	X	.3970	62	.0046	.4016	55		1 15/64	1.2344	65	available		
1/2–13	27/64	.4219	78	.0047	.4266	73	1 3/8–12	1 9/32	1.2813	87			
	7/16	.4375	63	.0047	.4422	58		1 19/64	1.2969	72	Reaming		
1/2–20	29/64	.4531	72	.0047	.4578	65	1 1/2–6	1 5/16	1.3125	87	recommended		
9/16–12	15/32	.4688	87	.0048	.4736	82		1 21/64	1.3281	79			
	31/64	.4844	72	.0048	.4892	68		1 11/32	1.3438	72			
9/16–18	1/2	.5000	87	.0048	.5048	80		1 23/64	1.3594	65			
	33/64	.5156	65	.0048	.5204	58	1 1/2–12	1 13/32	1.4063	87			
5/8–11	17/32	.5313	79	.0049	.5362	75		1 27/64	1.4219	72			

	35/64	.5469	66	.0049	.5518	62
5/8–18	9/16	.5625	87	.0049	.5674	80
	37/64	.5781	65	.0049	.5831	58
3/4–10	41/64	.6406	84	.0050	.6456	80
	21/32	.6563	72	.0050	.6613	68
3/4–16	11/16	.6875	77	.0050	.6925	71
7/8–9	49/64	.7656	76	.0052	.7708	72
	25/32	.7812	65	.0052	.7864	61
7/8–14	51/64	.7969	84	.0052	.8021	79
	13/16	.8125	67	.0052	.8177	62
1"–8	55/64	.8594	87	.0059	.8653	83
	7/8	.8750	77	.0059	.8809	73
	57/64	.8906	67	.0059	.8965	64
	29/32	.9063	58	.0059	.9122	54
1"–12	29/32	.9063	87	.0060	.9123	81
	59/64	.9219	72	.0060	.9279	67
	15/16	.9375	58	.0060	.9435	52

Taper Pipe		Straight Pipe	
Thread	Drill	Thread	Drill
1/8–27	R	1/8–27	S
1/4–18	7/16	1/4–18	29/64
3/8–18	37/64	3/8–18	19/32
1/2–14	23/32	1/2–14	47/64
3/4–14	59/64	3/4–14	15/16
1–11 1/2	1 5/32	1–11 1/2	1 3/16
1 1/4–11 1/2	1 1/2	1 1/4–11 1/2	1 33/64
1 1/2–11 1/2	1 47/64	1 1/2–11 1/2	1 3/4
2–11 1/2	2 7/32	2–11 1/2	2 7/32
2 1/2–8	2 5/8	2 1/2–8	2 21/32
3–8	3 1/4	3–8	3 9/32
3 1/2–8	3 3/4	3 1/2–8	3 25/32
4–8	4 1/4	4–8	4 9/32

(Standard Tool Co.)

Taper Pin and Reamer Sizes

Length of Pin	Reamer															
	6/0	5/0	4/0	3/0	2/0	0	1	2	3	4	5	6	7	8	9	10
3/8	50	44	38	32	29											
1/2	51	45	39	33	30	27										
5/8	52	46	41	34	30	27	21									
3/4	1/16	47	42	7/64	1/8	9/64	5/32	16	13/64	15/64						
1		49	44	37	31	29	25	11/64	9	1	I	P	W			
1 1/4					32	30	26	19	10	2	H	O	V	29/64		
1 1/2						*1/8	*9/64	20	3/16	7/32	G	5/16	V	29/64	35/64	43/64
1 3/4						*31	*29	*5/32	14	3	F	N	U	29/64	35/64	21/32
2						*33	*30	*25	*16	4	E	N	U	7/16	35/64	21/32
2 1/4						*7/64	*1/8	*26	*19	*11/64	D	M	23/64	7/16	17/32	21/32
2 1/2						*37	*31	*28	*21	*13/64	C	M	T	7/16	17/32	41/64
2 3/4						*40	*33	*29	*5/32	*8	*B	L	S	27/64	17/32	41/64
3						*42	*35	*30	*24	*11	*1	9/32	11/32	27/64	33/64	41/64
3 1/4									*26	*3/16	*2	J	R	27/64	33/64	5/8
3 1/2										*14	*2	*I	Q	Z	33/64	5/8
3 3/4										*16	*3	*H	*Q	Z	1/2	5/8
4										*17	*4	*G	*P	13/32	1/2	39/64
4 1/4										*19	*5	*F	*O	Y	1/2	39/64
4 1/2												*1/4	*O	X	31/64	39/64
4 3/4												*D	*5/16	*25/64	31/64	19/32
5												*C	*N	*W	31/64	19/32
5 1/4												*B			15/32	19/32
5 1/2															*15/32	37/64
5 3/4															*15/32	37/64
6															*29/64	37/64

* Hole sizes too small to admit taper pin reamers of standard length. Special, extra-length reamers are required for these cases.

Hardness Conversions

Brinell Indentation Diameter, (mm)	Brinell Hardness Number		Rockwell Hardness Number		Rockwell Superficial Hardness Number (Superficial Diamond Penetrator)			Tensile Strength (Approximate) 1000 psi
	Standard Ball	Tungsten-carbide Ball	B Scale	C Scale	15 N Scale	30 N Scale	45 N Scale	
2.45	—	627	—	58.7	89.6	76.3	65.1	347
2.50	—	601	—	57.3	89.0	75.1	63.5	328
2.55	—	578	—	56.0	88.4	73.9	62.1	313
2.60	—	555	—	54.7	87.8	72.7	60.6	298
2.65	—	534	—	53.5	87.2	71.6	59.2	288
2.70	—	514	—	52.1	86.5	70.3	57.6	274
2.75	—	495	—	51.0	85.9	69.4	56.1	264
2.80	—	477	—	49.6	85.3	68.2	54.5	252
2.85	—	461	—	48.5	84.7	67.2	53.2	242
2.90	—	444	—	47.1	84.0	65.8	51.5	230
2.95	429	429	—	45.7	83.4	64.6	49.9	219
3.00	415	415	—	44.5	82.8	63.5	48.4	212
3.05	401	401	—	43.1	82.0	62.3	46.9	202
3.10	388	388	—	41.8	81.4	61.1	45.3	193
3.15	375	375	—	40.4	80.6	59.9	43.6	184
3.20	363	363	—	39.1	80.0	58.7	42.0	177
3.25	352	352	—	37.9	79.3	57.6	40.5	170
3.30	341	341	—	36.6	78.6	56.4	39.1	163
3.35	331	331	—	35.5	78.0	55.4	37.8	158
3.40	321	321	—	34.3	77.3	54.3	36.4	152
3.45	311	311	—	33.1	76.7	53.3	34.4	147
3.50	302	302	—	32.1	76.1	52.2	33.8	143
3.55	293	293	—	30.9	75.5	51.2	32.4	139
3.60	285	285	—	29.9	75.0	50.3	31.2	136
3.65	277	277	—	28.8	74.4	49.3	29.9	131
3.70	269	269	—	27.6	73.7	48.3	28.5	128
3.75	262	262	—	26.6	73.1	47.3	27.3	125
3.80	255	255	—	25.4	72.5	46.2	26.0	121
3.85	248	248	—	24.2	71.7	45.1	24.5	118
3.90	241	241	100.0	22.8	70.9	43.9	22.8	114
3.95	235	235	99.0	21.7	70.3	42.9	21.5	111
4.00	229	229	98.2	20.5	69.7	41.9	20.1	109
4.05	223	223	97.3	—	—	—	—	104
4.10	217	217	96.4	—	—	—	—	103
4.15	212	212	95.5	—	—	—	—	100
4.20	207	207	94.6	—	—	—	—	99
4.25	201	201	93.8	—	—	—	—	97
4.30	197	197	92.8	—	—	—	—	94
4.35	192	192	91.9	—	—	—	—	92
4.40	187	187	90.7	—	—	—	—	90
4.45	183	183	90.0	—	—	—	—	89
4.50	179	179	89.0	—	—	—	—	88
4.55	174	174	87.8	—	—	—	—	86
4.60	170	170	86.8	—	—	—	—	84
4.65	167	167	86.0	—	—	—	—	83
4.70	163	163	85.0	—	—	—	—	82
4.80	156	156	82.9	—	—	—	—	80
4.90	149	149	80.8	—	—	—	—	73
5.00	143	143	78.7	—	—	—	—	71
5.10	137	137	76.4	—	—	—	—	67
5.20	131	131	74.0	—	—	—	—	65
5.30	126	126	72.0	—	—	—	—	63
5.40	121	121	69.0	—	—	—	—	60
5.50	116	116	67.6	—	—	—	—	58
5.60	111	111	65.7	—	—	—	—	56

(Carpenter Steel Co.)

Conversion Table: US Conventional to SI Metric

When You Know ⬇	Multiply By:		To Find ⬇
	Very Accurate	**Approximate**	
Length			
inches	* 25.4		millimeters
inches	* 2.54		centimeters
feet	* 0.3048		meters
feet	* 30.48		centimeters
yards	* 0.9144	0.9	meters
miles	* 1.609344	1.6	kilometers
Weight			
grains	15.43236	15.4	grams
ounces	* 28.349523125	28.0	grams
ounces	* 0.028349523125	.028	kilograms
pounds	* 0.45359237	0.45	kilograms
short ton	* 0.90718474	0.9	tonnes
Volume			
teaspoons		5.0	milliliters
tablespoons		15.0	milliliters
fluid ounces	29.57353	30.0	milliliters
cups		0.24	liters
pints	* 0.473176473	0.47	liters
quarts	* 0.946352946	0.95	liters
gallons	* 3.785411784	3.8	liters
cubic inches	* 0.016387064	0.02	liters
cubic feet	* 0.028316846592	0.03	cubic meters
cubic yards	* 0.764554857984	0.76	cubic meters
Area			
square inches	* 6.4516	6.5	square centimeters
square feet	* 0.09290304	0.09	square meters
square yards	* 0.83612736	0.8	square meters
square miles		2.6	square kilometers
acres	* 0.40468564224	0.4	hectares
Temperature			
Fahrenheit	* 5/9 (after subtracting 32)		Celsius

* = Exact

Conversion Table: SI Metric to US Conventional

When You Know ⬇	Multiply By:		To Find ⬇
	Very Accurate	**Approximate**	
Length			
millimeters	0.0393701	0.04	inches
centimeters	0.3937008	0.4	inches
meters	3.280840	3.3	feet
meters	1.093613	1.1	yards
kilometers	0.621371	0.6	miles
Weight			
grains	0.00228571	0.0023	ounces
grams	0.03527396	0.035	ounces
kilograms	2.204623	2.2	pounds
tonnes	1.1023113	1.1	short tons
Volume			
milliliters		0.2	teaspoons
milliliters	0.06667	0.067	tablespoons
milliliters	0.03381402	0.03	fluid ounces
liters	61.02374	61.024	cubic inches
liters	2.113376	2.1	pints
liters	1.056688	1.06	quarts
liters	0.26417205	0.26	gallons
liters	0.03531467	0.035	cubic feet
cubic meters	61023.74	61023.7	cubic inches
cubic meters	35.31467	35.0	cubic feet
cubic meters	1.3079506	1.3	cubic yards
cubic meters	264.17205	264.0	gallons
Area			
square centimeters	0.1550003	0.16	square inches
square centimeters	0.00107639	0.001	square feet
square meters	10.76391	10.8	square feet
square meters	1.195990	1.2	square yards
square kilometers		0.4	square miles
hectares	2.471054	2.5	acres
Temperature			
Celsius	* 9/5 (then add 32)		Fahrenheit

* = Exact

DECIMAL CONVERSION CHART

FRACTION		INCHES	M/M	FRACTION		INCHES	M/M
	1/64	.01563	.397		33/64	.51563	13.097
1/32		.03125	.794	17/32		.53125	13.494
	3/64	.04688	1.191		35/64	.54688	13.891
1/16		.0625	1.588	9/16		.56250	14.288
	5/64	.07813	1.984		37/64	.57813	14.684
3/32		.09375	2.381	19/32		.59375	15.081
	7/64	.10938	2.778		39/64	.60938	15.478
1/8		.12500	3.175	5/8		.62500	15.875
	9/64	.14063	3.572		41/64	.64063	16.272
5/32		.15625	3.969	21/32		.65625	16.669
	11/64	.17188	4.366		43/64	.67188	17.066
3/16		.18750	4.763	11/16		.68750	17.463
	13/64	.20313	5.159		45/64	.70313	17.859
7/32		.21875	5.556	23/32		.71875	18.256
	15/64	.23438	5.953		47/64	.73438	18.653
1/4		.25000	6.350	3/4		.75000	19.050
	17/64	.26563	6.747		49/64	.76563	19.447
9/32		.28125	7.144	25/32		.78125	19.844
	19/64	.29688	7.541		51/64	.79688	20.241
5/16		.31250	7.938	13/16		.81250	20.638
	21/64	.32813	8.334		53/64	.82813	21.034
11/32		.34375	8.731	27/32		.84375	21.431
	23/64	.35938	9.128		55/64	.85938	21.828
3/8		.37500	9.525	7/8		.87500	22.225
	25/64	.39063	9.922		57/64	.89063	22.622
13/32		.40625	10.319	29/32		.90625	23.019
	27/64	.42188	10.716		59/64	.92188	23.416
7/16		.43750	11.113	15/16		.93750	23.813
	29/64	.45313	11.509		61/64	.95313	24.209
15/32		.46875	11.906	31/32		.96875	24.606
	31/64	.48438	12.303		63/64	.98438	25.003
1/2		.50000	12.700	1		1.00000	25.400

Grade Marking for Bolts

Bolt Head Marking	SAE = Society of Automotive Engineers ASTM = American Society for Testing and Materials	Bolt Material	Minimum Tensile Strength in Pounds per Square Inch (psi)
No Marks	SAE Grade 1 SAE Grade 2 Indeterminate quality	Low-carbon steel Low-carbon steel	65,000 psi
2 Marks	SAE Grade 3	Medium-carbon steel, cold worked	110,000 psi
3 Marks	SAE Grade 5 ASTM – A 325 Common commercial quality	Medium-carbon steel, quenched and tempered	120,000 psi
Letters BB	ASTM – A 354	Low-alloy steel or medium-carbon steel, quenched and tempered	105,000 psi
Letters BC	ASTM – A 354	Low-alloy steel or medium-carbon steel, quenched and tempered	125,000 psi
4 Marks	SAE Grade 6 Better commercial quality	Medium-carbon steel, quenched and tempered	140,000 psi
5 Marks	SAE Grade 7	Medium-carbon alloy steel, quenched and tempered, roll-threaded after heat treatment	133,000 psi
6 Marks	SAE Grade 8 ASTM – A 345 Best commercial quality	Medium-carbon alloy steel, quenched and tempered	150,000 psi

Machine Screw and Cap Screw Heads

	Size	A	B	C	D
Fillister Head	#8	.260	.141	.042	.060
	#10	.302	.164	.048	.072
	1/4	3/8	.205	.064	.087
	5/16	7/16	.242	.077	.102
	3/8	9/16	.300	.086	.125
	1/2	3/4	.394	.102	.168
	5/8	7/8	.500	.128	.215
	3/4	1	.590	.144	.258
	1	1 5/16	.774	.182	.352
Flat Head	#8	.320	.092	.043	.037
	#10	.372	.107	.048	.044
	1/4	1/2	.146	.064	.063
	5/16	5/8	.183	.072	.078
	3/8	3/4	.220	.081	.095
	1/2	7/8	.220	.102	.090
	5/8	1 1/8	.293	.128	.125
	3/4	1 3/8	.366	.144	.153
Round Head	#8	.297	.113	.044	.067
	#10	.346	.130	.048	.073
	1/4	7/16	.183	.064	.107
	5/16	9/16	.236	.072	.150
	3/8	5/8	.262	.081	.160
	1/2	13/16	.340	.102	.200
	5/8	1	.422	.128	.255
	3/4	1 1/4	.526	.144	.320
Hexagon Head	1/4	.494	.170	7/16	
	5/16	.564	.215	1/2	
	3/8	.635	.246	9/16	
	1/2	.846	.333	3/4	
	5/8	1.058	.411	15/16	
	3/4	1.270	.490	1 1/8	
	7/8	1.482	.566	1 5/16	
	1	1.693	.640	1 1/2	
Socket Head	#8	.265	.164	1/8	
	#10	5/16	.190	5/32	
	1/4	3/8	1/4	3/16	
	5/16	7/16	5/16	7/32	
	3/8	9/16	3/8	5/16	
	7/16	5/8	7/16	5/16	
	1/2	3/4	1/2	3/8	
	5/8	7/8	5/8	1/2	
	3/4	1	3/4	9/16	
	7/8	1 1/8	7/8	9/16	
	1	1 5/16	1	5/8	

Standard System of Marking
for Threading Tools

Taps, dies, and other threading tools will be marked with the nominal size, number of threads per inch, and the proper symbol to identify the thread form. These symbols are in agreement with the A.S.A. B1-7-1949 *Standard on Nomenclature, Definitions, and Letter Symbols for Screw Threads.*

The markings for the British Threads are the fully abbreviated form based on the data in the British Standard Institute Specification No. 84-1940.

Symbols Used for American Threads

Symbol	Reference	Symbol	Reference
NC	American National Coarse Thread Series	NPSI	Dryseal American (National) Intermediate Internal Straight Pipe Thread
NF	American National Fine Thread Series		
NEF	American National Extra Fine Thread Series	NPSL	American (National) Internal Straight Pipe Thread for Locknut Connections (Loose Fitting Mechanical Joints)
N	American National 8, 12, and 16 Thread Series (8N, 12N, 16N)		
NH	American (National) Hose Coupling and Fire Hose Coupling Threads	NPSM	American (National) Internal Straight Pipe Thread for Mechanical Joints (Free Fitting)
NM	National Miniature Screw Thread	NPTR	American (National) Internal Taper Pipe Thread for Railing Joints (Mechanical Joints)
NGO	American (National) Gas Outlet Thread		
NS	American Special Thread (60° Thread Form)	AMO	American Standard Microscope Objective Thread
NPT	American (National) Taper Pipe Thread	ACME C	Acme Screw Thread—Centralizing Type
NPTF	Dryseal American (National) Taper Pipe Thread	ACME G	Acme Screw Thread—General Purpose Type
PTF	Dryseal SAE Short Internal Taper Pipe Thread	STUB ACME	Stub Acme Threads
ANPT	Military Aeronautical Pipe Thread Specification MIL-P-7105	N. BUTT	National Buttress Screw Thread
NPS	American (National) Straight Pipe Thread	V	A 60° "V" Thread with Truncated Crests and Roots. The Theoretical "V" Form is usually flatted several thousandths of an inch to the users specifications.
NPSC	American (National) Straight Pipe Thread in Pipe Couplings		
NPSF	Dryseal American (National) Fuel Internal Straight Pipe Thread	SB	Manufacturers Stovebolt Standard Thread
NPSH	American (Standard) Straight Pipe Thread for Hose Couplings and Nipples	STI	Special Threads for Helical Coil Wire Screw Thread Inserts

Symbols Used for British Threads

Symbol	Reference	Symbol	Reference
BSW	British Standard Whitworth Coarse Thread Series	BSPP	British Standard Pipe (Parallel) Thread
BSF	British Standard Fine Thread Series	WHIT	Whitworth Standard Special Thread
BSP	British Standard Taper Pipe Thread	BA	British Association Standard Thread

Bent Shank Taper Taps
In addition to the regular marking, bent shank taper taps will be marked with the table number to which they are made.

Ground Thread Taps (Limit Numbers)
All standard Ground Thread Taps made to Tables 327 and 329 will be marked with the letter G to designate Ground Thread. The letter G will be followed by the letter H to designate above basic (L below basic) and a numeral to designate the Pitch Diameter limits.

Example: G H3 Indicates a Ground Thread Tap with Pitch Diameter limits .0010 to .0015 in. over basic.

Pitch Diameter limits for Taps to 1″ diameter inclusive:

L1 = Basic to Basic minus .0005
H1 = Basic to Basic plus .0005
H2 = Basic plus .0005 to Basic plus .0010
H3 = Basic plus .0010 to Basic plus .0015

H4 = Basic plus .0015 to Basic plus .0020
H5 = Basic plus .0020 to Basic plus .0025
H6 = Basic plus .0025 to Basic plus .0030

(Morse Twist Drill & Machine Co.)

Screw Thread Elements for Unified and National Thread Form

Threads per Inch (n)	Pitch (p) $p = \dfrac{1}{n}$	Single Height Subtract from basic major diameter to get basic pitch diameter	Double Height Subtract from basic major diameter to get basic minor diameter	83 1/3% Double Height Subtract from basic major diameter to get minor diameter of ring gage	Basic Width of Crest and Root Flat $\dfrac{p}{8}$	Constant for Best Size Wire Also Single Height of 60° V-Thread	Diameter of Best Size Wire
3	.333333	.216506	.43301	.36084	.0417	.28868	.19245
3 1/4	.307692	.199852	.39970	.33309	.0385	.26647	.17765
3 1/2	.285714	.185577	.37115	.30929	.0357	.24744	.16496
4	.250000	.162379	.32476	.27063	.0312	.21651	.14434
4 1/2	.222222	.144337	.28867	.24056	.0278	.19245	.12830
5	.200000	.129903	.25981	.21650	.0250	.17321	.11547
5 1/2	.181818	.118093	.23619	.19682	.0227	.15746	.10497
6	.166666	.108253	.21651	.18042	.0208	.14434	.09623
7	.142857	.092788	.18558	.15465	.0179	.12372	.08248
8	.125000	.081189	.16238	.13531	.0156	.10825	.07217
9	.111111	.072168	.14434	.12028	.0139	.09623	.06415
10	.100000	.064952	.12990	.10825	.0125	.08660	.05774
11	.090909	.059046	.11809	.09841	.0114	.07873	.05249
11 1/2	.086956	.056480	.11296	.09413	.0109	.07531	.05020
12	.083333	.054127	.10826	.09021	.0104	.07217	.04811
13	.076923	.049963	.09993	.08327	.0096	.06662	.04441
14	.071428	.046394	.09279	.07732	.0089	.06186	.04124
16	.062500	.040595	.08119	.06766	.0078	.05413	.03608
18	.055555	.036086	.07217	.06014	.0069	.04811	.03208
20	.050000	.032475	.06495	.05412	.0062	.04330	.02887
22	.045454	.029523	.05905	.04920	.0057	.03936	.02624
24	.041666	.027063	.05413	.04510	.0052	.03608	.02406
27	.037037	.024056	.04811	.04009	.0046	.03208	.02138
28	.035714	.023197	.04639	.03866	.0045	.03093	.02062
30	.033333	.021651	.04330	.03608	.0042	.02887	.01925
32	.031250	.020297	.04059	.03383	.0039	.02706	.01804
36	.027777	.018042	.03608	.03007	.0035	.02406	.01604
40	.025000	.016237	.03247	.02706	.0031	.02165	.01443
44	.022727	.014761	.02952	.02460	.0028	.01968	.01312
48	.020833	.013531	.02706	.02255	.0026	.01804	.01203
50	.020000	.012990	.02598	.02165	.0025	.01732	.01155
56	.017857	.011598	.02320	.01933	.0022	.01546	.01031
60	.016666	.010825	.02165	.01804	.0021	.01443	.00962
64	.015625	.010148	.02030	.01691	.0020	.01353	.00902
72	.013888	.009021	.01804	.01503	.0017	.01203	.00802
80	.012500	.008118	.01624	.01353	.0016	.01083	.00722
90	.011111	.007217	.01443	.01202	.0014	.00962	.00642
96	.010417	.006766	.01353	.01127	.0013	.00902	.00601
100	.010000	.006495	.01299	.01082	.0012	.00866	.00577
120	.008333	.005413	.01083	.00902	.0010	.00722	.00481

Note: Using the Best Size Wires, measurement over three wires minus Constant for Best Size Wire equals Pitch Diameter.

BOLT TORQUING CHART

METRIC STANDARD						SAE STANDARD/FOOT POUNDS					
Grade of Bolt	5D	.8G	10K	12K		Grade of Bolt	SAE 1 & 2	SAE 5	SAE 6	SAE 8	
Min. Tensile Strength	71,160 P.S.I.	113,800 P.S.I.	142,200 P.S.I.	170,679 P.S.I.		Min. Tensile Strength	64,000 P.S.I.	105,000 P.S.I.	133,000 P.S.I.	150,000 P.S.I.	
Grade Markings on Head	5D	8G	10K	12K	Size of Socket or Wrench Opening	Markings on Head					Size of Socket or Wrench Opening
Metric		Foot Pounds				Metric	U.S. Standard	Foot Pounds			U.S. Standard
Bolt Dia.	U.S. Dec. Equiv.					Bolt Head	Bolt Dia.					Bolt Head	Nut
6mm	.2362	5	6	8	10	10mm	1/4	5	7	10	10.5	3/8	7/16
8mm	.3150	10	16	22	27	14mm	5/16	9	14	19	22	1/2	9/16
10mm	.3937	19	31	40	49	17mm	3/8	15	25	34	37	9/16	5/8
12mm	.4720	34	54	70	86	19mm	7/16	24	40	55	60	5/8	3/4
14mm	.5512	55	89	117	137	22mm	1/2	37	60	85	92	3/4	13/16
16mm	.6299	83	132	175	208	24mm	9/16	53	88	120	132	7/8	7/8
18mm	.709	111	182	236	283	27mm	5/8	74	120	167	180	15/16	1.
22mm	.8661	182	284	394	464	32mm	3/4	120	200	280	296	1-1/8	1-1/8

Cutting Fluids for Various Metals

Aluminum and its Alloys	Kerosene, kerosene and lard oil, soluble oil
Plastics	Dry
Brass, Soft	Dry, soluble oil, kerosene and lard oil
Bronze, High Tensile	Soluble oil, lard oil, mineral oil, dry
Cast Iron	Dry, air jet, soluble oil
Copper	Soluble oil, dry, mineral lard oil, kerosene
Magnesium	Low viscosity neutral oils
Malleable Iron	Dry, soda water
Monel Metal	Lard oil, soluble oil
Slate	Dry
Steel, Forging	Soluble oil, sulfurized oil, mineral lard oil
Steel, Manganese	Soluble oil, sulfurized oil, mineral lard oil
Steel, Soft	Soluble oil, mineral lard oil, sulfurized oil, lard oil
Steel, Stainless	Sulfurized mineral oil, soluble oil
Steel, Tool	Soluble oil, mineral lard oil, sulfurized oil
Wrought Iron	Soluble oil, mineral lard oil, sulfurized oil

EIA and AIA National Codes for CNC Programming

Preparatory (G) Functions	
G word	**Explanation**
G00	Denotes a rapid traverse rate for point-to-point positioning.
G01	Describes linear interpolation blocks; reserved for contouring.
G02, G03	Used with circular interpolation.
G04	Sets a calculated time delay during which there is no machine motion (dwell).
G05, G07	Unassigned by the EIA. May be used at the discretion of the machine tool or system builder. Could also be standardized at a future date.
G06	Used with parabolic interpolation.
G08	Acceleration code. Causes the machine, assuming it is capable, to accelerate at a smooth exponential rate.
G09	Deceleration code. Causes the machine, assuming it is capable, to decelerate at a smooth exponential rate.
G10-G12	Normally unassigned for CNC systems. Used with some hard-wired systems to express blocks of abnormal dimensions.
G13-G16	Direct the control system to operate on a particular set of axes.
G17-G19	Identify or select a coordinate plane for such functions as circular interpolation or cutter compensation.
G20-G32	Unassigned according to EIA standards. May be assigned by the control system or machine tool builder.
G33-G35	Selected for machines equipped with thread-cutting capabilities (generally referring to lathes). G33 is used when a constant lead is sought, G34 is used when a constantly increasing lead is required, and G35 is used to designate a constantly decreasing lead.
G36-G39	Unassigned.
G40	Terminates any cutter compensation.
G41	Activates cutter compensation in which the cutter is on the left side of the work surface (relative to the direction of the cutter motion).
G42	Activates cutter compensation in which the cutter is on the right side of the work surface.
G43, G44	Used with cutter offset to adjust for the difference between the actual and programmed cutter radii or diameters. G43 refers to an inside corner, and G44 refers to an outside corner.
G45-G49	Unassigned.
G50-G59	Reserved for adaptive control.
G60-G69	Unassigned.
G70	Selects inch programming.
G71	Selects metric programming.
G72	Selects three-dimensional CW circular interpolation.
G73	Selects three-dimensional CCW circular interpolation.
G74	Cancels multiquadrant circular interpolation.
G75	Activates multiquadrant circular interpolation.
G76-G79	Unassigned.
G80	Cancel cycle.
G81	Activates drill, or spotdrill, cycle.
G82	Activates drill with a dwell.
G83	Activates intermittent, or deep-hole, drilling.
G84	Activates tapping cycle.
G85-G89	Activates boring cycles.
G90	Selects absolute input. Input data is to be in absolute dimensional form.

EIA and AIA National Codes for CNC Programming (continued)

Preparatory (G) Functions

G word	Explanation
G91	Selects incremental input. Input data is to be in incremental form.
G92	Preloads registers to desired values (for example, preloads axis position registers).
G93	Sets inverse time feed rate.
G94	Sets inches (or millimeters) per minute feed rate.
G95	Sets inches (or millimeters) per revolution feed rate.
G97	Sets spindle speed in revolutions per minute.
G98, G99	Unassigned.

Miscellaneous (M) Functions

M word	Explanation
M00	Program stop. Operator must cycle start in order to continue with the remainder of the program.
M01	Optional stop. Acted upon only when the operator has previously signaled for this command by pushing a button. When the control system senses the M01 code, machine will automatically stop.
M02	End of program. Stops the machine after completion of all commands in the block. May include rewinding of tape.
M03	Starts spindle rotation in a clockwise direction.
M04	Starts spindle rotation in a counterclockwise direction.
M05	Spindle stop.
M06	Executes the change of a tool (or tools) manually or automatically.
M07	Turns coolant on (flood).
M08	Turns coolant on (mist).
M09	Turns coolant off.
M10	Activates automatic clamping of the machine slides, workpiece, fixture, spindle, etc.
M11	Deactivates automatic clamping.
M12	Inhibiting code used to synchronize multiple set of axes, such as a four-axis lathe that has two independently operated heads or slides.
M13	Combines simultaneous clockwise spindle motion and coolant on.
M14	Combines simultaneous counterclockwise spindle motion and coolant on.
M15	Sets rapid traverse or feed motion in the + direction.
M16	Sets rapid traverse or feed motion in the − direction.
M17, M18	Unassigned.
M19	Oriented spindle stop. Stops spindle at a predetermined angular position.
M20-M29	Unassigned.
M30	End of data. Used to reset control and/or machine.
M31	Interlock bypass. Temporarily circumvents a normally provided interlock.
M32-M39	Unassigned.
M40-M46	Signals gear changes if required at the machine; otherwise, unassigned.
M47	Continues program execution from the start of the program, unless inhibited by an interlock signal.
M48	Cancels M49.
M49	Deactivates a manual spindle or feed-override and returns to the programmed value.

EIA and AIA National Codes for CNC Programming (continued)

Miscellaneous (M) Functions

M word	Explanation
M50-M57	Unassigned.
M58	Cancels M59.
M59	Holds the rpm constant at its value.
M60-M99	Unassigned.

Other Address Characters

Address character	Explanation
A	Angular dimension about the X-axis.
B	Angular dimension about the Y-axis.
C	Angular dimension about the Z-axis.
D	Can be used for an angular dimension around a special axis, for a third feed function, or for tool offset.
E	Used for angular dimension around a special axis or for a second feed function.
H	Fixture offset.
I, J, K	Centerpoint coordinates for circular interpolation.
L	Not used.
O	Used on some N/C controls in place of the customary sequence number word address N.
P	Third rapid traverse code—tertiary motion dimension parallel to the X-axis.
Q	Second rapid traverse code—tertiary motion dimension parallel to the Y-axis.
R	First rapid traverse code—tertiary motion dimension parallel to the Z-axis (or to the radius) for constant surface speed calculation.
U	Secondary motion dimension parallel to the X-axis.
V	Secondary motion dimension parallel to the Y-axis.
W	Secondary motion dimension parallel to the Z-axis.

Powder Metallurgy Processes

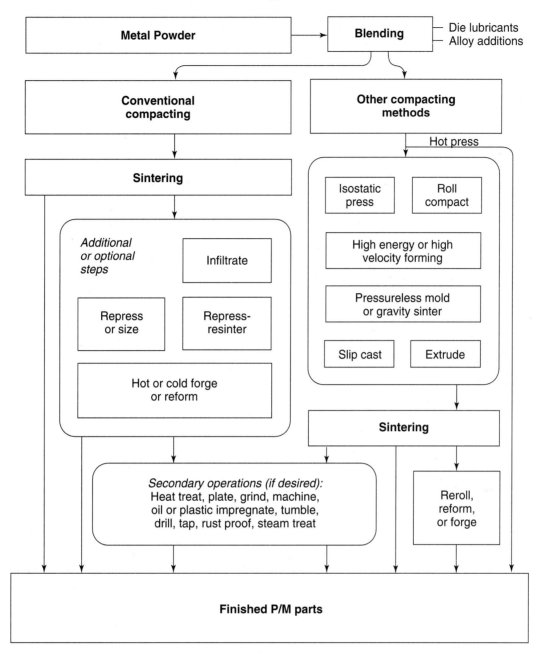

Sintering   Sintering is a solid-state phenomenon in which powdered metal particles become metallurgically bonded below the melting point of the metal. No adhesives or cements are used.

Infiltration   Pores of P/M parts are filled with a lower-melting-point metal such as copper-based alloy. When the part is sintered, the infiltrant material melts and penetrates into the P/M part by capillary action.

Coining and sizing   Basically, this operation involves repressing sintered parts in a die similar to the original compacting die.

Impregnation   The pores of the P/M part are filled with a lubricant or other nonmetallic such as plastic resin. This may be done by means of a vacuum or by soaking. The part then becomes self-lubricating (or pressure-tight if resin is used).

Master Chart of Welding and Allied Processes

Atomic hydrogen welding.......	AHW
Bare metal arc welding............	BMAW
Carbon arc welding.................	CAW
-gas................................	CAW-G
-shielded.........................	CAW-S
-twin...............................	CAW-T
Electrogas welding.................	EGW
Flux cored arc welding............	FCAW

Coextrusion welding.............	CEW
Cold welding........................	CW
Diffusion welding.................	DFW
Explosion welding.................	EXW
Forge welding......................	FOW
Friction welding...................	FRW
Hot pressure welding............	HPW
Roll welding........................	ROW
Ultrasonic welding................	USW

Dip soldering........................	DS
Furnace soldering.................	FS
Induction soldering...............	IS
Infrared soldering.................	IRS
Iron soldering.......................	INS
Resistance soldering..............	RS
Torch soldering.....................	TS
Ultrasonic soldering..............	USS
Wave soldering.....................	WS

Flash welding........................	FW
Projection welding.................	PW
Resistance seam welding.....	RSEW
-high frequency.............	RSEW-HF
-induction......................	RSEW-I
Resistance spot welding.......	RSW
Upset welding......................	UW
-high frequency.............	UW-HF
-induction......................	UW-I

Arc spraying.........................	ASP
Flame spraying.....................	FLSP
Plasma spraying..................	PSP

Flux cutting...........................	FOC
Metal powder cutting............	POC
Oxyfuel gas cutting..............	OFC
-oxyacetylene cutting......	OFC-A
-oxyhydrogen cutting.....	OFC-H
-oxynatural gas cutting..	OFC-N
-oxypropane cutting.......	OFC-P
Oxygen arc cutting..............	AOC
Oxygen lance cutting...........	LOC

Gas metal arc welding..............	GMAW
-pulsed arc......................	GMAW-P
-short circuiting arc............	GMAW-S
Gas tungsten arc welding..........	GTAW
-pulsed arc......................	GTAW-P
Plasma arc welding..................	PAW
Shielded metal arc welding........	SMAW
Arc stud welding.....................	SW
Submerged arc welding............	SAW
-series...............................	SAW-S

Block brazing...........................	BB
Diffusion brazing......................	CAB
Dip brazing.............................	DB
Exothermic brazing...................	EXB
Flow brazing............................	FLOW
Furnace brazing.......................	FB
Induction brazing......................	IB
Infrared brazing........................	IRB
Resistance brazing...................	RB
Torch brazing...........................	TB
Twin carbon arc brazing............	TCAB

Electron beam welding.....	EBW
-high vacuum.............	EBW-HV
-medium vacuum.......	EBW-MV
-nonvacuum..............	EBW-NV
Electroslag welding..........	ESW
Flow welding..................	FLOW
Induction welding............	IW
Laser beam welding.........	LBW
Percussion welding..........	PEW
Thermite welding.............	TW

Air acetylene welding......	AAW
Oxyacetylene welding.....	OAW
Oxyhydrogen welding.....	OHW
Pressure gas welding.....	PGW

Air carbon arc cutting........	CAC-A
Carbon arc cutting............	CAC
Gas metal arc cutting........	GMAC
Gas tungsten arc cutting...	GTAC
Plasma arc cutting...........	PAC
Shielded metal arc cutting.	SMAC

Electron beam cutting........	EBC
Laser beam cutting............	LBC
-air.............................	LBC-A
-evaporative...............	LBC-EV
-inert gas...................	LBC-IG
-oxygen.....................	LBC-O

Diagram circles: Arc welding (AW); Solid-state welding (SSW); Brazing (B); Soldering (S); Welding processes; Other welding; Resistance welding (RW); Allied processes; Oxyfuel gas welding (OFW); Thermal spraying (THSP); Allied processes; Oxygen cutting (OC); Thermal cutting (TC); Arc cutting (AC); Other cutting

(American Welding Society)

Grinding Wheel Markings

32A46-H8VBE

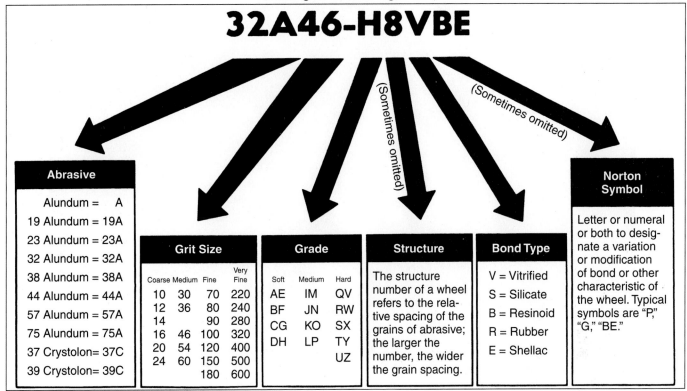

Abrasive		
Alundum =		A
19 Alundum =		19A
23 Alundum =		23A
32 Alundum =		32A
38 Alundum =		38A
44 Alundum =		44A
57 Alundum =		57A
75 Alundum =		75A
37 Crystolon=		37C
39 Crystolon=		39C

(Sometimes omitted)

(Sometimes omitted)

Grit Size			
Coarse	Medium	Fine	Very Fine
10	30	70	220
12	36	80	240
14		90	280
16	46	100	320
20	54	120	400
24	60	150	500
		180	600

Grade		
Soft	Medium	Hard
AE	IM	QV
BF	JN	RW
CG	KO	SX
DH	LP	TY
		UZ

Structure

The structure number of a wheel refers to the relative spacing of the grains of abrasive; the larger the number, the wider the grain spacing.

Bond Type

V = Vitrified
S = Silicate
B = Resinoid
R = Rubber
E = Shellac

Norton Symbol

Letter or numeral or both to designate a variation or modification of bond or other characteristic of the wheel. Typical symbols are "P," "G," "BE."

(Norton Co.)

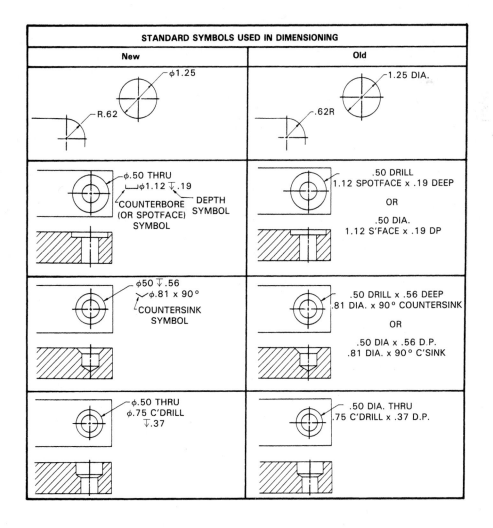

STANDARD SYMBOLS USED IN DIMENSIONING	
New	**Old**
φ1.25 / R.62	1.25 DIA. / .62R
φ.50 THRU / φ1.12 .19 — COUNTERBORE (OR SPOTFACE) SYMBOL — DEPTH SYMBOL	.50 DRILL / 1.12 SPOTFACE x .19 DEEP / OR / .50 DIA. / 1.12 S'FACE x .19 DP
φ50 .56 / φ.81 x 90° — COUNTERSINK SYMBOL	.50 DRILL x .56 DEEP / .81 DIA. x 90° COUNTERSINK / OR / .50 DIA x .56 D.P. / .81 DIA. x 90° C'SINK
φ.50 THRU / φ.75 C'DRILL / .37	.50 DIA. THRU / .75 C'DRILL x .37 D.P.

Quality control is a vital part of producing most machined products. This technician is using sophisticated spin testing equipment to check superprecision bearings for use in machine tools. (The Torrington Co.)

Glossary of Terms

A

A-axis: An angle that specifies the rotary motion of a machine tool or slide around the X axis, such that a right-handed screw advanced in the positive (+) *A* direction would be advanced in the positive (+) *X* direction.

Abrasive: A material that penetrates and cuts a material that is softer than itself. It may be natural (emery, corundum, diamond) or artificial (silicon carbide, aluminum oxide).

Abrasive machining: A process that uses an abrasive, in wheel or belt form, to remove material from a workpiece.

Absolute coordinates: Values of X, Y, and Z coordinates used in designating a point in space in the absolute system (a system in which all coordinate locations are measured from a fixed location on a machine table or from an absolute zero point established by a programmer or machine designer).

Absolute dimension: A dimension expressed with regard to the origin of a coordinate axis, but not necessarily coinciding with an absolute zero point.

Absolute positioning: Measures all tool movement from a fixed point of origin, or zero point. Compare with *Incremental positioning*.

Absolute zero: A temperature equal to –459.69°F or –273.16°C.

Accandec: A means of accelerating and decelerating feed rates to smooth starts and stops when an operation is under numerical control and when feed rate values are changed.

Acceptance test: A test to determine performance, capability, and conformity of software or hardware to design specifications.

Accuracy: Conformity of an indicated value to a value accepted as a standard.

Accurate: Made within the tolerances allowed.

Acme thread: Similar in form to a square thread in that the top and bottom of the thread is flat; however, the sides have a 29° included angle. It is used for feed and adjusting screws on machine tools.

Acquisition: A function that obtains information from memory locations or data files for use in data manipulation or handling.

Active storage: Data storage locations that hold data being transformed into motion.

Actual size: The measured size of a part after it is manufactured.

Actuator: A device that responds to a signal from a controller and performs an action.

Acute angle: An angle of less than 90°.

Adaptive control: A system that automatically and continuously monitors on-line performance of a manufacturing operation by measuring one or more variables of activity, comparing measured quantities with other measured quantities, and adjusting or modifying activity to meet production standards.

Addendum: That portion of a gear tooth that projects above or outside the pitch circle.

Address: The means of identifying information or location in a control system.

Adhesion: The process of joining materials by using an agent that sticks to both workpieces.

Adhesion process: Metal-joining method, such as soldering or brazing, that forms a bond between parts by filling joints with material that has a melting point lower than the base metal. Also, any method used to join materials with a glue or other bonding agent.

Adhesives: Materials that provide one of the newer ways to join metals and keep threaded fasteners from vibrating loose. In some applications, the resulting joints are stronger than the metal itself. Adhesive-bonded joints do not require costly and time-consuming operations such as drilling, countersinking, and riveting.

Align: To adjust to given points.

All-hard blade: One of two types of power hacksaw blade (the other is a flexible-back blade). It is best for straight, accurate cutting under a variety of conditions.

Allowance: The limits permitted for satisfactory performance of machined parts.

Alloy: A mixture of two or more metals fused or melted together to form a new metal.

Alphanumeric code: A system of instructions that consists of numbers, letters of the alphabet, and other symbols.

Aluminum Association Designation System: Every alloy is given an identifying number that consists of a four-digit code, followed by a dash and a temper designation, such as 7075-T6. The temper designation indicates the degree of hardness of the alloy.

American National Standards Institute (ANSI): An association that serves as a clearinghouse for nationally coordinated voluntary standards for fields ranging from information technology to building construction. Standards are established for such areas as definitions, terminology, symbols, materials, performance characteristics, procedures, and testing methods.

American National Thread System: The common thread form used in the United States, characterized by the 60° angle formed by the sides of the thread.

Amplifier: A device or system that provides an output different in magnitude from the control output.

Analog: Related to data presented in variable physical quantities, such as a dial indication of pressure.

Angle plate: A precisely made tool of cast iron, steel, or granite used to hold work in a vertical position for layout or machining. Faces are at right angles and may have slotted openings for easier mounting or clamping of the work to the machine tool table.

Annealing: The process of heating metal to a given temperature and cooling it slowly to remove stresses and induce softness.

Anode: A positive electrode.

Anodizing: A process for applying an oxide coating to aluminum. It is done electrolytically in an acid solution with equipment similar to that used for electroplating. The technique can be varied to produce a light-colored, porous coating that can be dyed in a variety of colors to a harder and nonporous coating for protection against corrosion.

Apron: A covering plate or casting that encloses and protects a mechanism. Also, a portion of the lathe carriage that contains gears, clutches, and levers for moving the carriage by hand and power feed.

Arbor: A shaft or spindle for holding cutting tools.

Arc: A discharge of electricity through a gas. In a gasoline engine, arcing (or sparking) takes place at the spark plug gap when the ignition coil fires to ignite the fuel mixture. After prolonged operation, the spark plug electrodes will be eroded by the action of the electric arcs. It is the basis of EDM.

Assembly: A unit fitted together from manufactured parts. A machine tool may comprise several assemblies.

Automated control: A control system that simultaneously monitors the quality of the product. It is able to start, stop, and sequence production.

Automatic feed: The capability of a machine tool to advance a tool into the workpiece mechanically.

Automatic screw machine: A variation of the lathe that was developed for high-speed production of large numbers of small parts. The machine performs a maximum number of operations, either simultaneously, or in a very rapid sequence.

Automatic System for Positioning of Tools (AUTOSPOT): A general-purpose computer program used in preparing instructions for numerical control positioning and straight-cut systems.

Automatically Programmed Tools (APT): A computer programming system describing parts illustrated on a design, in a sequence of statements, part geometry, cutter operations, and machine tool capabilities. It is used for turning, point-to-point, and multiaxis milling.

Automation: An industrial technique in which mechanical labor and mechanical control are substituted for human labor and human control. It is an extension and refinement of mass production.

Axis: A real or imaginary line that is equally distant from the surface or sides of something. Also, a point of reference.

B

B-axis: An angle that specifies the rotary motion of a machine tool or slide around the Y axis.

Back gears: Gears fitted to belt-driven machine tools to increase the number of spindle speeds. They are used to slow the spindle speed of a lathe for cutting threads, knurling, and making heavy roughing cuts.

Backlash: Lost motion (play) in moving parts, such as a thread in a nut, caused by looseness.

Band machining: A widely employed technique that makes use of a continuous saw blade. Chip removal is rapid, and accuracy can be held to close tolerances, eliminating or minimizing many secondary machining operations.

Base metal: A pure metallic element; the principal metal of an alloy.

Batch processing: A manufacturing operation in which a designated quantity of material is manufactured. Also, a method of processing jobs so that each job is completed before the next one is started.

Bed: One of the principal parts of a machine tool. It contains V-ways or bearing surfaces that support and guide the work of the cutting tool.

Bench grinder: A grinding machine that has been mounted on a bench or table. The grinding wheels mount directly onto the motor shaft. Normally, one wheel is coarse, for roughing, and the other is fine, for finish grinding. Compare with *Pedestal grinder*.

Beryllium: A metal that weighs almost 80% less than steel, yet offers virtually equal strength. Although brittle, it is easy to machine and is used when weight is a critical factor, such as in missiles and aircraft. It is also used in nuclear reactors. Beryllium is considered an "exotic" metal.

Bevel: An angle formed by a line or surface that is not at a right angle to another line or surface.

Bevel gear: A toothed wheel employed to change the angular direction of power between shafts. The teeth are either straight or curved.

Bill of materials: A listing of the numbers, names, materials, and quantities of the parts specified on a set of working drawings.

Binary code: A code in which each allowable position has one of two possible states. For example, it can be expressed as 1 (one) or 0 (zero), on or off, etc.

Bistable: Elements that have two output possibilities and will hold a given condition until switched.

BIT: An abbreviation for "binary digit." It is a single character of a language employing exactly two distinct characters.

Blanking: A stamping operation in which a die is used to shear or cut a desired shape from flat sheets or strips of metal.

Blind hole: A hole that does not go completely through the workpiece.

Block: In programming, binary words grouped together as a unit to provide complete information for a cutting operation. Also, one or more rows of punched holes in a tape separated from other words by an end-of-block character.

Blowhole: A hole produced in a casting when gases are entrapped during the pouring operation.

Bomb out: Complete failure of a computer routine, resulting in the need to restart or reprogram the computer. Also referred to as a "crash."

Boring: An internal machining operation in which a single-point cutting tool is used to enlarge a hole.

Boring mill: A huge machine capable of turning and boring work with diameters as large as 40′ (12 m). Work that is too large or too heavy to be turned in a horizontal position is machined on a vertical boring mill.

Box jig: See *Jig*.

Brazing: Joining metals by fusion of nonferrous alloys that have melting temperatures above 800°F (427°C) but lower than the metals being joined.

Brinell hardness: A measure of the resistance of a material to being indented. The Brinell test is an index of such factors as body hardness of the metal, machinability, uniformity of grade, and temper after heat treatment.

Briquetting: The process of compressing powdered metal into the final shape before sintering.

Brittleness: Characteristics that cause metal to break easily. In some respects, it is the opposite of toughness.

Broach: A long, multi-tooth cutting tool with three kinds of teeth shaped to give a desired surface: rough, semifinished, and finished.

Broaching: A manufacturing process for machining flat, round, and contoured surfaces, both internal and external. A broach is pushed or pulled across the work, with each tooth removing only a small portion of the material. Cutting a keyway is typically a broaching process.

Buffing: The process of bringing out the luster in metal using cloth wheels (usually cotton or muslin disks sewn together) and a polishing compound. Proper wheel speed depends upon the size of the wheel.

Bug: A flaw or defect in program code, or in the design of a computer, that renders the program incapable of performing the objectives for which it was written.

Burnishing: The process of finishing a metal surface by compressing the surface. It is often done by tumbling the work with steel balls.

Burr: A sharp edge remaining on metal after cutting, stamping, or machining. A burr can be dangerous if not removed.

Bushing: A bearing for a revolving shaft. Also, a hardened steel tube used on jigs to guide drills and reamers.

Byte: A sequence of binary digits operated upon as a single unit. A byte is composed of 8 bits, or binary digits.

C

C-axis: An angle that specifies the rotary motion of a machine tool or slide around the Z axis.

Calibration: The adjustment of a device so that output is within a designated tolerance for specific input values.

Cam: A rotating or sliding element that, because of the curvature of its driving surface, imparts complicated motions to followers or driven elements of the machine tool.

Canned cycle: In numerical control, a set of operations preset in hardware or software and initiated by a single command. Several operations are performed in a predetermined sequence; the function ends with a return to the beginning condition.

Carbon content: Refers to the amount of carbon contained in steel. Carbon content is measured in points or percentage (100 points equal 1%).

Carburizing: A process that introduces carbon to the surface of steel by heating the metal to just below its melting temperature in contact with carbonaceous solids, liquids, or gases. The piece is held at that temperature for a predetermined time and then quenched.

Cartesian Coordinate System: The basis of numerical control programming, it is a system that defines the direction and distance the work and/or tool must move. The position of any point can be defined with reference to a set of axes (X, Y, and Z) that are at right angles to each other.

Case hardening: A process of surface-hardening iron-based alloys so that the surface layer or case is made substantially harder than the interior or core. Typical case hardening processes are carburizing, cyaniding, and nitriding.

Casting: An object made by pouring molten metal into a mold.

Cathead: A sleeve or collar that fits over out-of-round or irregular-shaped work, permitting it to be supported in a steady rest. The work is centered by using adjusting screws located around the circumference of the collar.

Cathode: A negative electrode.

Cathode-ray tube (CRT): A vacuum tube in which a stream of electrons is projected onto a fluorescent screen to produce a luminous spot. The cathode-ray output in a computer displays information in graphic form or by character representation.

Cell: A manufacturing unit, usually comprised of one or two machines with computer-controlled and automated work-handling equipment.

Center finder: A device used in a drilling machine to locate the center of round stock. Also known as a "wiggler."

Center line: A line used to indicate an axis of a symmetrical part. It consists of a series of long and short dashes.

Centerless grinding: A technique in which a workpiece is not supported between centers; rather, it is positioned on a work support blade and fed automatically between a regulating or feed wheel and a grinding wheel.

Central processing unit (CPU): The heart of a computer; it handles memory and computational functions.

Centrifugal casting: A method for casting metals in which a mold is rotated during pouring and solidification of the metal.

Cermets: Composites of ceramic and metal that are widely used for high-temperature applications. Their resistance to high temperatures and wear make them ideal for use in super high-speed cutting tools.

Channel: A longitudinal row of holes punched along the length of paper tape. Standard numerical control tape has eight channels. Also known as "tracks" or "levels."

Character: An elementary mark, such as a letter or digit, used to represent data.

Chaser: A thread-cutting tool that fits into a die head used on a turret lathe or screw machine. It is usually a hardened steel plate with several correctly pitched teeth cut into it. Three or four chasers are used in a die head.

Chasing threads: Cutting threads on a machine tool.

Chatter: Vibration caused by a cutting tool springing away from the work. Chatter produces small ridges on machined surfaces.

Chemical blanking: Total removal of metal from certain areas by chemical action. It is used to produce small, intricate, ultrathin parts.

Chemical cutting fluids: Cooling and lubricating liquids that contain no oil. Because they are not actually fluids (graphite, mica, and white lead are examples), a wetting agent is often added to provide lubricating qualities. See *Cutting fluid.*

Chemical machining: A technique in which chemicals, usually in an aqueous solution, are employed to etch away selected portions of metal to produce accurately contoured parts. Chemical machining falls into two categories: chemical milling and chemical blanking.

Chemical milling: A technique for machining metal to exacting tolerances by chemical action. It is used to remove metal from large surface areas, such as reducing the weight of sheet metal parts.

Chip: A small piece of semiconductor material on which electronic components, such as integrated circuits, are formed. Also, a section of metal that is removed as a workpiece is being machined.

Chip breaker: A small groove cut into the face of a lathe tool near the cutting edge to break chips into small pieces.

Chipless machining: A metalworking technique in which a series of dies replaces the usual cutting tools on the lathe, drill press, or milling machine. Material is usually in coil form and is referred to as "wire." The process is sometimes called cold heading or cold forming.

Chuck: A device to hold work or cutting tools on a machine tool.

Circular interpolation: A mode that reduces the number of calculations required on a numerical control machine to program a curved surface. The circular interpolator in the machine control unit will automatically compute the necessary number of intermediate points to describe the circular cut. It also generates the electronic signals that will run the servos and guide the cutting tool in making the cut.

Circular pitch: The distance from the center of one gear tooth to the center of the next tooth, measured along the pitch circle.

Classes of fits: Standard working tolerances for thread accuracy, indicated by the last number on a thread description. Fits for inch-based threads are: Class 1, loose fit; Class 2, free fit; Class 3, medium fit; and Class 4, close fit.

Clearance: Distance by which one object is separated from another object.

Climb milling: A milling technique in which the teeth of a cutting tool advance into the work in the same direction as the feed. Also known as "climb cutting."

Clockwise: From left to right in a circular motion; the direction in which hands on a clock rotate.

Closed jig: See *Jig.*

Closed loop system: A numerical control system in which a reference signal is compared with a position signal generated by a monitoring (feedback) unit on a machine tool. The comparison is used to adjust the machine tool to reduce the difference to zero. An imaginary loop is formed by data flow. Also known as "feedback control system."

CNC: See *Computer numerical control.*

CO_2 laser: A device that produces a narrow beam of light that can be used to cut, weld, or heat-treat materials.

Coated abrasive: A finishing material consisting of an abrasive adhered to a backing such as paper or cloth.

Code: A system of organized symbols (bits) representing information in a language that can be understood and handled by a control system.

Coining: A metal-forming process that creates two-sided parts by squeezing the metal between a pair of dies.

Cold circular saw: A tool that uses a circular, toothed blade capable of producing very accurate cuts. Large cold circular saws can be used to sever round metal stock up to 27" (675 mm) in diameter.

Cold heading: See *Chipless machining.*

Color harden: A hardening technique usually done for appearance only.

Color temper: Reheating hardened steel and observing color changes to determine a quenching temperature to obtain the desired hardness.

Column: A vertical shaft designed to bear axial loads in compression.

Column and knee milling machine: So named because of the parts that provide movement to the workpiece. It consists of a column that supports and guides the knee in vertical (up and down/Z-axis) movement, and a knee that supports the mechanism for obtaining table movements. These movements are transverse (in and out/Y-axis) and longitudinal (back and forth/X-axis).

Combustible materials: Solids, liquids, or gases that are capable of burning. Combustible materials are classified into four categories: Class A fires involve ordinary combustible materials (paper, wood, textiles); Class B fires involve flammable liquids and grease; Class C fires involve electrical components; and Class D fires involve flammable metals, such as magnesium and lithium.

Command: A signal from a machine control unit (MCU) initiating one step in a complete computer program.

Composites: A family of materials that consists of two or more different but complementary substances that are physically combined.

Compound rest: A slide in the lathe located above a base cross-slide. The upper slide can be revolved to any required angular position.

Computer graphics: Graphs, charts, and/or drawings generated by a computer. They are displayed on a video screen or printed by a plotter or printer.

Computer numerical control (CNC): A system in which a program is used to precisely position tools and/or the workpiece and to carry out the sequence of operations needed to produce a part.

Computer-aided design (CAD): Use of computers to aid in designing a product.

Computer-integrated manufacturing (CIM): A computer system that controls a sequence of manufacturing operations involving several machines.

Concave surface: A curved depression in the surface of an object.

Concentric: Having a common center.

Cone pulley: A one-piece pulley having two or more diameters.

Continuous casting: A casting technique in which ingot is continuously solidified and withdrawn while it is being poured, so its length is not determined by mold dimensions.

Continuous path system: An operation in which the rate and direction of the relative movement of machine members are under continuous control, so the machine travels through a designated path at a specified rate without pausing.

Contour: The outline of an object.

Contouring control system: A numerical control system that generates a contour by guiding a machine or cutting tool along a path resulting from coordinated, simultaneous motion along two or more axes.

Control: A signal received at a system input and used as intelligence to produce a modification in output.

Controller: A device through which commands are introduced and manipulated to compute, encode, and store data; produce readouts; and process computation and output. In computer numerical control, it is also known as the machine control unit (MCU).

Convex surface: A surface of an object that curves outward.

Coolant: A fluid or gas used to cool the cutting edge of a tool to prevent it from burning up during machining operations.

Coordinate measuring machine (CMM): Used in quality control testing, it was devised to make measurements electronically. It can be programmed to check thousands of individual reference points on an object against specifications.

Core: A body of sand or other material that is formed to a desired shape and placed in a mold to produce a cavity or opening in a casting.

Counterbore: Enlarging a hole to a given depth and diameter.

Counterclockwise: From right to left in a circular motion; the opposite of the direction in which hands on a clock rotate.

Countersink: Chamfering a hole to receive a flat-head screw.

Creep grinding: A surface grinding operation that is often performed in a single pass with an unusually large depth of cut.

Cross slide: A part of a machine tool that permits the carriage to make transverse tool movements.

Cryogenics: The study and development of extremely low temperature processes, techniques, and equipment.

Cursor: The movable pointer used by a CRT operator to indicate where entries or actions are to take place.

Cutter compensation: The process of taking into account the difference in radius between a cutting tool and a programmed numerical control operation, in order to achieve accuracy.

Cutter offset: The difference between a part surface and the axial center of a cutter or cutter path during a machining operation. Also, a numerical control feature that enables a machine operator to use an oversized or undersized cutter.

Cutting fluid: A liquid used to cool and lubricate a cutting tool to improve the quality of the surface finish. There are four basic types of cutting fluids: mineral oils, emulsifiable (water-based) oils, chemical and semichemical fluids, and gaseous fluids.

Cyaniding: The introduction of carbon and nitrogen simultaneously into a ferrous alloy by heating while in contact with molten cyanide; usually followed by quenching to produce a hardened case.

Cyanoacrylate quick setting adhesive: A bonding agent known by such names as Eastman 910™, Super Glue™, and Crazy Glue™. It is used to hold matching metal sections together while they are being machined.

Cycle: A sequence of operations repeated regularly. Also, the time necessary for one sequence of operations to occur.

D

Data: Information that is input to a computer system and then processed so it can be output in a sensible form. It usually consists of numbers, letters, or symbols that refer to or describe an object, idea, condition, situation, or other type of information.

Dead band: Range through which an input can be varied without initiating a response.

Dead center: A stationary center used to hold work on a lathe.

Debug: In numerical control, a process of detecting, locating, and removing software errors and hardware problems that cause malfunctions in a computer.

Deburr: The process of removing, through various finishing techniques, sharp, raised edges produced by the machining process.

Decarburizing: The process of removing carbon from metals.

Deceleration distance: The calculated distance for decreasing the speed of movement along an axis to avoid overshooting a position.

Dedendum: The portion of a gear tooth between pitch circle and root circle; it is equal to addendum plus clearance.

Demagnetizing: The process of removing the magnetism from a piece that had been held in a magnetic chuck.

Depth of cut: Refers to the distance the cutter is fed into the work surface. The depth of cut varies greatly with lathe condition, material hardness, speed, feed, amount of material to be removed, and whether it is to be a roughing or finishing cut.

Dial indicators: An instrument used for centering and aligning work on machine tools, checking for eccentricity, and inspecting. There are two types of indicators: balanced indicators take measurements on either side of a zero line, while continuous indicators read from zero in a clockwise direction.

Diametral pitch: The ratio of the number of gear teeth to the number of inches of pitch diameter.

Diamond dressing tool: A device used to evenly wear away a grinding wheel face and keep it from becoming so loaded or glazed it cannot cut freely.

Diamond-edge band: A specially designed band machining tool for cutting material that is difficult or impossible to cut with a conventional toothed blade. The diamonds are only on the front edge of the band, where the cutting is accomplished.

Die: A tool used to cut external threads. Also, a tool used to impart a desired shape to a piece of metal.

Die casting: A casting process in which molten metal is forced under pressure into a permanent mold.

Die cavity: A hollow space inside a die where metal solidifies to form a casting.

Die chaser: See *Chaser*.

Die stock: A handle used to hold and rotate a threading die over the workpiece.

Dielectric fluid: A liquid (usually a light mineral oil) used to form a nonconductive barrier between the electrode and the work at the arc gap in EDM.

Digitizing: The process of converting a scaled, but nonmathematical, drawing into coordinate numerical control locations for programming purposes.

Dip coating: Applying a finishing material by completely immersing the part or product.

Direct numerical control (DNC): Use of a shared computer to program, service, and log a process such as a machine tool cutting operation. Part program data are distributed via data lines to machine tools. Compare with *Distributed numerical control*.

Disk: A random access storage component of a computer to store numerical control programs or other data.

Distributed numerical control (DNC): A control system in which individual numerical control machine tools are connected to a central command computer. Compare with *Direct numerical control.*

Dividing head: A machine tool attachment for accurate spacing of holes, slots, gear teeth, and flutes. When geared to a table lead screw, it can be used to machine spirals.

Dog: A device on the side of a machine tool worktable used to turn off the automatic feed mechanism or to reverse travel.

Draft: Clearance on a pattern that allows easy withdrawal of the pattern from the mold.

Drift: A tapered piece of flat steel used to separate tapered shank tools from sleeves, sockets, or machine tool spindles.

Drill point gage: An instrument used to ensure a drill is correctly sharpened.

Drill press: A machine that rotates a drill against stationary material with sufficient pressure to cause the drill to penetrate the material. It is primarily used for cutting round holes.

Drill rod: A carbon steel rod accurately and smoothly ground to size. It is available in a large range of sizes.

Drill template: The simplest form of an open jig, it consists of a plate with holes to guide the drill. Also known as a "plate jig."

Drilling: Cutting or enlarging a hole by use of a cutting tool sharpened on its point.

Drive fit: Using force or pressure to fit two pieces together.

Drop forging: A forming operation, usually done under impact, that compresses metal in dies.

Dry abrasive cutting: Employs a rotary abrasive wheel, without the use of a liquid coolant, for rapid, less-critical cutting. Compare with *Wet abrasive cutting.*

Dual dimensioning: A system that employs both the U.S. Conventional system of fraction or decimal dimensions and SI Metric dimensions on the same drawing.

Ductility: A property of metal that permits permanent deformation by hammering, rolling, and drawing without breaking or fracturing.

Dump: To remove all or part of the contents of a computer storage device. Also, printed copy resulting from an operation.

Dwell: A timed delay of programmed duration; for example, a hydraulic cushion at the end of each stroke in a grinding operation.

E

Eccentric: Not on a common center. Also, a device that converts rotary motion into a reciprocating (back-and-forth) motion.

Eddy-current test: A nondestructive test in which impedance (resistance) changes in a coil brought near a metal product indicate the presence of flaws.

Edit: To modify the format of a program or to alter data input or output by inserting or deleting characters.

Electrical discharge machining (EDM): A process by which an electric spark is used to erode away small amounts of material from a workpiece. EDM is used to machine soft or extremely hard materials that are difficult to work by conventional machining.

Electrical discharge wire cutting (EDWC): A metal-removing process in which a small-diameter wire electrode is fed through a workpiece, causing the material to be removed by spark erosion.

Electrochemical machining (ECM): A process in which a stream of electrolyte (typically salt water) is pumped at high pressure through a gap between the positively charged work and the negatively charged tool (electrode). The current passing through the gap removes material from the work by electrolysis, duplicating the shape of the electrode tool as it advances into the metal.

Electrode: A metal rod through which electricity flows. In EDM, it is comparable to the cutting tool on a conventional machine tool.

Electrohydraulic forming: A process in which high-voltage electrical energy is discharged from a capacitor bank into a thin wire or foil suspended between two electrodes immersed in water. As the wire or foil is vaporized by the electric current discharge, the vapor products expand, converting the electrical energy to hydraulic energy. The shock wave forms the metal against the die. It is a variation of explosive forming.

Electrolysis: Producing chemical reactions by passing an electric current through an electrolyte. The method is used in ECM.

Electrolyte: An electrically conductive fluid. It is used in ECM.

Electromachining: Processes in which electrical energy is applied directly to remove metal by erosion. Mechanical forces have no influence on the processes, as they do in conventional machining techniques. EDM and ECM are electromachining processes.

Electronic beam machining (EBM): A metal-removing process in which heat is produced by a sharply focused electron beam.

Electroplating: A plating process accomplished by passing an electric current from an anode (usually made of plating material) through an electrolyte to the workpiece (the cathode).

Emery: A natural (not manufactured) abrasive for grinding and polishing.

End of block (EB): A character that represents the end of a line or block of information in a numerical control program.

Erosion: In EDM, the wearing away of small amounts of material from a metal by an electric spark.

Etchant: A strong alkaline solution used in the chemical milling process. A prepared part is immersed in the solution, and the resulting chemical action removes the desired metal.

Expansion fit: A process in which a part is placed in liquid nitrogen or dry ice until it shrinks enough to fit into its mating part. Interference develops between the fitted pieces as the cooled piece expands. It is the reverse of shrink fit.

Explosive forming: Shaping metal parts in dies using a controlled explosion to generate forming pressure.

External thread: A screw thread cut on an outside surface.

F

Fabricate: To assemble or make from component parts.

Face milling: Machining large, flat surfaces parallel to the cutter face.

Faceplate: A disk fixed to the spindle of a lathe, used to attach the workpiece.

Facing: In lathe work, cutting across the end of a workpiece to create a flat surface.

Fastener: Any device used to hold two objects or parts together, such as bolts, nuts, screws, pins, keys, rivets, and chemical bonding agents or adhesives.

Fatigue: The tendency for metal to break or fracture under repeated or fluctuating stresses.

Feedback: Information returned from the output of a machine or process for use as input in subsequent operations.

Ferrous: An alloy containing a significant amount of iron.

File band: A band machine accessory used to obtain a smooth, uniformly finished surface. A series of small file segments make up the file band; the individual units interlock and form a continuous file.

Fillet: Curved surface that connects two surfaces that form an angle.

Finishing teeth: The teeth on a broach that provide the smoothest finish on the surface.

Fit: Clearance or interference between two mating parts. There are several classes of fits.

Fixture: A device for holding work rigidly while machining operations are performed. It does not guide the cutting tool.

Flame hardening: A method of surface hardening steel by rapidly heating the surface with the flame of an oxyacetylene torch, then quenching.

Flash: A fin of excess metal along the mold joint line of a casting, occurring between mating die faces.

Flask: In foundry work, a wooden or metal form consisting of a cope (top portion) and a drag (bottom portion) used to hold molding sand.

Flexible manufacturing system (FMS): An arrangement of machines and a connecting transport system under the control of a central computer that allows processing of several workpieces simultaneously.

Flexible shaft grinder: A small, portable, hand-held unit that performs many grinding jobs, from light deburring to die-polishing, and is powered by electricity or air.

Flexible-back blade: A blade used on metal cutting saws when safety requirements demand a shatterproof tool. These blades are also used for cutting odd-shaped work if there is a possibility of the work coming loose in the vise.

Floppy disk: A flexible plastic disk coated with magnetic oxide and used to store computer data. Also known as a diskette.

Flow: Quantity of a fluid passing a point per unit of time. In pneumatics, this is commonly represented in cubic feet per minute (cfm).

Fluorescent penetrant inspection: A nondestructive testing technique in which an oil-based penetrant is sprayed onto the work and drawn into every crack and flaw. The surface is rinsed with a solvent to remove the excess penetrant. After developing, the surface is viewed under ultraviolet light where defects appear fluorescent.

Flute: A groove machined in a cutting tool to facilitate easy chip removal and to permit cutting fluid to reach the cutting point.

Flux: Fusible material used in brazing and welding to dissolve and aid in removal of oxides and other undesirable substances.

Fly cutter: A single-point tool fitted in a lathe arbor; it is inexpensive to make but relatively inefficient because only one point does the cutting.

Follower rest: Similar to a steady rest except it provides support directly in back of the cutting tool and follows along during the cut. Compare with *Steady rest*.

Foot-pound: The US Conventional measurement unit for torque.

Force fit: When interference between two mating parts is sufficient to require force to press the pieces together, and the parts are considered permanently assembled.

Forge: To form metal with heat and/or pressure.

Form grinding: A cutting operation in which the grinding wheel is shaped to produce the required contour on the work.

Forming: A method of working sheet metal into useful shapes by pressing or bending.

Free fit: A fit that is used when tolerances are liberal. Clearance is sufficient to permit a shaft to run freely without binding or overheating when properly lubricated.

Friction saw: A metal-cutting tool with a blade that may or may not have teeth. The saw operates at very high speeds (20,000 surface feet or 6000 meters per minute) and actually melts its way through the metal.

Fused disposition modeling (FDM)™: A rapid modeling technique that produces fully functioning prototypes. When made of acrylonitrile butadiene styrene, the parts can be installed and run for proof that a design works. The FDM process forms three-dimensional objects from CAD-generated solid or surface models. A temperature-controlled head extrudes thermoplastic material layer by layer. The designed object emerges as a three-dimensional part without tooling.

G

G function: A preparatory code on a program tape to indicate a special function. The letter *g* is lowercase when shown on a printout. For example, g90 indicates that all the following coordinate locations are expressed in absolute dimensions.

Gage: A tool used for checking metal parts to determine whether they are within specified limits.

Gage blocks: See *Jo-block*.

Gaging: To check parts with various gages to determine whether the pieces are made within specified tolerances.

Gang milling: Using two or more milling cutters to machine several surfaces at one time.

Garbage: Erroneous, faulty, unwanted, or extraneous data in a computer or numerical control program.

Gaseous fluid: A type of cutting fluid; compressed air is the most commonly used.

Gate: The point where molten metal enters a mold cavity.

Gears: Toothed wheels that transmit rotary motion from one shaft to another shaft without slippage.

Geometric dimensioning and tolerancing (GDT): The control of the size of the features of a part and the allowances (either oversize or undersize) to achieve interchangeable manufacturing.

Gib: A wedge-shaped strip that can be adjusted to maintain a proper fit of the movable surface of a machine tool.

Graduate: To divide into equal parts by engraving or cutting lines or graduations into metal.

Graduations: Lines that indicate measurement points on tools and machine dials.

Gravity feed: Moving materials from one location to another using the force of gravity.

Grinding: An operation that removes material by rotating an abrasive wheel or belt against the work.

Gun drilling: A process used to drill deep, accurately sized holes.

H

H code: A code that calls out a specific fixture offset value that may be stated for one, two, or three axes. It is designated by the letter *h* and followed by two digits.

Half nut: A mechanism used to lock the lathe carriage to the lead screw to cut threads.

Hardened steel square: A square with true right angles, inside and outside. It is accurately ground and lapped for straightness and parallelism and comes in sizes up to 36″ (910 mm). It is recommended when extreme accuracy is required.

Hardening: Heating and quenching certain iron-based alloys to produce a hardness superior to that of untreated material. Hardening causes metals to have a higher resistance to penetration and abrasion.

Hardness testing: Techniques to determine the degree of hardness of heat-treated material.

Headstock: On a lathe, the structure that contains the spindle to which various work-holding attachments are fitted.

Heat treatment: Careful application of a combination of heating and cooling cycles to a metal or alloy in a solid state to bring about certain desirable properties, such as hardness or toughness.

Helical gears: Gears with teeth cut at some angle other than a right angle to the gear face, permitting two or more teeth to be engaged at all times. Operation is smoother and less noisy than that of spur gears.

Helix: A path generated by a point as it advances at a fixed rate on the surface of a cylinder, such as screw threads or flutes on a twist drill.

High-energy-rate forming (HERF): A metal-forming technique involving the release of a high-energy source, such as electrical, pneumatic-mechanical, or explosives.

Hob: A special gear cutter designed to cut gear teeth on a continuous basis.

Hobbing: Cutting gear teeth with a hob. The gear blank and hob rotate together in mesh during the cutting operation.

Home position: A fixed location in the basic coordinate axes of a machine tool. It is usually the point in the process when tools are fully retracted, permitting any necessary changes.

Honeycomb: Sections of thin material bonded together to form a structure that is similar in appearance to the wax comb that bees create to store honey. The structure gives existing metals greater strength and rigidity while reducing weight.

Honing: The process of removing a relatively small amount of material from a surface by means of abrasive stones to obtain a desired finish or an extremely close dimensional tolerance.

Horizontal band saw: Frequently referred to as a cutoff machine, it is a tool with a long, continuous blade that moves in only one direction. Cutting is continuous, and the blade can run at very high speeds because it rapidly dissipates the cutting heat.

Horizontal spindle milling machine: A category of milling machine where the cutter is fitted onto an arbor mounted in the machine on an axis parallel with the worktable. Multiple cutters may be mounted on the spindle for some operations.

Hydroabrasive process: A water-jet process that includes abrasive particles in the water.

Hydrodynamic machining (HDM): A computer-controlled cutting process that was developed to shape composites. The technique uses a 55,000 psi (379 000 kPa) water jet to cut complex shapes with minimal waste. Depending upon the material, tolerances can be held to ±0.004″ (0.1 mm). Also known as water-jet cutting.

I

ID: Abbreviation for inside diameter.

Idler gear: A gear situated between a driving gear and a driven gear to transfer motion, without any change of direction or of gear ratio.

Impact machining: An ultrasonic technique using a shaped cutting tool that oscillates about 25,000 times per second to pound a slurry of fine, abrasive particles against the work. The technique works best on hard, brittle materials (glass, quartz, silicon, and carbides) and is ineffective on soft materials, such as aluminum and copper. Also known as "slurry machining."

Impedance: The total opposition to alternating current, including resistance, capacitance, and inductance. Also known as "electrical impedance."

In-line transfer system: A chain or belt conveyor that moves a pallet containing a parts-holding device or fixture between operations.

Incremental positioning: Measuring each tool movement with reference to the prior (last) tool position. Compare with *Absolute positioning.*

Independent chuck: A chuck in which each jaw can be moved independently of the other jaws.

Indexable insert cutting tools: Widely used for turning and milling operations, the inserts are manufactured in a number of shapes and sizes for different turning geometries. As an edge dulls, the next edge is rotated into position until all edges are dulled; the inserts are then discarded.

Indexing: The process of providing equal spacing of holes and slots on the periphery of a cylindrical piece using a dividing or indexing head.

Indicator: A sensitive instrument capable of measuring slight variations when testing the trueness of the work, machines, or machine attachments.

Induction hardening: The rapid heating of a part surrounded by a high-frequency electric coil to above its critical temperature for a few seconds, followed by rapid cooling.

Input: Transfer of information by an appropriate medium into a computer or machine control unit.

Inserted tooth cutter: A milling cutter with replaceable teeth.

Inspection: Measuring and checking of finished parts to determine whether they meet specifications.

Integrated circuit (IC): A tiny single-structure assembly of electronic components containing many circuits and functions.

Interchangeable: Refers to a part that has been made to specific dimensions and tolerances and is capable of being fitted in place of an identically made part.

Interface: A functioning connection between two elements or components within a system.

Internal cut: A precision sawing operation on a vertical band machine in which the band is threaded through a hole drilled in the piece, welded, and the work maneuvered along the prescribed line.

Internal grinding: A cutting operation done to secure a fine surface finish and accuracy on inside diameters. Work is mounted in a chuck and rotated. During the grinding operation, the revolving grinding wheel moves in and out of the hole.

Internal threads: Screw threads cut on the inside surface of piece. Internal threads are made on the lathe with a conventional boring bar and a cutting tool sharpened to the proper shape.

International Standards Organization (ISO): The agency responsible for establishing and publishing manufacturing standards worldwide.

International System of Units (SI): The metric system of weights and measures. Abbreviated SI (Systeme International). See *SI Metric*.

Interpolation: A method of determining machine movements between two or three coordinate points.

Investment casting: A casting method that involves making a wax, plastic, or frozen mercury pattern, surrounding it with a wet refractory material, then melting or burning out the pattern after the material has dried and set. Molten metal is then poured into the resulting mold cavity.

J

Jarno taper: A standard taper of 0.600 inches per foot used on machine tools.

Jig: A device that guides a cutting tool and aligns it to the workpiece so that all parts produced are uniform and within specifications. An open (plate) jig consists of a plate with holes to guide the drill. It fits over the work. A box (closed) jig encloses the work and is used when holes must be drilled in several directions.

Jo-block (Johansson block): Precisely made steel blocks used by industry as a standard of measurement. They are made in a range of sizes and with a dimensional accuracy of ± 0.000002 (two millionths) inch, with a flatness and parallelism of ±0.000003 (three millionths) inch. Also called "gage blocks."

Just-in-time manufacturing (JIT): A system in which materials are delivered at the time they are needed for production, rather than being held in inventory. Its purpose is to expend the fewest resources to produce the final product.

K

Key: A small piece of metal embedded partially in the shaft and partially in the hub to prevent rotation of the gear or pulley on the shaft.

Keyway: The slot or recess in a shaft that holds the key.

Knee: The structure that supports the saddle and table of a column and knee-type milling machine.

Knife-edge blade: A vertical band machine tool employed to cut material that would tear or fray if machined by a conventional blade. Such materials include sponge rubber, cork, cloth, corrugated cardboard, and rubber.

Knurling: The process of impressing diamond or straight-line patterns onto a metal surface by rolling with pressure to improve the appearance and provide better grip. The rolls are called knurls.

L

Laminated object manufacturing (LOM)™*:* A new technology for the rapid generation of models, prototypes, accurate patterns, and molds. This state-of-the-art, three-dimensional modeling method uses inexpensive solid sheet material, such as paper, plastic, and composites, to form the desired designs. Molds made from the composite materials can be used in direct tooling applications for such processes as injection molding, vacuum forming, and sand casting.

Land: The metal left between flutes or grooves in drills, reamers, taps, and other cutting tools.

Lapping: The process of finishing surfaces with a very fine abrasive, like diamond dust or abrasive flours.

Lard oil: A cutting oil made from animal fat, often mixed with mineral oil to improve lubrication.

Laser-beam machining: A process that uses an intense source of coherent light energy as a cutting tool.

Lathe: A machine in which a workpiece in a work-holding device is rotated while a stationary cutting tool is forced against it. Some operations performed on a lathe include turning, boring, facing, thread cutting, drilling, and reaming.

Lathe dog: A device for clamping work so that it can be machined between centers.

Lay: Describes the direction of the predominant tool marks, grain, or pattern of surface roughness.

Lay out: To locate and scribe points for machining and forming operations.

Layout dye: A coating applied to metal to make layout lines more visible.

Lead: The distance a nut will advance on a screw in one revolution.

Lead screw: A long precision screw on the front of the lathe bed that moves the tool carriage when cutting threads.

Lip clearance: The amount that the surface of the point of a twist drill is relieved back from the lips (the cutting edges of the drill).

Live center: A center used to hold and rotate work on a lathe.

Longitudinal: Lengthwise.

Lubricating: Having the quality of reducing friction and cutting forces.

M

M function: A miscellaneous function in a numerical control program, designated by the letter *m*, followed by a two-digit number. It indicates such functions as coolant on/off or spindle rotation direction.

Machinability: The ease or difficulty of machining as it relates to the hardness of the material to be cut.

Machinability index: A reference that indicates the degree of ease or difficulty of machining a particular material. The index is based on the machining characteristics of a common steel, which is given an index of 100. Magnesium alloy has a machinability index of 500 to 2000 and is relatively easy to machine. In contrast, tool steel has a machinability index of 34 and is difficult to machine.

Machine bolt: A type of fastener employed to assemble parts that do not require close tolerances. Bolts are manufactured with square and hexagonal heads, in diameters ranging from 1/2" to 3". The nuts are similar in shape to the bolt head and are usually furnished with the machine bolts. Tightening the nut produces a clamping action to hold the parts together.

Machine control unit (MCU): A system of hardware and software used to control the operation of a computer numerical control machine.

Machine reamer: See *Reamer.*

Machine tools: That class of machines which, taken as a group, can reproduce themselves.

Machinist: A person who is skilled in the use of machine tools and is capable of making complex machine setups.

Magnaflux™: See *Magnetic particle inspection.*

Magnetic chuck: A device that uses a magnetic field to hold work for grinding.

Magnetic forming: A metal-shaping process that uses an insulated induction coil wrapped around or placed within the work. As very high momentary currents are passed through the coil, an intense magnetic field is developed. This causes the work to collapse, compress, shrink, or expand depending on the design of the coil. Coil location depends upon whether the metal is to be squeezed inward or bulged outward. The coil is shaped to produce the desired shape in the work. Also known as "electromagnetic forming" or "magnetic pulse forming."

Magnetic particle inspection: A nondestructive inspection technique that makes use of a magnetic field and magnetic particles to detect and locate flaws on or near the surface of ferromagnetic (iron-based) materials.

Major diameter: The largest diameter of a thread measured perpendicular to the axis.

Malleability: A property of metal that determines its ease in being shaped when subjected to mechanical working such as forging or rolling.

Mandrel: A slightly tapered, hardened steel shaft that supports work machined between centers.

Manipulator: In automated manufacturing, the articulated "arm" of the robot. The end of the arm is fitted with a "wrist" capable of angular and/or rotational motion.

Manual data input (MDI): A feature of a machine control unit that allows a programmer or operator to enter data directly to the unit rather than through an outside storage medium such as punch tape, disk, or computer memory.

Manufacturing: The use of machines, tools, and processes to convert raw materials into new products.

Marform: A drawing process that forms metal sheet by using a movable steel punch and a rubber-headed ram.

Match plate: A plate on which matching patterns for metal casting are mounted or integrated. It is used to facilitate molding.

Megahertz (MHz): A unit of frequency equal to one million hertz (cycles per second).

Menu: A list of functions appearing on a video display terminal that indicates the possible operations that design or controlling equipment can perform.

Mesh: To engage gears to form a working contact.

Metal spraying: A metal-coating technique in which a metal wire or powder is heated to its melting point and droplets are sprayed by air pressure to produce the desired coating on the work surface.

Metrology: The science that deals with systems of measurement.

Microinches: Millionths of an inch, shown as XX μin.

Micrometer caliper: A precision tool capable of measuring to 0.001" (0.01 mm). When fitted with a Vernier scale, it will read to 0.0001" (0.002 mm). Also known as a "mike."

Micrometers: Millionths of a meter, shown as XX μm.

Microsecond: One millionth of a second.

Milling: Removing metal with a rotating cutter on a milling machine.

Milling machine: A machine that removes metal from work by means of a rotary cutter.

Millisecond: One thousandth of a second.

Mineral oils: Cutting fluids best suited for light-duty (low speed, light feed) operations where high levels of cooling and lubrication are not required.

Minor diameter: The smallest diameter of a screw thread, measured across roots and perpendicular to axis. Also known as "root diameter."

Mist coolant: Cutting fluid applied on a band machine by flooding in mist form. It is used for high-speed sawing of free machining nonferrous metals. It is also employed when tough, hard-to-machine materials are cut.

Miter gears: Right angle bevel gears having the same number of teeth. They are used to transmit power through shafts at right angles to each other.

Morse taper: A standard taper of approximately 5/8 inch per foot, used on lathe centers and drill shanks.

Multiple-spindle drilling machine: A power-driven machine with multiple drilling heads, allowing several operations to be performed without changing drills.

Music wire: A carbon steel wire used to manufacture springs.

N

Nanosecond: One billionth of a second.

NC: Abbreviation for National Coarse series of screw threads. Also the abbreviation for numerical control.

Necking: Machining a groove around a cylindrical shaft.

Newton meter (N·m): The SI Metric measurement unit for torque.

NF: Abbreviation for the National Fine series of screw threads.

Nitriding: A case-hardening technique in which a ferrous alloy is heated in an atmosphere of ammonia or in contact with a nitrogenous material to produce surface hardness by absorption of nitrogen. Quenching is not necessary.

Noncorrosive: Will not wear away material gradually by chemical action. Mineral cutting oils, for instance, are noncorrosive cutting fluids.

Nonferrous: Not containing iron.

Normalizing: A process in which ferrous alloys are heated to approximately 100°F above a critical temperature range and cooled slowly in still air at room temperature to relieve stresses that may have developed during machining, welding, or forming operations.

Numerical control (NC): A system (composed of a control program, a control unit, and a machine tool) that controls the actions of the machine through coded command instructions.

Numerical data: Information expressed by a set of numbers or symbols.

O

Obtuse angle: An angle of more than 90° and less than 180°.

Occupational Safety and Health Act (OSHA): A law passed by Congress in 1970 to establish standards for occupational safety and health. Employers and employees have the duty to comply with all rules and regulations established by OSHA.

OD: Abbreviation for outside diameter.

Off-center: Eccentric, not accurate.

Offset tailstock method: A method for machining external tapers on a lathe. Jobs that can be turned between centers may be taper-turned by this technique.

Oil hardening: Using a mineral oil as a quenching medium in heat-treatment or surface hardening of certain alloys.

Open jig: See Jig.

Open-loop system: A control system that has no means for comparing output with input for machine control purposes.

Optical comparator: A gaging system for inspection and precise measurement of small parts and sections of larger parts.

Out-of-true: Term meaning not on center, eccentric, out of alignment.

P

Pallet: The support base, frame, or tray used on an in-line material transfer system.

Pedestal grinder: Similar to a bench grinder, except the grinding machine is on a pedestal (base) fastened to the floor. It is larger than a bench grinder.

Peening: The mechanical working of metal by use of hammer-like blows.

Peripheral milling: A milling operation that is done when the surface being machined is parallel with the periphery of the cutter.

Permanent mold: A mold, usually made of metal, used for repeated production of similar castings.

Pickling: Removing stains and oxide scales from metal surfaces by immersion in acid baths.

Pinion: The smaller of two mating gears.

Pitch: The distance from a point on one thread or gear tooth to the corresponding point on the next thread or tooth.

Pitch diameter: The diameter of the pitch circle of a gear.

Plain protractor: An angle-measuring tool used in layout work when angles do not need to be laid out or checked to extreme accuracy. The head is graduated from 0° to 180° in both directions for easy reading.

Plain turning: Turning in which the entire length of the piece is machined to a specified diameter.

Planer-type surface grinder: A machine tool that makes use of a reciprocating motion to move the worktable back and forth under the grinding wheel.

Plaster mold casting: A casting process, used primarily with aluminum, in which plaster molds are used in place of sand molds for a better surface finish.

Plate jig: See Jig.

Plunge grinding: Grinding method in which work is mounted between centers and rotated while in contact with the grinding wheel. The area being ground is no wider than the wheel face.

Point-to-point system: See Position control system.

Polishing: A process that uses a fine abrasive and a wax or oil compound to improve the surface finish on a workpiece.

Position control system: A positioning system in which controlled motion is required only to reach a given end point, with no path control during the transition from one point to the next. Also known as a "point-to-point system".

Position sensor: A device for measuring a position and converting this measurement into a form convenient for transmission. Also known as a "position transducer."

Pot broaching: A manufacturing process used for machining flat, round, and contoured surfaces, both internal and external, in which the tool is stationary, and the work is pulled through it.

Powder metallurgy (P/M): A technique used to shape parts from metal powders.

Precision: The degree of refinement with which an operation is performed or measured; held to close tolerances.

Precision grinding: A finishing operation in which a minute amount of material is removed with each pass of the grinding wheel to generate a smooth, accurate surface.

Precision microgrinder: A small, portable hand grinder that performs many grinding jobs, from light deburring to die-polishing, and is powered by electricity or air.

Preparatory function: In CNC programming, a command-changing mode of operation or control, such as from positioning to contouring or calling for a repeat cycle of machine.

Press fit: A class of fit where interference between mating parts is sufficient to require force to press the pieces together.

Printout: A printed sheet containing computer program data.

Profilometer: An electronic instrument for measuring surface roughness.

Programmer: In CNC, there are two important types of programmers: A parts programmer is a person who writes instructions for a computer to act upon to develop a specific program of operation (for example, turning a shaft to the desired diameter). A computer programmer is a person who develops routines that give a computer basic intelligence to act upon instructions that have been prepared by a parts programmer.

Protective clothing: Clothing that is worn in a machine shop to protect the body. Safety glasses and hearing protectors are two of the most important articles because shop areas produce both noise and flying chips. Other protective clothing includes steel-toed shoes, lead aprons, caps or hairnets, and respirators.

Pull broach: See *Broach.*

Pyrometer: A temperature-measuring device, originally designed to measure high temperatures; however, some are now used in any temperature range.

Q

Quenching: A rapid cooling of heated metal by contact with fluids or gases to impart hardness to the material.

Quick return: A mechanism on some machine tools that can be engaged to rapidly move the worktable to its starting point during a noncutting cycle.

Quill: A steel tube in the head of some machine tools that encloses the bearings and rotating spindle on which are mounted cutting tools. It is geared to a handwheel and/or lever that is used to raise or lower the rotating cutting tool on the work surface. A quill can be locked in position.

R

Rack: A flat strip of metal with teeth designed to mesh with those of a gear wheel, as in "rack and pinion." It is used to change rotary motion to reciprocating motion.

Raker set: A three-tooth saw set in which one tooth is angled toward the left, another one straight, and the next one angled toward the right, alternating continuously along the length of the blade. Raker set is recommended for cutting large solids or thick plate and bar stock.

Ram: Part of a machine that moves back and forth and carries a cutting tool.

Rapid traverse: A machine tool mechanism that rapidly repositions the workpiece for the next cut.

Rate of feed: The table movement speed on a milling machine.

Readout: A visual display of data.

Ream: To finish a drilled hole with a reamer.

Reamer: A cutting tool used to enlarge, smooth, and size a drilled hole by removing a small amount of metal.

Reciprocating hand grinder: A tool used to finish dies.

Red hardness: In metals, the quality of remaining hard when red hot.

Reference line: A layout line from which all measurements are made. Also known as a baseline.

Relief: Clearance provided around a cutting edge by removal of tool material.

Rework: The process of repairing defective parts, assemblies, or entire products. Rework adds no value to the product and is a form of waste.

Right angle: An angle of 90°.

Riser: A reservoir of molten metal provided to compensate for contraction of cast metals as they solidify.

Robot: A programmable, multifunctional manipulator designed to move materials, parts, tools, or specialized devices through variable programmed motions to perform a variety of tasks.

Rockwell hardness: A measure of the hardness of a material. The depth of the indentation of either a steel ball or a specially designed diamond cone penetrator under a prescribed load is the basis for the test.

Roller burnishing: A cold-working, metal-finishing operation that, rather than removing metal, compresses or "irons out" the peaks of the surface into the valleys.

Rolling: A process of forming and shaping metal by passing it through a series of driven rollers.

Root diameter: The smallest diameter of a thread.

Rotary table: A milling machine attachment consisting of a round table with T-shaped slots, rotated by means of a handwheel actuating a worm and worm gear.

Roughing: Rapid removal of surplus stock when the quality of the surface finish is not important.

Roughing teeth: The teeth on a multitoothed broach that remove the largest amount of material, producing a roughly finished surface. Subsequent teeth will produce a semi-finished and finished surface.

Row: In numerical control, a path perpendicular to the edge of a paper or plastic tape, along which information may be stored by presence or absence of holes or magnetized areas.

Runner: A channel through which molten metal flows from the sprue to the casting and risers.

S

Safe edge: A file edge without teeth.

Safety equipment: Tools or equipment that help prevent or mitigate accidents in the potentially dangerous environment of a machine shop. Machine guards, fire extinguishers, eye wash stations, vacuum dust collectors, power switches (for locking off equipment), and brushes (for removing machine chips) are a few examples.

Sand-mold casting: A process that involves pouring molten metal into a cavity that has been formed in a sand mold. The resulting parts usually must be machined to final dimensions and finish.

Sandblasting: A method used to clean castings and metals by blowing sand against them under very high pressure.

Scale: Surface oxidation on metals caused by burning, oxidizing, or cooling. The oxides usually form as loose scale.

Scale drawings: Drawings made other than actual size (1:1). A drawing made one-half size would have a scale of 1:2. A scale of 2:1 would mean that the drawing is twice the size of the actual part.

Scleroscope: A testing device that drops a hammer onto the test piece; a measurement of the resulting bounce or rebound of the hammer is used to determine hardness.

Scrap: Excess material generated during machining.

Scraping: Removing minute portions of a wearing surface to achieve a precision fit and finish not attainable by ordinary filing techniques.

Scribe: To draw a line on a metal workpiece with a sharp pointed tool (a scriber).

Semichemical cutting fluids: Cooling and lubricating liquids that may have a small amount of mineral oil added to improve the fluids' lubricating qualities. Semichemical cutting fluids incorporate the best qualities of both chemical and emulsifiable (water-based) cutting fluids.

Sensors: Electrical devices that receive varied types of information and transmit feedback to a computer or other control device.

Sequence number: Identifies the relative location of blocks or groups of blocks in a program.

Servomechanism: In EDM, the drive unit that accurately controls electrode movement and maintains the correct distance between the work and the electrode as machining progresses.

Set up: The positioning of a workpiece, attachments, and cutting tools on a machine tool.

Setover: The distance a lathe tailstock is offset from the normal centerline of the machine. It is a method of taper-turning.

Setscrews: Semipermanent fasteners that are used for such applications as preventing pulleys from slipping on shafts, holding collars in place on assemblies, and positioning shafts.

Shim: A thin piece of sheet metal used to provide proper clearance between mating parts.

Shrink fit: A type of fit in which an outer member is expanded by heating to permit insertion of an inner member, in turn, obtaining a tight fit as the outer member cools and shrinks. The fit is tight and considered permanent.

SI Metric: The metric system of weights and measures. (*SI* stands for the French words Systeme International.)

Side milling cutter: A type of milling cutter with cutting edges on the circumference and on one or both sides. The cutters are made in solid form or with inserted teeth.

Single-point cutting tool: A cutting tool with one face and one cutting edge.

Sintering: Forming a coherent bonded mass by heating metal powders without melting; it is used mostly in powder metallurgy.

Skills standards: Developed by the National Tooling and Machining Association, with the aid of the metalworking industry, these are the industry's requirements for skilled workers and the basis for industry-recognized certification obtained through performance testing.

Slab broach: A broaching operation using a flat toothed strip that is usually held (singly or in groups) in a slotted fixture.

Slip bushings: Inserts used to guide the drills when a combination of open and box jigs must be used to perform several operations. The slip bushings are removed for subsequent operations such as reaming, tapping, countersinking, counterboring, or spot facing.

Slitting: An operation in which a thin cutter or rotary knife is used to cut sheet metal into narrow strips.

Slotting: Similar to slitting, except that the cut is made only part way through the work. The slot in a screw head is an example of slotting.

Smart tooling: Involves the use of cutting tools and work-holding devices that can be readily reconfigured to produce a variety of shapes and sizes within a given part family. This makes it economically feasible to manufacture products in smaller lot sizes.

Society of Automotive Engineers (SAE): The organization that devised standards used to classify steels.

Software: The entire set of programs, procedures, and related documentation associated with a computer system.

Soldering: Joining metals with a nonferrous filler metal that has a melting point lower than the base metals. It is normally carried out at temperatures below 800°F (427°C).

Spline: A series of grooves cut lengthwise around a shaft or hole.

Spot facing: Machining a circular spot on the surface of a part to furnish a flat bearing surface for mounting a bolthead or nuthead.

Sprue hole: An opening in a mold into which molten metal is poured.

Spur gear: A wheel with teeth that run straight across the gear face, perpendicular to the sides. It is the most commonly used gear.

Square: A tool used to check 90° (square) angles. It is also used for laying out lines that must be at right angles to a given edge or parallel to another edge.

Staking: Joining two parts by upsetting metal at their junction.

Standard: An accepted base for a uniform system of measurement and quality.

Statistical Process Control (SPC): A quality control method used to reduce the number of rejects and lower production costs. Production is periodically measured and analyzed to determine if adjustments should be made.

Steady rest: A support for long, thin workpieces that keeps the work from springing or bending away from the cutting tool. The rest also reduces "chatter" when long shafts are machined. Compare with *Follower rest.*

Steel rule: A measuring tool, available in at least three basic types of graduations: fractional inch, decimal inch, and metric.

Stellite: An alloy of cobalt, chromium, and tungsten used to make high-speed cutting tools.

Stereolithography: A rapid prototyping technique that allows designers to quickly generate three-dimensional models or prototypes of parts with relatively inexpensive materials. The stereolithography process uses a computer-guided, low-power laser beam to harden a liquid polymer plastic into the programmed shape. The three-dimensional hard plastic models can be studied to determine whether they are the best solution to a design problem.

Stop: A device on a machine tool worktable used to turn off the automatic feed mechanism or to reverse travel. Also known as a "dog."

Straddle milling: Using two or more milling cutters to perform several milling operations simultaneously.

Straight set: A two-tooth saw set in which one tooth is angled to the side and the other one straight, alternating continuously along the length of the blade. Straight set is recommended for materials like aluminum and magnesium.

Straight-cut system: A system in which controlled cutting action occurs only along a path parallel to linear, circular, or other machine ways.

Straightedge: A precision tool for checking the accuracy of flat surfaces.

Strain: A measure of change in the shape or size of a body, compared to the body's original shape or size.

Stress: Intensity, at a point in a body, exerted by an external force.

Stress-relieving: A process for removing internal stresses that have developed in parts that have been cold worked, machined, or welded. Steel parts are heated to 1000°F to 1200°F (547°C to 660°C), held at this temperature one hour or more per inch of thickness, and then slowly air- or furnace-cooled.

Stretch forming: Shaping metals by applying tension to stretch the heated sheet or part, wrapping it around a die, and then cooling it.

Super finish: A finish in which surface irregularities have been reduced to a few millionths of an inch to produce an exceptionally smooth and long-wearing surface.

Surface gage: A scribing tool used to check whether a part is parallel to a given surface.

Surface plate: An iron or granite plate, ground or lapped to a smooth flat surface, and used for precision layout and inspection.

Surface roughness standards: A series of small plates with varying degrees of surface roughness that allow a machinist or inspector to compare finishes by sight and touch.

T

T function: A code identifying a tool select command in a program. As with *m* and *g* codes, the *t* also appears lowercase on a printout.

Tailstock: A movable lathe fixture that mounts on ways to support work between centers. It can be fitted with tools for drilling, reaming, and threading.

Tang: Flats or tongue machined on the end of tapered shanks. Tang fits into a slot in the mating part and prevents the taper from rotating in the mating part. Also, the part of a file that fits into a handle.

Tantalum: A ductile metal capable of withstanding temperatures ranging from 2500°F to 4000°F (1372°C to 2206°C). It is used in making surgical tools, pen points, and electronic equipment, and is being used increasingly in space-age technology applications.

Tap: A tool used to cut internal threads.

Taper: A piece that uniformly increases or decreases in diameter to assume a wedge or conical shape.

Taper attachment: A guide attached to a lathe and used to accurately cut internal and external tapers.

Tapping: Forming an internal screw thread in a hole or other part by means of a tap. Also, opening the pouring hole of a melting furnace to remove molten metal.

Temper: The hardness and strength of a rolled metal.

Tempering: Heating hardened steel to just below its critical temperature, then slowly cooling the steel to reduce brittleness and toughness.

Template: A pattern or guide used for layout design.

Tensile strength: The maximum stress and strain a material can bear.

Tension: A stretching or pulling force.

Thread: To cut a screw thread.

Thread rolling: Applying a thread to a bolt or screw by rolling it between two grooved die plates, only one of which is in motion, or between rotating circular dies.

Thread-cutting stop: A device used to stop the thread-cutting operation on a lathe so the tool can be removed from the work after each cut and repositioned before the next cut is started.

Threaded fastener: A bolt or similar device that uses the wedging action of the screw thread to clamp parts together. To achieve maximum strength, a threaded fastener should screw into its mating part at least a distance equal to one and one-half times the thread diameter.

Three-tooth rule: In a properly selected power hacksaw blade, at least three teeth must come in contact with the work.

Three-wire method of measuring threads: To accurately measure thread size, three wires of a specific diameter are fitted into screw threads, and a micrometer measurement is made over the wires. A mathematical formula provides the information necessary to calculate the correct measurement.

Titanium: A metal used for applications that require a material to be lightweight, high-strength, corrosion-resistant, and high-temperature resistant. It weighs only about half as much as steel, yet is almost as strong.

Tolerance: A permissible deviation from a basic dimension.

Tool crib: A room or area in a machine shop where tools and supplies are stored and dispensed as needed.

Tool function: A command in a computer program identifying a tool and calling for its selection. The actual tool change may be initiated by a separate tool-change command.

Tool post: Mounts the cutting tool on the carriage of the lathe.

Toolroom: An area or department where tools, jigs, fixtures, and dies are manufactured.

Tooth form: The shape of the tooth on a band machine saw. There are three basic forms: standard, skip, and hook.

Tooth rest: A device that quickly and accurately positions the teeth of a gear cutter.

Torque: The amount of turning or twisting force applied to a threaded fastener or part. It is measured in force units of foot-pounds (ft.-lbs.) or the SI Metric equivalent, newton meters (N•m). Torque is the product of the force applied times the length of the lever arm.

Track: See *Row.*

Train: A series of meshed gears.

Transducer: A device that converts an input signal into an output signal of a different form.

Transverse: Movement at a right angle to the main direction of travel. An in and out movement along the Y axis.

True: On center.

Tumbler gears: Idle gears in a gear train used to reverse the rotation of the driven gear.

Tungsten carbide: The hardest human-made metal (almost as hard as diamond). The metal is shaped by molding tungsten, carbon, and cobalt powders under heat and pressure.

Turret lathe: A lathe equipped with a six-sided tool holder (turret) that holds multiple tools and can be revolved to present the appropriate tool for a particular operation.

Twist drill: A common drill made by forging or milling rough flutes and then twisting them to a spiral shape. After twisting, the drills are milled and ground to approximate size. Finally, they are heat-treated and ground to exact size.

U

US Conventional: The "English" system of weights and measures used in the United States. Compare with *SI Metric.*

Ultrasonic machining: A machining process in which sound waves above the audible range propel an abrasive for use as a metal-cutting tool.

Ultrasonic testing: Techniques that make use of sound waves above the audible range to detect cracks and flaws in almost any kind of material that is capable of conducting sound. Sound waves may also be employed to measure the thickness of the same materials from one side.

Ultraviolet light: Also known as "black light," it is employed in conjunction with a fluorescent dye for parts inspection. Any defects show because the dye flows into and remains in the flaws.

Unified Thread System: A standard thread form. The threads are identified as UNF (Unified National Fine) and UNC (Unified National Coarse). Fasteners using this thread series are interchangeable with fasteners using the American National Thread System.

Universal chuck: A chuck on which all jaws move simultaneously at a uniform rate to automatically center round or hexagonal stock.

Universal tool and cutter grinder: A grinding machine designed to support cutters (primarily milling cutters) while they are sharpened to specified tolerances. Special attachments permit straight, spiral, and helical cutters to be sharpened accurately. Other attachments enable the machine to be adapted to all types of internal and external cylindrical grinding.

V

V-block: A square or rectangular steel block with a 90° V-groove through the center, provided with a clamp for holding round stock for drilling, milling, and laying out operations.

V-ways: Portions of machine tool beds that are raised and shaped like an inverted V; they act as bearing surfaces, guiding and aligning the movable portion of the machine.

Variability: Inconsistency that occurs in manufacturing a product.

Vents: Narrow openings in molds that permit gases generated during metal pouring to escape.

Vernier caliper: A precision measuring instrument, used for both inside and outside measurements, that is accurate to 1/1000" (0.001") and 1/50 mm (0.02 mm).

Vernier protractor: An angle-measuring tool used in layout work when angles must be extremely accurate. With this tool, angles of 1/12 of a degree (5 minutes of arc) can be precisely measured.

Vertical milling attachment: A mechanism that can be attached to a horizontal milling machine to allow such operations as end and surface milling, radius and cam milling, drilling, reaming, boring, and cutting slots and keyways.

Vertical spindle milling machine: A type of milling machine in which the cutter is normally perpendicular (at a right angle) to the worktable. On many vertical spindle machines, the spindle can be tilted to perform angular cutting operations.

Vitreous enamel: A glass layer that has been fused to sheet or cast iron surfaces, forming an extremely hard coating that is smooth and easy to clean. Also known as porcelain.

W

Washer: A flat, ring-shaped piece used to improve the binding ability of a screw fastener. It also prevents surface marring.

Water-jet cutting: See *Hydrodynamic machining (HDM)*.

Waviness: A quality of a machined surface that takes the form of smoothly rounded peaks and valleys. Waviness is of greater magnitude than roughness.

Wavy set: A saw tooth set in which several teeth are angled to the right and several to the left, alternating continuously along the blade. Wavy set is recommended for work with varying thicknesses, such as pipe, tubing, and structural materials.

Webster hardness tester: A portable device for testing the hardness of materials such as aluminum, brass, copper, and mild steel. It can be used on assemblies that cannot be brought into the laboratory, or to test a variety of shapes that other testers cannot check, such as extrusions, tubing, or flat stock. The tester's dial indicator reading is converted to the Rockwell hardness scale by referring to a conversion chart furnished with the tool.

Wet abrasive cutting: Uses a rotary abrasive wheel with a liquid coolant to produce a fine surface finish and permit cutting to close tolerances. The cuts are burn-free and have few or no burrs. Compare with *Dry abrasive cutting*.

Wheel dresser: A tool for cleaning, resharpening, and restoring the mechanical accuracy of the cutting faces of grinding wheels.

Wiggler: See *Center finder*.

Work envelope: The volume of space defined by the reach of a robot's arm in three dimensions.

Work hardening: The increase in hardness that develops in metal as a result of cold forming.

Working drawing: A drawing or drawings that give a machinist the necessary information to make and assemble a mechanism.

Wringing fit: A fit that is practically metal to metal. It is used with gage blocks and requires a twisting motion to assemble.

X

X axis: Machine movement to left or right, in relation to the operator.

X-ray inspection: A nondestructive testing technique that has become a routine step in acceptance of parts and materials.

Y

Y axis: Machine movement toward or away from the operator.

Yield point: The point of stress or strain at which a material fractures.

Z

Zero point: The point from which all dimensions are referenced in an absolute positioning system.

Acknowledgments

The author expresses his sincere thanks to the many organizations and manufacturers who cooperated so generously in supplying the technical information and many of the photographs used in this textbook. Any omissions from the following list are purely accidental.

Adept Technology, Inc.; Alden Corp.; Aloris Tool Co., Inc.; Aluminum Company of America; American Foundrymen's Society; American National Standards Institute; American Welding Society; Armstrong Bros. Tool Co.; Armstrong-Blum Mfg. Co.; Association for Manufacturing Technology; Autodesk, Inc.; AVCO Aerostructures Div., AVCO Corp.; Baldor; Bell Helicopter Textron/Boeing Helicopters; Bethlehem Steel Co.; Bird-Johnson Company; Black and Decker; Boeing Co.; Boston Gear Co.; Bridgeport Machines, Inc.; Broaching Machine Specialties; Brown & Sharpe Mfg. Co.; Buick Div. of GMC; Burgess-Norton Mfg. Co.; Bystronic, Inc.; Carboloy, Inc.; Central Foundry Div. Of GMC; Challenge Machinery Co.; Charmilles Technologies; Chicago-Latrobe; Chick Machine Tool, Inc.; Chrysler Corp.; Cincinnati Lathe & Tool Co.; Cincinnati Milacron; CITCO Div., Western Atlas, Inc.; Clausing Industrial, Inc.; Coated Abrasive Manufacturers Institute; Cogsdill Tool Products, Inc.; Columbian Vise and Mfg. Co.; Commander Die & Machine Co.; Compositek Corporation; Computervision; Cross Co.; Dapra Corporation; Darex Corp.; Defiance Machine and Tool Co.; Deka-Drill, South Bend Lathe; Delcam International; Delco Moraine Div. of GMC; Deneb Robotics, Inc.; DoALL Co.; Driv-lok, Inc.; duMont Corp.; Dumore Co.; DuPont Co.; Eljer Plumbingware; Emco-Maier Corp.; Emergency Vehicles, Inc.; ENCO Mfg. Co.; Engis Corporation; EROWA Technology, Inc.; Euclid; Evans & Sutherland; Everett Industries, Inc.; Ex-Cell-O Corp.; EZFeatureMILL–Engineering Geometry Systems; Facit; Fanuc Robotics North America, Inc.; Federal Products Co.; Flow International Corp.; Ford Motor Co.; French National Railroads; General Dynamics Corp.; General Motors Corp.; Giddings & Lewis, Inc.; Greenfield Tap & Die; Grumman Aerospace Corp.; Hamilton Standard Div., United Technologies; Hamilton Watch Co.; Hammond Machinery Builders.; Bill Hannan; Hardinge, Inc.; Harig Div. of Bridgeport Machines Inc.; Harrison/REM Sales, Inc.; Hartel Cutting Technologies, Inc.; Heidenhain Corp.; Heli-Coil Corp.; Helisys, Inc.; Hewlett-Packard Marketing Communications; Hexcel Corp.; Hirata Corporation of America; Honeywell, Inc., Industrial Automation and Control; Hougen Manufacturing , Inc.; Howmet Corp.; Hughes Electro-optical and Data Systems Group; Hurco Manufacturing Co.; Hydramat, Inc.; Ingersoll Rand; Ingersoll-Rand Co.; Waterjet Systems; Interlen Products Corporation; Iscar Metals, Inc.; Jacobs Mfg. Co.; Jergens, Inc.; Jet Equipment & Tools; C. E. Johannson Co.; Johnson Gas Appliance Co.; Jones & Lamson Machine Co.; Jones & Shipman, Inc.; Jorgensen Conveyors, Inc.; Justright Manufacturing Company; Kaiser Tool Co., Inc.; Kearney & Trecker Corp.; Kelsey-Hayes; Kennametal, Inc.; Kesel/JRM International, Inc.; Komet of America, Inc.; L.W. Chuck Co.; LaPoint Machine Tool Co.; K.O. Lee Co.; LeBlond Makino Machine Tool Co.; Light Beam Technology, Inc.; Light Machines Corp.; Lindberg Steel Treating Co.; Lockheed-Martin; Lockwood Products, Inc.; Loctite Corp.; Lufkin Rule Co.; Lumonics Corp.; Magnaflux Corp.; Magnus Chemical; Maho Machine Tool Corp.; Marposs Corp.; Master Lock Co.; Mazak Corp.; Metal Powder Industries Federation; Metco, Inc.; Microphoto, Inc.; MIFCO McEnglevan Industrial Furnace Co., Inc.; Millersville University; Minitool Inc.; Mitsubishi Materials USA Corporation; Mitutoyo/MTI Corp.; Morse Tool Co.; Motoman; MTS Systems Corp.; Nachi Robotics; NASA; National Broach & Machine Co.; National Machinery Co.; National Machine Systems; National Machine Tool Builders' Assoc.; NEYTECH; Nicholson File Co.; Northrop-Grumman Corp.; Norton Co.; NSK America; O.S. Walker Co., Inc.; Okuma America Corporation; Optica; Optical Gaging Products, Inc.; Osborn Mfg. Co.; Parker-Kalon; Parlec, Inc.; Peter Wolters of America, Inc.; Pfauter-Maag Cutting Tools; Philadelphia Gear Corp.; Polygon Company; Pontiac Div., General Motors Corp.; Precision Castparts Corp.; Procyon Machine Tools; Radyne; Reishauer Elgin; Renishaw, Inc.; Republic-Lagun Machine Tool Co.; Revolution Tool Company; Rofin-Sinar, Inc. ; Rohm and Haas; Rush Machinery, Inc.; The Ryan Company ; Sandvik Coromant Co.; W.J. Savage Co.; William L. Schotta, Millersville University; Sears, Roebuck and Company–Orland Park Retail Store, Orland Park, Illinois; Shakeproof Div., Illinois Tool Works, Inc.; Sharnoa Corp.; Sharp Industries, Inc.; Shore Instrument and Manufacturing Co., Inc.; Simmons Machine Tool Corp.; South Bend Lathe Corp.; South Shore Tool and Development Corp.; SpeedFam; Standard Tool Co.; L. S. Starrett Co., Athol, Massachusetts; Stratasys, Inc.; Sundstrand Corp.; Sunnen Products Company; Surf-Chek; Surfware; 3D Systems; 3M Company; Taft-Pierce Co.; Threadwell Manufacturing Co.; Tinius Olsen Testing Machine Co.; Tornos-Bechler S.A.; Toshiba Machine Co., America; Tri-Tool, Inc.; U.S. Air Force Thunderbirds; U.S. Amada, Ltd.; U.S. Army; U.S. Navy; Union Carbide Corp.; Union Carbide Corp.; Unison Corp.; USM Corp., Fastener Group; Valenite, Inc.; Waldes Kohinoor, Inc.; Bob Walker; Warner & Swasey Co.; Webster Instrument, Inc.; Weldon Machine Tool, Inc.; W.F. Wells; Westech Automation Systems; Westech Products Group/Gantrex Machine Tool Loaders; Westinghouse Electric Corp.; J.H. Williams and Co.; Willis Machinery and Tools Corp.; John Winter; Wilson Instruments/Instron Corp.; Wilson Mech. Inst. Div., American Chain and Cable Co., Inc.; Wilton Corp.; WMW Machinery Company, Inc.; Yukiwa Seiko USA, Inc.; Carl Zeiss, Inc.

Index